AMORPHOUS AND LIQUID SEMICONDUCTORS

AMORPHOUS AND LIQUID SEMICONDUCTORS

Edited by

J. Tauc

Brown University
Providence
Rhode Island, U.S.A.

PLENUM PRESS • LONDON AND NEW YORK • 1974

Library of Congress Catalog Number: 73-81491

ISBN 0-306-30777-4

Copyright © 1974 by Plenum Publishing Company Ltd

Plenum Publishing Company Ltd
4a Lower John Street
London W1R 3PD
Telephone 01-437-1408

U.S. Edition published by
Plenum Publishing Corporation
227 West 17th Street
New York, New York 10011

TYPESET BY H CHARLESWORTH & CO LTD, HUDDERSFIELD

MADE AND PRINTED IN GREAT BRITAIN BY
THE GARDEN CITY PRESS LIMITED
LETCHWORTH, HERTFORDSHIRE SG6 1JS

Preface

Solid state physics after solving so successfully many fundamental problems in perfect or slightly imperfect crystals, tried in recent years to attack problems associated with large disorder with the aim to understand the consequences of the lack of the long-range order. Semiconductors are much more changed by disorder than metals or insulators, and appear to be the most suitable materials for fundamental work.

Considerable exploratory work on amorphous and liquid semiconductors was done by the Leningrad School since the early fifties. In recent years, much research in several countries was directed to deepen the understanding of the structural, electronic, optical, vibrational, magnetic and other properties of these materials and to possibly approach the present level of understanding of crystalline semiconductors. This effort was stimulated not only by purely scientific interest but also by the possibility of new applications from which memory devices in the general sense are perhaps the most challenging.

The research met with serious difficulties which are absent in crystals. The theorists have to learn how to live without the Bloch theorem, without the mathematical simplicity introduced by symmetry and long range order; they have to struggle with inhomogeneous systems and with contradictory and changing experimental results. The experimentalists in turn are plagued with the difficulty of mastering the reproducible preparation of materials, with experimental data curves which are typically less structured and sharp (how much nicer it is to have graphs with sharp peaks than very broad bands), and with the vague generalities and loosely defined concepts of the present theories. Therefore the progress has been slow and most of the fundamental questions are still open to discussion. It reminds one of the situation in which semiconductor research was during the thirties.

Nevertheless, just as the basic ideas for the understanding of crystalline semiconductors were available in the pre-germanium era, very probably most of the present interpretations have a sound foundation and will survive as basic concepts for the future more detailed and better founded work. The recent book on the Electronic Properties in Non-crystalline Materials by Mott and Davis presents an ingenious attempt to correlate the observed electronic effects in disordered solids on the basis of a few concepts derived from some theoretical considerations and generalized with a sharp physical intuition.

In our book we do not try to give a unified picture of the field. A few experts in the field were invited to discuss their respective subfields. They were asked to describe the basic methods, their merits, limitations and achievements. This approach is hopefully useful for readers working or intending to work in the field. Of course, each contributor necessarily includes some of his personal preferences in the interpretation and meaning of his results. Actually these views are not too diverging, but no attempt was made to unify them.

A large class of amorphous solids – glasses – have been studied extensively over many decades from the point of view of their thermodynamical and structural properties. The first two chapters deal with this aspect which is considered essential for any deeper understanding of the nature of the amorphous state. The inherent metastability of the amorphous state compared to the crystalline state is responsible for many discrepancies in the experimental results. It should be understood not only for improving the reproducibility of the measurements but also because the control of the structural parameters is the heart of such applications of amorphous semiconductors which cannot be produced by devices constructed of crystalline materials. Indeed, most kinds of memory devices are based on the ability of glasses to exist in different structural states.

The third chapter is an introduction to the theoretical treatment of the electronic states in highly disordered systems. This chapter briefly summarizes some of the basic ideas of an important theoretical school in this field. It starts with simple consideration but later difficult fundamental problems are treated in a rather sophisticated way.

After dealing in some detail with the optical and transport properties of amorphous semiconductors, we include a chapter on some of their applications in practical devices. We believe that it is of general interest to the reader to learn about the physical principles of various devices based on amorphous semiconductors. It should also be a stimulating reading on a field whose future depends much on further ingenuity in applied research. This remark does not mean that a deeper understanding of the fundamentals should be underestimated. It is well known what a difference it made in the crystalline semiconductor technology in the fifties compared with the thirties.

The last chapter deals with liquid semiconductors. They are closely related to amorphous semiconductors but have many interesting new features.

J. Tauc

Providence

Contributors

B. G. Bagley — Bell Laboratories Incorporated, Murray Hill, New Jersey 07974, U.S.A.

Morrel H. Cohen — James Franck Institute, University of Chicago, Chicago, Illinois 60637, U.S.A.

E. N. Economou — Department of Physics and Center for Advanced Studies, University of Virginia, Charlottesville, Virginia 22901, U.S.A.

J. E. Enderby — Department of Physics, University of Leicester, Leicester, England.

Karl F. Freed — James Franck Institute, University of Chicago, Chicago, Illinois 60637, U.S.A.

H. Fritzsche — Department of Physics and the James Franck Institute, University of Chicago, Chicago, Illinois 60637, U.S.A.

R. Grigorovici — Institute of Physics of the Rumanian Academy of Sciences, Bucharest, Rumania.

E. S. Kirkpatrick — Thomas J. Watson Research Centre, P.O. Box 218, Yorktown Heights, New York 10598, U.S.A.

J. Tauc — Division of Engineering and Department of Physics, Brown University, Providence, Rhode Island 02912, U.S.A.

Contents

Chapter 1

The Nature of the Amorphous State

B. G. Bagley

Bell Laboratories, Incorporated, Murray Hill, New Jersey 07974

1.1 INTRODUCTION

This chapter is a brief introduction to the general concepts and definitions of the amorphous state. Techniques for the characterization of amorphous materials and their transformations are reviewed. Also, structural and chemical changes which can lead to inhomogeneities, and the detection of these inhomogeneities, are discussed. Because of the difficulty in completely characterizing amorphous materials, one must be aware of the possible structures, transformations, and inhomogeneities which can occur in order to better interpret measured properties.

More extensive reviews of various aspects of the subject are as follows: amorphous semiconductors, Kolomiets (1964) and Pearson (1964a); glasses, Weyl and Marboe (1962, 1964, 1967) and Rawson (1967); glass formation and glass transition theory, Turnbull and Cohen (1960) and Turnbull (1965a, 1965b, 1969); phase transformations, Turnbull (1956); and spinodal decomposition, Cahn (1968).

1

1.2   THE AMORPHOUS STATE

1.2.1   Definitions

Prior to the early 1900's, solids were designated amorphous if they had a formless fracture surface (currently termed conchoidal fracture). With the advent of X-rays and their use in characterizing crystalline materials, it was found (with only a few exceptions) that those materials which were characterized as amorphous by fractography, were also noncrystalline as evidenced by X-ray diffraction. That is, they did not exhibit the sharp reflections associated with crystalline materials, but instead, exhibited a few broad halos (Figure 1.1). Amorphous and non-crystalline are, in present terminology, synonymous. Thus we can structurally classify materials either as crystalline or non-crystalline (amorphous). A crystal is defined as a substance consisting of atoms arranged in a pattern that repeats periodically in three dimensions (Barrett (1952)). Any material which does not meet this criterion of a periodically repeated pattern of atoms is non-crystalline. The negative nature of the definition of amorphous leads to difficulties, for whether we classify a material as crystalline or not depends on a definition of how much long range order constitutes a crystal, the experimental techniques used to determine the degree of order, and possibly some intuition or prejudice. This is discussed further in Section 1.2.4 and in Chapter 2.

In addition to a structural classification, we can also classify materials according to their mechanical (flow) behavior. Thus solids are defined (Condon, 1954) as materials whose shear viscosities exceed $10^{14.6}$ poise, while fluids (liquids and gasses) have shear viscosities of less than $10^{14.6}$ This arbitrary division point (viscosity equal to $10^{14.6}$) corresponds to a

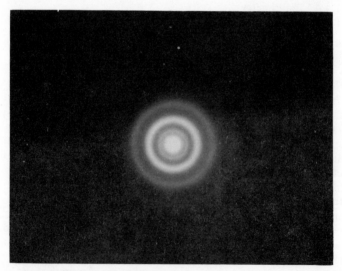

Fig. 1.1   Transmission electron diffraction pattern obtained from a vapor deposited thin film of amorphous As_2Se_3 $2 As_2Te_3$.

relaxation time of one day. That is, a material is considered to be solid if there is no permanent change in its shape upon the application of a small force for one day (Turnbull 1969).

A distinctive class of amorphous solids are glasses, which are defined (Morey, 1954) as amorphous solids obtained from the melt (solidification of a liquid). This distinction between glasses and other amorphous solids may appear pedantic. However, it is intuitively pleasing to consider an amorphous solid (glass) obtained by cooling a liquid to be both structurally and thermodynamically related to that liquid. The connection between an amorphous solid produced, for example, by extensive plastic deformation of a crystal, and other states of the material is more tenuous.

1.2.2   Glass Formation and the Glass Transition

Upon cooling a liquid below its melting point it will either crystallize, or form a glass. During crystallization, the viscosity, entropy, volume, and internal energy change discontinuously, and the transformation is first order. In glass formation, however, these properties change continuously, although the change may be rapid in the vicinity of the glass transition temperature (T_g). Figure 1.2 depicts the temperature dependence of the volume of a material which can either form a glass or crystallize. The thermal expansion and heat capacity of the glass are generally observed to be close to that of the crystal.

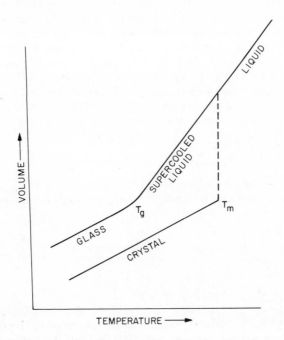

Fig. 1.2   Schematic of a volume-temperature dependence for a material in various states (after Jones (1956)).

Glass formation is not a rare phenomenon. Many materials form glasses in relatively large masses (greater than 1 gm) and at moderate cooling rates (greater than $1°C/min$) (see Rawson (1967) and Weyl and Marboe (1962, 1964, 1967) for review). There are examples of glass formers from each of the bonding types; covalent (As_2S_3), ionic $(KNO_3-Ca(NO_3)_2)$, metallic (Pd_4Si), van der Waal's (o-terphenyl) and Hydrogen $(KHSO_4)$. Glasses can be prepared which span the whole range of electrical conductivity. As examples; dielectric (SiO_2), semiconducting $(As_2Se_3.2As_2Te_3)$ and metallic (Pd_4Si).

Central to our understanding of the nature of glasses, is an explanation of the wide range of glass forming tendency. At the one extreme, for example, is As_2S_3 which is such a good glass former that it is virtually impossible to grow crystals from the melt, and at the other are the monatomic metals which have not yet been prepared as glasses using the most drastic quenches obtainable.

According to the free volume model of the liquid state developed by Cohen and Turnbull (1959) (also Turnbull and Cohen (1961)), all liquids would undergo a glass transition to an amorphous solid except for the intervention of crystallization. Thus, according to this viewpoint, glass formation becomes a question of preventing crystallization, and the explanation of why certain materials are good glass formers lies in the reasons for slow crystallization. Turnbull and Cohen (1958, 1960) have treated in detail the conditions whereby crystallization can be bypassed. Glass formation becomes more probable the greater the cooling rate, the smaller the sample volume, and the slower the crystallization rate.

The crystallization of an amorphous material proceeds by the processes of nucleation and growth, and the crystallization rate is suppressed by reducing either (or both) of these processes. We discuss first the kinetics of nucleation. Turnbull and Cohen, using nucleation theory and a number of assumptions, arrive at Eq. (1.1) (see Turnbull (1969)) as a reasonable upper limit to the nucleation frequency as a function of undercooling.

$$I = 10^{32} \exp \left[-\frac{16\pi\alpha^3\beta}{3T_r(\Delta T_r)^2} \right] \qquad (1.1)$$

Here I is the nucleation frequency (in $cm^{-3}\ sec^{-1}$), $T_r = T/T_m$ and $\Delta T_r = (T_m-T)/T_m$ (T_m is the melting point). α and β are dimensionless parameters defined as:

$$\alpha = \frac{(NV_c^2)^{1/3}\sigma}{\Delta H} \qquad (1.2)$$

$$\beta = \frac{\Delta H}{RT_m} \qquad (1.3)$$

where N is Avagadro's number, V_c the molar volume of the crystal, σ the liquid-crystal interfacial tension, and ΔH the enthalpy of melting. In Figure 1.3 the nucleation frequency as a function of undercooling Eq. (1.1)

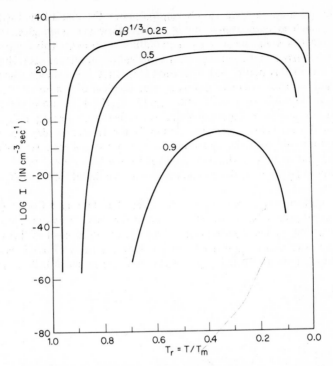

Fig. 1.3   Dependence of the homogeneous nucleation frequency (I) on undercooling for various values of $\alpha\beta^{1/3}$ (after Turnbull (1969)).

is shown for various values of $\alpha\beta^{1/3}$. The two extremes of glass forming ability are evident in the figure. It would be experimentally impossible to suppress the crystallization of liquids for which $\alpha\beta^{1/3}$ is small (< 0.25). Likewise for $\alpha\beta^{1/3} > 0.9$ homogeneous nucleation is so infrequent that these liquids would be excellent glass formers. A determination of β is experimentally accessible by simple calorimetry and values are generally between 1 and 10. Because of the difficulty in a direct measurement of σ, the determination of α is not easy. For many simple liquids a value for α of at least 1/2 seems appropriate (Turnbull (1969)). Thus $\alpha\beta^{1/3} \gtrsim \frac{1}{2}$. From Figure 1.3 we see that a liquid for which $\alpha\beta^{1/3} = 0.5$ will have infrequent nucleation events down to rather large undercoolings (down to $T_r \approx 0.8$). However, continued undercooling leads to nucleation frequencies as high as 10^{25} cm^{-3} sec^{-1}. Present experimental techniques would not allow one to quench a liquid which has this high a nucleation frequency without the material undergoing crystallization.

The pre-exponential of Eq. (1.1) contains a factor η^{-1}, where η is the viscosity. In Eq. (1.1), and therefore Figure 1.3, η was taken as 10^{-2} poise and independent of temperature (this is consistent with Eq. (1.1) being an upper bound approximation). This value for η and the assumption of temperature independence closely approximates the conditions for simple systems, such as monoatomic liquid metals. However good glass formers have viscosities which are strongly temperature dependent. This temperature dependence has a marked effect on the nucleation frequency, as is shown in Figure 1.4. In this figure, the nucleation frequency is calculated using $\alpha\beta^{1/3}$ = 0.5 and with the viscosity having a value (typical of a number of simple molecular liquids) of $10^{-3.3}$ exp $[3.34\ T_m/(T-T_g)]$ poise (Turnbull (1969)). We see from Figure 1.4 that the effect of increasing T_g/T_m is to decrease the nucleation frequency, sharpen the frequency dependence on temperature, and shift the maximum frequency to smaller undercooling, all of which aid in glass formation. Indeed, we see that even if $\alpha\beta^{1/3}$ = 0.5, there would be no difficulty in cooling to a glass a liquid for which T_g/T_m = 2/3.

The importance of the T_g/T_m ratio led Turnbull and Cohen (1960, 1961) to propose a thermodynamic criterion for glass formation. They have suggested that T_g scales with cohesive energy, and in support of this suggestion they note that for a given molecular type the ratio $kT_g/\Delta H_v$ is nearly constant (ΔH_v is the enthalpy of vaporization per molecule). Because glass

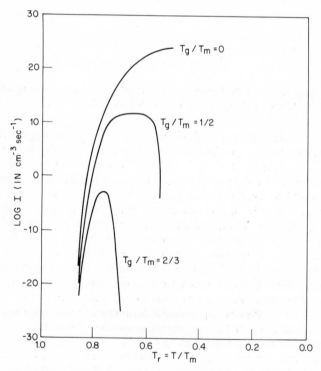

Fig. 1.4   Dependence of the homogeneous nucleation frequency (I) on undercooling for various values of the ratio T_g/T_m. $\alpha\beta^{1/3}$ is taken as 0.5 and the viscosity is assumed to vary as $10^{-3.3}$ exp $[3.34\ T_m/(T-T_g)]$ (after Turnbull (1969)).

formation is enhanced the larger is the fraction T_g/T_m, and $kT_g/\Delta H_v$ is approximately constant, Turnbull and Cohen (1961) suggested that the glass forming tendency is greater the smaller is the ratio $kT_m/\Delta H_v$. In a multi-component system, if the cohesive energy is only weakly dependent on composition, then T_g will also vary little with composition and glass formation will be enhanced for those compositions with the lowest liquidus temperatures, i.e. near eutectic compositions. The basic validity of this thermodynamic criterion is demonstrated by noting that there indeed are systems (ionic and metallic, for example) in which glasses can be formed only in eutectics, that is, the compositions where T_m is the most depressed.

The preceding discussion has focused on the kinetics of nucleation. Since growth follows nucleation, if nucleation is prevented there will be no crystallization. However, even if nucleation occurs, either homogeneously or heterogeneously on impurity particles, glasses can still be formed if the growth rate (u) is slow. Turnbull, (1969), considering growth rate kinetics and heat transport, arrives at the equation for the growth rate of a flat interface:

$$u = \frac{10^{10}\lambda\beta f}{\eta} \quad \frac{\Delta T'}{1+K} \tag{1.4}$$

where

$$K = 10^{10}R\frac{\lambda\beta^2}{V_c}\frac{\delta f}{\kappa\eta} \tag{1.5}$$

and λ is the distance the interface moves for each molecular interfacial jump, f is the fraction of crystal surface sites available for molecular attachment, R is the gas constant, κ is the thermal conductivity, and $\Delta T'$ is the reduced temperature undercooling to a sink a distance δ away. β, η and V_c have been previously defined. For $K \gg 1$

$$u = \frac{V\kappa\Delta T'}{\beta R\delta} \tag{1.6}$$

and the interface growth kinetics are determined mainly by heat transport, a condition unsatisfactory for good glass formation.

For $K \ll 1$,

$$u = \frac{10^{10}\lambda\beta f\Delta T'}{\eta} \tag{1.7}$$

and the growth kinetics are determined mostly by molecular transport across the interface, a situation conducive to glass formation.

For most materials $K \sim 2 \times 10^9$ $(f\delta/\kappa\eta)$ and for an upper bound calculation we take f = 1 and δ = 0.1 cm. Thus, for K to be less than 1 for a good thermally conducting (metallic) liquid, the viscosity (η) must be greater than ~ 10 poise. Likewise, for the nonmetallic less thermally conducting liquids, η must be greater than $\sim 10^3$ poise. As an indication of how slow growth is in a glass, at T_g ($\eta = 10^{15}$ poise) Eq. (1.4) predicts an upper bound (maximum growth rate) of approximately 10 nm per day (Turnbull (1969)).

It has long been recognized that good glass formation is associated with a high viscosity at the melting point (SiO_2, for example, has a viscosity of 10^7 poise at $T_m = 1710°C$). This had led to the 'geometrical' theories of glass formation, the most well known of which is Zachariasen's (1932) coordination rules for oxide glasses. Weyl and Marboe (1962, 1964, 1967) and Rawson, (1967) have reviewed in detail the geometrical theories. As a description for glass formation, the crystallization kinetic approach of Turnbull and Cohen and the geometrical theories are complementary. The kinetic approach formulates the problem in terms of experimental conditions (sample volume and cooling rate) and 'macroscopic' material parameters (η, σ, f, T_g/T_m, etc.). The geometrical theories, on the other hand, are an attempt, using crystal chemistry, to correlate the molecular constituents and structure to physical properties, concentrating on those properties which most affect glass formation. For example, this approach would contend that a theory which explains why SiO_2 is so viscous at its melting point also explains why it is a glass former. The chemical and structural aspects of semiconducting glass formation have been recently discussed by de Neufville (1972).

We now discuss the property changes which occur when crystallization is avoided and we cool a liquid to form a glass. For most materials (pure SiO_2 and GeO_2 are exceptions), on cooling the liquid there is a sharp (over a narrow temperature range) but continuous change in the slope of the specific volume, i.e. a change in the thermal expansion coefficient (Figure 1.2). Accompanying this change in thermal expansion, there is a marked, but again continuous, change in specific heat. The changes in both thermal expansion and specific heat are from values characteristic of the liquid, to values near those of the corresponding crystalline solid. The temperature at which the properties change abruptly is denoted the glass transition temperature, T_g (Figure 1.2).

As the temperature of a fluid is lowered, the viscosity increases. Another common definition of T_g is the temperature at which the fluid becomes solid, with two designations having wide acceptance, T_g is the temperature at which (1) $\eta = 10^{14.6}$ poise (Condon, 1954) or (2) $\eta = 10^{13}$ poise (Parks and Gilky (1929).

While most physical properties manifest some change in the vicinity of T_g, the electrical conductivity of glassy semiconductors is an exception. Vengel and Kolomiets (1957) observed no change in either the temperature dependence or absolute value of the electrical conductivity upon cooling through the glass transition of several glassy chalcogenides from the As_2Se_3 − As_2Te_3 system.

For a property which manifests a change at the glass transition (specific volume for example), the experimentally determined value below T_g is dependent on the sample thermal history. For example, for faster cooling rates, the glass is observed to have a greater specific volume. In addition, for measurements performed while the sample is cooling. T_g is observed to decrease with decreasing cooling rate. Property measurements made while the temperature is increasing through the glass transition, are dependent not only on the heating rate, but also on the cooling rate with which the glass was initially formed. When the heating and cooling rates are identical, hysteresis effects are not experimentally observed. These effects are illustrated in Figure 1.5 (for specific volume) and Figure 1.6 (for specific heat).

These kinetic effects clearly indicate that the glass is relaxing towards a more stable state. The two possibilities for this state, are (1) the crystalline state, as suggested by Kauzman (1948) or (2) an ideal glassy state in internal metastable equilibrium, as suggested by Gibbs and Di Marzio (1958) and Cohen and Turnbull (1964). Turnbull (1971) and Turnbull and Polk (1971) have extensively reviewed the evidence supporting each of these two possibilities.

In the same way that many properties of the crystalline state are viewed as being determined by imperfections in an ideal, perfect, crystal, it is useful to view the properties of glasses as influenced by 'imperfections' in an ideal,

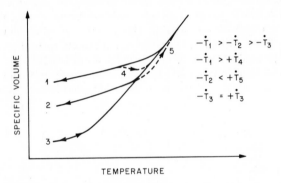

Fig. 1.5　The effect on the specific volume of various cooling $(-\dot{T})$ and heating $(+\dot{T})$ rates through the glass transition.

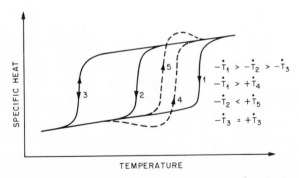

Fig. 1.6　The effect on the specific heat of various cooling $(-\dot{T})$ and heating $(+\dot{T})$ rates through the glass transition

perfect, glass. As examples of such imperfections we have 'dangling bonds', 'excess volume', composition and density fluctuations and internal stresses.

Gibbs and Di Marzio (1958) and Cohen and Turnbull (1964) have suggested that this ideal glass will have zero entropy at $T = 0\ °K$. This is in contrast to the disorder view of entropy and the Lewis and Gibson (1920) suggestion that glasses will always contain a residual entropy. Experimental resolution of these differing viewpoints has not yet occurred. Simon (1930) observed that the residual entropy of glassy glycerol at absolute zero was 5 cal/mole deg. However, Oblad and Newton (1937) demonstrated that the residual entropy value was dependent upon the measurement rate, and equilibrating for longer times reduced the observed entropy. The concept of the ideal glassy state would suggest that even larger equilibrating times are needed as the temperature is reduced, in order to remove the 'imperfections' and thereby reduce the entropy to an immeasurable value. Zeller and Pohl (1972) have recently observed a contribution to the low temperature specific heat of glass which varies linearly with temperature. They suggest that this linear contribution may be a general characteristic of the glassy state as it was observed in all the glasses they examined, including both organic and inorganic materials. These authors also observed that the low temperature thermal conductivity of glassy materials varies with temperature as $T^{1.8}$. These experimental results have been interpreted by Fulde and Wagner (1971), and Anderson, Halperin and Varma (1972).

Over a broad temperature range the molecular transport properties of the fluid are very well described by the empirical Fulcher (1925) equation. For the fluidity (φ),

$$\phi = 1/\eta = \phi_o \exp\left[-\frac{A}{T-T_o}\right]; \tag{1.8}$$

for diffusion

$$D = T\,D_o \exp\left[-\frac{A'}{T-T'_o}\right]; \tag{1.9}$$

and for ionic conduction,

$$J = J_o \exp\left[-\frac{A''}{T-T''_o}\right]; \tag{1.10}$$

A, A', A'', T_0, T'_0, T''_0, φ_0, D_0, and J_0 are all experimentally determined constants and T is the temperature in $°K$. For measurements on the same material and where the transport of the same molecular species is rate controlling for all three processes, it has been observed that,

$$A \cong A' \cong A'' \text{ and } T_o \cong T'_o \cong T''_o.$$

It has also been observed that for 'spherical molecules', D and φ in the fluid state are related by a generalized Stokes-Einstein relation, $D = BT\,\varphi$. B is a constant with a value often near to the Stokes-Einstein value of $(k/3\pi a_0)$ where k = the Boltzmann constant and a_0 is the molecular diameter.

Theory must account for these properties of fluids, their temperature dependence, the glass transition, and the properties of the amorphous solid. The two most widely accepted theories are the entropy theory, as formulated by Gibbs and Di Marzio (1958), Gibbs (1960), and Adam and Gibbs (1965), and the free-volume theory as developed by Eyring (1936), Fox and Flory (1950), Williams, Landel and Ferry (1955), Cohen and Turnbull (1959), and Turnbull and Cohen (1961, 1970).

The entropy theory is the result of a statistical mechanical calculation based on a quasi-lattice model. The configurational entropy (S_c) of a polymeric material was calculated as a function of temperature by a direct evaluation of the partition function (Gibbs and Di Marzio (1958). The results of this calculation are that, (1) there is a thermodynamically second order liquid to glass transformation at a temperature T_2, and 2), the configurational entropy in the glass is zero; i.e. for $T > T_2$, $S_c \Rightarrow O$ as $T \Rightarrow T_2$, and for $T < T_2$, $S_c = 0$. This transformation is not experimentally observed, however, for practical cooling rates result in a kinetically limited glass transition at T_g, a temperature higher than T_2. Infinitely slow cooling rates would be required for the direct observation of the second-order glass transition. Adam and Gibbs (1965) extended this earlier work to include molecular transport by considering cooperative atomic rearrangements, thermally activated over an energy barrier. As an expression relating shear viscosity and configurational entropy, they derive,

$$\eta = \eta_0 \exp(E/T\,S_c), \qquad\qquad (1.11)$$

where η_0 and E are constants. The Adam-Gibbs theory is applicable to all materials; i.e., it is not restricted to polymers alone.

According to the free volume theories, molecular transport in all fluids, including 'simple' fluids, is determined by the magnitude and temperature dependence of the free volume; the free volume being that part of the total volume not 'occupied' by the molecules and thus available for redistribution over the entire sample volume. Below the glass transition the free volume is approximately zero, temperature independent, and distributed uniformly over all the molecules; major fluctuations which could lead to a redistribution of free volume over the sample volume do not occur. Atomic transport in the glass is determined by 'solidlike' diffusive displacements which require a cooperative motion of several molecules to provide sufficient free volume locally for the diffusive step to occur. Thermal expansion in the glass is determined by the anharmonicity in the atomic pair potentials, just as in a crystalline solid. As the temperature is increased, there is, at T_g, a rapid increase in free volume, an appearance of configurational entropy, and the material undergoes the glass transition. This free volume can be redistributed over the entire sample volume with changes in potential energy no larger than kT. A decreasing density fluctuation which increases the potential energy locally (over the nearest neighbor coordination sphere) is offset by a locally increased density fluctuation which decreases the energy elsewhere in the sample. In this fluid state, atomic transport is determined by an increasing proportion of 'gaslike' diffusive displacements and the self-diffusion coefficient (D) is given by the relationship (Turnbull and Cohen (1970));

$$D = 1/3 \; \bar{u} \; d\left[v^* + (v_f/p) \right] \exp \; (-p \, v^*/v_f) \tag{1.12}$$

where $\bar{u}$ is the gas kinetic velocity, d is proportional to the diffusive displacement, v^* is a critical volume required before a diffusive jump can occur, v_f is the free volume, and p is an overlap factor ($\frac{1}{2} < p < 1$).

The kinetic nature of the experimentally observed glass transition (at T_g) is ever present. For any reasonable cooling rate, a temperature will always be reached where equilibrium cannot be maintained and the properties of the solid (including specific volume and configuration entropy) will be determined by its thermal history. This kinetic nature of the glass transition led to the concept of the 'fictive' temperature, as the temperature at which the glass would be in thermodynamic equilibrium (Tool (1964)). The fictive temperature of a material thus depends upon each sample's thermal history.

1.2.3 Preparation

Liquids with small crystallization kinetic constants can form glasses directly from the melt (Section 1.2.2), and for glass formation and stability there is extensive data on many glasses. Often, however, there are no data for a particular composition of interest. Differential scanning calorimetry and differential thermal analysis with programmed cooling rates are ideal techniques for measuring glass stability. One determines the fastest cooling rate necessary to avoid crystallization, which is detected thermally as an exothermic reaction. Such a study has been made on the As_2Se_3-As_2Te_3 system, and the increasing glass stability as selenium is substituted for tellurium was clearly demonstrated (Bagley and Blair (1970)).

If the composition is dictated for reasons other than stability, then the crystallization kinetic constants are also determined, and the only variables are the cooling rate and sample size. It is desirable to cool as slowly as possible to avoid stresses and to allow an approach to the ideal glassy state. Care must always be exercised, however, that partial crystallization and/or phase separation have not occurred and gone undetected. The care with which resulting glasses must be examined for inhomogeneities is determined by the sensitivity of the properties of interest to the presence of such inhomogeneities.

For a good glass former such as As_2S_3, the crystallization kinetics are so slow so as to allow very slow cooling of a large mass and still obtain a glass. Indeed, it is very difficult to grow As_2S_3 crystals from the melt. As the possibility for crystallization increases, the liquid must be quickly cooled through the region of maximum crystallization rate. As_2Se_3 is a less stable glass than As_2S_3 and must be cooled more quickly from the melt in order to form a homogeneous glass free of microcrystals. As the kinetics increase further, there reaches a point at which the bulk material (a few grams) cannot be cooled directly from the melt fast enough to form a homogeneous glass. As_2Te_3 is an example of such a material.

If a very fast quench (into iced brine, for example) of a small bulk sample is not fast enough to avoid crystallization, then techniques experimentally more difficult must be used. 'Splat' cooling is the most drastic quench obtainable on a liquid which can result in a solid still properly termed a glass. In the gun technique of splat cooling, a small amount (milligrams) of material is shot onto a thermally conducting substrate (Duwez and Willens (1963)). This spreads the liquid out and forces good thermal contact with the substrate resulting in a high quenching rate. Cooling rates of about $10^7°C/sec$ have been obtained by this technique (Predecki et al. (1965)). Duwez and coworkers have investigated the properties of tellurium based amorphous semiconductors prepared by splat cooling. Interestingly, they observed that pure tellurium, even at this very fast quench rate, could not be quenched into the glassy state This is in contrast to selenium, which is a good glass former and can be prepared in bulk by relatively slow cooling rates. Luo and Duwez (1963), however, observed that tellurium containing 10–30 at .% gallium, 10–30 at .% indium, or 10–25 at .% germanium could be obtained in the amorphous state. The structures (Luo, 1964), crystallization behavior (Willens (1962)), transport properties (Tsuei (1966)), and Mossbauer spectra, (Tsuei and Kankeleit (1967)), of these binary systems were investigated. The piston and anvil splat cooling technique (Pietrokowsky, 1962), in which a small drop is smashed between two thermally conducting surfaces, yields only powders when quenching the brittle amorphous semiconductors.

There is a class of techniques for preparing amorphous solids in which the liquid state is bypassed completely. These include electrodeposition and the various techniques for deposition from the vapor state, i e. vacuum deposition, sputtering, and decomposition of gaseous compounds by radio frequency discharges. There are several reasons one might choose one of these techniques for the preparation of amorphous materials. First: with a proper choice of deposition parameters (substrate temperature and deposition rate) very high equivalent quenching rates can be obtained. Thus many amorphous materials which cannot be obtained as glasses by quenching the melt can be obtained by these techniques. Amorphous GeTe, for example, is not obtainable as a glass, but it can be obtained by vapor quenching. Its interesting properties have been extensively studied (Chopra and Bahl (1969), Betts et al. (1970), Bahl and Chopra (1969, 1970)). Secondly, there are times when these techniques must be employed because the desired amorphous solid is not directly connected with the equilibrium liquid state. Amorphous tetrahedrally bonded semiconducting germanium and silicon are examples. Both of these elements melt to a dense metallic liquid, the semiconducting liquid being experimentally inaccessible. Finally, these preparation techniques are ideal for the deposition of experimentally advantageous, technologically important, thin films.

Experimental aspects of vapor deposition have been extensively reviewed by Chopra (1969). Many amorphous semiconducting thin films have been prepared by the techniques of vacuum evaporation and cathodic sputtering. The most extensive work has been done in those materials which exhibit switching and memory effects (Chapter 6), and in the elemental semi-conductors, silicon and germanium. In spite of the large number of studies on vapor deposited amorphous Si and Ge, there remain several discrepancies and unresolved questions concerning their properties (see Chapters 4 and 5). The variation in properties is likely due to the variation in deposition para-meters (quality of vacuum, deposition rate, substrate temperature, etc.) and the discrepancies will be resolved with a more complete characterization of the resulting films.

Because of a preferential vaporization of one or more of the components, a difficulty often encountered in the vacuum evaporation of multi-component systems is maintaining composition control. An evaporated thin film (multicomponent) is very likely to have a different composition than the bulk starting material, and the problem is compounded by the difficulty in obtaining an accurate chemical analysis on the film. Better composition control is obtained by flash evaporation or sputtering.

Chittick, Alexander and Sterling (1969) and Chittick (1970) have des-cribed a technique for the preparation of amorphous silicon and germanium by using a radio-frequency glow discharge decomposition of silane and germane gases. Based upon the estimated amount of oxygen in the system (10 ppm), and the absence of the strong silicon-oxygen absorption in the infrared spectrum, these films are judged to be less contaminated by oxygen than films prepared by the other deposition techniques. These authors report the results of electrical, optical, photoconductive, and density measurements. They conclude that the properties of these films are qualitatively similar to those obtained on films prepared by evaporation, but quantitative distinctions must await more detailed characterization of the films produced by both techniques. LeComber and Spear (1970) have made detailed drift mobility and conductivity measurements on amorphous silicon prepared by this glow discharge technique

Electroplating is the deposition of a material by the passage of a current from electrodes through an electrolyte (see Lowenheim (1963) for a review). Whether this electrolyte is aqueous or not depends upon the desired electrodeposit. Germanium, for example, has thus far only been deposited from nonaqueous solutions. Fink and Dokras (1949) found that elemental germanium could be deposited from solutions of GeI_4 in various glycols. Szekely (1951) improved the process by deposition from a solution of $GeCl_4$ in propylene glycol at a current density of 40 amps/dm^2. Szekely's process has been used to electrodeposit amorphous germanium for thermal (Chen and Turnbull, 1969), small angle X-ray scattering (Chen, 1967), and infrared absorption (Tauc et al. (1970a)) studies. No other amorphous semi-conductors have been prepared by electrodeposition.

It should also be noted that crystalline solids can also be made amorphous by irradiation with neutrons or ions, or by massive plastic deformation. The structure (Large and Bicknell (1967), Moss et al. (1971)), optical reflectance (Kurtin *et al.* (1969)), electron spin resonance spectra (Crowder *et al.* (1970)), and Raman spectra (Smith *et al.* (1971)) of amorphous silicon obtained by ion bombardment are very similar to that observed in amorphous silicon obtained by evaporation or sputtering.

1.2.4 Structure

Amorphous materials are often defined operationally by their diffraction patterns. That is, the diffraction patterns consist of a few broad halos rather than sharp Bragg reflections. However, markedly different structures can lead to qualitatively the same diffraction patterns. A small crystallite size, strains, and imperfections, broaden the normally sharp crystalline reflections, and as the crystal size gets smaller a point is reached where the crystalline reflections overlap and the diffraction pattern appears like that for an amorphous material. The crystallite size at which the diffraction pattern appears amorphous has been discussed by Germer and White (1941), and Piggott (1966), and is approximately four unit cells in diameter for monatomic close packed materials. Thus, diffraction cannot distinguish between a random arrangement of atoms, as suggested by Bernal (1959, 1960) for monatomic systems and Zacharison (1932) for polyatomic systems, and a microcrystalline one, where each microcrystallite contains approximately two hundred atoms. Indeed, because of this, microcrystalline models for the amorphous state had been proposed for polyatomic systems by Valenkov and Porai-Koshits (1937) and for monatomic systems by Mott and Gurney (1938). The crystallite size which will yield an amorphous diffraction pattern will differ from material to material, but it can be estimated from the observed crystalline diffraction pattern (Piggott (1966)). The microcrystalline structure can be further complicated when there is the possibility of its also being polyphase. Of course, any of the infinite number of structures between the continuous random packing and the microcrystalline will also yield amorphous diffraction patterns. In view of the variety of preparation techniques and materials, there is no reason to believe that all amorphous solids are characterized by just one of these structures exclusively. Also, there is no reason to conclude that vapor or electro-deposition cannot produce the ideal random structures often associated with solids obtained from the melt (i.e. glasses).

There are times when it is important to be able to distinguish between these various structures. The preparation techniques such as vapor deposition, electrolytic deposition, and splat cooling are yielding amorphous semiconductors not obtainable in bulk from the melt. While these materials could have the continuous random structure, experimental evidence also indicates the other structures are possible. For example if the substrate temperature for vapor deposition or splat cooling is lowered continuously, one observes finer and finer crystallite sizes until the diffraction pattern finally becomes that for an amorphous solid. This strongly suggests, of course, that these amorphous solids may be microcrystalline and one would like to distinguish between the structures experimentally. It is often observed that the properties of amorphous semiconducting thin films are very sensitive to the preparation technique and thermal history (see Chapters 4 and 5). It is important to be able to distinguish whether or not these differences are due to different structures, and whether the thermal effects are due to structure changes (as examples: crystallite coarsening, relaxation to a more stable random structure or a reduction in the number of 'broken bonds').

We shall see in the balance of this section and in Chapter 2 that subtle differences in amorphous structure cannot be characterized by the present resolution of experimental techniques normally used for structural investigations. It may turn out that, rather than determining the structure and correlating properties with it, the properties (such as optical) may be a more sensitive tool from which the structure must be inferred.

What techniques can experimentally distinguish between the various structures? We can distinguish the microcrystalline polyphase structure from the others by electron microscopy or small angle X-ray scattering (if the various phases differ in electron density), or NMR or Mossbauer (if the phases differ in their spectra). Distinguishing between the more homogeneous structures, the continuous random arrangement and the microcrystalline single phase, is much more difficult however. When the crystallite size is about 10–20 Å the number of atoms in the boundaries is comparable to the number of atoms in the interiors of the crystallites. The nearest neighbor coordination of these interfacial atoms is quite distorted and may appear very much like that of the random structure. Thus techniques which look at the nearest neighbor coordination, such as NMR and Mossbauer, will not be able to resolve the structures. Also, when the crystallite size is as small as 10–20 Å, the crystals and their boundaries are difficult to detect by electron microscopy. Rudee and Howie (1972) have recently reported resolving 14 Å microcrystallites in amorphous Ge but the interpretation of their results is not conclusive (Chandari et al. (1972), Berry and Doyle (1973)). And, finally, at the present time the detailed atomic arrangements for the various models are not well enough defined and large-angle diffraction techniques (electron, X-ray, and neutron) are not accurate enough to definitively resolve the various structures experimentally, although progress is being made in this direction (see Chapter 2).

Small angle X-ray scattering, being sensitive to long range electron density variations, is in principle able to distinguish between homogeneous and inhomogeneous atomic arrangements, and thus between the dense random and microcrystalline models. Because of this, and the observation of a strong small angle X-ray scattering from isolated crystallites, it has been frequently concluded that an absence of small angle X-ray scattering invalidates a crystallite model. Warren (1937) was the first to present these arguments. He noted that the X-ray diffraction pattern of vitreous SiO_2 obtained by Warren, Krutter and Morningstar (1936) was similar to what would be obtained from 7.7Å crystallites of cristobalite. He did not, however, observe small angle scattering and thus concluded; 'the absence of small angle scattering means that the glass is a continuous medium with no discrete particles or breaks in the scheme of bonding'. While small angle scattering should be sensitive to the inhomogeneous excess volume (void space) distribution, Bienenstock and Bagley (1966) have presented an upper bound calculation which demonstrates that, for the appropriate values of crystallite size and density, it is very difficult experimentally to distinguish between the models, and conclusions based on earlier work are unwarranted. The calculation did not give support to one model or another but demonstrated that, in general, the technique is not definitive. Experimentally the results are even more difficult to interpret because a microcrystalline material will have small angle scattering due to Bragg double reflection, and the random model will have scattering due to density fluctuations.

When direct experimental techniques fail to distinguish between the structural models, we may call upon more intuitive arguments. Turnbull (1962) has pointed out that the resistance of liquids and glasses to nucleation implies something about their structure. If an amorphous material contained microcrystalline regions, there would be no nucleation barrier, and the crystallization kinetics would be limited only by grain growth. Thus, at temperatures just below the melting point, slow crystallization kinetics would indicate nucleation was required. This argument is most applicable to glasses which do indeed crystallize slowly upon heating to just below T_m. In most of these glasses, once nucleation occurs growth proceeds rapidly. Thus it is difficult to argue that these materials were originally microcrystalline.

Turnbull (1965b) has also proposed a distinction between the structures based upon a density deficit argument. In a dense random structure there are no structural discontinuities and the density is determined only by the density of the random packing. Bernal and Mason (1960) and Scott (1960) have shown experimentally that the density for the dense random packing of hard spheres is 86% that of close packing and Polk (1971) has demonstrated that the density for random packing in a tetrahedrally bonded system is only 2% less than the corresponding crystal. In the microcrystallite model, however, there is a structural discontinuity at the crystallite boundary which can markedly reduce the density. In a hard sphere system if we assume the boundaries to consist of 1/2 of a monolayer, the density of a solid consisting of crystallites four atoms in diameter would be 12–17% less dense than the density of the corresponding bulk single crystals. For a microcrystalline size three atoms in diameter the density would be 15–23% less (Bagley et al. (1968)). These crystallite diameters are about the size at which the diffraction pattern becomes amorphous. The corresponding calculation for a tetrahedrally bonded system has not, as yet, been done. Rather than having a large density deficit, however, the actual densities of most glasses are only a few percent less than that of the corresponding crystal. The experimental situation for the vapor deposited thin films is not as clear, however, where density deficits or from 0% to 27% have been reported for amorphous germanium. (Clark (1967), Donovan et al. (1969, 1970) Chopra and Bahl (1970) and Light (1969)), and a deficit of from 3.5% to 27% for amorphous silicon (Mogab and Block (1971)).

Turnbull (1965a) has also pointed out that the mode of the amorphous to crystalline transformation yields information about the amorphous phase structure. It is a consequence of the Cohen-Turnbull theory (1959) of the amorphous state that there is a basic discontinuity, both structural and thermodynamic, between the amorphous and crystalline states. This implies that the amorphous to crystalline transformation must occur by nucleation and growth. A microcrystalline solid, on the other hand, without the nucleation barrier, is expected to transform analogous to a grain growth process. The characteristic of this type of process is that many grains grow at the expense of a few, the driving force being the reduction of interfacial energy. Thus, the average crystallite size of the microstructure increases continuously, and the diffuse diffraction pattern sharpens continuously, as the transformation proceeds. In nucleation and growth the microstructure and the diffraction pattern are those of crystals in an unchanged amorphous matrix. Thus, observations on the mode of transformation may give insight as to the original structural state. The dendritic crystallization observed by Willens (1962) in the Ge-Te system is a good example of a nucleation and growth transformation. The mode of transformation argument has been used by Chopra and Bahl (1970) and Moss and Graczyk (1969) to infer that the structures of their vapor deposited amorphous films were random. Care must be exercised in the use of this criterion, however, since copious nucleation and subsequent growth would appear like crystallite coarsening. Likewise, the transformation of microcrystalline polyphase materials, and microcrystalline materials with decreased boundary mobilities due to impurity poisoning, could appear as nucleation and growth. It is likely, however, that in this case the amorphous matrix would still suffer grain coarsening characteristics. Also, one could conceive of an amorphous material consisting of microcrystals of one structure in which, by nucleation and growth, crystals of a different structure grew.

1.3   INHOMOGENEITIES IN GLASS

1.3.1   Sources

It is convenient to classify the sources of inhomogeneities in glass into two groups. This classification is useful for indicating how the inhomogeneities arise, and how we can eliminate them when they occur. The first group are those inhomogeneities which arise because of the intrinsic properties of the material. The intrinsic inhomogeneities include density fluctuations, structural inhomogeneities (crystals in a glassy matrix) and compositional inhomogeneities (phase separation and crystallites with a composition different than the matrix). These intrinsic inhomogeneities will be discussed in more detail.

The second group of inhomogeneity sources are extraneous to the material itself and are the result of the preparation procedure. These extraneous sources include bubbles or voids, incomplete reaction of starting materials, contamination from the container, and motes (insoluble or slowly soluble particulate material) introduced through the starting material.

The preparation of a homogeneous glass is not trivial. There are some sources of inhomogeneities which can be removed in obvious ways. Proper selection of container material, the use of cleaner mote-free starting materials, and the selection of compositions which preclude phase separation are examples. However, the preparation procedures for avoiding some inhomogeneities are sometimes conflicting. As examples; a fast quench to avoid crystallization introduces excess volume which results in larger density fluctuations and inversely, using a very slow cooling rate to maximize the density and minimize strains increases the likelihood of nucleating crystals. Likewise, the higher temperature and longer reaction times which may be required for a complete reaction of starting materials and the removal of voids also increases the possibility of contamination from the crucible. In cases such as these, the actual preparation technique is always a compromise.

1.3.2 Density Fluctuations

In this section we consider the density fluctuations which occur in a dense (far from the liquid-gas critical point) single component system. Such density fluctuations play a role in optical properties (Section 1.4 and Chapter 4) and are thought to play a role in transport properties (Chapter 5). Composition fluctuations in multicomponent systems are considered in Section 1.3.4.

In the liquid, the mean squared value for the density fluctuation ($\Delta\rho$) for the total system (sample) is given by the equation,

$$< (\Delta\rho)^2 > = -\frac{kT\rho^2}{V^2} \left(\frac{\partial V}{\partial P}\right)_T = \frac{kT\rho^2}{V} K_T \tag{1.13}$$

where ρ is the density, V the volume, P the pressure, and K_T the isothermal compressibility (see Münster (1969) for an extensive review of fluctuation theory). We see, from Eq. (1.13), that experimentally resolvable density fluctuations vanish for macroscopic sample sizes.

The density fluctuation of the total sample is the superposition of local fluctuations in a dynamic, time dependent, equilibrium. We now consider, as a subsystem of the total liquid sample, a fixed (but arbitrary) volume ΔV which contains a variable number of molecules. The local fluctuations are considered random with no correlation between the various ΔV's. This subsystem can be considered as a grand canonical ensemble in equilibrium with its surroundings (the balance of the sample). Then,

$$< (\Delta\rho)^2 > = -\frac{kT\rho^2}{(\Delta V)^2} \left(\frac{\partial \Delta V}{\partial P}\right)_T \tag{1.14}$$

The mean squared value for the density fluctuation increases as smaller subsystem volumes (ΔV) are specified. There is, however, a lower limit to the ΔV that can be considered. ΔV must be large enough to give meaning to the thermodynamic quantities used, and to make a connection to experiment through macroscopic parameters such as the isothermal compressibility. This lower limit on ΔV is typically $10^6 - 10^8$ Å.[3]

As the temperature of the liquid is lowered towards that of the glass transition, as noted previously, there will be a temperature at which the liquid cannot attain thermodynamic equilibrium (metastable with respect to the crystal) in the time scale of typical experimental conditions For example, a fluid with a viscosity of 10^{13} poise and shear modulus (G) of 10^{10} dynes/cm^2 has a relaxation time (τ) of 10^3 seconds ($\eta = G\tau$) and would require experimental times in excess of this to achieve equilibrium. When equilibrium is no longer maintained (with decreasing temperature) the density fluctuations, including those due to excess volume, become irreversibly time dependent, and the formalism becomes very complex.

There are two hypothetical limiting cases of interest. In one, an infinitely slow cooling rate maintains thermodynamic equilibrium to the ideal glass, and the equilibrium formalism is applicable. In the other a fluid in equilibrium (at its fictive temperature) is quenched infinitely fast to a temperature low enough so that no molecular transport occurs. In this case, what were dynamic fluctuations in time becomes static fluctuations in space. The most elementary treatment of this glass is then as a thermodynamic system with one additional parameter, the fictive temperature. In an actual experiment, of course, relaxations take place and the state of the system is dependent upon its entire thermal history and requires many parameters for its definition. Detailed discussion of the use of irreversible thermodynamics for the study of relaxation processes in liquids and glasses is contained in reviews by Davies (1956, 1960).

1.3.3 Crystallization
As noted in Section 1.2.2, crystallization proceeds by a two-step process consisting of the formation of a nucleus and its subsequent growth.

Nucleation
For monatomic systems, the steady state nucleation frequency per unit volume in a clean system (homogeneous nucleation), neglecting strain energy (valid for the crystallization of a fluid), is given by the equation (Turnbull and Fisher (1949)),

$$I = K_v \exp \left[- \frac{(\Delta G^* + \Delta G_a)}{kT} \right] . \tag{1.15}$$

Here ΔG^* is the critical activation free energy which, for a spherical nucleus and isotropic liquid-crystal interfacial tension (σ) and the liquid-crystal free energy difference per unit volume (ΔG_v), is given by,

$$\Delta G^* = \frac{16 \pi \sigma^3}{3 (\Delta G_v)^2} \tag{1.16}$$

ΔG_a is the free energy activation barrier for an atom to cross the liquid-crystal interface, and the kinetic constant, K_v, is given approximately by the value,

$$K_v \simeq N_\varrho \nu_o \qquad (1.17)$$

where N_ϱ is the number of molecules per unit volume in the liquid state and ν_o is an atomic attempt frequency (often approximated by the Debye frequency).

The application of these equations to the group of materials classed as amorphous semiconductors is complicated by the variety of structures involved (Chapter 2). The possible structures vary from the linear polymeric group VI elements (S, Se), through those partially crosslinked by the introduction of group IV and/or V elements, to the completely crosslinked three dimensional network structures such as Ge or $GeSe_2$. Hoffman and Lauritzen (1961) have treated in detail the nucleation aspects of polymeric materials with specific model calculations for ΔG^* and σ.

According to classical nucleation theory, the nucleation energy barrier, ΔG^*, (Equation 1.16) arises from the need to create a crystal-liquid interface of area ΔA and the resulting energy increase $\sigma \Delta A$. Thus the magnitude of σ is a measure of nucleation resistance. A number of authors (Turnbull (1950a, 1950b, 1965a), Jackson and Chalmers (1956), Skapski (1956), Hoffman and Lauritzen (1961), Ewing (1971)) have derived correlations between the liquid-crystal interfacial tension and the enthalpy of melting (ΔH). One of these is (see Hoffman and Lauritzen (1961)),

$$\frac{\sigma}{\Delta H} = e \, d_o \qquad (1.18)$$

where e is a constant and d_o is the crystalline lattice spacing. Experimentally, σ is inferred from a determination of the nucleation frequency (or the temperature of maximum undercooling) and the use of Eq. (1.15). Maximum undercooling is achieved by the use of a technique in which the material is finely dispersed to avoid heterogeneous nucleation (Turnbull, 1950c). By use of this technique Turnbull (1950b) observed e to be about 0.5 for metals, Thomas and Stavely (1952) observed $e \approx 0.3$ for nonpolymeric molecular liquids, and for organic polymers $e \approx 0.1$ (Turnbull's experimental work cited in Hoffman and Lauritzen (1961)).

Recently, Crystal (1970) has studied in detail the crystallization of selenium. By application of the Hoffman-Lauritzen (1961) theory of polymer crystallization, Crystal obtains values for the liquid-crystal interfacial tension of selenium of 9.5 ergs/cm^2 for the lateral face and 337 ergs cm^2 for the chain folded face. He also finds $e \approx 0.1$ to be a valid approximation for this inorganic polymer.

We consider now heterogeneous nucleation, which is a nucleation 'catalyzed' by the presence of foreign material. Important examples of such catalyzing materials are impurity particles (motes) and container walls. Theoretical analysis (see Turnbull (1956) for review) results in the equation (analogous to Eq. 1.15) for the heterogeneous nucleation frequency,

$$I_h = K_h \exp \left[- \frac{\Delta G_h^* + \Delta G_a}{kT} \right] . \qquad (1.19)$$

For heterogeneous nucleation on a flat surface,

$$\Delta G_h^* = \Delta G^* \; \frac{(2 + \cos \theta) \; (1 - \cos \theta)^2}{4} \tag{1.20}$$

where θ is the nucleus-catalytic surface contact angle. Thus the presence of the nucleating surface serves to reduce the energy barrier by reducing the interfacial energy contribution, and as the contact angle approaches zero, the energy barrier disappears. Analogous to Eq. (1.17), the preexponential for heterogenous nucleation (K_h) is given by,

$$K_h \approx N_s \nu_o, \tag{1.21}$$

where N_s is the number of atoms per unit area in contact with the catalytic surface sites.

Another source of nucleating sites are crystalline regions of the parent material entrapped in micro cavities in the container walls or in motes (Turnbull (1950d)). For a cylindrical microcavity of radius r_c the crystalline phase will be stable for all cavities for which

$$r_c \le \frac{2 \, \sigma \cos \theta}{\Delta G_v} \tag{1.22}$$

Superheating above T_m is required to melt these entrapped crystalline nuclei. There is a direct correspondence between the ability to undercool and the extent of prior superheating, until the undercooling is large enough for surface heterogeneous nucleation (Eq. 1.19) to intervene.

Growth
We now consider the growth, without a change in chemical composition, of a crystalline nucleus from its liquid. Viewing crystal growth as an activated process over an interfacial energy barrier ($\Delta G_a'$), Hillig and Turnbull (1956) (see also Turnbull (1956), and Turnbull and Cohen (1960)), using rate theory, obtain the equation for growth rate (u),

$$u = f \lambda \nu_o \exp \left(- \frac{\Delta G_a'}{kT} \right) \left[1 - \exp \left(\frac{V_m \Delta G_v}{kT} \right)' \right] \tag{1.23}$$

where V_m is the molecular volume and all other terms have been previously defined. $\Delta G_a'$ is experimentally inaccessible, so to allow comparison with experiment, Turnbull and Cohen (1960) assumed that crystal growth and diffusion are determined by the same kinetic process. Thus using

$$D \approx \nu_o \lambda^2 \exp \left[- \frac{\Delta G_a'}{kT} \right] \tag{1.23}$$

and the Stokes-Einstein relation we have an expression relating crystal growth to fluidity,

$$u = \frac{fkT\phi}{3\pi\lambda^2} \left[1 - \exp \left(\frac{V_m \Delta G_v}{kT} \right) \right]. \tag{1.24}$$

The fraction of available crystal interface sites, f, is presumed to increase (approaching 1) with decreasing temperature.

Growth of the linear polymeric materials (chains) is complicated by the added thermodynamic parameters introduced by the possibility of chain folding. Lauritzen and Hoffmann (1960), and Hoffman and Lauritzen (1961), have developed in great detail a theory for the growth of polymeric materials. Their theory accounts for the experimental observations that polymeric materials grow with a lamellar morphology, having a lamellar thickness that decreases with decreasing temperature. The lamellar thickness (of the order of 100 Å) is much smaller than the molecular length, with the molecules being chain folded within the lamellae.

Photo-Induced Crystallization

Light can produce a marked enhancement of the crystallization kinetics over that obtained from purely thermal effects (Dresner and Stringfellow (1968), Paribok-Aleksandrovich (1970), Feinleib et al. (1971), Pearson and Bagley (1971)). This effect has been used to write holograms (Brandes et al. (1970)) and discrete images (Feinleib et al. (1971)) with lasers, and the optical memory aspects of this transformation are treated in Chapter 6.

Keneman (1971) and Omachi and Igo (1972) have reported writing high-efficiency holograms in vapor deposited thin films of As_2S_3. Because the crystallization kinetics (without a composition change) of As_2S_3 are so slow, the photo induced change is not likely a thermally induced crystallization. Two other mechanisms have been proposed: one is the photo decomposition $As_2S_3 \overset{h\nu}{\to} 2As + 3S$ (Berkes et al. (1971), (1973)) and the other a photo induced structural change from a molecular glass to a more stable cross-linked network (Moss et al. (1973)). In a 7 mole % As-93 mole % S glass, the photo induced change results in a microstructure of pure rhombic sulfur microcrystals within an arsenic sulfur glass matrix, the same end result as that induced thermally (Pearson and Bagley (1971)). Indeed, in the dilute arsenic glass, the transformation is analogous to that which occurs in pure sulfur, in which the first stage of crystallization is the precipitation of rhombic sulfur microcrystals in a polymeric sulfur glass matrix.

Light irradiation often produces both photo and thermal effects which are difficult to resolve. Feinleib et al. (1971), argue that the very fast crystallization rate (crystallized with $1-16\ \mu$ sec light pulses) obtained by the light irradiation of an amorphous $Te_{81}\ Ge_{15}\ Sb_2\ S_2$ alloy cannot be accounted for by purely thermal effects, there being also a strong photo effect which they attribute to the creation of a high density of excess electron-hole pairs. The high speed of the crystallization in this composition was confirmed by Hamada et al. (1972) using a double light pulse technique; the second pulse being a sampling pulse. They observe the irradiated spot of the thin film to be crystallized in $50\ \mu$ sec. However, from their studies on amorphous thin films in the Te-Ge-As system, Weiser et al. (1973) conclude that the light induced crystallization is a purely thermal effect.

Dresner and Stringfellow (1968) studied in detail the photo-enhanced growth aspect of the crystallization of amorphous selenium. For an incident mean photon energy of about 3 eV they observe the growth rate to vary linearly with light intensity for low intensities, and to vary as the square root of the intensity and finally saturate at high intensities (Figure 1.7). In these experiments the thermal effect is small; the calculated temperature rise at maximum light intensity is only 3×10^{-2} °C. They demonstrated

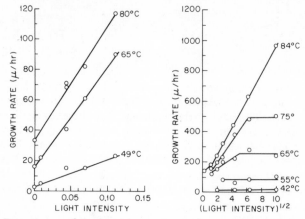

Fig. 1.7   Dependence of the crystal growth rate in amorphous selenium on light intensity and temperature (after Dresner and Stringfellow (1968)).

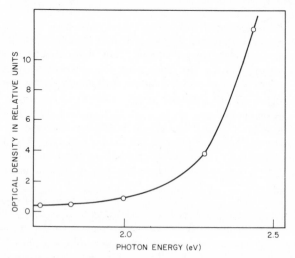

Fig. 1.8   Optical density of amorphous selenium as a function of the incident light photon energy (after Paribok-Aleksandrovich (1970)).

that the growth enhancement is the result of electron-hole pairs being produced in the amorphous phase, and by applying an electric field during irradiation, they also demonstrated that the kinetics are controlled by the flux of holes towards the interface. They point out that broken Se-Se bonds, analogous to mobile holes, would promote growth by reducing the interfacial energy barrier, i.e., $\Delta G'_a$ in Eq. (1.23), and that the hole-electron pair annihilation energy is also available to aid crystallization.

Recently Kim and Turnbull (1973) have studied the thermal and photo-induced crystallization of amorphous Se. They observe a fast thermal crystallization rate in pure Se and conclude that impurities play an important role when photo-enhanced effects are observed.

Paribok-Aleksandrovich (1970) determined the effect of photon energy on the crystallization of amorphous selenium. He observed a photo-enhancement for photon energies exceeding the Se-Se bond strength of 1.8 eV (Cottrell (1958)) $-$ 1.9 eV (Pauling (1960)), becoming extensive when $h\nu$ exceeds 2.2 eV (Figure 1.8).

Light can affect the nucleation as well as the growth aspect of crystallization. The thermodynamic barrier to nucleation ΔG^* (Eqs. (1.15) and (1.16)) is infinite at T_m and decreases with increasing undercooling (ΔG_V, Eq. (1916) increases from zero with increasing undercooling). At large undercoolings ΔG_a (Eq. (1.15)) dominates. Thus the nucleation frequency goes through a maximum with increasing undercooling. If, as would be expected, the photo enhancement served to lower the kinetic barrier, then the nucleation frequency would both increase and shift to a larger under-cooling. In the (unlikely) limit of ΔG_a going to zero, then the maximum nucleation frequency occurs at T_m ($^\circ$K)/3. Ovshinsky and Klose (1971) observed a photo-enhancement of the nucleation frequency in a thin film of unstated (proprietary) composition. They observe a photo plus thermal surface nucleation frequency of 5×10^6 cm^{-2} whereas the thermal nucleation alone was 10^4 cm^{-2}.

Electric Field Effects

The work of Uttecht *et al.* (1970) demonstrates that the electric field can play an important role in crystallization. These authors observe that the crystallization of an As$_{11}$Te$_7$Ge$_2$ alloy with the application of an electric field always nucleated at the positive voltage electrode and grew towards the negative. If the voltage was reversed before a complete crystalline filament was formed, then the already formed filament was observed to recede and a new filament observed to nucleate at the now positive electrode.

Feinleib *et al.* (1971) have suggested that the electric field enhances crystallization kinetics by producing excess electron-hole carriers; there being a direct analogy to the photo-enhanced crystallization discussed above.

In addition to excess carrier effects, van Roosbroeck and Casey (1972) have noted that for these relaxation semiconductors, the recombination front is established at the anode, which thus is where the voltage drop, and therefore heating, occurs. The kinetics are locally enhanced because of this local, additional, thermal effect.

There is yet another effect that can play a role in the observed polarity dependence. If crystallization requires the rejection of a solute, then nucleation is aided in areas depleted of solute and continued growth requires, and becomes limited by, the diffusion of the rejected solute ahead of the moving interface. Thus the kinetics would be enhanced if the field gave the rejected solute an increased mobility (ionic conduction). The importance of this process is indicated by two observations of Uttecht *et al.* (1970). First, they observe that in the As$_{11}$Te$_7$Ge$_2$ alloy the chemical composition of the crystalline filament is different from the glass from which it grows; and second, they observe that many times the filament, upon approaching the negative electrode, dips below the surface to make contact. This second observation is consistent with the notion that the easiest path for the filament would be to go around, rather than through, the built-up concentrations of rejected solute. This effect would not occur in materials which crystallize without a composition change.

1.3.4 Phase Separation

Phase separation is the unmixing of an initially homogeneous multi-component material into two or more amorphous phases.* The region of the phase diagram (temperature and composition are the usual variables) in which these amorphous phases coexist is called the immiscibility gap; which can be either thermodynamically stable as in the Se-Sb system (Figure 1.9, 55–85 atom percent Sb; above 571°C), or thermodynamically metastable with respect to crystallization (sub-liquidus) as is also exhibited by Se-Sb alloys (Figure 1.9, dotted lines). For the composition range 10–30 atom percent Sb (Figure 1.9) we also see the inflected or 'S' shaped liquidus curve which often signals a subliquidus immiscibility gap (Roy (1962)).

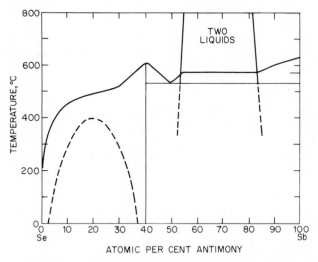

Fig. 1.9   Phase diagram for the Se-Sb system (after Berkes and Myers (1971) and Myers and Berkes (1972)). Metastable immiscibility is delineated with a dashed line.

The occurrence of liquid immiscibility is determined by phase equilibrium thermodynamics, as developed and reviewed by Gibbs (1961). The driving potential for the unmixing process is a reduction in the system free energy. That is, phase separation can occur whenever the free energy for the polyphase system is lower than that for the homogeneous single phase.

---

*Crystallization is also a phase separation. However, current terminology distinguishes between phase separation (amorphous phase immiscibility) and crystallization.

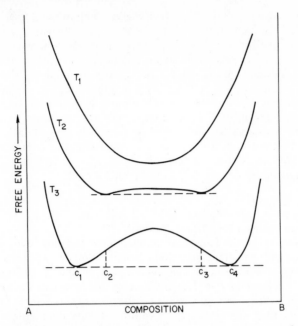

Fig. 1.10 Variation of free energy with composition and temperature for a system which can experience phase separation. $T_1 > T_2 > T_3$.

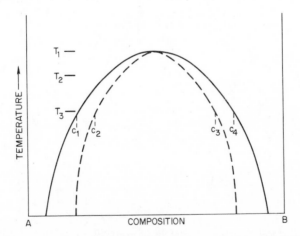

Fig. 1.11 Schematic phase diagram for a binary system having a miscibility gap. The phase boundary is shown solid while the spinodal is dashed. The labeled temperatures and compositions correspond to those in Figure 1.10.

Figure 1.10 depicts hypothetical free energy-composition curves for three selected temperatures for a binary system at constant pressure. The immiscibility gap for this same (hypothetical) system is shown by the solid curve in Figure 1.11 where the phase boundary is the locus of points of lowest free energy (Figure 1.10). We use a construction due to Gibbs (1961) wherein the compositions of the two equilibrium phases are given by a common tangent drawn between the two minima on the free energy curve. For compositions in the immiscibility gap, the free energy difference between the homogeneous and polyphase (equilibrium) systems is given by the difference between the free energy curve and this common tangent (Figure 1.10). Also shown (dotted) on Figure 1.11 is the spinodal (Gibbs, 1961) which corresponds to the locus of points for which

$$(\partial^2 G/\partial C^2)_{T,P} = 0 \qquad\qquad (1.25)$$

(G is the free energy and C the composition).

The spinodal curve divides the immiscibility gap into two regions of differing stability and transformation (unmixing) mode and kinetics. In the outer region, between the spinodal and the phase boundary, the system is in metastable equilibrium. The system is stable against small fluctuations in composition, which increase the system free energy, but metastable with respect to the final state. However, as with the density fluctuations there are small composition fluctuations which are in a state of internal dynamic equilibrium. This is illustrated in Figure 1.12 wherein small fluctuations about composition C_5 are seen to raise the free energy, whereas the equilibrium state consists of two phases of compositions C_1 and C_4. In this region the unmixing proceeds by nucleation and growth, requiring the formation of an interface and therefore introducing an activation barrier for the creation of the second liquid phase nucleus. However, upon approaching the spinodal the interface becomes progressively more diffuse and the activation barrier approaches zero (Cahn and Hilliard (1959)).

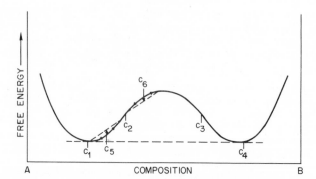

Fig. 1.12 Free energy-composition curve for temperature T_3 of Figures 1.10 and 1.11 illustrating how composition fluctuations at C_6, within the spinodal (C_2-C_3), lower the system free energy, while small fluctuations outside the spinodal at C_5 increase the free energy.

Within the spinodal, the system is thermodynamically unstable; a small composition fluctuation will serve to lower the free energy. This is illustrated in Figure 1.12 wherein it is seen that composition fluctuations about composition C_6 lower the free energy. Once a composition fluctuation is formed, reversion to the homogeneous phase is improbable as this would then increase the free energy. Furthermore, we see from Figure 1.12 that for compositions within the spinodal there is a path of continuously decreasing free energy all the way to the final state (compositions C_1 and C_4). Thus, within the spinodal, there is no activation barrier and the kinetics are limited only by the diffusion process. Hillert (1961) and Cahn (1961) have examined in detail the stability conditions within the spinodal. For an incompressible isotropic binary solution of constant molar volume, and for composition fluctuations of the form (C_o is the initial composition);

$$C - C_o = a \cos bx \tag{1.26}$$

it is found that only fluctuations whose wave number (b) are less than a critical value are unstable. This critical value is given by

$$b_c = \left(\frac{-(\partial^2 F/\partial C^2)_{C_o}}{2K_o} \right)^{1/2} \tag{1.27}$$

where F is the Helmholz free energy density and K_o is the gradient energy density coefficient (assumed positive).

For conditions requiring nucleation and growth for the unmixing, the transformation kinetics are described by Eqs. (1.15), (1.16), (1.17), and (1.23) recognizing that the parameters now represent liquid-liquid rather than liquid-crystal values. That is, ΔG_a and $\Delta G'_a$ represent the liquid-liquid interfacial barriers, σ the liquid-liquid interfacial tension, ΔG_v the homogeneous liquid to phase separated liquid free energy difference per unit volume for the separating component, N_ϱ the number of phase separating molecules per unit volume, and V_m the molecular volume of the separating species.

Within the spinodal, the transformation mode (termed spinodal decomposition) and kinetics have been theoretically treated by Hillert (1961) and Cahn (1961, 1965). This theoretical development is based on obtaining a solution to the diffusion equation with the concentration gradients determined by thermodynamic constraints. For the initial stages of the unmixing, when the composition gradients are small, the derived diffusion equation is;

$$\frac{\partial C}{\partial t} = M \left(\frac{\partial^2 F}{\partial C^2} \right)_{C_o} \nabla^2 C - 2 M K_o \nabla^4 C, \tag{1.28}$$

where M is the diffusion mobility, which must have a positive value, and t is the time. The solution to this equation is;

$$C(r, t) - C_o = \exp[R'(b)t] \cos(b \cdot r) \tag{1.29}$$

where the amplification factor $(R'(b))$ is given by,

$$R'(b) = -M\left(\frac{\partial^2 F}{\partial C^2}\right)_{C_o} b^2 - 2MKb^4. \tag{1.30}$$

Composition fluctuations can grow only when $R'(b)$ is positive, which occurs only within the spinodal region and for $b < b_c$. $R'(b)$ has a rather sharp maximum at;

$$b_M = b_c/\sqrt{2} = \left(\frac{-\left(\frac{\partial^2 F}{\partial C^2}\right)_{C_o}}{4K_o}\right)^{\frac{1}{2}} \tag{1.31}$$

Thus spinodal decomposition proceeds as follows (Cahn (1961, 1965)); while all wave numbers may be present initially, the transformation kinetics are quickly dominated by those wave numbers (b_M) corresponding to the maximum kinetic amplification factor (Eq. 1.31). Thus the spatial composition will be a superposition of sine waves of nearly fixed wavelength $(2\pi/b_M)$ but with random orientations, phases, and amplitudes. For volume fractions greater than 0.15 ± 0.03 the microstructure will have a three dimensional interconnected morphology. For second phase volume fractions less than 0.15, the microstructure will consist of isolated particles (Cahn (1965)). Similar morphologies can also result from nucleation and growth processes (Haller (1965)).

In a liquid immiscibility gap there is no elastic (strain) energy contribution to the system energy. Also, because of an easy atomic accommodation at the interface, liquid-liquid interfacial energies are generally small. These two effects (small strain and interfacial energies) have two important consequences. First, it is kinetically difficult to prevent the unmixing process. For both the nucleation and growth and spinodal decomposition processes, the kinetic barriers are small or non-existent; the separation kinetics are limited only by diffusion. However, diffusion is sluggish in a glass. Thus for an immiscibility gap whose consolute temperature is below T_g, unmixing may be suppressed. Likewise strain energy effects suppress the spinodal (Cahn (1962)) and therefore become important if the unmixing occurs in a rigid, glassy, material.

The second important consequence is that, once unmixed, the second phase can exist as a very fine dispersion (often tens of Angstrom units), there being a negligible driving force for coarsening. Thus phase separation may be difficult to detect.

The presence of phase separation should always be considered a possibility. Morral (1968) has shown that the thermodynamic tendency for phase separation increases as the number of components increases. Possible exceptions are at compositions with simple stoichiometric ratios, e.g. As_2Se_3. While there is extensive experimental evidence documenting phase separation in oxide systems (Levin (1970)), similar experimental evidence in the non-oxide semiconducting glasses is still very meager (Myers and Berkes (1972)).

Because of the multiplicity of phases (any of which may be more fluid or more undercooled than the homogeneous system) and the extensive interfacial boundary, after phase separation one (or more) of the phases is likely to be more unstable with respect to crystallization than was the initially homogeneous system. Thus phase separation can result in enhanced crystallization kinetics. Cahn (1969) also describes conditions in which crystallization must be preceded by phase separation; the crystallization being thermodynamically prohibited until phase separation occurs.

1.4   DETECTION AND CHARACTERIZATION OF INHOMOGENEITIES

For purposes of detection it is convenient to classify inhomogeneities into three size regions; less than 10 Å, larger than 1 μm, and the intermediate range, 10 Å–1 μm.

The less than 10 Å region properly belongs in the realm of structural analysis, which is treated in detail in Chapter 2. On an atomic scale, all materials could be considered structurally and (if polyatomic) compositionally inhomogeneous. Thus a rather arbitrary dividing line is required, and 10 Å is convenient. In this region techniques such as large angle X-ray, neutron, and electron scattering, and techniques which examine nearest neighbor coordination, such as electron spin resonance, nuclear magnetic resonance, Mössbauer spectroscopy, and optical spectroscopy, all can provide useful structural information.

For inhomogeneity sizes larger than 1 μm, direct microscopic examination is convenient. Samples can be freshly fractured or cut and polished, and either etched or unetched. By using light and dark field microscopy and polarimetry microcrystals can be detected, and the presence of Schlieren patterns indicate composition and density fluctuations as shown by variations of the refractive index. Optical microscopy can be used for surface examination, or bulk examination by transmission through thin films or thin sections prepared by cutting and polishing. Optical microscopy was used by Dresner and Stringfellow (1968) for studying the crystallization kinetics of selenium (see Section 1.3.3), Shappirio et al. (1970) for characterizing the microstructure of glasses in the Ga-As-Te-Ge system, and by a number of investigators (Pearson and Miller (1969)), Stocker (1969, 1970) Uttecht et al. (1970), Tanaka et al. (1970), Kikuchi et al. (1970), Haberland and Kehrer (1970) and Champness and Armitage (1971)) for documenting the filamentary and second phase nature of the electrical memory effect (see Chapter 6). Infra-red transmission microscopy is available for examining those bulk glasses which are opaque in the visible. Vaško's (1968, 1969) investigations of the homogeneity of nonoxide chalcogenide glasses by infra-red microscopy clearly demonstrates the usefulness of this technique. His results, documenting the extent of inhomogeneity of many glasses thought to be single phase, graphically emphasizes the care required for the preparation of homogeneous glasses.

Electron microprobe analysis (see Birks (1971) for review) is a suitable technique for characterizing chemical inhomogeneities with a spatial resolution down to ≈ 1 μm. In this technique, a focused beam of electrons is used to excite, in the sample, characteristic x-rays whose wavelengths and intensities are determined by a spectrometer. Elements whose atomic numbers are 5 or greater can be detected with commercial microprobes. Uttecht *et al.* (1970) and Sie (1970) have used microprobe analysis to demonstrate that the crystalline memory filament which forms in a glass of composition $As_{55}Te_{35}Ge_{10}$ has a different composition ($As_{39}Te_{57}Ge_4$) than that of the glass matrix, thus requiring atomic diffusion for the filament formation.

Scanning electron microscopy (see Thornton (1968) for review) is an experimental technique used to reveal surface topography with a spatial resolution down to 100 Å. With this technique an image of the surface is generated by secondary electrons emitted by the surface after irradiation by a primary electron beam. Imaging of the backscattered primary electron beam can also be used to reveal surface details. This technique has been used by Armitage *et al.* (1970) and Guntersdorfer (1971) to demonstrate the damage that can occur during the switching of Te-As-Ge-Si alloys. A scanning electron micrograph of the fractured surface of a $As_4Ge_{15}Te_{81}$ glass (Takamori *et al.* (1971)) showed that crystallization was surface-nucleated. Ovshinsky and Klose (1971) used scanning electron microscopy to determine the crystallite sizes photo-induced in a glass.

The 10 Å–1 μm size region is a difficult one in which to detect inhomogeneities. Direct techniques include the observation of the inhomogeneities by electron microscopy, and the observation of radiation (optical, X-ray or electron) scattered by the inhomogeneities In a less direct approach, the presence of inhomogeneities is inferred from their anticipated effect on measured properties; thermal, optical, and electrical (discussed in Chapter 5) for example. These techniques are now briefly reviewed.

1.4.1 Electron Microscopy

Electron microscopy is a powerful direct experimental technique. Using electron microscopy information can be obtained on the presence of inhomogeneities, and on their shapes, sizes, size dispersion and number density. The experimental and theoretical aspects of this technique have been reviewed by Hirsh *et al.* (1965). There are two methods of observation. In the first, the topography of the sample surface is replicated and it is this replica, and not the sample which is then examined in the electron microscope. Generally carbon is used as the replicating material and shadowing at an angle with heavy elements (Pt) is used to accentuate the surface relief. The resolution limit is about 50 Å due to a microstructure in the replica. As the sample itself is not examined, a diffraction pattern is not obtained. The sample surface can be either etched or unetched. An unetched surface will reveal cracks, voids, and polyphase microstructures if the various phases have differing fracture properties. For etching the surface, care must be exercised in the choice of etchant, as etching artifacts are easily introduced. The etch must reliably attack, at different rates, the often subtle composition changes accompanying phase separation, or the structural (and compositional) changes accompanying partial crystallization. For inhomogeneity sizes larger than 100–200 Å this technique overlaps with scanning electron microscopy, which also examines surface topography but which does not require the preparation of a replica. These techniques were used by Pearson (1964), Roy and Caslavska (1969), Phillips *et al.* (1970), and Bunton (1971) to demonstrate that all of the glasses they examined, which included many diverse chalcogenide glasses, were polyphase and inhomogeneous. Pearson's observations led him to suggest that the Hall effect-Seebeck effect sign anomaly was due to the presence of second phases (Pearson (1964)).

The second method of electron microscopic examination is by direct transmission through the sample. With this method the interior of the sample is examined (not surface topography) and diffraction patterns can be obtained. As electrons can penetrate only about 1000 Å of material, thin sections of the bulk must be prepared. This can be done by crushing brittle materials and examining the feathered edges (Prebus and Michener (1954)), cutting and polishing (Doherty and Leombruno (1964)), Seward *et al.* (1967), chemical thinning (James and McMillan (1968)), or ion milling (Bach (1970), Barber (1970)). If they can be removed from the substrate, vapor deposited thin films can be examined directly

The presence of microcrystals can be detected in the microstructure using both light and dark field techniques, and in the diffraction pattern where the crystalline reflections will be superimposed on the amorphous pattern. Phase separation can be detected in the microstructure if the two phases differ in electron density. In general, to be easily observed in transmission the inhomogeneities should be larger than about one tenth of the sample thickness. Practical electron microscopic resolutions are about 10 Å. Thus, the detection of inhomogeneities in the 10–50 Å size range by transmission electron microscopy is difficult, but the detection of larger sizes is relatively easy. Care must be used to insure that the observed inhomogeneities are representative of the bulk glass and are not the result of the examination process itself. For example, thermal and electron damage effects can manifest themselves as changes in the microstructure (i.e. induced phase separation or crystallization) which were not originally present in the bulk glass. This is a difficulty not encountered in using the replica technique.

Transmission electron microscopy was used by Shappirio *et al.* (1970) to characterize the structural homogeneity of their glasses; by Bagley and Northover (1970) to study thermally induced transformations in chalcogenide thin films; by Feinleib *et al.* (1971) to characterize the photo induced phase changes in a chalcogenide thin film; and by Donovan and Heinemann (1971) to document the presence of voids in amorphous germanium (see Figure 5.37, Chapter 5).

1.4.2   Small Angle X-Ray Scattering

Radiation is scattered when it passes through an inhomogeneous medium. X-rays are scattered by fluctuations in electron density. The scattering angle and the size of the scattering inhomogeneity are inversely related. Thus for atomic dimensions and x-ray wavelengths (1.5 Å) the scattering occurs at the large angles ($5-180°2\theta$) normally associated with structural investigations (Chapter 2). As the size of the scattering center (inhomogeneity) increases, measurements must be made at smaller angles ($< 5°2\theta$), closer to the direct beam. Commercial instruments are available which can measure scattered intensities down to about 15 seconds of arc, corresponding to sizes of up to 2 μm and thus overlapping with optical microscopy. For the lower limit, inhomogeneities as small as 10 Å can be detected. The theory, experimental techniques, and applications of small angle scattering have been reviewed by Guinier *et al.* (1955) and Brumberger (1967).

Scattering experiments are performed in a transmission geometry. The optimum sample thickness is the inverse of the linear absorption coefficient. For amorphous semiconductors this optimum thickness is generally in the range $10-100$ μ. Care must be exercised to avoid, or remove by corrections, instrumental and surface scattering, but any resulting bulk scattering is then due to the presence of electron density inhomogeneities. As an example of the information obtained by small angle X-ray scattering Figure 1.13

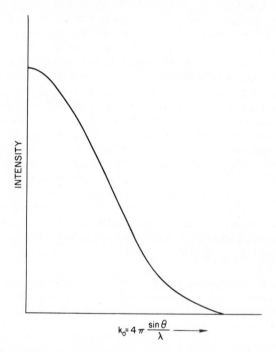

$k_o = 4\pi \frac{\sin\theta}{\lambda}$ ——→

Fig. 1.13 The angular (θ) dependence in terms of the variable k_o, of the radiation scattered by a dilute dispersion of homogeneously sized particles.

depits a hypothetical scattering curve for a dilute dispersion of homogeneously sized scatterers. At the small scattering k_o $(= 4\pi \frac{\sin \theta}{\lambda})$ values, Guinier has demonstrated that the scattered intensity (I_x) has a Gaussian shape:

$$I_x = I_o \exp \left[\frac{-k_o{}^2 R_G^2}{3} \right]$$ (1.32)

where R_G is the Guinier radius and is the square root average of the distance of each atom from the center of mass of the scattering particle. R_G is a unique parameter obtained from the scattering curve and is independent of the shape of the particles. For spherical particles with radius R_S,

$$R_G^2 = 3/5 \ R_S^2$$ (1.33)

At the larger k_o values, the scattered intensity is related to another parameter of interest, the inhomogeneity-matrix total interfacial area.

If there is a dispersion of sizes, or shapes, or if the number density is large enough that there is scattering interference between particles, then interpretation of the data is more difficult (see reviews cited). In the general case, the interpretation is based on a theoretical development of Debye and Bueche (1949) (also see Debye (1960)) in which the scattered intensity is written in terms of a correlation function $(c(r))$ for the electron density fluctuations $(\delta \rho_e)$;

$$I_x \text{ (in electron units)} = \left\langle \left(\frac{\delta \rho_e}{\rho_e} \right)^2 \right\rangle V \int_o^\infty c(r) \frac{\sin k_o r}{k_o r} \ 4\pi r^2 dr$$ (1.34)

Here ρ_e is the electron density and V the irradiated volume. All of the structural information available from an X-ray scattering experiment is contained in the correlation function $c(r)$, which is obtained by a Fourier transformation of Eq. (1.34).

A recent application of small angle X-ray scattering is for the determination of phase separation kinetics. Rundman and Hilliard (1967) noted the correspondence between the composition fluctuations corresponding to Cahn's solution to the diffusion equation and the small angle scattering intensity which results from these composition (i.e. electron density) fluctuations (Eq. 1.34). They obtain the relation,

$$I_x(k_o,t) = I(k_o,0) \exp [2 \ R'(k_o)t]$$ (1.35)

Here $R'(k_o)$ is the amplification factor of Eq. (1.30) $(b = k_o)$, t the time, and $I(k_o,0)$ the initial intensity. Thus $R'(k_o)$ can be determined by measuring the time evolution of the scattered X-ray intensity.

Kinetic studies of phase separation in $Na_2O–SiO_2$ glasses by Andreev et al. (1970a, 1970b) and Tomozawa et al. (1970) gave qualitative agreement with theory. In Figure 1.14 is shown the results of Andreev et al. (1970a) for the time evolution of the scattered intensity as a function of k_o. These

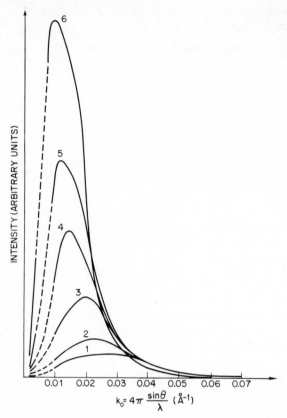

Fig. 1.14 Intensity scattered from a phase separated sodium silicate glass. $k_o < 0.004$
Å^{-1} are light scattering results and $k_o > 0.01$ are X-ray scattering results.
Glass was heat treated at $530°C$ for 1 hr. (1), 10 hrs. (2), 16 hrs. (3), 25 hrs.
(4), 30 hrs. (5) and 42.5 hrs. (6) after Andreev *et al.* (1970a)).

authors used light scattering to obtain data at small k_o and X-ray scattering
to obtain data at large k_o.

The scattering from randomly oriented non-spherical inhomogeneities is
isotropic. Oriented scatterers, however, result in an anisotropic scattering.
Cargill (1972) observed an anisotropy in the small angle X-ray scattering
from evaporated amorphous germanium films. From his results he concludes
that the amorphous films ($\sim$7 μm thick) contain 1–2 volume % of rod
shaped voids whose dimensions are 22 Å $\times$ 46 Å in the plane of the film and
2200 Å normal to the film plane.

Moss and Graczyk (1969) have observed a small angle scattering of
electrons (another radiation suitable for diffraction) from a 100 Å thick
film of vapor deposited amorphous silicon. Upon annealing the film, the
small angle scattered intensity is reduced. From their results they conclude

that the amorphous film has a density 10–15% lower than that of the crystalline material, and that this density deficit is distributed as voids (the 'Swiss cheese' model of Ehrenreich and Turnbull (1970)). Moss and Graczyk also point out that the presence of internal voids could account for the surface states observed by Brodsky and Title (1969).

1.4.3 Optical Properties

In the late 1800's it was recognized that light will be scattered in a medium which contains variations in the dielectric constant (Rayleigh (1964)). Since then, the theoretical and experimental aspects of light scattering have been comprehensively developed (see recent reviews by Fabelinskii (1968) and Kerker (1969)).

For large inhomogeneities (lower limit about $\lambda/10$) the scattered intensity has an angular dependence from which structural information can be obtained. This information is analogous to that obtained from small angle X-ray scattering; i.e., the Guinier radius, interparticle distances, and the interfacial area. Because of the large difference in wavelength between X-rays (≈ 1 Å) and light (≈ 1 μm), however, light scattering yields these results only for inhomogeneities several orders of magnitude (or more) larger in size than for the equivalent X-ray characterization. For isolated (non-interfering) particles smaller than $\lambda/10$, the scattering is isotropic (Rayleigh scattering) so detailed structural information cannot be obtained; although of course, the presence of scattering indicates the presence of a nonhomogeneous refractive index.

Light scattering has been used extensively to characterize polyphase structures in the visibly transparent oxide glasses (see, for example, Andreev et al. (1970a)). Indeed, upon extensive phase separation, the normally transparent oxide glasses suffer a marked change in their optical properties, becoming turbid and opalescent. As yet, there have not been any similar studies (requiring, of course, infra-red wavelengths) on the semiconducting glasses.

In some experiments (see Tauc et al. (1970b), for example), the parameter of interest is the absorption coefficient, which must be extracted from a measurement of the total loss. In such a case, the contribution to the total loss due to scattering must be either accounted for by supplementary scattering measurements, or eliminated by preparing a homogeneous sample.

Galeener (1971a) has recently developed a theory for the effect of very small voids (a few tens of Angstrom units) on optical properties. He finds that the presence of these small voids introduces new resonance peaks (void resonances) into the imaginary part of the effective dielectric constant. He suggests (Galeener (1971b)) that the structure in the uv dielectric constants of evaporated amorphous germanium observed by Donovan et al. (1970b) is consistent with void resonances from a connected network of dislike microcracks about 6 Å in width. Using transmission electron microscopy, Donovan and Heinemann (1971) have observed such a microstructure in amorphous Ge films deposited on substrates whose temperatures are $150°C$ or lower (see Figure 5.37, Chapter 5). However, Cargill (1972) concludes that his small angle X-ray scattering data (see Section 1.4.2) are inconsistent with Galeener's cracklike void model for amorphous Ge.

1.4.4 Calorimetry

Calorimetry can be used to detect the enthalpy and specific heat changes which accompany structural transitions Accurate data can be easily obtained through use of one of the commercially obtainable scanning calorimeters or differential thermal analyzers. The calorimeters determine transition temperatures and provide quantitative data on reaction enthalpies, heat capacities, and heat capacity changes Differential thermal analysis, however, only provides quantitative data on the reaction temperatures. The scanning character of each of these techniques provides a 'thermal spectrum' of characteristic temperatures, glass transition temperatures and crystalline melting points for example. Scanning calorimetry also provides a spectrum of characteristic energies such as the enthalpies of melting. In Figure 1.15 is shown a scanning calorimetric trace (increasing temperature) for $As_2Se_3 \cdot 2 As_2Te_3$ (Bagley and Bair (1970)). As calorimetry only provides energy and specific heat changes, subsidiary structural investigations are required to characterize the nature of the changes. Shown in Figure 1.15 are the glass transition at 136°C (scale expanded for clarity), a crystallization exotherm and the melting endotherm at 323°C. Often the thermal spectra are more complex than that shown in Figure 1.15 and detailed interpretation is then more difficult.

A large amount of information can be obtained from thermal spectra. The observation of a glass transition indicates the presence of a glassy phase, and two glass transitions (Moynihan et al. (1971)) indicates phase separation with the two phases having different glass transition temperatures. de Neufville (1972) has suggested that phase separation is also indicated when the glass transition temperature is invariant with composition. Červinka et al. (1970) have demonstrated the glassy nature of ternary alloys (based on $CdAs_2$) quenched from the melt by observing, thermally, the glass transition and subsequent crystallization. Hrubý and Štourač (1971a, 1971b) have

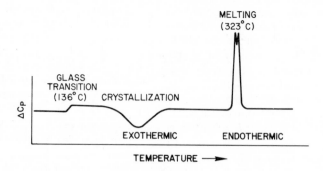

Fig. 1.15 Differential scanning calorimetric trace of an $As_2Se_3 \cdot 2 As_2Te_3$ glass. Temperature scan was from low temperature to high.

made similar observations to demonstrate the glassy nature of the two binary compositions $CdAs_2$ and As_2Te_3.

Differential thermal analysis was used to delineate the glass forming regions in the Ge-Te system (Takamori *et al.* (1970)) and the Ge-Te-As system (Savage (1971)). Myers and Felty (1967) and de Neufville (1972) have thermally determined the composition dependence of the glass transition temperature for several multicomponent chalcogenide glasses, from which they have obtained information on the chemistry and structure of these materials.

Pearson and Bagley (1971) used scanning calorimetry to demonstrate that α sulfur microcrystals can be photoprecipitated from arsenic-sulfur glasses. The presence of these crystals was evidenced by the observation of a melting endotherm as the temperature was scanned through the sulfur melting point. Finally, a number of authors (Fritzsche and Ovshinsky (1970), Bagley and Bair (1970), Phillips *et al.* (1970), Pinto (1971), and Johnson and Quinn (1971)) have used calorimetric techniques to demonstrate that the mechanism for memory switching (Chapter 6) is the crystallization and reversion of the glass.

We conclude this section on the detection and characterization of inhomogeneities, and this chapter, with a warning: Inhomogeneities can be easily (and sometimes inadvertently) introduced into amorphous solids, and their presence can both affect physical properties and be difficult to detect.

REFERENCES

Adam, G. and Gibbs, J. H., (1965), *J. Chem. Phys.* **43**, 139.
Anderson, P. W., Halperin, B. I. and Varma, C. M., (1972), *Phil. Mag.* **25**, 1.
Andreev, N. S., Boiko, G. G. and Bokov, N. A., (1970a) *J. Non Crystalline Solids,* **5**, 41.
Andreev, N. S. and Porai-Koshits, E. A., (1970b), *Discuss. Faraday Soc.,* **50**, 135.
Armitage, D., Brodie, D. E. and Eastman, P. C., (1970), *Canadian J. Physics,* **48**, 2780.
Bach, H., (1970), *J. Non-Crystalline Solids,* **3**, 1.
Bagley, B. G., Chen, H. S. and Turnbull, D., (1968), *Mat. Res. Bull.,* **3**, 159.
Bagley, B. G. and Bair, H. E., (1970), *J. Non-Crystalline Solids,* **2**, 155.
Bagley, B. G. and Northover, W. R., (1970), *J. Non-Crystalline Solids,* **2**, 161.
Bahl, S. K. and Chopra, K. L., (1969), *J. Appl. Phys.,* **40**, 4940.
Bahl, S. K. and Chopra, K. L., (1970), *J. Appl. Phys.* **41**, 2196.
Barber, D. J., (1970), *J. Materials Sci.,* **5**, 1.
Barrett, C. S., (1952), Structure of Metals, 2nd ed. p. 1. McGraw Hill Book Co., New York.
Berkes, J. S., Ing, S. W. Jr. and Hillegas, W. J., (1971), *J. Appl. Phys.,* **42**, 4908.
Berkes, J. S. and Myers, M. B., (1971), *J. Electrochem. Soc.,* **118**, 1485.
Berkes, J. S., Short, J. M. and Johnson, K. J., (1973), Fifth International Conference on Amorphous and Liquid Semiconductors, Garmisch-Partenkirchen, Germany.
Bernal, J. D., (1959), *Nature,* **183**, 141.
Bernal, J. D., (1960), *Nature,* **185**, 68.
Bernal, J. D. and Mason, J., (1960), *Nature,* **188**, 910.
Berry, M. V. and Doyle, P. A., (1973), *J. Phys. C: Solid State Phys.,* **6**, 46.
Betts, F., Bienenstock, A. and Ovshinsky, S. R., (1970), *J. Non-Crystalline Solids,* **4**, 554.
Bienenstock, A. and Bagley, B. G., (1966), *J. Appl. Phys.,* **37**, 4840.
Birks, L. S., (1971), Electron Probe Microanalysis; Chemical Analysis Vol. 17, 2nd ed. Wiley-Interscience, New York.

Brandes, R. G., Laming, F. P. and Pearson, A. D., (1970), *Applied Optics*, **9**, 1712.
Brodsky, M. H. and Title, R. S., (1969), *Phys. Rev. Lett*, **23**, 581.
Brumberger, H., editor, (1967), Small Angle X-Ray Scattering, Gordon and Breach, New York.
Bunton, G. V., (1971), *J. Non-Crystalline Solids*, **6**, 72.
Cahn, J. W., (1961), *Acta Met.*, **9**, 795.
Cahn, J. W., (1962), *Acta Met.*, **10**, 907.
Cahn, J. W., (1965), *J. Chem. Phys.*, **42**, 93.
Cahn, J. W., (1968), *Trans. AIME*, **242**, 166.
Cahn, J. W., (1969), *J. Amer. Ceram. Soc.*, **52**, 118.
Cahn, J. W. and Hilliard J. E., (1959), *J. Chem. Phys.*, **31**, 688.
Cargill, G. S., (1972), *Phys. Rev. Lett.* **28**, 1372.
Červinka, L., Hrubý, A., Matyáš, M., Šimecek, T., Škácha J., Štourač, L., Tauc, J. and Vorlíček, V., (1970), *J. Non-Crystalline Solids*, **4**, 258.
Champness, C. H. and Armitage, D., (1971), Proc. Second. Int. Conf. on Conduction in Low-Mobility Materials, p. 339. Taylor and Francis, London.
Chaudhari, P., Graczyk, J. F. and Charbnau, H. P., (1972), *Phys. Rev. Lett.*, **29**, 425.
Chen, H. S., (1967), Ph.D. Thesis, Division of Engineering and Applied Physics, Harvard University, Cambridge, Mass.
Chen, H. S. and Turnbull, D., (1969), *J. Appl. Phys.*, **40**, 4214.
Chittick, R. C., (1970), *J. of Non-Crystalline Solids* **3**, 255.
Chittick, R. C., Alexander, J. H. and Sterling, H. F., (1969), *J. Electrochem. Soc.*, **116**, 77.
Chopra, K. L., (1969), Thin Film Phenomena, McGraw Hill Book Co. New York.
Chopra, K. L. and Bahl, S. K., (1969), *J. Appl. Phys.* **40**, 4171.
Chopra, K. L. and Bahl, S. K., (1970), *Phys. Rev.* **B1**, 2545.
Clark, A. H., (1967), *Phys. Rev.*, **154**, 750.
Cohen, M. H. and Turnbull, D., (1959), *J. Chem. Phys.*, **31**, 1164.
Cohen, M. H. and Turnbull, D., (1964), *Nature*, **203**, 964.
Condon, E. U., (1954), *Amer. J. Phys.*, **22**, 43.
Cottrell, T. L., (1958), The Strengths of Chemical Bonds, 2nd ed. p. 258. Butterworths, London.
Crowder, B. L., Title, R. S., Brodsky, M. H., and Pettit, G. D. (1970), *Appl. Phys. Lett.*, **16**, 205.
Crystal, R. G., (1970), *J. Polymer Science* **A-2 8**, 1755.
Davies, R. O., (1956), *Reports on Progress in Physics*, **19**, 326.
Davies, R. O., (1960), in Non-Crystalline Solids, Chap. 9, p. 232. (V. D. Frechette, editor), J. Wiley, New York.
Debye, P., (1960), in Non-Crystalline Solids, Chap. 1, p. 1. (V. D. Frechette, editor), J. Wiley, New York.
Debye, P. and Bueche, A. M., (1949), *J. Appl. Phys.*, **20**, 518.
de Neufville, J. P., (1972), *J. Non-Crystalline Solids*, **8/10**, 85.
Doherty, P. E. and Leombruno, R. R., (1964), *J. Amer. Ceram. Soc.*, **47**, 368.
Donovan, T. M., Spicer, W. E. and Bennett, J. M., (1969), *Phys. Rev. Lett.*, **22**, 1058.
Donovan, T. M., Ashley, E. J. and Spicer, W. E., (1970a), *Physics Letters*, **32A**, 85.
Donovan, T. M., Spicer, W. E., Bennett, J. M. and Ashley, E. J., (1970b), *Phys. Rev.*, **B 2**, 397.
Donovan, T. M. and Heinemann, K., (1971), *Phys. Rev. Lett*, **27**, 1794.
Dresner, J. and Stringfellow, G. B., (1968), *J. Phys. Chem. Solids*, **29**, 303.
Pol Duwez, and Willens, R. H., (1963), *Trans. AIME 227*, 362.
Ehrenreich, H. and Turnbull, D., (1970), *Comments on Solid State Physics*, **3**, 75.
Ewing, R. H., (1971), *J. Crystal Growth*, **11**, 221.
Eyring, H., (1936), *J. Chem. Phys.*, **4**, 283.
Fabelinskii, I. L., (1968), Molecular Scattering of Light, Plenum Press, New York.
Feinleib, J., de Neufville, J., Moss, S. C. and Ovshinsky, S. R., (1971), *Appl. Phys. Lett.*, **18**, 254.
Fink, C. G. and Dokras, V. M., (1949), *J. Electrochem. Soc.*, **95**, 80.

Fox, T. G. and Flory, P. J., (1950), *J. Appl. Phys.*, **21**, 581.
Fritzsche, H. and Ovshinsky, S. R., (1970), *J. Non-Crystalline Solids*, **2**, 148.
Fulcher, G. S., (1925), *J. Amer. Ceram. Soc.*, **77**, 3701.
Fulde, P. and Wagner, H., (1971), *Phys. Rev. Lett.*, **27**, 1280.
Galeener, F. L, (1971a), *Phys. Rev. Lett.*, **27**, 421.
Galeener, F. L., (1971b), *Phys. Rev. Lett.*, **27**, 1716.
Germer, L. H. and White, A. W., (1941), *Phys. Rev.*, **60**, 447.
Gibbs, J. H., (1960), in Modern Aspects of the Vitreous State, (J. D. Mackenzie, editor), Vol. 1, p. 152. Butterworths, London.
Gibbs, J. H. and Di Marzio, E. A., (1958), *J. Chem. Phys.*, **28**, 373.
Gibbs, J. W., (1961), Scientific Papers, Vol. 1. Dover New York.
Guinier, A., Fournet, G., Walker, C. B. and Yudowitch, K. L., (1955), Small Angle Scattering of X-Rays, J. Wiley, New York.
Guntersdorfer, M., (1971), *J. Appl. Phys.* **42**, 2566.
Haberland, D. R. and Kehrer, H. P., (1970), *Solid-State Electronics*, **13**, 451.
Haller, W., (1965), *J. Chem. Phys.*, **42**, 686.
Hamada, A., Kurosu, T., Saito M., and Kikuchi, M., (1972), *Appl. Phys. Lett.*, **20**, 9.
Hillert, M., (1961), *Acta Met.*, **9**, 525.
Hillig, W. B. and Turnbull, D., (1956), *J. Chem. Phys.*, **24**, 914.
Hirsch, P. B., Howie, A., Nicholson, R. B., Pashley, D. W. and Whelan, M. J., (1965), Electron Microscopy of Thin Crystals, Butterworths London.
Hoffman, J. D. and Lauritzen, J. I. Jr., (1961), *J. Resh. Nat. Bur. of Standards*, **65A**, 297.
Hrubý, A. and Štourač, L., (1971a), *Mat. Resh. Bull.*, **6**, 247.
Hrubý, A. and Štourač, L., (1971b), *Mat. Resh. Bull.*, **6**, 465.
Ioffe, A. F. and Regel, A. R., (1960), *Progr. in Semiconductors*, **4**, 237.
Jackson, K. A. and Chalmers, B., (1956) *Can. J. Phys.*, **34**, 473.
James, P. F. and McMillan, P. W., (1968), *Phil. Mag.*, **18**, 863.
Johnson, R. T. Jr., and Quinn, R. K., (1971), *Solid State Comm.*, **9**, 393.
Jones, G. O., (1956), Glass, Methuen, London.
Kauzman, W., (1948), *Chem. Rev.*, **43**, 219.
Keneman, S. A., (1971), *Appl. Phys. Lett.*, **19**, 205.
Kerker, M., (1969), The Scattering of Light and Other Electromagnetic Radiation, Academic Press, New York.
Kikuchi, M., Iizima, S., Sugi, M. and Tanaka, K., (1970), *Supp. to J. Japan Soc. Åppl. Phys.*, **39**, 203.
Kim, K. S. and Turnbull, D., (1973), to be published.
Kolomiets, B. T., (1964), *Phys. Stat. Sol.*, **7**, 359; ibid 713.
Kurtin, S., Shifrin, G. A. and McGill, T. C., (1969), *Appl. Phys. Lett.*, **14**, 223.
Large, L. N. and Bicknell, R. W., (1967), *J. Materials Science*, **2**, 589.
Lauritzen, J. I. Jr. and Hoffman, J. D., (1960), *J. Resh. Nat. Bur. of Standards*, **64A**, 73.
LeComber, P. G. and Spear, W. E., (1970), *Phys. Rev. Lett.*, **25**, 509.
Levin, E. M., (1970), in Phase Diagrams: Materials Science and Technology, (A. M. Alper, editor), Vol. III, p. 144. Academic Press New York.
Lewis, G. N. and Gibson, G. E., (1920), *J. Amer. Chem Soc.*, **42**, 1529.
Light, T. B., (1969), *Phys. Rev. Lett.*, **22**, 999.
Lowenheim, F. A., editor, (1963), Modern Electroplating, 2nd ed. J. Wiley, New York.
Luo, H. L., (1964), Ph.D. Thesis California Institute of Technology Pasadena, Calif.
Luo, H. L. and Pol Duwez, (1963), *Appl. Phys. Letters*, **2**, 21.
Mogab, C. J. and Block, R. G., (1971), unpublished work.
Morey, G. W., (1954), The Properties of Glass, 2nd ed. p. 28. Reinhold Publ. Co. New York.
Morral, J. E., (1968), Ph.D. Thesis, Department of Metallurgy and Materials Science, Mass. Instit. of Tech., Cambridge, Mass.
Moss, S. C., Flynn, P. and Bauer, L. O., (1971), *Bulletin Amer. Phys. Soc.*, **16**, 1392
Moss, S. C. and Graczyk, J. F., (1969), *Phys. Rev. Letters* **23**, 1167.

Moss, S. C., de Neufville, J. P. and Ovshinsky, S. R., (1973), *Bull. Amer. Phys. Soc.,* 18, 389.

Mott, N. F. and Gurney, R. W., (1938), *Rept. Progr. Phys.,* 5, 46.

Moynihan, C. T., Macedo, P. B., Aggarwal, I. D. and Schnaus, U. E., (1971), *J. Non-Crystalline Solids,* 6, 322.

Münster, A., (1969), Statistical Thermodynamics Vol. 1, Chap. III and Section 5.14. Springer-Verlag, Berlin.

Myers, M. B. and Felty, E. J., (1967), *Mat. Resh. Bull,* 2, 535.

Myers, M. B. and Berkes, (1972), *J. Non-Crystalline Solids,* 8/10, 804.

Oblad, P. G. and Newton, R. F., (1937), *J. Amer. Chem. Soc,* 59, 2495.

Ohmachi, Y. and Igo, T., (1972), *Appl. Phys. Lett.,* 20, 506.

Ovshinsky, S. R. and Klose, P. H., (1971), in 1971 SID International Symposium Digest of Technical Papers, p. 58. Lewis Winner New York.

Paribok-Aleksandrovich, I. A., (1970), *Soviet Physics-Solid State,* 11, 1631.

Parks, G. S. and Gilky, W. A., (1929), *J. Phys. Chem.,* 33, 1428.

Pauling, L., (1960), The Nature of the Chemical Bond, 3rd ed. p. 85. Cornell University Press, Ithaca.

Pearson, A. D., (1964a), in Modern Aspects of the Vitreous State, (J. D. Mackenzie editor), Vol. 3 p. 29. Butterworths, London.

Pearson, A. D., (1964b), *J. Electrochem. Soc.,* 111, 753.

Pearson, A. D. and Miller, C. E., (1969), *Appl Phys. Lett,* 14, 280.

Pearson, A. D. and Bagley, B. G., (1971), *Mat. Res. Bull,* 6, 1041.

Phillips, S. V., Booth, R. E. and McMillan, P. W., (1970), *J. Non-Crystalline Solids,* 4, 510.

Pietrokowsky, P., (1962), *J. Scient. Instr.,* 34, 445.

Piggott, M. R., (1966), *J. Appl. Phys.,* 37, 2927.

Pinto, R., (1971), *Thin Solid Films,* 7, 391.

Polk, D. E., (1971), *J. Non-Crystalline Solids,* 5, 365.

Prebus, A. F. and Michener, J. W., (1954), *Indust. and Engr. Chem.,* 46, 147.

Predecki, P., Mullendore, A. W., and Grant, N. J., (1965), *Trans.,* AIME 233, 1581.

Rawson, H., (1967),, Inorganic Glass-Forming Systems, Academic Press, New York.

Rayleigh, J. W. S. Lord, (1964), Scientific Papers, Dover, New York.

Roy, R., (1962), in Symposium on Nucleation and Crystallization in Glasses and Melts, (M. K. Reser, G. Smith, and H. Insley, editors), p. 39. Amer. Ceram. Soc., Columbus, Ohio.

Roy, R. and Caslavska, V., (1969), *Solid State Comm.* 7, 1467.

Rudee, M. L. and Howie, A., (1972), *Phil. Mag.* 25, 1001.

Rundman, K. B. and Hilliard, J. E., (1967), *Acta. Met.,* 15, 1025.

Savage, J. A., (1971), *J. Materials Sci.,* 6, 964.

Scott, G. D., (1960), *Nature,* 188, 908.

Seward, T. P., Uhlmann, D. R., Turnbull, D. and Pierce, G. R., (1967), *J. Amer. Ceram. Soc.,* 50, 25.

Shappirio, J. R., Eckart, D. W. and Cook, C. F., Jr., (1970), *J. Non-Crystalline Solids,* 2, 217.

Sie, C. H., (1970), *J. Non-Crystalline Solids,* 4, 548

Simon, F. E., (1930), *Ergebn. Exact. Naturwiss* 9, 244.

Skapski, A. S., (1956), *Acta Met.,* 4, 576.

Smith, J. E. Jr., Brodsky, M. H., Crower, B. L., Nathan, M. I. and Pinczuk, A., (1971), *Phys. Rev. Lett,* 26, 642.

Stocker, H. J., (1969), *Appl. Phys. Lett.,* 15, 55.

Stocker, H. J., (1970), *J. Non-Crystalline Solids,* 2, 371.

Szekely, G., (1951), *J. Electrochem. Soc.,* 98, 318.

Takamori, T., Roy, R. and McCarthy, G. J., (1970), *Mat. Resh. Bull.,* 5, 529.

Takamori, T., Roy, R. and McCarthy, G. J., (1971), *J. Appl. Phys.,* 42, 2577.

Tanaka, K., Iizima, S., Sugi, M. and Kikuchi, M., (1970), *Solid State Comm.,* 8, 75.

Tauc, J., Abraham, A., Zallen, R. and Slade, M., (1970a), *J. Non-Crystalline Solids,* 4. 279.

Tauc, J., Menth, A., and Wood, D. L., (1970b), *Phys. Rev. Lett.,* **25**, 749.

Thomas, D. G. and Stavely, L. A. K., (1952), *J. Chem. Soc.,* 4569.

Thornton, P. R., (1968), Scanning Electron Microscopy, Applications to Materials and Device Science, Chapman and Hall, London.

Tomozawa, M., MacCrone, R. K. and Herman, H., (1970), *Physics and Chem. of Glasses,* **11**, 136.

Tool, A. Q., (1964), *J. Amer. Ceram. Soc.,* **29**, 240.

Tsuei, C. C., (1966), Ph.D. Thesis, California, Institute of Technology, Pasadena, Calif.

Tsuei, C. C. and Kankeleit, E., (1967), *Phys. Rev.,* **162**, 312.

Turnbull, D., (1950a), in Thermodynamics in Physical Metallurgy p. 282., American Society for Metals, Cleveland.

Turnbull, D., (1950b), *J. Chem. Phys,* **18**, 769.

Turnbull, D., (1950c), *J. Chem. Phys.* **18**, 768.

Turnbull, D., (1950d), *J. Chem. Phys.,* **18**, 198.

Turnbull, D., (1956), in Solid State Physics-Advances in Research and Applications, (F. Seitz and D. Turnbull editors), Vol. 3, p. 225. Academic Press, New York.

Turnbull, D., (1962), *J. Phys. Chem.,* **66**, 609.

Turnbull, D., (1965a), in Physics of Non-Crystalline Solids (J. A. Prins editor), p. 41. North Holland Publ. Co., Amsterdam.

Turnbull, D., (1965b), in Liquids: Structures Properties, Solid Interactions (Thomas J. Hughel editor), p. 6. Elsevier Publ. Co., Amsterdam.

Turnbull, D., (1969), *Contemp. Phys.,* **10**, 473.

Turnbull, D., (1971), in Solidification, p. 1. American Society for Metals, Metals Park, Ohio.

Turnbull, D. and Cohen, M. H., (1958), *J. Chem. Phys.,* **29**, 1049.

Turnbull, D. and Cohen, M. H., (1960), in Modern Aspects of the Vitreous State (J. D. Mackenzie editor), Vol. 1., p. 38. Butterworths, London.

Turnbull, D. and Cohen, M. H., (1961), *J. Chem. Phys,* **34**, 120.

Turnbull, D. and Cohen, M. H., (1970), *J. Chem. Phys.,* **52**, 3038.

Turnbull, D. and Fisher, J. C., (1949), *J. Chem. Phys.,* **17**, 71.

Turnbull, D. and Polk, D., (1972), *J. Non-Crystalline Solids* **8/10**, 19

Uttecht, R., Stevenson, H., Sie, C. H., Griener, J. D. and Raghavan, K. S., (1970), *J. Non-Crystalline Solids,* **2**, 358.

Valenkov, N. and Porai-Koshits, E., (1937), *Z. Krist.,* **95**, 195.

Van Roosbroeck, W. and Casey, H. C. Jr., (1972), *Phys. Rev.,* B**5**, 2154.

Vaško, A., (1968), *Mat. Resh. Bull.,* **3**, 209.

Vaško, A., (1969), in The Physics of Selenium and Tellurium (W. C. Cooper editor), p. 241. Pergamon Press, New York.

Vengel, T. N. and Kolomiets, B. T., (1957), *Soviet Physics-Technical Physics,* **2**, 2314.

Warren, B. E., (1937), *J. Appl. Phys.,* **8**, 645.

Warren, B. E., Krutter, H. and Morningstar. O., (1936), *J. Am. Ceram. Soc.,* **19**, 202.

Weiser, K., Gambino, R. J. and Reinhold, J. A. (1973). *Appl. Phys. Lett.,* **22**, 48.

Weyl, W. A. and Marboe, E. C., (1962), The Constitution of Glasses, A Dynamic Interpretation, Vol. 1, Fundamentals of the Structure of Inorganic Liquids and Solids, Interscience Publishers, New York.

Weyl, W. A. and Marboe, E. C., (1964), The Constitution of Glasses, A Dynamic Interpretation, Vol. II Part One, Constitution and Properties of Some Representative Glasses, Interscience Publishers, New York.

Weyl, W. A. and Marboe, E. C., (1967), The Constitution of Glasses, A Dynamic Interpretation, Vol. II: Part Two, Constitution and Properties of Some Representative Glasses, Interscience Publishers, New York.

Willens, R. H., (1962), *J. Appl. Phys.,* **33**, 3269.

Williams, M. L., Landel, R. F. and Ferry J. D., (1955), *J. Amer. Chem. Soc.,* **77**, 3701.

Zachariasen, W. H., (1932), *J. Am. Chem. Soc.,* **54**, 3841.

Zeller, R. C. and Pohl, R. O., (1971), *Phys. Rev.,* B **4**, 2029.

Chapter 2

Structure of Amorphous Semiconductors

R. Grigorovici

Institute of Physics of the Roumanian Academy of Sciences, Bucharest

2.1 INTRODUCTION

When a few decades ago solid state physics emerged as a discipline in its own right it centered its interest on crystals. Among the different reasons one certainly was that crystallography offered the physicist a ready-made formalism describing the symmetry properties of periodic crystal lattices. Only the glasses presented too much technical interest to be completely pushed aside.

Liquid metals, like mercury, and solid amorphous semiconductors, like selenium, were trivial examples of non-crystalline electronic conductors. The first attempt to finding a correlation between the structure of amorphous and liquid semiconductors and their electric properties was initiated by Joffe (1947) and culminated in the rule of Joffe and Regel (1960). More recently many efforts have been devoted to finding theoretical treatments able to predict the energy spectrum of the electrons in amorphous semiconductors and to describe the optical and transport properties of these materials. However any such calculation implies the knowledge of their structure. It is the aim of this contribution to review the present state of this knowledge (see also Grigorovici, 1969a; Turnbull and Polk, 1972; Grigorovici, 1973).

2.2 SEMICONDUCTING PROPERTIES AND STRUCTURE

2.2.1 The Crystalline Lattice and the Chemical Bond

There is an obvious correlation between the prevailing type of chemical bond which links to each other the atoms of a solid and the resulting crystal lattice (Pauling, 1960; Wells, 1963; Krebs, 1968). Here we are concerned with semiconductors only and, therefore, covalent bonds are at least strongly predominant in all of them. It is therefore useful to analyse the correlation between a particular type of covalent bond and the resulting crystal lattice.

The simplest case is that of the quadrivalent elements. In all of them, except Pb, at least one of the polymorphs involves four hybrid sp^3 orbitals with tetrahedral symmetry. The 4 electrons which occupy these orbitals stem from the s^2p^2 ground state of the atoms by promotion into a sp^3 state. There might be also some d and f contributions to the hybridization which confine the electrons into narrower orbitals and destroy the rotational symmetry of the sp^3 orbitals. The two-atoms configuration with the dihedral angle $\delta = 0$ (Figure 2.1) is called eclipsed and its binding energy is slightly less than that of the staggered configuration with $\delta = 2\pi/3$.

The crystal which corresponds to the highest binding energy is the diamond lattice, in which all atoms are linked in the staggered configuration (atoms 1 and 2 in Figure 2.2). This lattice is the high temperature and high pressure polymorph of C, the normal temperature and pressure polymorph of Si and Ge, and the low temperature form of Sn.

Similar structures can be found in many $A^{IV}B^{IV}$, $A^{III}B^{V}$, and $A^{II}B^{VI}$ semiconducting compounds, while the character of the bond becomes increasingly ionic. In these latter compounds two related crystal structures

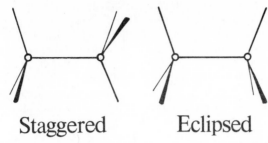

Fig. 2.1.   Staggered and eclipsed configurations in a tetrahedral bond.

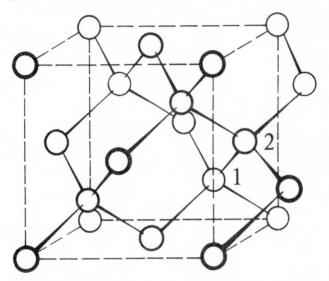

Fig. 2.2.   The diamond lattice. Atoms 1 and 2 in staggered configuration (after Pauling (1960)).

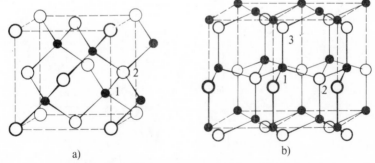

a)   b)

Fig. 2.3.   a) The sphalerite lattice (Zn, black spheres; S, white spheres). Atoms 1 and 2 in staggered configuration.
b) The wurtzite lattice. Atoms 1 and 2 in staggered, atoms 1 and 3 in eclipsed configuration (after Pauling (1960)).

are encountered. The first one, called sphaleritic (Figure 2.3a), is the analogue of the diamond lattice, the A^{II} and B^{VI} atoms being tetrahedrally surrounded by 4 B^{VI} and A^{II} atoms, respectively, in the staggered configuration. The second form, called wurtzitic, differs from the former one only by a quarter of the two-atom configurations being eclipsed instead of staggered (atoms 1 and 3 in Figure 2.3b). Some $A^{IV}B^{IV}$ and $A^{II}B^{VI}$ compounds like SiC or ZnS exist in numerous polytypes differing from one another by the alternation over different periods of diamond- and wurtzite-like packings.

$A^{II}B^{IV}C_2^V$ and $A^{II}B^{II}C_2^{VI}$ compounds (chalcopyrites) are chemically and structurally close analogues of $A^{III}B^V$ and $A^{II}B^{VI}$ compounds. Because of the difference between the A-C and B-C distances the tetrahedra are slightly distorted and, if the A and B atoms form ordered sublattices, the crystals are tetragonal rather than cubic. In some $A^{II}B^{IV}C_2^V$ compounds, e.g. in ZnSnAs$_2$, order-disorder transitions take place at a few degrees below the melting point, leading to a tetragonal-cubic transformation.

Another compound with tetrahedral coordination is CdAs$_2$ (Červinka, Hosemann and Vogel, 1970). By the transfer of two electrons, one from every As atom, Cd changes its electron configuration from $4d^{10}5s^2$ into $4d^{10}5s^2 5p^2$, while As changes from $3d^{10}4s^2 4p^3$ into $3d^{10}4s^2 4p^2$, i.e. the configurations of Sn and Ge, respectively. In the CdAs$_2$ crystal each Cd atom is surrounded by 4 As atoms, each As atom by 2 Cd and 2 As atoms (Figure 2.4). The $A^{II}B^{IV}C_2^V$ crystal CdGeAs$_2$ is related to the CdAs$_2$ crystal. The additional Ge atoms alternate with the Cd atoms along the c-axis and

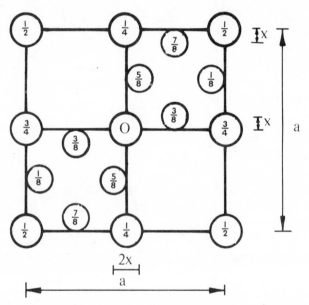

Fig. 2.4.   The CdAs$_2$ lattice (Cd, big circles; As, small circles). $a = 7.96$ Å; $c = 4.67$ Å; $x = 0.48$ Å (after Cervinka *et al.* (1970)).

occupy the hitherto empty centers of the distorted As tetrahedra (B sites). It must, however, be mentioned that the truly tetravalent Ge atoms lie at the center of nearly undistorted tetrahedra, while the distortion around the Cd and especially the As atoms is much higher (Vaipolin, 1972), thus revealing the imperfect sp³ character of their bonds.

In the above-mentioned crystals all orbitals are provided with just the number of electrons necessary to achieve pairing by overlapping. There is therefore no resonance between the bonds and, in consequence, all these materials are insulators or semiconductors.

An alternative bond between two A^{IV} atoms is the double bond. In this case the condition of maximal overlapping of the orbitals participating in

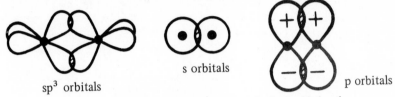

sp³ orbitals   s orbitals   p orbitals

Fig. 2.5. Overlapping of atomic orbitals in the double bond.
a) Hybrid sp³ orbitals;
b) s and p orbitals.

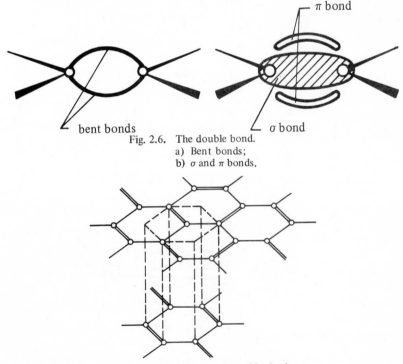

π bond

bent bonds   σ bond

Fig. 2.6. The double bond.
a) Bent bonds;
b) σ and π bonds.

Fig. 2.7. The hexagonal graphite lattice.

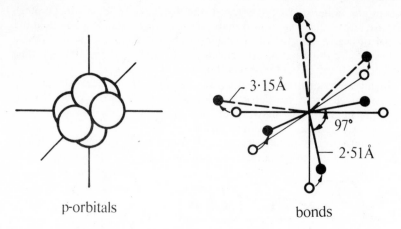

p-orbitals   bonds

Fig. 2.8. *p* orbitals and first- (——) and second-neighbour (- - -) bonds in As (after
Krebs (1966). o → ·: shift of atoms due to *s* hybridization of *p* bonds.

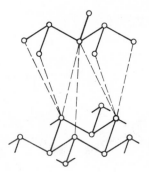

Fig. 2.9.   The α-As lattice. First- (——) and second-neighbours (- - -) bonds.

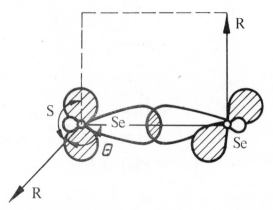

Fig. 2.10.   Staggered configuration in Se_2R_2 molecule (R = radical). Orbitals containing paired electrons are shaded. ϑ = bond angle; δ = dihedral angle.

the double bond might be expressed in two different ways: by bent bonds or by δ- and π-type orbitals (Figure 2.5). In both cases free rotation of one atom against the other about the double bond is again hindered, but the highest binding energy now corresponds to a coplanar configuration (Figure 2.6).

The corresponding crystal is the graphite lattice (Figure 2.7). It consists of parallel hexagonal networks of C atoms in which the double bonds resonate between the different possible configurations, so that every C-C bond acquires a one-third double-bond character. The layers are linked to each other by weak van der Waals forces, the distance between them being more than twice the distance between the atoms within the layer. Consequently graphite is a highly anisotropic semimetal; its electrical conductivity is much higher perpendicularly to the c-axis of the hexagonal crystal than parallel to it. The high conductivity is attributed to the π-electrons of the double bonds.

The stable polymorphs of As and Sb have to be considered here in spite of being semimetals when crystalline, as they display semiconducting properties when amorphous. The fundamental state is a $s^2 p^3$ one. Three symmetrical p bonds are therefore fully developed. Nevertheless some sp hybridization occurs, followed by the asymmetrization and some angular modification in the bonds (Figure 2.8). Three of the bonds become stronger and shorter and the angle between them increases from 90 to 97° in As and to 95°30′ in Sb; therefore the coordination number in both As and Sb is 3. The other 3 bonds are weaker and longer, by 25% in As and by 15% in Sb. A layer structure results (Figure 2.9) in which each layer consists of a network of flattened tetrahedra with the atoms at their corners.

In Se and Te the fundamental state is a $s^2 p^4$ one. Thus the s orbital and one of the 3 p orbitals are completely filled. Again sp hybridization leads to a pronounced asymmetrization of the remaining two p orbitals. There is obviously a strong repulsion between the completely filled p orbitals. In a $Se_2 R_2$ molecule (Figure 2.10; filled orbitals shaded) this repulsion will force the Se atoms into a staggered configuration. Therefore the remaining two orbitals directed towards the R radicals will tend to form a dihedral angle δ of 90° and the same angle ϑ with the Se-Se bond. In fact δ and ϑ are near these values, but they depend also on the substituents and their interaction. The potential barrier opposing the rotation about the Se-Se bond is 5–6 kcal/mole at $\delta = 0°$ and 2–3 kcal/mole at $\delta = 180°$.

Group VI elements like S, Se and Te also form hexahalides ($A^{VI} B_6^{VII}$) and $A_2^{VI} B_{10}^{VII}$ molecules (Figure 2.11) with strictly octahedral symmetry about the A^{VI} atom. Some d hybridization must be postulated in order to form he 6 necessary symmetrically oriented and only half-filled highly assymetrical $d^2 sp^3$ orbitals. The corresponding crystals are trigonal Se and Te, and α- and β-monoclinic Se.

Trigonal Se is the stable polymorph of Se. It consists of closely packed, infinite and parallel helicoidal chains with trigonal symmetry (Figure 2.12). The values of the dihedral and bond angles in the chain are $\delta = 102°$ and $\vartheta = 105°$; the bond length is 2.32 Å. Four next-nearest neighbours lie at 3.46 Å

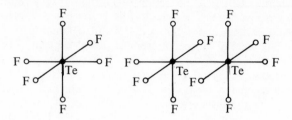

Fig. 2.11. Structure of TeF$_6$ and Te$_2$F$_{10}$ molecules.

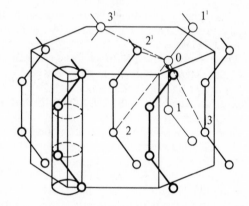

Fig. 2.12. Structure of trigonal Se. First-neighbour bonds (2.32 Å): 0–1,1′. Second-neighbour bonds (3.46 A): 0 - - - 2, 2′, 3, 3′.

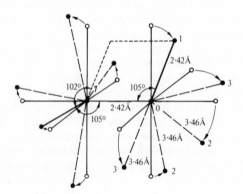

Fig. 2.13 Two-atoms configuration in trigonal Se. First- (——) and second-neighbour (- - -) bonds. o → ·: shift of atoms from ideal d^2sp^3 configuration (see Figure 2.11 b).

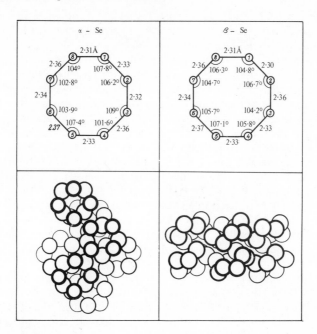

Fig. 2.14. 8-fold rings and their spatial arrangement in α- and β-monoclinic Se (after Wyckoff (1964)).

opposite to the bond angle (2 coplanar and 2 in a plane perpendicular to that of the angle). The ratio $3.46/2.32 = 1.49$ is too low to assign these four bonds a purely van der Waals character. The steric arrangement (Figure 2.13) suggests rather a distorted octahedral symmetry (see Figure 2.11). Another strong argument favouring this latter interpretation is the structural change taking place when passing to Te and Po, the heavier A^{VI} atoms. In Te $\delta = 100°$; $\vartheta = 102°$; $r_1 = 2.86$ and $r_2 = 3.47$ Å; thus $r_2/r_1 = 1.21$, while the Po lattice is a simple cubic one with $\delta = \vartheta = 90°$ and $r_2/r_1 = 1$. The Po simple cubic structure is also found in $(Au,Ag,Cu)_{30}Te_{70}$ alloys.

The other polymorphs of Se are α- and β-monoclinic Se. Both consist of closely packed, slightly distorted and puckered 8-fold rings. In α-Se the bond angle ϑ varies between $101.6°$ and $109°$, the bond length between 2.31 and 2.37 Å; in β-Se the corresponding values are 104.2 to 107.1 and 2.30 to 2.37 Å, respectively (Figure 2.14). The mean dihedral angle in a perfectly regular 8-fold puckered ring with $\vartheta = 105°$ is $\delta = 102.2°$ in excellent agreement with the δ value in trigonal Se. The only difference with the trigonal chain is that in the ring δ changes its sign at every bond, while in the chain its sign is the same over the whole length of the chain as well as in all chains of the crystal. The next-nearest neighbours in both α- and β-Se lie at about 3.8 Å ($r_2/r_1 \simeq 1.61$), but different Se atoms have different environments. These are sure indications that one deals with two different close packings of ring molecules rather than of atoms.

In Te $\vartheta = 102°$. To form a regular 8-fold ring δ should be $105.2°$, which differs significantly from the value $100°$ in the trigonal crystal. Indeed no polymorph of Te based on 8-fold rings is known.

If no restrictions are put on the value of the dihedral angle, other crystalline polymorphs of Se and Te could be imagined, while conserving the bond lengths and angles of both the first- and second-nearest neighbours of the trigonal crystals. In particular plane zig-zag chains with $\delta = 180°$ can be packed together extremely well. But no such crystalline polymorph is known. All these facts show how sensitive the crystal structure is to small differences in the lattice energy.

Of interest are also some $A^{IV}B^{VI}$ compounds, e.g. GeTe. By transferring one electron from the Te to the Ge atom, both get a similar electronic configuration (Ge: $4s^2 4p^3$; Te: $5s^2 5p^3$) and thus three unpaired p electrons each. No wonder 6-fold coordination occurs in the GeTe crystal which has a slightly distorted NaCl structure. At $400°C$ even this distortion disappears. SnTe and PbTe present the NaCl structure at even lower temperatures, while for GeS and GeSe only the distorted NaCl structures are known. There is no crystalline SiO.

The last element of interest to us is boron. Though a group III element, its chemical behaviour is rather similar to that of Si because its three valence electrons are unable to fill all its four orbitals ($2s2p^3$). This electron deficiency also lies at the root of its peculiar structural properties. Its compounds with H, the boranes, range from B_2H_6, through B_6H_{10}, to $B_{10}H_{14}$. While in B_2H_6 a slightly distorted tetrahedral symmetry is easily recognizable (Figure 2.15), in B_6H_{10} the first signs of an asymmetrical 6-fold ligancy appear. In $B_{10}H_{14}$ the molecule nearly closes into a shell in which each B atom is linked to 5 nearest neighbours within the shell and one outer H atom.

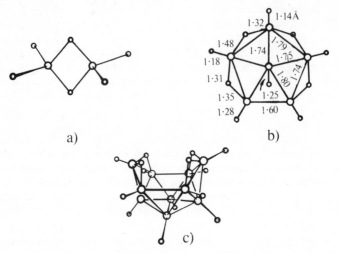

Fig. 2.15.   Structure of borane molecules: B_2H_4; B_6H_{10}; $B_{10}H_{14}$ (B, big circles; H, small circles (after Wells (1962)).

There are three B polymorphs: α- and β-rhombohedral and tetragonal boron. In all of them the essential structural unit is a nearly regular icosahedron containing twelve 6-fold coordinated B atoms 1.73 to 1.79 Å apart. In α-B the outer bonds either point directly towards a B atom in a neighbouring icosahedron or split into two weaker bonds pointing each towards B atoms in different icosahedra (3-center or δ-type bond). In β-B some icosahedra are merged with each other, while in tetragonal B some supplementary 4-fold coordinated B atoms are linked tetrahedrally to four B atoms in four different icosahedra (Figure 2.16). The two types of 4- and 6-fold ligancy encountered in B_2H_6 and $B_{10}H_{14}$ are easily recognizable.

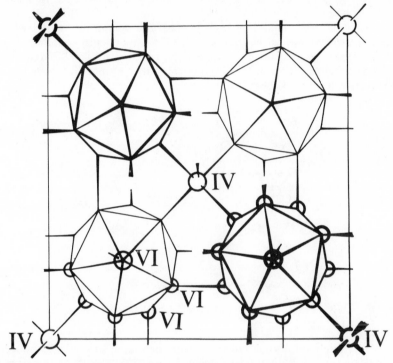

Fig. 2.16.   Unit cell of tetragonal B. VI – 6-fold coordinated atoms; IV – 4-fold co-ordinated atoms (after Pauling (1960)).

Semiconducting compounds in which the valencies of the atoms combine in such a way as not to give the same number of valence electrons per atom confer the different atoms present in the lattice different ligancies.

In an $A_4^{III}B_3^{IV}$ compound like N_4Si_3 the four sp^3 orbitals of the Si atoms are only half filled. Therefore each Si atom is surrounded tetrahedrally by four N atoms. Two of the five valence electrons of the N atoms are paired in one of the sp^3 orbitals. The other three half-filled ones form the edges of a trigonal pyramid with the N atom at the summit and three Si atoms at the corners of the base.

In an $A^{III}B^{VI}$ compound like InSe an electron will pass from the Se to the In atom, giving them a $4s^2 4p^3$ and a $5s^2 5p^2$ configuration, respectively. Obviously the same symmetries as in $N_4 Si_3$ must appear in the lattice, only that now every In atom is surrounded by three Se and one In atom, every Se atom by three In atoms.

In an $A_2^{III}B_3^{VI}$ compound like $In_2 Se_3$ only two of the Se atoms are able to get rid of one of their p electrons, becoming 3-valent, while the third Se atom is able to bind only two In atoms in an angular configuration. Both sp^3 and sp hybridization occur and while the ligancy of In is four and the Se atoms are bound tetrahedrally, 2/3 of the Se atoms have three In atoms, 1/3 of them only two In atoms as their nearest neighbours. The average ligancy of Se is therefore 2.67 in $In_2 Se_3$.

In $A_2^V B_3^{VI}$ compounds like $As_2 Se_3$, sp hybridization is present in both As and Se. The $As_2 Se_3$ lattice combines the 3-fold pyramidal coordination of the nearest Se neighbours of the As atoms with the 2-fold angular coordination of the nearest As neighbours of the Se atoms in a single lattice with layer structure (Figure 2.17).

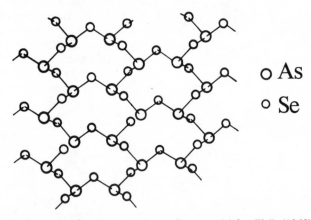

o As

o Se

Fig. 2.17.   Structure of one layer in the $As_2 Se_3$ crystal (after Wells (1962)).

All examples of crystals quoted hitherto have been selected because they are related to some amorphous semiconductor. The selection is therefore a narrow one. It might however suffice to show that, if the crystal structure of a given element or compound could not quite be predicted from the type of chemical bond present in related molecules, it can nevertheless be understood on this basis, at least in so far as covalent bonds are concerned.

2.2.2   Long-range and Short-range Order

The crystal lattice is an idealized description of the spatial arrangement of its atoms. The idealization consists in the basic assumption that the arrangement is strictly periodical in space, i.e. that spatial limitation, thermal vibrations and defects always existing in real crystals are overlooked.

The usual description of the lattice is obtained by applying to a given elementary cell a certain translation group. The symmetry operations which transform all translation vectors of the group into themselves consist in rotations, inversions and reflections. Only 1-, 2-, 3-, 4-, and 6-fold rotation and rotation-inversion axes are compatible with translation groups. The symmetry groups define the 7 crystal systems, while the 14 space lattices corresponding to the translation symmetries are known as Bravais lattices. Every real crystal lattice can be described as a superposition of parallel Bravais lattices. Therefore the symmetry operations of the whole crystal lattice are more numerous than those of the translation group. They also include screw axes and glide-reflection planes. All these operations form the spatial group of the crystal which gives a complete description of its lattice and characterizes the long-range order in the crystal.

Nearly all the information we have about the long-range order in crystals results from diffraction experiments with X-rays, electrons and neutrons. The diffraction pattern of a crystal is easily obtained by way of its reciprocal lattice. Each of its points corresponds to a set of equidistant planes in the real lattice which reflect the incident X-ray beam in a well defined direction. This direction is given by Ewald's construction (Figure 2.18). Let us take the point O as the origin of the reciprocal lattice and suppose that the crystal is irradiated in a given direction $\overline{PO}$. Let us take the length of $\overline{PO}$ equal to $1/\lambda$ and draw a sphere of radius $1/\lambda$ about P. There will be scattering of the incident radiation only if this sphere touches one of the points of the reciprocal lattice, e.g. Q, and the scattered wave will travel exactly in the direction $\overline{OQ}$.

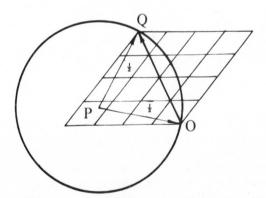

Fig. 2.18.   Ewald's construction. PO – incident; OQ – diffracted radiation.

If a high number of small crystals are used as a target and the incident radiation is monochromatic, Ewald's condition will be satisfied in a certain fraction of these crystals and, because of the rotational symmetry of the construction about $\overline{PO}$, the diffraction pattern will consist of rings centered on the incident beam. This is the Debye-Scherrer diagram.

Any slight deviation from the ideal infinite periodical structure of the lattice due to thermal vibrations or to the finite size of the crystals perturbs the ideal periodicity of the planes associated with a certain point of the reciprocal lattice. As a consequence the size of the points gets finite and this leads to a broadening of the diffraction rings.

The question is now what sort of order, if any, is conserved in a solid semiconductor when the long-range order is lost altogether, not only slightly disturbed. Such a solid is called amorphous.

The atoms cannot be distributed in space at random as the shape of the diffraction pattern shows striking similarities with that of the Debye-Scherrer diagram of a sample consisting of very small crystals of the related element or compound. Taking into account the strong directional character of the covalent bond, it would be difficult to admit that a total structural disorder could exist in a metastable amorphous phase and not to admit that the atoms of this phase would be arranged in such a way as to satisfy at least a high number of chemical bonds. Therefore the order within the amorphous phase must show some similarity with that of the crystal lattice, but nevertheless be compatible with the loss of long-range order. This order extends only to a distance of 10 to 20 Å. This is the short-range order we would like to know and to describe as precisely as possible.

Direct information about the short-range order in both crystals and amorphous materials can be obtained by analysing the fine structure of the X-ray absorption spectrum beyond the absorption edge proper. This structure is due to the influence of the environing atoms on the absorbing one (see Section 2.4.1). The influence vanishes rapidly with increasing distance, so that the information one gets from the extended X-ray absorption fine structure (EXAFS) is confined to a few coordination spheres.

Hitherto three different approaches have been used in order to describe the short-range order of amorphous semiconductors:

– The microcrystal approach: it is supposed that perfect order reigns over a small distance, after which there is an abrupt change in the orientation of the next ordered domain;

– The perturbed crystal approach: it is supposed that a high number of more or less conventional structural defects are frozen into an ideal crystal;

– The continuous network approach: it is supposed that there are no discontinuities within the amorphous phase, its atoms forming a three-dimensional array of continuously interconnected atoms.

All these approaches are compatible with the general features of the diffraction patterns of amorphous semiconductors, because all of them lead to a progressive broadening of the points of the reciprocal lattice of the related crystal when moving away from its origin. An option can be made only on the base of a thorough analysis of high quality structural data and on arguments from other types of measurement.

2.2.3 The Description of the Short-range Order in Amorphous Semiconductors

As mentioned above, the diffraction pattern of an amorphous solid shows rotational symmetry about the incident beam. This is due to the fact that the irradiated volume of the sample comprises a very high number of small, randomly oriented and more or less ordered domains. The EXAFS is recorded in the regularly transmitted beam itself. It is therefore impossible to deduce the three-dimensional short-range order of an amorphous solid from its diffraction pattern or its EXAFS. Only a one-dimensional description of this order results in the form of a radial distribution function (RDF) which yields the atom density as a function of the distance from an arbitrarily chosen atom. Peaks in the RDF indicate by their position and area the radius of and the number of atoms in a given coordination sphere (see Sections 2.3.2 and 2.3.3). The isotropy of the diffraction pattern reflects only the macroscopic isotropy of the investigated solid, but has nothing to do with the isotropy of the short-range order. Especially in amorphous semiconductors with their low coordinations the local short-range order cannot be but highly anisotropic.

Yet, an analysis of the shape of successive peaks in the RDF provides more than just one-dimensional information. In a tetrahedrally connected glass, for instance, fluctuations in the interatomic distance show up in the shape of the first peak, fluctuations in bond angles in the second peak, and fluctuations in dihedral angles both in the area and the shape of the third peak.

However, between the short-range order and the RDF there is not a one-to-one correspondence: the three-dimensional short-range order cannot be inferred from the diffraction pattern or the EXAFS of an amorphous solid, but a given model can agree or not with the corresponding RDF, the diffraction pattern or the EXAFS.

Therefore the only way towards describing the three-dimensional short-range order in amorphous semiconductors is the construction of structural models.

All approaches in modelling use more or less the related crystal lattices as their starting point. Modelling based on the microcrystal and perturbed crystal approach lets the short-range order of the related crystal essentially unchanged. The macroscopic isotropy is achieved by random orientation of the crystals. The continuous network approach however starts more or less openly from pair and higher correlation functions. Such a model must have a more pronounced statistical character: the spatial distribution of the atoms will differ somewhat from one region to another, showing both statistically varying orientations and local radial and angular distortions.

2.3   THE DIFFRACTION PATTERN OF AN AMORPHOUS SOLID

2.3.1   General Features
It was shown in Section 2.2.3 that the diffraction patterns of amorphous solids are in many respects similar to that of a microcrystalline powder.

Let us firstly admit that the amorphous solid consists indeed of a great number of perfectly ordered and finite clusters of atoms oriented at random. Within each cluster which contains N atoms of different types i there are a certain number of randomly oriented interatomic distances x_{ij}. There is also no correlation between the positions of the atoms contained in different clusters. The intensity distribution of the scattered X-rays as a function of $s = (2 \sin \vartheta)/\lambda$ will be given by

$$I_N(s) = \sum_i f_i^2 + \sum_1^N {}_i \sum_1^N {}_j f_i f_j \frac{\sin 2\pi s x_{ij}}{2\pi s x_{ij}} \quad , \tag{2.1}$$

where f_i are the atomic scattering factors of the different atoms and $x_{ij} = |x_{ij}|$. This is the so-called Debye formula (Furukawa, 1962). If all atoms are equal

$$I_N(s) = \phi_s N f^2 = N f^2 \left[1 + \frac{2}{N} \sum_n \frac{\sin 2\pi s x_n}{2\pi s x_n} \right] \quad , \tag{2.2}$$

where ϕ_s is called the interference function and n is the numerical index attributed to a given interatomic distance $[n_{max} = N(N-1)/2]$. Eq. (2.2) can be used also for clusters of molecules or atom groups by replacing the atomic scattering factor f by the structure factor of the molecule or group, F. The diffraction pattern based on a certain microcrystal model can now be confronted with the experimental results. This method, sometimes called 'correlation method', is valuable especially by the information it conveys about the shortest interatomic distances in a certain model and is easy to manipulate even if more than one type of atom is present in the structure.

One might renounce from the very beginning to any vectorial description of the short-range order, by simply using the probability $P(r)$ of finding an atom on a sphere of radius r.

If all atoms are identical

$$\phi(s) = \frac{I(s)}{N f^2} = 1 + \frac{2}{sv} \int_0^\infty r \left[P(r) - 1 \right] \sin (2\pi sr) \, dr \tag{2.3}$$

$$= 1 + \int_0^\infty 4\pi r^2 \left[\rho_a(r) - \rho_0 \right] \frac{\sin 2\pi sr}{2\pi sr} \, dr. \tag{2.4}$$

Here N is the number of atoms contained in the sample; $\rho_0 = 1/v$ is the average atom density, v being the average volume occupied by one atom: $\rho_a(r) = \rho_0 P(r)$ is the local density. $4\pi r^2 \rho_a(r)$ is called the atomic radial distribution function (ARDF). It gives the number of atoms contained within a spherical shell of radius r and unit thickness centered on an arbitrary atom (see e.g. Figure 2.30). Sometimes, in order to avoid the steep increase in slope at high r and better define the positions of the peaks of $\rho_0(r)$, the differential atomic radial distribution function (DARDF) $4\pi r^2 [\rho_a(r) - \rho_0]$ is used (see e.g. Figure 2.27).

When more than one type of atom is present in the structure, different types of RDFs are used (Filipovitch, 1955; Paalman and Pings, 1963).

If the investigated sample contains N_i atoms of different species $i = 1,2,...m$; Z_i is the number of electrons contained in the atom of type i; and $\epsilon_{in}(r)$ defines the electron density about the center of the n-th atom of type i taken over the whole sample, then the diffraction pattern is described by the intensity distribution

$$I(s) = I(0) + \int_0^\infty 4\pi r^2 \sum_1^m N_i Z_i \left[\epsilon_i(r) - \epsilon_0\right] \frac{\sin 2\pi sr}{2\pi sr} \, dr. \qquad (2.5)$$

Here $\epsilon_i(r) = (1/N_i Z_i) \sum_1^{N_i} Z_i \epsilon_{in}(r)$ is the mean electron density over the whole sample about the centers of all atoms of type i and ϵ_0 is the mean electron density. $I(0)$ is the scattered intensity at the center of the diffraction pattern due only to the geometry of the sample and its macroscopic properties; it is inaccessible to measurement. $4\pi r^2 \sum_1^m N_i Z_i \epsilon_i(r)$ is the true electronic radial distribution function.

In order to eliminate the influence of the finite distribution of electrons within a given atom, each atom of type i is considered to be a single punctiform scattering center of scattering factor f_i given by

$$f_i^2(s) = 4\pi \int_0^\infty \epsilon_i(r) * \epsilon_i(-r) r^2 \, \frac{\sin 2\pi sr}{2\pi sr} \, dr = 4\pi \int_0^\infty \overset{2}{\epsilon_i}(r) r^2 \frac{\sin 2\pi sr}{2\pi sr} \, dr, \quad (2.6)$$

where $\overset{2}{\epsilon_i}(r)$ symbolizes the convolution product between $\epsilon_i(r)$ and $\epsilon_i(-r)$ within the atom of type i. If these atoms would scatter independently, the scattered intensity

$$I_g(s) = \sum_1^m N_i f_i^2(s) \qquad (2.7)$$

would be due exclusively to the electronic distribution within them. Therefore $I(s) - I_g(s) - I(0)$ represents the contribution to the diffraction pattern of the interatomic interference only. The corresponding radial distribution function $4\pi r^2 \sum_1^m N_i [Z_i \epsilon_i(r) - \overset{2}{\epsilon_i}(r)]$ is called 'electronic' (ERDF), but is used to characterize in fact the atomic radial distribution in the amorphous solid.

One can also try to localize the scattering centers within the centers of the atoms from the very beginning. This is achieved by superposing the partial interference functions

$$\phi_{ij}(s) = 1 + \int_0^\infty 4\pi r^2 \rho_0 \left[\frac{\rho_{ij}}{c_j \rho_0} - 1\right] \frac{\sin 2\pi sr}{2\pi sr} \, dr, \qquad (2.8)$$

where ρ_{ij} is the density distribution of atoms of type j about an atom of type i; $c_j = N_j/N$ is the concentration of atoms of type j and ρ_0 is the mean atom density. The total interference function

$$\phi(s) = \frac{I(s)-I(0)-N[\bar{f^2}-(\bar{f})^2]}{N(\bar{f})^2} = 1 + \int_0^\infty 4\pi r^2 \,[\rho(r)-\rho_0]\, \frac{\sin 2\pi sr}{2\pi sr}\, dr, \quad (2.9)$$

where $\bar{f}$ and $\bar{f^2}$ are the mean and mean square atomic scattering factors, respectively;

$$\rho(r) = \Sigma_i \,\Sigma_j \,c_i f_i f_j \,\frac{\rho_{ij}}{(\bar{f})^2} \quad\quad\quad (2.10)$$

is the weighted atomic radial density and, in consequence,

$$\phi(s) = \Sigma_i \Sigma_j \,\frac{c_i \,c_j \,f_i \,f_j}{(\bar{f})^2}\, \phi_{ij}(s)\,. \quad\quad\quad (2.11)$$

The corresponding so-called atomo-electronic radial distribution function (AERDF) is $4\pi r^2 \,\rho(r)$.

Eq. (2.10) implies that $f_i f_j/(\bar{f})^2$ is independent of s, which is certainly not exactly true, but proves to be a useful approximation. More often this approximation takes the form of assigning each type of atom a constant effective number of electrons $K_i = f_i/f_e$. Here f_e is the scattering factor of an hypothetical electron which scatters in all directions as much power as an average atomic electron within an array of N independently scattering atoms, so that

$$I_t = N \,\Sigma_i \,c_i \,K_i^2 \,f_e^2 = N \,\Sigma_i \,c_i \,Z_i^2 \,f_e^2\,. \quad\quad\quad (2.12)$$

In this case Eq. (2.9) can be rewritten as

$$\phi(s) = 1 + N f_e^2 \int_0^\infty 4\pi r^2 \,[\Sigma_i \Sigma_j c_i K_i K_j \,\rho_{ij}(r) - (\Sigma_i c_i K_i)^2 \rho_0] \frac{\sin 2\pi sr}{2\pi sr}\, dr. \quad (2.13)$$

In neutron diffraction f_i has to be replaced by the scattering amplitude b_i for neutrons, linked to the scattering cross-section σ_i by $\sigma_i = 4\pi b_i^2$.

The most obvious consequence of using AERDFa instead of ERDFs is the narrowing of the peaks related to the different coordination spheres. Unfortunately because of the increase of $1/f_e^2$ with increasing s in the interference function (2.9), experimental errors and uncertainties in the numerical evaluation of K_i introduce spurious effects when calculating AERDFs from experimental data (see Section 2.3.2). On the contrary the intensity function $I(s)-I_g(s)-I(0)$ tends towards zero with increasing s, thus avoiding such effects.

2.3.2 The Atomic Radial Distribution Function in Elements

The RDF is obtained relatively simply by a Fourier transformation of the interference function (2.4) if only one type of atoms is present in the solid:

$$4\pi r^2 \,\rho_a(r) = 4\pi r^2 \,\rho_0 + 8\pi r \int_0^\infty s\,[\phi(s)-1]\sin(2\pi sr)\,ds \quad\quad\quad (2.14)$$

$$= 4\pi r^2 \,\rho_0 + 8\pi r \int_0^\infty s\,i(s)\sin(2\pi sr)\,ds \quad\quad\quad (2.15)$$

where

$$i(s) = \frac{I(s)}{Nf^2} - 1 \qquad (2.16)$$

and ρ_0 is the mean atom density of the material.

The $i(s)$ function has to be calculated from the experimental diffraction pattern $I(s)$. This calculation involves:

— corrections determined by the absorption of the radiation in the sample, the geometry of the experimental arrangement, the polarization of the scattered radiation and the presence of incoherent scattering;

— the knowledge of the mean atom density ρ_0 and of the angular dependence of the atomic scattering factor f;

— normalization of $I(s)$ so as to fit both the total scattered power and the angular dependence of $I(s)$ at high diffraction angles to that given by the same number N of independently scattering atoms, Nf^2.

The necessary data and procedures can be found e.g. in the International Tables for X-ray Crystallography and in Klug and Alexander (1954). In a first approximation the values of f for X-rays are about proportional to Z^2 and independent of λ. In a better approximation f is dependent on λ due to slight dispersion corrections. For electrons the absolute values of f are much higher than for X-rays thus restricting electron diffraction measurements to very thin films only. The angular dependence is steeper than for X-rays and the dependence on Z follows at low angles ϑ a $Z^{1/3}$ law, making the detection of light atoms in the presence of heavier ones easier by electron diffraction. Elimination of unelastically scattered electrons from the diffracted beam and corrections for multiple scattering are essential in the evaluation of electron diffraction data. In exchange the scanning of an electron diffraction pattern can be achieved within seconds, while a precision X-ray measurement takes in general hours.

Neutrons are scattered by the atomic nucleus. The scattered intensity is spherically symmetrical. The dependence of the scattering cross-section σ on the atomic mass number A follows in general an $A^{1/3}$ law, but big fluctuations in σ occur from nucleus to nucleus in a seemingly random manner due to resonance and magnetic scattering.

The calculated RDF reflects reality only partially. In particular the Fourier transformation (2.14–2.15) can yield the real RDF only if the integrals are extended to infinity, while the Bragg angle ϑ is necessarily limited to $\pi/2$ and thus s to $s_{max} = 2/\lambda$. The limitation of the integration to s_{max} transforms a δ-shaped maximum of the real RDF situated at $r=a$ and corresponding to a coordination sphere of N atoms into a

$$\sin 2\pi s_{max}\,(r-a)/2\pi s_{max}\,(r-a)$$

function. The result is a central peak of height $h_{max} = 2\,s_{max}N$, width (at h=0) $\Delta = 1/s_{max}$ and area $A = 1.18\,N$. It is surrounded by secondary maxima and minima whose overlapping with other central or secondary peaks can simulate the existence of maxima without any physical meaning or falsify the positions or the areas below the central peaks. If there are also static or dynamic (thermal) fluctuations in r about the mean value a which follow, say, a Gaussian distribution, the shape of the peak will be the result of a convolution product between the two broadening functions.

Other artefacts might be introduced into the RDF by errors of measure and by the use of incorrect correction and normalizing factors. This happens especially at high values of s where the difference between $I(s)$ and Nf^2 is relatively small and therefore (see Eq. 2.16) $i(s)$ becomes very imprecise. These errors are further exaggerated by multiplication by s (see Eq. 2.15). The result is the appearance in the RDF of spurious maxima and minima without physical significance.

The termination error and spurious effects can be got rid of after Kaplow, Strong and Averbach (1965) by an iteration procedure. It starts with removing meaningless ripples from the RDF (e.g. between $r = 0$ and the first peak), proceeds by alternative transformations of RDFs into $I(s)$ functions extended far beyond s_{max} and vice-versa and by adequate alterings of both functions until a good match is obtained between the extended and the experimental $I(s)$ over the range $0 < s < s_{max}$. The last extended $I(s)$ is then used to get a final RDF_f free from errors, but consistent with experiment. The peaks of RDF_f will be broadened only due to fluctuations in r, the corresponding $DRDF_f$ will start from $r = 0$ as a parabola given by $-4\pi\rho_o r^2$ (see Figure 2.27). Simpler alternative procedures consist in multiplying $I(s)$ by an appropriate artificial temperature factor which suppresses the ripples in the RDF, but also diminishes its resolution, or in extending $i(s)$ analytically beyond s_{max}, if $i(s)$ shows below s_{max} the form of a damped periodic function.

2.3.3 The Radial Distribution Function in Compounds

The Fourier transformations analogous to Eqs. (2.14) to (2.16) can be written starting from Eqs. (2.5) through (2.13):

$$\text{ERDF} = 4\pi r^2 \sum_1^m N_i \left[Z_i \epsilon_i(r) - \bar{\epsilon_i}(r) \right] = 4\pi r^2 \, \Sigma_i N_i Z_i \epsilon_0 + 8\pi r \int_o^\infty s \left[I - I(0) - I_g \right]$$

$$\sin (2\pi sr) \, ds; \tag{2.17}$$

$$\text{AERDF} = 4\pi r^2 \frac{\Sigma_i \Sigma_j c_i f_i f_j \rho_{ij}(r)}{(\bar{f})^2} = 4\pi r^2 \, \rho_o + 8\pi r \int_o^\infty s \frac{I - I(0) - N\bar{f}^2}{N(\bar{f})^2}$$

$$\sin (2\pi sr) \, ds \tag{2.18}$$

or

$$= 4\pi r^2 \sum_1^m \sum_1^m c_i c_j K_i K_j \rho_{ij}(r) = 4\pi r^2 \left[\sum_1^m c_i K_i \right]^2 \rho_o + 8\pi r \int_o^\infty s \frac{I - I(0) - N \Sigma_i c_i f_i^2}{N f_e^2}$$

$$\sin (2\pi sr) \, ds. \tag{2.19}$$

The relative advantages of using the ERDF or AERDF have already been discussed in Section 2.3.1. A comparison between both types of RDF may reveal the artificial character of some parasitic peaks in the AERDF.

To extract structural information from both the ERDF and the AERDF firstly their peaks must be attributed to a certain pair of atoms. This can be done for the first peak by supposing that the radius of the first coordination sphere corresponds to the sum of the covalent radii of the involved atoms. If more than two types of atoms are involved, the peak may be broadened by partial overlapping of narrower ones corresponding to different atom pairs. The analogy between the amorphous and the crystalline phases can also be used, but in general not farther than the second coordination sphere.

Much more information can be gained in this respect by recording diffraction patterns of the same amorphous material by using two different radiations, e.g. X-rays and neutrons (see e.g. Henninger and Buschert, 1967). As shown in Section 2.3.2 the effective number of electrons K_i for X-rays varies monotonously with Z, while the scattering amplitudes b_i for neutrons vary irregularly with the nuclear species. If only two types of atoms are present and experimental conditions are selected in such a manner that, for instance, $K_i > K_j$, but $b_i > b_j$, then the peaks in which $i-i$ pairs are predominant will be greatly enhanced in the X-ray RDF relative to the neutron RDF.

A complete solution for a binary compound or alloy can be given in the approximation of the AERDF if one disposes of three different diffraction patterns of the same material taken in conditions under which the two types of atoms have as different scattering factors f_1 and f_2 as possible for the incident radiations. This can be done by using alternatively X-rays of different wave lengths, electrons, and neutrons and/or by changing the isotopic composition of the sample. If one admits that the partial interference functions (see Eq. 2.8) do not depend on composition (c_1 and c_2), which seems to be true to a first approximation in some alloys, the three measurements can be done with the same radiation, but on three different compositions.

In the above-mentioned cases all three partial interference functions and, in consequence, the three partial RDFs, namely $4\pi r^2 \rho_{11}(r)$, $4\pi r^2 \rho_{12}(r)$ and $4\pi r^2 \rho_{22}(r)$ which characterize the structure completely can be obtained (Steeb and Hezel, 1966). The fourth partial RDF, namely $4\pi r^2 \rho_{12}(r) = (c_1/c_2) \, 4\pi r^2 \rho_{12}(r)$ is not independent.

Supposing that we are able to attribute to each peak of the AERDF the type of atom pairs involved, the area below a given peak will be

$$A = \Sigma_i \Sigma_j c_i K_i K_j n_{ij}, \tag{2.20}$$

where n_{ij} is the number of atoms of type j surrounding one atom of type i.

If the different types of pairs give overlapping maxima, then

$$A = c_1 K_1^2 n_{11} + c_2 K_2^2 n_{22} + c_1 K_1 K_2 n_{12} + c_2 K_1 K_2 n_{21}. \tag{2.21}$$

Supplementary hypotheses have to be made to separate the four different contributions. In alloys, for instance, one might be allowed to admit that

$$n_{11} + n_{12} = n_{22} + n_{21} = c_1 n_1 + c_2 n_2, \tag{2.22}$$

where n_1 and n_2 are the first coordination numbers in the pure elements. Eqs. (2.21) and (2.22) together with

$$c_1 n_{12} = c_2 n_{21} \qquad (2.23)$$

not only determine the four partial coordination numbers, but also allow for the short-range degree of ordering to be found.

2.4   THE EXTENDED X-RAY ABSORPTION FINE STRUCTURE

X-ray absorption beyond the K, L, . . . edges is due to the ejection of photoelectrons of increasing energy which are scattered by the surrounding atoms. This scattering introduces variations with energy of the value of the dipole transition matrix between the intial electron state and its final state and, consequently, of the X-ray absorption coefficient as seen in Figure 2.19. In spite of many previous attempts (see Azaroff, 1963) a satisfactory theory of the extended X-ray absorption fine structure (EXAFS) has been developed only recently (Sayers, Lytle and Stern, 1970). As a corollary, the EXAFS has been shown by Sayers, Stern and Lytle (1971) to provide valuable information about the short-range order in both crystals and amorphous materials.

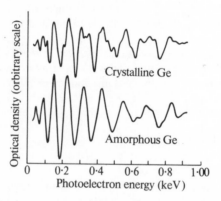

Fig. 2.19.   Oscillatory part of the optical density for X-rays beyond the main absorption edge (EXAFS) in crystalline and amorphous Ge (after Sayers, Stern, and Lytle (1971)).

2.4.1   General Features

The oscillatory photoelectric cross section written in the dipole approximation of radiative transitions is

$$\sigma = 4\pi^2 \; \alpha \; h\nu \; |\vec{r}_{if}|^2 \; N(E), \qquad (2.24)$$

where α is the fine structure constant, $h\nu$ the photon energy, $N(E)$ is the density of the final states and, under simplifying assumptions, the dipole matrix element $|\vec{r_{ij}}|^2$ is given by

$$M = \int \psi_i^* (\vec{r}) \; \vec{r} \; \psi_f (\vec{r}) \; dr. \tag{2.25}$$

For the K edge ψ_i may be represented by the atomic $1s$ wave function, and the problem consists in calculating the final photoelectron wave function ψ_f, including scattering.

In this calculation allowance had to be made for the following facts:

– The final state has p symmetry as imposed by the dipole selection rule $\Delta l = \pm 1$. This eases the extrapolation of ψ_f, calculated just outside the ionized atom represented by a muffin-tin potential, to near its nucleus, thus obtaining the result of the interference of the outgoing wave with the waves scattered back from the surrounding neutral atoms represented by δ-function potentials. When the waves add constructively, σ increases and vice-versa.

– $N(E)$ is supposed to be that of free electrons in spite of the existence of energy gaps, because scattering is weak and the gaps are narrow compared with the photon energy scale which extends from 30 to 1000 eV.

– As the photoelectron proceeds farther from its origin, its wave function ψ_f loses coherence with the initial wave function ψ and tends to form an eigenstate, thus no more producing scattered waves. This is accounted for by an exponential decay of ψ_f from its origin which is estimated to be so rapid as to make only 3 to 7 coordination spheres be felt in the EXAFS.

The final formula shows that the oscillatory part χ (K) *of the EXAFS* depends in principle not only on the number of atoms N_j in a given coordination shell of radius r_j, but also on their spatial arrangement. However the terms generated by this dependence are utterly negligible and so, finally, $\chi (k)$ is given by

$$\chi(k) = -k f (k) \; \Sigma_j \; [N_j \exp (-\gamma r_j)/r_j^2] \; \exp (-\sigma_j^2 k^2 /2) \sin [2kr_j + 2\eta (k)].$$

$$\tag{2.26}$$

Here $k = 2\pi/\lambda$, $f(k)$ is the usual electron scattering factor, γ describes the decay of ψ_f, σ_j is the mean square amplitude of the relative displacements of the atoms in the j-th shell around r_j and $\eta(k)$ is the phase shift of the photoelectron caused by the potential of the absorbing atom. This formula is in excellent agreement with experiments on crystalline metals (Cu, Fe) and semiconductors (Ge).

2.4.2 The Radial Distribution Function in Elements

Eq. 2.26 has the form of a simple sum of damped sine waves. The first coordination shell determines its main features; the following ones contribute to its details. This suggest the possibility of inverting it in order to obtain a RDF. This has been done by taking the Fourier transform of

$$-\chi(k)\,k^{-1}\,f(k)^{-1}:$$

$$G(r) = -(2/\pi)^{1/2} \int_{0}^{\infty} \chi(k)\,k^{-1}\,f(k)^{-1}\,\sin\,[2\,kr + 2\eta\,(k)]\,\mathrm{d}k \qquad (2.27)$$

$$= \Sigma_j\,[N_j\exp\,(-\gamma\,r_j)/r_j^2\,\,\sigma_j]\,\exp\,[-2(r-r_j)^2/\sigma_j^2] + \Delta(r),$$

which, if $\Delta(r)$ can be neglected, as it proved to be, gives the function $G(r)$ in the form of a sum of Gaussian peaks. Their area is no more proportional to N_j because of the damping factor γ, but comparison between the known RDF of the related crystal and its $G(r)$ allows for the determination of both γ and $\eta(k)$ and thus the evaluation of N_j in the amorphous material. The width of the peaks is determined by instrumental broadening, termination effects and by real fluctuations in r. Specific results on a-Ge films are shown in Figure 2.20.

2.4.3   The Radial Distribution Function in Compounds and Alloys

The theory of EXAFS has not yet been extended beyond elements. Obviously it will be there that the advantages of EXAFS will show up, because the environments of atoms of different species will be recorded separately, something one diffraction experiment alone cannot achieve. Therefore EXAFS promises to provide valuable complementary information about the short-range order of amorphous semiconducting compounds and alloys, Sayers, Lytle and Stern (1972).

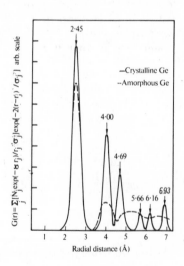

Fig. 2.20.   The radial distribution function $G(r)$ of crystalline and amorphous Ge obtained by Fourier transformation of the data of Figure 2.19 (after Sayers, Stern, and Lytle (1971)).

2.5   THE NEAREST NEIGHBOUR CONFIGURATION

2.5.1   Joffe and Regel's Rule

Joffe and Regel (1960) stated the empirical rule that a molten or amorphous semiconductor retains its semiconducting properties in spite of the destruction of the long-range order if only the non-crystalline phase conserves the short-range order present in the related semiconducting crystal. The short-range order was characterized by the number of atoms in and the radius of the first coordination sphere. There is at present not a single known exception to this rule. For the time being quantum theory is not yet able to handle condensed systems composed of many atoms, but lacking spatial periodicity. Thus at present only Joffe and Regel's rule is able to predict to some extent the properties of a given amorphous semiconductor. Theory will however finally master the present difficulty and then a quantitative description of the short-range order will be essential for its confrontation with experiment.

It is with this in mind that in the present chapter (2.5) only the nearest neighbour configuration will be used to classify these materials (Table 2.1). A more detailed treatment of the short-range order in amorphous semiconductors including, at least in principle, the whole more or less ordered region around an arbitrary atom is given in the next chapter (2.6).

2.5.2   The 2-Fold Configuration (chain and ring structures)

a-Se and perhaps also a-Te can be obtained by quenching their melt, certainly both by vacuum deposition on cold substrates. The first peak of the RDF of a-Se shows that in all samples, even if obtained or annealed differently, the mean interatomic distance always lies between the mean value in monoclinic Se and that in trigonal Se, as shown by Krebs and Schultze-Gebhardt (1955); Richter and Gommel (1957); Richter and Herre (1958); Andreyevski, Nabitovitch and Voloshchuk (1960); Krebs and Steffen (1964); Henninger, Buschert and Heaton (1967); Kaplow, Rowe and Averbach (1968); Richter (1972). The area of this peak always corresponds to 2 atoms; its width shows that in a-Se the static fluctuations in r_1 are lower than those in monoclinic Se (± 0.04 Å), if thermal broadening is taken into consideration. All these facts suggest that the chain and ring structures are present in both molten and solid a-Se, a-Te has shown recently to contain short Te chains (Ichigawa (1973)). In accordance with Joffe and Regel's rule, a-Se and a-Te are both semiconductors.

2.5.3   The 3-Fold Configuration (layer structures)

Amorphous As and Sb can be obtained by vacuum deposition, by chemical precipitation, and electrolytically as thin films. In all varieties of a-As and a-Sb the 3-fold coordination of the crystalline phases is exactly conserved (Krebs and Schultze-Gebhardt (1955); Krebs, Schultze-Gebhardt and Thees (1955); Richter and Gommel (1957); Krebs and Steffen (1964), Breitling (1972)) $r_{1_{am}}$ is 0.3 to 0.7% smaller than $r_{1_{cr}}$.

TABLE 2.1 *Amorphous Semiconductors*

1. Elements

Group	Symbol	Configuration	
		cryst.	amorph.
III	B	6 and 4	~ 6
IV	C	3 or 4	3 and 4
	Si,Ge	4	4
V	As,Sb	3	3
VI	Se	2	2
	Te	2	2

2. Compounds

Type	Formula	Configuration	
		cryst.	amorph.
IV-IV	SiC	IV: 4-IV	–
III-V	GaP,GaAs,GaSb,InSb	(III,V): 4(V,III)	(III,V): 4(V,III)
II-IV-V_2	CdGe(P,As)$_2$,ZnSnAs$_2$	V: 2.II + 2.IV (II,IV): 4.V	V: 2.II + 2.IV (II,IV): 4.V
IV-VI	SiO	–	(IV,VI): 2(VI,IV)
	(Ge,Sn,Pb) (Se,Te)	(IV,VI): 6(VI,IV)	IV: 4.VI; VI: 2(IV,VI)
III-VI	(Ga,In) (Se,Te)	III: 3.VI + 1.IV VI: 3.III	III: ~ 3.5.VI + ~ 1.IV VI: ~ 3.5.III
III$_2$-VI$_3$	(Ga,In)$_2$(Se,Te)$_3$	III: 4.VI VI: 2.67.III	III: 4.VI VI: 2.67.III
V$_4$-IV$_3$	N$_4$Si$_3$	IV: 4.III III: 3.IV	IV: 4.III III: 3.IV
V$_2$-VI$_3$	(As,Sb,Bi)$_2$(S,Se,Te)$_3$	V: 3.VI VI: 2.V	V: > 3.VI VI: > 2.V

3. Alloys

Type	Composition	Related crystals
II-X$_x$-V$_2$	CdX$_x$As$_2$ X=Ge,Si,Sb,Tl,Mg,Al, Ga,In	CdAs$_2$; CdGeAs$_2$
II-V$_x$	Mg(Sb,Bi)$_x$	Mg$_3$(Sb,Bi)$_2$
IV$_x$-VI$_{1-x}$	Ge$_x$(S,Se,Te)$_{1-x}$	Ge; S,Se,Te; Ge(S,Se,Te)
III$_x$VI$_{1-x}$	(Ga,In)$_x$Te$_{1-x}$	Te; (Ga,In)Te; (Ga,In)$_2$Te$_3$
	Tl$_2$Te$_x$	Tl$_2$Te$_3$; TlTe; Tl$_5$Te$_3$
V$_x$-VI$_y$	As$_x$S$_y$	As; S; As$_2$S$_3$
IV$_x$-V$_2$-VI$_{3 \pm y}$	(Ge,Si)$_x$As$_2$Te$_{3\pm y}$	As$_2$Te$_3$; GeTe
(I + I)$_3$-VI$_7$	(Cu + Au)$_3$Te$_7$	Te; (Cu,Ag,Au)$_x$Te$_{1-x}$

Note: (III,V) means: III or V group elements
      (III V) means: III and V group elements
      :4.IV means: are linked to four IV group atoms

While however crystalline As and Sb are semimetals, a-As and a-Sb have been found to be semiconductors (Krebs (1952); Moss (1959)), though their properties are far from being known in much detail. Baumann (1966) however obtained superconducting a-Sb films by evaporating Sb atoms from SbAu alloys instead of Sb_n molecules from pure Sb. The short-range order in this form of a-Sb has not yet been investigated.

2.5.4 The 4-Fold Configuration (tetrahedral structures)

Amorphous Si, Ge, SiC, GaP, GaAs, GaSb and InSb can be obtained by vacuum evaporation, by cathodic sputtering, electrolytically, and by r.f. decomposition of gaseous compounds as thin films. By releasing the hydrostatic pressure of about 140 kbar at $100°C$ from a sample of Ge_{IV}, a high pressure polymorph with b.c.c. lattice, a glassy form of Ge was obtained by Bates (1966), but its short-range order is not yet known.

The RDFs of the above-mentioned amorphous materials show a first peak whose area corresponds to $N = 4 \pm 0.1$ (Richter and Breitling (1958), on Ge and Si; Mikolaichuk and Dutchak (1964), on GaSb; Coleman and Thomas (1967), on Si; Grigorovici and Mănăilă (1968), on Ge; Light and Wagner (1968), on Ge; Moss and Graczyk (1969), on Si, Graczyk and Moss (1970), on Si; Breitling (1972), on Ge; Shevchik and Paul (1972), on Ge; Shevchik (1972), on Ge, GaP, GaAs and GaSb; Temkin, Connell and Paul (1972) on GeSn alloys; Shevchik, Lamin and Tejeda (1973) on GeSi alloys; Sayers, Stern and Lytle (1971), on Ge by EXAFS). The maximum of the first peak lies between 0.02 and 0.07 Å above the crystalline interatomic distances; the higher values were provided by X-ray data obtained photographically, the lower ones by X-ray diffractometry, by electron diffraction and by EXAFS.

An analysis of the shape of the first peak of the RDF of a-Ge and a-Si yielded static standard deviations $\sqrt{\overline{u^2}}$ from the mean interatomic distance lying in a-Ge between 0.06 Å (Shevchik and Paul (1972)) and 0.12 Å (Grigorovici, Mănăilă and Vaipolin (1968)) and at 0.09 Å in a-Si (Graczyk and Moss (1970)). It must however be emphasized that the shape of the first peak is rather sensitive to deposition conditions and annealing (Richter and Breitling (1958); Breitling (1972); Moss, Flynn and Bauer (1973)). It is also interesting to note that while a-GaAs yields an AERDF quasi-identical to that of a-Ge, with which it is isoelectronic, a-GaP and a-GaSb present much broader first peaks (Shevchik (1972)). Taking into account that the covalent radii of Ga and As are nearly equal, while those of Ga, P, and Sb differ, this broadening implies that 'wrong' Ga-Ga, P-P, and Sb-Sb pairs are present in the amorphous phase, in accordance with Raman scattering measurements by Wihl, Cardona and Tauc (1972).

Tetrahedral coordination is also present in amorphous $A^{II} B_x^{IV} C_2^V$ alloys and in the compounds with X = O and X = 1. If the crystal is heavily tetragonally distorted, i.e. (2a-c) 0.10 (A^{II} heavier than B^{IV}, e.g. A^{II} = Cd; B^{IV} = Si, Ge; C^V = P, As), bulk glasses can be obtained by quenching the melt of the compound (Vaipolin, Osmanov and Rud (1965); Goriunova (1965) for $CdGeAs_2$ and $CdGeP_2$; (Ugai, Ziubina and Aleinikova (1968); Hrubý and Štourač (1971)) for $CdAs_2$, or that of alloys like $CdGe_x As_2$, where x could vary between 0 and 1.3 (Jensen (1964); Hrubý and Štourač (1969); Červinka et al. (1970); Hrubý and Honserova (1972); Vaipolin, Kuzmenko and Osmanov, (1972)). Other elements like Si, Sb, Tl, Mg, Al, Ga and In play a similar role over narrower ranges. If A^{II} is nearly equal in mass or lighter than B^{IV}, $2a - c$ diminishes in the crystals, reaching zero, e.g. in $ZnSnAs_2$. The corresponding amorphous phase can be obtained by vacuum deposition as thin films (Manaila and Popescu (1971)), but not by quenching the melt.

Structurally only the $CdGe_xAs_2$ system and a-$CdGeP_2$ have been investigated hitherto. In spite of forming a homogeneous amorphous phase, $CdGe_xAs_2$ splits by crystallization into two phases: $CdAs_2$ and $CdGeAs_2$. After Červinka, Hosemann, and Vogel (1970) vitrification of $CdAs_2$ is eased by the Ge atoms occupying at $x < 0.2$ the empty spaces of the $CdAs_2$ lattice (Figure 2.4) and at $x > 0.2$ the empty centers of the As tetrahedra (B-sites), as in the $CdGeAs_2$ crystal.

The introduction of Ge and Si atoms does not disturb the redistribution of electrons in $CdAs_2$ by which each atom gets 4 unpaired electrons. Therefore the same factors, i.e. differences in covalent radii and in deformability of the tetrahedra combined with the partially ionic character of the bonds which determine the tetragonalization of the crystal (Borshchevskii *et al.* (1967)), must be supposed to determine vitrification too. Analysing the shape of the first peak of the RDFs of bulk a-$CdGeP_2$ and a-$CdGeAs_2$, Grigorovici, Mănăilă, and Vaipolin (1968) and Mănăilă and Popescu (1971) have shown that the difference in length between the P-Cd and P-Ge (2.55 vers. 2.33 Å) and the As-Cd and As-Ge (2.65 vers. 2.55 Å) bonds is at least partially conserved in the amorphous phase.

2.5.5   The 6-Fold Configuration (icosahedral structures)

The structure of a-B obtained by the reduction at $1000-1200°C$ of BCl_3 with H_2 has been investigated by X-ray diffraction by Badzian (1967). The RDF shows a series of well defined peaks. The first two lie at 1.8 and 3.1 Å and contain 6.6 and 17.5 atoms, respectively. These values compare well with those of α-rhombohedral B crystals: $r_1 = 1.71$ to 1.79 Å; $N_1 = 5.5$; $r_1' = 2.03$ Å; $N_1' = 0.5$; $r_2 = 2.9$ to 3.3 Å; $N_2 = 17$. It is nevertheless difficult to assess exactly which of the polymorphs of B is most narrowly related to a-B. Because tetragonal B is a high temperature polymorph and a-B-crystallizes above $1200°C$ into β-rhombohedral B, i.e. at the same temperature at which the $\alpha \to \beta$ transition of B occurs, Badzian concludes that a-B is a frozen stage in this transition.

2.5.6   Mixt Configurations

3-4 configurations
- Amorphous C is obtained either in thin films by vacuum deposition or in film and bulk form by pyrolysis of certain organic compounds. Structural investigations by Franklin (1950); Richter, Breitling, and Herre (1961); Kakinoki *et al.* (1960); Noda and Inagaki (1964); and Furukawa (1964) showed that the first peak of the RDF lies somewhere between 1.42 and 1.54 Å, its area corresponds to a coordination number between 3 and 4, and its width is rather large (Figure 2.21). Its particular position, area, and shape

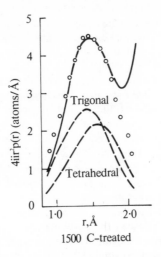

Fig. 2.21. First peak of the RDF of vitreous carbon treated at 1500°C. ———
experimental curve; o o o sum of trigonal and tetrahedral components
(after Noda and Inagaki (1964)).

depends on the way the sample has been obtained or annealed. The above
facts strongly suggest the existence of a mixt 3–4 configuration in a-C.

— Amorphous $A^{III}B^{VI}$ compounds, like a-InSe, show the first maximum of
the AERDF split into two peaks at 2.60 and 3.15 Å (Tatarinova, and
Kazmazovskaya, 1961). If the two peaks are attributed to In-Se and In-In
pairs, every Se atom is found to be surrounded by 3.46 In atoms, every In
atom by 3.46 Se and 0.95 In atoms against 3 and 3 + 1, respectively, in
crystalline InSe. These results show that the short-range order present in the
InSe crystal is conserved in detail in the amorphous phase.

— Amorphous N_4Si_3 is obtained as thin films by the decomposition in a
glow discharge of a mixture of ammonia and silane in the ratio 3:1. The first
peak of the AERDF at 1.7 Å corresponds to the N-Si distance in the crystal.
On this basis Coleman and Thomas (1967) conclude that a-N_4Si_3 contains
the same Si:4N and N:3Si configurations as the crystal.

Mixt 2–3–4–6 configurations
As shown in 2.2.1 mixt 2–3 configurations are found in $A_2^YB_3^{YI}$ crystals.
Amorphous phases of binary or pseudo-binary compounds with A = As, Sb,
Bi and B = S, Se, Te are obtained either by vacuum deposition
(Andreyevski, Nabitovitch, and Voloshchuk (1961)) or by quenching the
melt, especially if small quantities of other elements like Ga, Ge or I are
added (Kolomiets (1964), Hilton, Jones, and Brau (1966)). The thermal
properties of Ge-doped As_2Se_3 (Kolomiets, Pajasova, and Štourač (1965);
Štourač, Kolomiets, and Shilo, (1968)) indicate that, in accordance with
Myuller's (1966) views, the vitrification of As_2Se_3 by addition of Ge is

favoured by the intercalation of the added atoms between the normal structural units of crystalline As_2Se_3, strengthening the chemical bonds between these units, but creating microdomains whose structural symmetry is incompatible with the symmetry of the host lattice.

As an example, structural analysis of a-As_2Se_3 by Porai-Koshits and Vaipolin (1963) shows an increase in the coordination number by 10 to 20% both of Se atoms around As atoms and vice-versa compared with the crystal. The layer structure of the crystal is thought to be conserved in a-As_2Se_3, but the layers are supposed to be distorted and welded to one another in some points. Similar results have been obtained by Andreyevski, Nabitovitch, and Voloshchuk (1961) and Young and Thege (1971) on a series of a-$A_2^{V}B_3^{VI}$ films. Recently Tsushihashi and Kawamoto (1971) discussed how large numbers of excess As or S atoms can be included into the As_2S_3 glass structure by formation of 'wrong' As-As and S-S links, respectively. The inclusion of Ag and Ge atoms has been investigated structurally by Mănăilă and Popescu (1972). While Ag atoms break up the As-S rings leading to early crystallization, Ge atoms enter the glass structure in high quantities being accommodated, at least at low concentrations, between the layers. However, the conservation of the layer structure itself is doubtful (Grigorovici (1973)) because the second order As-As and S-S distances are not present in the RDF.

Of great interest are at present Ge- and Si-doped As_xTe_y glasses and films because of their threshold switching properties (Ovshinsky, 1966, 1968). Unfortunately until recently structural analysis was quite powerless in face of ternary alloys. Structural investigations by X-ray diffraction due to Hilton et al. (1966) on a $Ge_{15}As_{45}Te_{40}$ glass, for instance, yielded an AERDF with two peaks at 2.50 and 4.02 Å. The position of the first peak corresponds to an average of all interatomic distances which range from 2.43 Å (Ge-Ge) to 2.86 Å (Te-Te). Thus little can be inferred about the structure.

Another group of amorphous semiconductors are the $A^{IV}B^{VI}$ compounds and alloys. a-SiO, which is a much investigated material with large technical uses, a-SnTe, a-PbTe and the whole series of a-Ge_xTe_{1-x} and a-Ge_xSe_{1-x} alloys can be obtained by vacuum evaporation or sputtering. Bulk glassy Ge_xTe_{1-x} and Ge_xSe_{1-x} alloys can be obtained by splat-cooling the melt or even slower quenching if relatively small quantities of P, As, S, Si, or I are added to the melt. This is most easily achieved around the compositions close to an eutecticum (Hilton, Jones, and Brau, 1966; Feltz et al., 1971), e.g. for $x \simeq 0.15$ in the Ge-Te system.

AERDFs of some Ge-Te alloys (bulk samples with $x = 0.11$; splat-cooled samples with $x = 0.25$; thin films with $x = 0.5$, $x = 0.66$, and $x = 0.72$) have been obtained by Luo and Duwez (1963), by Bienenstock, Betts, and Ovshinsky (1969) and by Dove, Heritage, Chopra and Bahl (1970). They show a first peak shifting from 2.64 to 2.70 Å with decreasing x and broadening asymmetrically around $x = 0.5$; its area corresponds to $N = 2$ at $x \simeq 0.1$ and to $N \simeq 4$ at $x \simeq 0.5$ (Figure 2.22). A second peak lies for all values of x at 4.2 Å. A small shoulder at 5.1 Å in the $x = 0.11$ and $x = 0.72$ alloys becomes a small but well defined peak in the $x = 0.66$ alloy.

Thus, as expected, a 4-fold coordination of Ge or Te atoms around the Ge atoms prevails at high x values ($2.50 < r_1 < 2.70$ Å); a 2-fold coordination of Te or Ge atoms around the Te atoms prevails at low x values ($2.86 > r_1 > 2.70$ Å); and, perhaps, some 6-fold coordination of Ge atoms around Te atoms and vice-versa is present near $x = 0.5$ ($r_1 = 3.0$ Å as

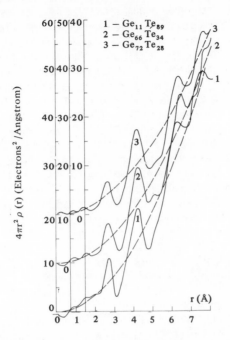

Fig. 2.22.   AERDF of Ge_xTe_{1-x} alloys (after Bienenstock, Betts, and Ovshinsky (1970)).

in crystalline GeTe). The second peak at 4.2 N corresponds to the second peak found in the RDF of both liquid Te and a-Ge and, indeed, shows no significant changes with x, while the sharp third peak which appears in the x = 0.66 alloy corresponds to the third coordination sphere in crystalline GeTe. More recent studies on a-GeSe and a-GeTe alloys by Fawcett *et al.* (1972) and Dove *et al.* (1972) by X-ray diffraction; by Betts *et al.* (1972) and Nicoteva *et al.* (1972) by combined X-ray and neutron diffraction; by Sayers, Lytle and Stern, (1972), by EXAFS, clearly favour structural models in which the Ge atoms are 4-fold, the Se and Te atoms 2-fold contained. However, models in which the Ge atoms are 3-fold, the Te atoms 3- and 2-fold coordinated cannot be completely dismissed neither on the basis of diffraction data, nor on considerations of chemical binding (Bienenstock (1973)).

The technologically important a-SiO has no crystalline analogue. While earlier structural studies by Brady (1959); Coleman and Thomas (1967); Lin and Joshi (1969) were rather inconclusive, a recent investigation by Yasaitis and Kaplow (1972) proved beyond doubt that a-SiO is no mixture of Si and SiO_2 crystallites, but has a structure of its own in which most probably both the Si and the O atoms are 2-fold coordinated and grouped in chains or rings. a-SnTe and a-PbTe are known to exist only in the form of thin films deposited at low temperature (Brown, Miller and Allgaier (1970)).

It should also be mentioned that Te-rich Ge_xTe_{1-x} alloys are model memory switching materials. They frequently show microheterogeneity in the glassy state (Phillips, Booth, and McMillan (1970); Feltz *et al.* (1972); Kinser *et al.* (1972)) and in a high electric field (Ovshinsky (1966), (1968)) or under strong illumination (Feinleib, de Neufville, and Moss (1970)) lower their resistance along a narrow path due to crystallization accompanied by phase separation (Sie, Dugan, and Moss (1971)).

Amorphous $A_2^{III}B_3^{VI}$ compounds have mixt 2-, 3- and 4-fold coordinations as in their cystals. In a-In_2Se_3 Andreyevski, Nabitovitch and Voloshchuk (1962) found that both r_1 and the two N_1 values correspond extremely well with those in the crystal. Te-rich a-$(Ga,In)_x Te_{1-x}$ $(0.1 < x < 0.3)$ alloys obtained by splat-cooling the melt (Luo (1964)) show however very low first coordination numbers $(1.3 < N_1 < 2.2)$, probably because of the high contribution of Te-Te pairs to diffraction.

A last example of mixt configurations are the amorphous $Te_{70}Cu_{25}Au_5$ alloys obtained by Duwez and Tsuei (1970) by splat-cooling their melt. The simple cubic structure found in crystalline $(Au,Ag)_{30}Te_{70}$ is converted around the above composition into an amorphous phase. Obviously the contribution of the Te-Te pairs is again highly predominant in the first peak of the AERDF obtained by X-ray diffraction and, consequently, $N_1 \simeq 2$. The amorphous alloy also displays the same quadrupole splitting as that found in Mössbauer experiments in crystalline Te, but not in cubic or monoclinic $AgTe_2$ (Tsuei and Kankeleit (1967)). It was therefore inferred that the amorphous alloy consists of randomly oriented Te chains, with the Cu and Au atoms distributed at random between the chains.

2.5.7 Molecular Configurations

A noteworthy configuration effect has been found in a-$Te_x Tl_2$ films by Ferrier, Prado, and Anseau (1971), though not by a structural investigation. The effect is absent in polycrystalline samples and takes place in the liquid phase only for x = 1. The electrical conductivity vs. composition plot shows sharp minima corresponding to the molecular species $Te_x Tl_2$ with x taking all integer values between 1 and 7. This suggests a corresponding unusual variation in the short-range order in a-TeTl alloys which has still to be confirmed structurally.

Recently molecular units have also been suggested to be the building blocks of as-deposited a-As_2S_3 and a-As_2Se_3 films (Moss, de Neufville and Ovshinsky (1973)). Also, some of the microcrystal models of a-Se discussed under 2.6.1 contain well defined molecular units, like different types of rings and chains.

2.6 MODELLING OF THE STRUCTURE OF AMORPHOUS SEMICONDUCTORS

2.6.1 The Microcrystal or Cluster Approach

The amorphous solid is thought to consist of crystallites of one of the real or potential polymorphs of the same element or compound. The crystallites are supposed to be small enough to explain the broadening of the diffraction maxima by a size effect.

Confrontation with experiment can be done either by calculating the diffraction pattern of randomly oriented and finite crystallites with the aid of Debye's formula (2.2) or by calculating the RDF of the crystallites.

In the first case a poor fit between model and experiment can be improved by ascribing the crystallites an appropriate shape or by distorting them conveniently.

In the second case both a certain increase of all interatomic distances by a constant factor is supposed to take place and fluctuations about these distances increasing with a certain power of *r* are admitted. Mostly the distribution of interatomic distances about the mean value is supposed to be Gaussian. In the case of lattices held together by bonds of different strengths, an appropriate anisotropy in both the increase of the interatomic distances and their fluctuations have often been used. Mixtures of crystallites of more than one polymorph, of non-existing but plausible crystallites and even of non-crystalline clusters have been used in order to achieve a better fit.

Examples

Silicon and *Germanium* present probably the simplest case of all, because only one stable polymorph is known to exist under normal conditions.

Moss and Graczyk (1969) recorded the diffraction pattern of a-Si films, using a scanning electron diffraction device and filtered 50 kV electrons. On the other hand they calculated the diffraction pattern of randomly oriented, normal Si crystallites of various sizes without being able to obtain an acceptable fit with the experimental results (Figure 2.23). The authors concluded that it is extremely unlikely that a-Si could consist of microcrystallites which have the structure of the bulk crystal.

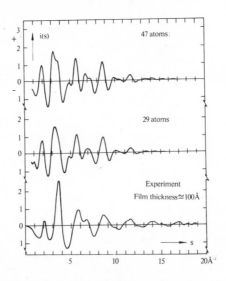

Fig. 2.23. The interference function *i(s)* of a-Si films compared with calculated *i(s)* functions of small Si crystallites of 29 and 47 atoms (after Moss and Graczyk (1969)); $s = (4 \pi \sin \vartheta)/\lambda$.

a-Ge films offer a similar picture. A comparison between the RDF of a-Ge and a calculated curve corresponding to a uniformly expanded and conveniently perturbed microcrystal model is shown in Figure 2.24. There is again some similarity between the two curves, but there are also some striking differences, especially the near disappearance of the third and fifth peaks of the calculated RDF in the experimental one.

Richter and Breitling (1958) succeeded in improving the fit between the calculated and the experimental first two peaks of the diffraction pattern by assuming that the crystallites are shaped like small platelets, only a few atom layers thick (Schichtpakete). To achieve random orientation they had

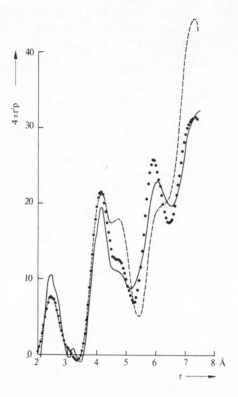

Fig. 2.24. The RDF of a-Ge. ——— experimental curve (after Richter and Breitling
(1958)); - - - calculated curve for microcrystalline Ge; . . . . . . calculated
curve for the model of Grigorovici and Mănăilă (1969, 1971); Grigorovici
(1970).

to be imbedded into a non-specified completely disordered matrix. The
distance between the atom layers had also to be increased for films
deposited at low temperatures by as much as 11% (Breitling (1972)). Nevertheless the model cannot account for the lack of the two peaks in the RDF
at 4.7 and 7.1 Å. In an attempt to explain this disagreement, the authors
became aware of the fact that the lacking maxima correspond to those
interatomic distances which change drastically when neighbouring tetrahedra are rotated about their common bond. No wonder wurtzite-type Si
crystals fit the RDF and I(s) of a-Si better than normal crystals (Rudee and
Howie (1972)).

   Their remark incited Coleman and Thomas (1967) and Grigorovici and
Mănăilă (1968) to look independently on possibilities of building small
three-dimensional clusters from atoms linked tetrahedrally to each other,
but not in the staggered two-atoms configuration exclusively present in the
diamond lattice. If exclusively eclipsed bonds, known to exist in the
wurtzite lattice, are used to linking together the atoms, 5-fold plane rings
are formed (Figure 2.25a). The presence of such rings has been detected by
Mader (1971) is small Ge particles deposited by electron beam evaporation
on NaCl substrates. They arise by multiple twinning along common [110]
axes situated parallel to the substrate. There is however a small angular

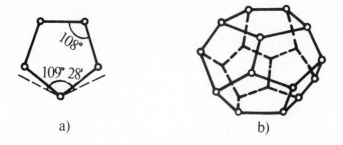

Fig. 2.25.   Tetrahedrally interconnected atoms in eclipsed configuration. a) Pentagonal
ring (the tetrahedral angle of 109°28′ is exaggerated in the drawing);
b) regular dodecahedron (amorphon).

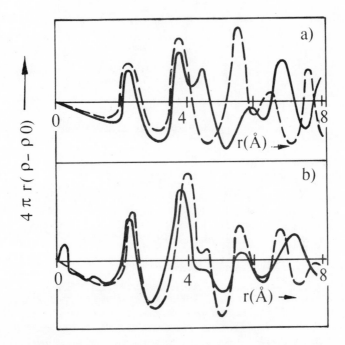

Fig. 2.26.   DARDF of a-Si.
a) DARDF of microcrystalline Si; - - - DARDF of the amorphonic structure.
b) —— experiment; - - - DARDF of a 60/40 mixture of microcrystals and
amorphons (after Coleman and Thomas (1967)).

misfit of 1°28′ per bond resulting from the difference between the pentagonal angle of 108° and the tetrahedral angle of 109°28′. Twelve 5-fold
rings can merge into a regular pentagonal dodecahedron or 'amorphon'
(2.25b). Two or more dodecahedra can merge along one of their 5-fold rings
and so on. But such mergers cannot continue indefinitely because of the
ever increasing angular misfits. Therefore only heavily distorted 5-fold rings
can be fitted into some high pressure Si and Ge polymorphs (Si III and Ge III).

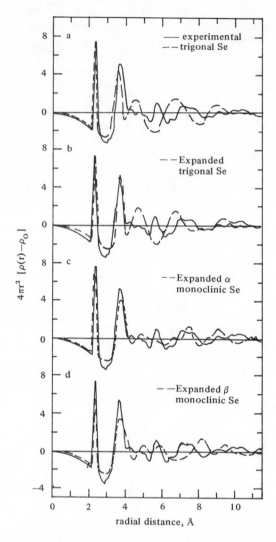

Fig. 2.27   DARDF of vacuum-deposited Se compared with that of unexpanded and
expanded trigonal, α- and β- monoclinic Se crystals, 13–14 Å in diameter
(after Kaplow, Rowe, and Averbach (1968)).

As predicted, clusters containing staggered and eclipsed bonds respectively, differ in their RDFs chiefly at 4.7 and 7.1 Å (Figure 2.26). The third coordination sphere, for instance, contains 12 atoms in the diamond lattice and none in the amorphon, while the first two coordination numbers remain unchanged. Therefore 60/40 and 50/50 mixtures of diamond-type and a-morphon-type clusters fit quite well the experimental RDFs of a-Si (Figure 2.26) and a-Ge, respectively.

 — *Selenium* obtained in the amorphous state in different ways (see Section 2.4.2) yields RDFs of different forms, except for the first narrow peak. A comparison of the DARDF of vacuum-deposited a-Se, for instance, with those of the three crystalline polymorphs (Figure 2.27) shows striking differences with all of them even if, as it has been done in order to obtain a better fit, the weaker bonds between the chains or rings have been expanded appreciably more than the stronger ones.

In the frame of the microcrystal approach a-Se is supposed to consist of a mixture of trigonal chains and rings in different ratios depending on the method of preparation, the thermal history and the author. While Henninger, Buschert, and Heaton (1967) prefer a random array of trigonal chains, Kaplow, Rowe, and Averbach (1968) stress the similarity of the RDF of a-Se with that of the monoclinic forms of Se, and Andreyevski, Nabitovitch, and Voloshchuk (1960), Krebs and Schultze-Gebhardt (1955) and Krebs and Steffen (1964) prefer a mixture of trigonal chains and 8-fold rings, without excluding rings containing even more atoms.

The real stumbling block in all these attempts are the peaks at 5.7 and 7.2 Å which are present in the RDFs of all types of a-Se, but absent or much lower in all crystalline polymorphs, expanded or not. Various other basic structural units like 6-fold rings (Richter and Herre (1958)), coplanar zig-zag chains (Richter and Breitling (1966)) or layer parcels of such chains (Richter (1972)) have been proposed in order to make this peak appear in the RDF of an hypothetic crystal lattice, sometimes with much success (Figure 2.28).

 — *Carbon* exists in a series of amorphous forms. Some of them are indeed microcrystalline ones, e.g. active charcoal (Richter, Breitling, and Herre (1956)). However a-C obtained by the pyrolysis of a polymer whose structure was investigated by Franklin (1950) did not fit into such a model. She

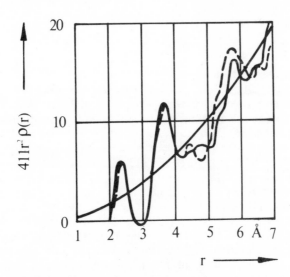

Fig. 2.28.   RDF of a-Se.—— experiment; - - - calculated RDF for crystallites composed of 6-fold rings (after Richter and Herre (1958)).

nevertheless proposed a microcrystal model in which very small graphite crystallites displaying an increased distance between the parallel layers of atoms were supposed to be embedded into a completely disordered matrix of unspecified structure. Similar difficulties were encountered later in the interpretation of the RDF of vacuum-deposited a-C films and of so-called vitreous carbon samples obtained by the pyrolysis of some cross-linked polymers.

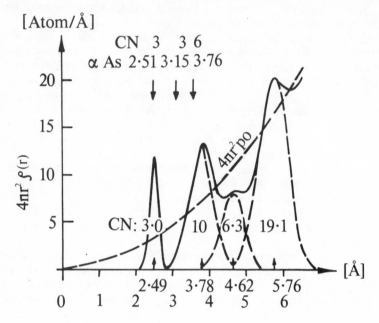

Fig. 2.29.   RDF of a-As. CN – coordination numbers. ↑: peaks in a-As; ↓: peaks in the α-As crystal (after Krebs (1969)).

—The RDFs of different amorphous forms of *arsenic* and *antimony*, while conserving the first neighbours configuration of their stable crystals, do not fit their RDFs (Figure 2.29). Richter and Gommel (1957) prefer to conserve the normal crystal lattice, but to shape the crystallites into small platelets oriented at random and also to admit statistical rotations about the bonds linking together the flattened tetrahedra within one atom layer and to change more or less the distance between these layers depending on preparation and annealing (Breitling (1972)). Krebs and Steffen (1964) suppose that a-As and a-Sb consist of a mixture of normal crystallites whose layers are however superposed in three different configurations and of abnormal crystallites displaying a structure similar to that of black phosphorus.

—In only one case the RDFs of amorphous semiconducting *alloys*, namely a-MgBi$_x$ and a-MgSb$_x$ films, have been interpreted tentatively by Ferrier and Herrell (1970) as being a mixture of very small crystallites of Mg$_3$X$_2$ and of X, where X stands for Mg, Bi or Sb, depending on which of them is in excess.

Merits and short-comings

The microcrystal approach presents the combined advantages of simplicity and versatility. It uses without alteration the whole traditional arsenal of crystallography. No wonder most of the modelling done in the earlier papers is openly or implicitely based on this approach. However acceptable fits between model and experiment could be achieved only by subjecting the crystallites without any thermodynamic justification to heavy distortions, by giving them peculiar shapes, or by admitting the presence of crystallites or atom clusters not encountered otherwise.

One is also forced to admit that the crystallites are 10–20 Å in size. The crystallites must be oriented at random in order to achieve the macroscopic isotropy of the amorphous solid. High-angle boundaries should therefore be present. The thickness of these often postulated 'completely disordered' regions which separate the crystallites from each other must be of the same order as the size of the crystallites themselves. It is difficult to admit that about half the volume of an amorphous semiconductor could be completely disordered without also admitting a high fraction of the covalent bonds being unsatisfied (Warren (1937)).

However the acceptor densities found by barrier capacity measurements in a-Ge and a-Si films (Grigorovici, Croitoru, Dévényi, and Teleman (1964); Grigorovici, Croitoru, Marina, and Năstase (1968)) lie between 10^{17} and 10^{19} cm^{-3}. The density of unpaired electrons as measured by ESR in a-Si, a-Ge, and a-SiC films (Ditina, Strahov, and Helms (1968); Brodsky and Title, (1969); Title, Brodsky, and Crowder (1970)) were also found to lie between 10^{17} and 10^{20} cm^{-3}, depending on preparation and annealing conditions. The ESR signal had all the characteristics of surface states in Si, Ge and SiC crystals and this fact has been interpreted as a proof for the existence of an inner surface attributed to the presence of voids. No ESR signal was recorded by the last authors in chalcogenide glasses, and the concentration of unpaired electrons was found to be below 10^{17} cm^{-3} from magnetic insceptibility measurements (Tauc *et al.* (1972)). Similar results were found for a-CdGeAs$_2$ (Di Salvo *et al.* (1973)). Spear (1960) estimated the density of broken bonds in a-Se to be of the order 10^{20} cm^{-3}.

All this means that, at its worst, only about one in thousand bonds are broken in amorphous semiconductors, which contradicts the microcrystal and also the non-crystalline cluster approach. Its uncontested usefulness certainly lies in offering some guidance when trying the other two approaches.

Electron microscope dark field images of amorphous thin films of Ge and different alloys obtained by the elimination of the central diffraction peak reveal discrete bright regions of about the expected size of 15 Å (Rudee (1971); Chaudhari, Graczyk and Herd (1972)). While Rudee (1972) saw in these observations the proof of the existence of microcrystals Chaudhari, Graczyk and Chabnau (1972) and Shevchik (1972) proved that continuous random network models of the type described below (2.6.3) provide sharp enough diffraction maxima in order to explain the appearance of the bright spots. If, however, part of the central diffraction peak and the first diffraction ring were allowed to contribute to the formation of the image, a few lattice fringes appeared within these areas (Rudee and Howie (1972); Rudee

(1972)). The fringe distances matched the corresponding III interplanar distance in crystalline Ge, a fact that seemed incompatible with diffraction by a random network model (Howie, Krivanek and Rudee (1973)), though Berry and Doyle (1973) contested this conclusion, too. Finally a compromise began to emerge: Rudee admitted the necessity of linking his wurtzite crystallites by a structure quite similar to a random network, while Chaudhari admitted more local order to exist in certain volumes than given by a truly random network.

2.6.2   The Perturbed Crystal Approach

Its principle consists in considering all amorphous phases as being quenched high-temperature crystal forms which contain a high number of inherent structural defects. In a very general form this approach has been formulated by Eckstein (1968). She proposes to consider all vitreous materials as being in a frozen non-equilibrium state. In order to account for the diversity of configurations present in the material, this state must be characterized not only by its 'phonon' temperature, but by a whole set of different 'configurational' temperatures, one for each degree of configurational freedom. It is rather difficult to see how this approach is to be used in a particular case. Some specific attempts have however been made on these lines.

Examples

An elaborate attempt was made on a-Se by Kaplow, Rowe, and Averbach (1968). They started by measuring very carefully the X-ray diffraction pattern of vacuum-deposited and cast a-Se. There were significant differences between the samples, depending on the method of preparation and the temperature of the substrate, but comparison with the model was based on the vacuum-deposited sample obtained at room temperature.

None of the RDFs of the three crystalline polymorphs of Se fitted the experimental RDF (Figure 2.27) or showed the peaks at 5.7 and 7.2 Å in their RDFs. So the authors formed spherical computer arrays of 100 atoms arranged alternatively in one of the three lattices of Se. Then the positions of the atoms were changed at random by a Monte Carlo procedure, each move being of the size of the amplitude of thermal vibrations. Only those movements were retained which improved the agreement between the experimental RDF and the RDF of the computer array. The moves were stopped when additional ones no longer improved the agreement.

Only 10^5 moves were necessary to reach this stage when starting with α-monoclinic Se (8-fold ring structure) (Figure 2.30). The necessary rms static displacement $\sqrt{u_s^2}$ was only 0.20 Å compared with the rms thermal displacement $\sqrt{u_{th}^2}$ of 0.29 Å existing in trigonal Se at room temperature. The highest displacement reached by one of the atoms was $\Delta u = 1.1$ Å. The mean bond angle ϑ_p in the perturbed crystal was $104 \pm 11.2°$ compared with bond angles lying between 101.6 and 109° in the crystal. When starting with β-Se (also an 8-fold ring structure) the results were similar ($\sqrt{u_s^2} = 0.16$ Å; $\Delta u = 1.1$ Å; $\vartheta_p = 100 \pm 7.4°$; $104.2 < \vartheta_{cr} < 107.1°$).

A visual inspection of the perturbed structure showed that the rings were only slightly distorted, but they appeared to be rather much tilted out as a whole off their normal position in the lattice. Bridges with trigonal chain symmetry have been formed between neighbouring rings which had opened slightly.

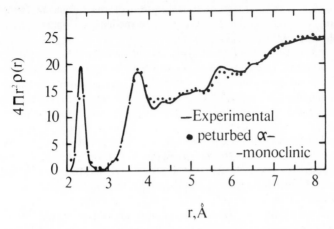

Fig. 2.30.   RDF of a-Se.── experiment (cf. Figure 2.27); . . . . RDF of perturbed α-Se
crystal (after Kaplow, Rowe, and Averbach (1968)).

When starting with the trigonal lattice (chain structure), not only much
more moves were necessary to reach the point of best agreement, but all
distortions were much higher ($\sqrt{\overline{u_s^2}} = 0.68$ Å; $\Delta u = 1.7$ Å; $\vartheta_p = 116 \pm 19.9°$;
$\vartheta_{cr} = 105°$). Visual inspection of the perturbed model showed heavily
distorted but still prevailing, roughly parallel chains. Bridges formed
between the chains by large atomic excursions looked very much like rings.

The authors conclude that a distorted ring structure seems to be closer to
the structure of a-Se than a distorted chain structure, but that probably a
mixture of both types of structural elements has to be admitted, a conclusion supported by IR and Raman spectra measurements on a-Se (Lucovsky
et al., (1967)).

−An attempt to describe a-Ge and a-Si as heavily distorted crystals was
made by Seeger and Chik (1969) in an abbreviated form. Noting that the
vast majority of the tetrahedral bonds remain intact in a-Ge and a-Si, the
authors attribute their lower density to a high number of strongly relaxed
mono- and divacancies, in most of which the lacking atom or pair of atoms
are however surrounded by 5 and 7 atoms, respectively. If this is indeed the
case, the acceptors so formed will be low-lying ones and will be filled with
electrons, not being therefore active. Only di- or multivacancies are
supposed to exist, as monovacancies are very mobile in Ge and Si crystals,
even below room temperature. The order of magnitude of the heat of
crystallization of a-Ge of 2.75 kcal/mole (Chen and Turnbull (1969)) is said
to be accounted for by this model. Annealing and recrystallization processes
might be due to the elimination of multiple vacancies by migration.

Merits and short-comings
In their attempt Kaplow, Rowe, and Averbach search after a definite answer
to a definite question: what is the slightest distortion to be given to a
certain crystal in order to fit the RDF of a related amorphous material?
They solve the problem without any physics being involved, but the final
answer is much less precise than the mathematical procedure: a-Se is
supposed to consist of an undefined mixture of distorted 8-fold rings and

trigonal chains. The preferential orientation of the initial crystallographic directions is conserved in the final stage; so the isotropy characteristic of an amorphous phase is not achieved.

The calculation has however the merit of yielding definite values for the rms static displacement of the atoms and the statistics of bond angles. These data allow for an evaluation of the heat of crystallization of a-Se. With a force constant $K_{Se} = 2.69.10^5$ dyne/cm and a static displacement $\sqrt{u_s^2} = 0.20$ Å on the one hand and the statistics of bond angles in the model of a-Se obtained when starting with α-monoclinic Se on the other, a heat of transformation of 9.33 kcal/mole results. Experimental values lie between 1.1 and 1.4 kcal/mole. The disagreement would be worse if starting from trigonal Se. So definitely the perturbations involved in the models of a-Se calculated by Kaplow, Rowe, and Averbach are much higher than the real ones.

Seeger and Chik's approach is less elaborate, but physically more meaningful. However more details must be known before being able to discuss its merits and short-comings.

2.6.3   The Continuous Network Approach
In this approach the amorphous solid is supposed to consist of an infinite, non-periodical, three-dimensional array of interlinked atoms in which the short-range order about each atom is imposed by the same characteristics of the chemical bond as in the crystal.

These two apparently contradictory conditions can be fulfilled simultaneously by intimately mixing two or more types of local orders known to exist in some crystalline polymorphs or polytypes and differing by the spatial arrangement of first- or second-order neighbours. In view of the sensitivity of crystal structure to small energy variations (see Section 2.2.1), this approach provides structural models with low configurational entropies, a general characteristic of glasses. Also fluctuations in bond lengths, bond angles, and dihedral angles are bound to appear (Grigorovici and Mănăilă (1969). If in such a model all bonds are satisfied, the network is called 'ideal'. It is called 'random' if the configurational parameters are allowed to vary at random according to a specified distribution function.

Modelling can be achieved by hand or by computer. The most important features of the modelling process concern the simulation of the spatial and energetic characteristics of the chemical bonds, the simulation of the build-up process (eventually its kinetics) and, if wanted, the simulation of randomness.

So far mostly tetrahedrally connected models have been built, all structural units being identical and all bonds equivalent. The only exception is a-C, where tetrahedral and trigonal connectivities had to be used alternatively.

In modelling by hand simulation of the chemical bond can be achieved by using ball- and spoke units which allow to a given extent for the variation of bond lengths and angles and free rotation about the bond. No attempt has been made hitherto to simulate a given ratio between the energy increases connected with these three distortions in order to minimize the configurational entropy. The obvious advantages of this type of units consists in the ease with which the location of atoms, the measurement of parameters and their fluctuations as well as the simulation of differences in covalent radii and connectivities between different atoms can be achieved. They do not allow for the simulation of higher order interactions.

Another type of units is obtained by dividing the related crystal lattice into space-filling polyhedra called Voronoi polyhedra, Wigner-Seitz, or proximity cells. Their faces are perpendicular to the middle of the lines which link the central atom to its first and higher-order neighbours. The cell has to be simplified so as to fit into all polymorphs used in modelling. Connectivity is achieved by touching corresponding faces exactly or nearly so. Rigid polyhedra admit only increases in bond length and correlate wrongly any change in bond angle with an increase in bond length. They correlate correctly but rather severely bond and dihedral angles. They allow for the simulation of higher-order neighbours interaction, but not for that of differences in covalent radii and connectivities between different atoms. Location of atoms and parameter fluctuations are difficult to evaluate.

The build-up process is mostly simulated by adding to a central unit successive 'correlation' layers which saturate the bonds let free in the precedent layer. It comes nearest to the conditions under which a film is deposited. In this phase rules have to be introduced in order to achieve 'ideality' or 'randomness'. An alternative way of building a model starts from a completely disordered system of atoms which are then shifted in small steps so as to occupy the correct distances to the nearest and next-nearest neighbours (Henderson (1971). It comes nearest to the process of quenching a melt.

Modelling by computer has to simulate all the above-mentioned build-up processes and conditions by an appropriate program.

Examples

−*Amorphous* and *vitreous carbon.* The first attempt of modelling the structure of a-C vacuum-deposited films using the continuous network approach was made by Kakinoki *et al.* (1960). Their experimental technique (electron diffraction, $s_{max} = 35$ Å^{-1}) yielded a carefully measured $I(s)$ curve (Figure 2.31) and a RDF of high resolution which showed two well defined but rather broad peaks at $r_1 = 1.50$ and $r_2 = 2.54$ Å. Using Debye's Eq. (2.2) the authors tried to match without success the whole measured $I(s)$ function by a calculated curve for a spherically symmetrical array of atoms in which only these two interatomic distances were present.

The authors succeeded however to fit well the whole $I(s)$ curve by using a mixture of the two graphite and diamond distances ($r_{1g} = 1.41$; $r_{2g} = 2.44$; $r_{1d} = 1.54$; $r_{2d} = 2.53$ Å) in about equal quantities in their hypothetical array of atoms (Figure 2.31). The rather broad first peak of the experimental RDF could then be decomposed into two narrower peaks whose widths were only slightly larger than $1/s_{max}$. The fluctuations of ∼0.11 Å which still had to be supposed to affect the first interatomic distance in a-C significantly exceed the thermal fluctuations in the crystals.

Finally Kakinoki *et al.* constructed a ball-and-spoke model of a-C in which half the atoms were in graphite-like, the other in diamond-like configuration (Figure 2.32). No special measures were taken to achieve randomness. Islands of equally connected atoms about 20 Å in size emerged and the whole array looked very similar to the ball-and-spoke model of an

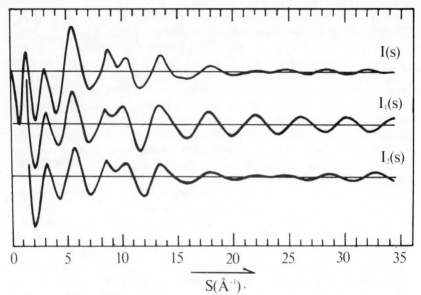

$$S(\text{Å}^{-1}).$$

Fig. 2.31. Diffraction pattern $I(s)$ of a-C film. $I_1(s)$ – calculated curve using only two interatomic distances: 1.50 and 2.54 Å; $I_2(s)$ – the same, but using four distances: 1.41; 1.54; 2.44 and 2.53 Å (after Kakinoki *et al.* (1960)); $s = (4\pi \sin \vartheta)/\lambda$.

Fig. 2.32. Model of a-C. o – trigonally and • – tetrahedrally interconnected atoms (after Kakinoki *et al.* (1960)).

oxide glass. Stresses resulted where these islands met and they were thought to reveal themselves in the supplementary fluctuations of interatomic distances found in the experimental RDF.

Later Noda and Inagaki (1964) and Furukawa (1964) interpreted the diffraction pattern of vitreous carbon – a glassy material obtained by the pyrolysis of certain cross-linked polymers – in a similar way (Figure 2.21). To account for the peculiar properties of vitreous C which, in spite of its low density, is chemically inert, impermeable to gases, and difficult to graphitize, Noda and Inagaki suggested that the tetrahedral bonds should serve only to link graphite-like islands to each other, while Kakinoki (1965) supposed that the trigonally and tetrahedrally bound islands should be linked to each other by heat resistant oxygen bridges.

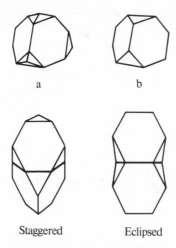

a   b

Staggered   Eclipsed

Fig. 2.33. Voronoi polyhedra of the diamond lattice (a – exact; b – simplified) in
staggered and eclipsed configuration (after Grigorovici (1969)).

—Amorphous germanium. The first continuous ideal network model of
a-Ge was devised by Grigorovici and Mănăilă (1969). They used a simplified
Voronoi polyhedron common to both the diamond and the wurtzite lattice
(Figure 2.33) as a building unit. Because of the rigid correlations between
bond length, bond angle and dihedral angle, constraints on the addition of
every new correlation layer were very high. Only two possible dihedral
angles ($0°$ and $60°$, eclipsed and staggered positions, Figure 2.33) emerged,
with only very small fluctuations about these values. The number of
possible configurations was therefore relatively small. So, instead of building
one very big model, the authors preferred to explore *all* possible ideal
configurations using indiscriminately staggered and eclipsed bonds. They are
listed in Table 2.2, each of the five correlation layers being characterized by
the number N of atoms it contains and by the numbers s and e of staggered
and eclipsed bonds satisfied during its addition. The formation of s and e
bonds was considered to be equally probable and the corresponding
statistical weights are listed in the last column. Each structure contained
around 150 atoms and consisted of three types of rings merged with each
other: 6-fold puckered 'chair' rings, 6-fold 'boat' rings, and 5-fold plane
rings. In the most representative structure (2/2; 10/3; 23/11; 42/26) the
ratio between 6- and 5-fold rings was 4:1. Obviously in a big model the
configurations listed in Table 2.2 are supposed to appear with a frequency
given by their statistical weights. No difficulty appeared in fitting new
correlation layers to the core of a given model. The total number of configurations per atom increased with the total number of atoms contained in the
model.
    Finally the atoms were located by computation. The fluctuations in
bond length and angles did not result clearly from the model. An average
RDF has been calculated (Grigorovici and Mănăilă (1971); Grigorovici
(1970)) using the statistical weights of Table 2.2, by admitting a linear
expansion of 3% for all bonds, fluctuations around a given coordination

TABLE 2.2

Correlation Layer									Statistical Weight × 10^6
I		II		III		IV		V	
s/e	N	s/e	N	s/e	N	s/e	N	N	
4/0	4	12/0	12	36/0	24	60/0	42	64	16
						54/6	42	66	61
						48/12	42	68	92
						42/18	42	70	61
						36/24	42	72	16
				33/3	24	60/0	43	65	489
						54/6	43	67	489
				30/6	24	60/0	44	66	1 465
				27/9	24	60/0	45	67	977
				24/12	24	60/0	46	68	244
		9/3	12	36/0	25	64/0	43	66	489
						57/7	43	66	489
						58/6	43	68	1 465
						51/13	43	68	1 465
						52/12	43	70	1 465
						45/19	43	70	1 465
						46/18	43	72	489
						39/25	43	72	489
				33/3	25	64/0	44	67	3 906
						57/7	44	67	3 906
		6/6	12	32/6	26	52/12	43	68	5 860
						46/18	43	70	11 719
						40/24	43	72	5 860
		3/9	12	24/18	27	45/21	42	67	7 813
						39/27	42	69	7 813
		0/12	12	12/36	28	28/36	40	60	3 906
									62 500
3/1	4	12/0	12	30/6	24	60/0	42	64	6 945
						54/6	42	66	6 945
				27/9	24	60/0	43	65	6 945
						54/6	43	67	6 945
						55/7	43	63	41 667
				24/12	24	50/14	44	66	41 667
				21/15	24	45/21	45	68	13 889
		9/3	12	30/6	25	56/8	43	66	31 250
						49/15	43	66	31 250
				27/9	25	56/8	44	67	31 250
						49/15	44	67	31 250
									250 000
2/2	4	10/3	12	26/8	24	54/14	43	64	93 748
				23/11	24	46/26	44	61	187 494
				20/14	24	38/38	45	58	93 748
									375 000
1/3	4	6/9	12	18/15	24	42/24	42	64	125 000
				15/18	24	27/45	43	65	125 000
									250 000
0/4	4	0/18	12	0/36	24	0/54	36	52	62 500 / 62 500
									1 000 000 / 1 000 000

sphere increasing as $\sqrt{r}$ and a termination effect corresponding to the s_{max} used by Richter and Breitling (1958). Good agreement with their RDF for a freshly deposited a-Ge film was obtained (Figure 2.24, dotted curve). Obviously the fluctuations in r were exaggerated in the first peak, which was too low, and subevaluated in the second peak, which was too high.

The compatibility of both the preservation of the tetrahedral symmetry of the first neighbourhood of each atom and the loss of long-range order is due to the 5-fold symmetry of some of the rings. The odd number of atoms contained in 5-fold rings also explains the appearance of 'wrong' pairs which was detected in a-$A^{III}B^{V}$ films (see Section 2.5.4), making it compelling.

The energy per atom of a cluster of atoms corresponding to one of the configurations listed in Table 2.2 and a given number of correlation layers could also be calculated, by attributing to a staggered bond the binding energy of 1.633 eV equal to that in the Ge crystal and by lowering this value correspondingly if the bond was eclipsed, affected by angular misfits or increased in length. For an unsaturated bond at the surface of the cluster the binding energy is zero. The result of such a calculation is displayed in Figure 2.34 for a few selected ideal structures and 1 to 4 correlation layers. The most important feature was that small clusters which contained predominantly eclipsed bonds lay at lower energies because of the smaller number of free faces. The most probable configuration of Table 2.2 (2/2; 10/3; 23/11; 46/26) scarcely differs in energy from the crystallite over the whole range. Combined with the fact resulting from inspection of Table 2.2 that structures with many s bonds allow easily for the addition of e bonds, while structures which contain many e bonds favour the addition of new e bonds, the formation of an amorphous phase by successively arriving atoms

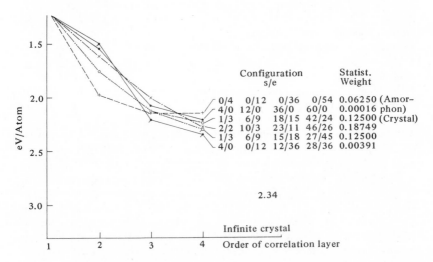

Fig. 2.34. Energy of tetrahedrally bound atom clusters of different structures (s/e – ratio between staggered and eclipsed bonds) and sizes (N – number of correlation layers) (after Grigorovici and Mănăilă (1969)).

gets a thermodynamical explanation. In bigger clusters the curves cross again until finally the diamond structure will correspond to the lowest energy.

According to this model crystallization of a-Ge can be achieved only by breaking successively all bonds and by rebuilding the whole structure, i.e. it must have and indeed has all the characteristics of a reconstructive phase transformation. The heat of crystallization evaluated in a calculation based on the energy gained by transforming all eclipsed bonds in the above-mentioned statistically weighted structure into staggered ones, by bringing all atoms nearer to each other, and by healing the angular misfits (Grigorovici and Mănăilă (1970)) leads to a value of 3.08 kcal/mole which compares favourably with the experimental value of 2.75 kcal/mole found by Chen and Turnbull (1969).

A second attempt of modelling by hand an ideal continuous a-Ge or a-Si network having a random character and maximal density was made by Polk (1971). He used ball- and -spoke units which allowed only for an increase in bond length, for bending of bonds and for free rotation about the bonds. Randomness and maximal density were achieved by choosing, when adding new atoms, that configuration which reduced strain by bending and elongation of bonds to a minimum. Implicitly this reduced the configurational entropy. The model consisted essentially of merged 6- and 5-fold rings in the ratio 4:1. Often but not always could it be decided if a 6-fold ring was chair- or boat-like because of the relatively strong distortions. Strain did not build up when the model grew to 440 units. The central region comprising about 300 units had a density 92.9% of that of the crystal; densities in three partial regions of increasing radius were 90.6, 96.7, and 91.9%, respectively, proving the homogeneity of the model. Supposing that the increase of 1.3% in the interatomic distance compared to that in the crystal was an artefact, the density of ideal random a-Ge would be only 2% less than that of the Ge crystal. This compares favourably with the lowest densities found in highly annealed a-Ge films by Donovan et al. (1970) or in a-Si films if the volume of the voids was eliminated (Moss and Graczyk (1969)). The standard deviation in the first interatomic distance was rather too low (0.03 Å) compared with the experimental values ranging from 0.06 to 0.12 Å (see Section 2.6.2); that of the second interatomic distance which corresponds to the elementary tetrahedron edge, translated into bond angle deviations, was $10°$ with a cut-off at $20°$. This compares better with experimental values of 10 to $12°$ (Moss and Graczyk (1969); Shevchik and Paul (1972)). Some of the units of the third correlation layer entered the second coordination sphere.

The RDF of the model was given in the form of a histogram (Figure 2.35) and compared with the RDF of a-Si by Moss and Graczyk (1969) as far as the 4-th coordination sphere of the crystal only. The third peak of the RDF is shifted to higher r than in the experimental curve (see also Figure 2.24), but not more than it was often found, e.g. by Richter and Breitling (1958), namely around 5 Å in a-Ge. This shift was attributed to the distortion by flattening of the 6-fold chair- or boat-like rings.

Polk's random network model confirmed many essential features of the Grigorovici-Mănăilă continuous network model like the mixing of merged 5- and 6-fold rings in about the same ratio, the overlap of correlation layers on the r scale, the possibility to extend the model without limits, etc. Due to the peculiarities of the ball-and-spoke units he used, the increase in the

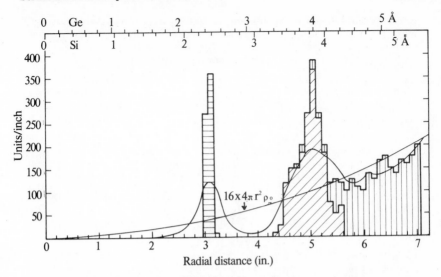

Fig. 2.35. The histogram of an ideal random continuous network model of a-Ge or a-Si (after Polk (1971)). Horizontally shaded area — first correlation layer; diagonally shaded area — second correlation layer; vertically shaded area — third- and higher order correlation layers. Full curve — RDF of a-Si (after Moss and Graczyk (1969)).

mean interatomic distance he got was still overrated, though less than in the proximity cell model, but due to the bias contained in the highest possible density goal, the fluctuations around this distance were underrated. The model thus reproduces, with some exaggeration, the experimental conclusion that the elementary tetrahedron is little distorted radially and much more angularly (see e.g. Figure 2.20), something the proximity cell model was not able to simulate. There was however no statistics of dihedral angles in order to confirm or to infirm the preference for staggered and eclipsed configurations postulated by Grigorovici and Mănăilă. Also no comparison with experiment was offered between 6 and 8 Å, though the RDF of both a-Si and a-Ge show characteristic structures in this region. However correct density values were obtained. Thus Polk's model constitutes a definite progress in continuous network modelling of tetrahedrally connected elemental amorphous semiconductors.

Modelling by computer was achieved only very recently. Shevchik and Paul (1972) follow essentially Grigorovici and Mănăilă's as well as Polk's correlation layer procedure with the difference that an atom added to the existing core is free to search for optimal bonding conditions only until it is linked by two bonds, when it becomes frozen. These conditions simulate rapid deposition of a film onto a cold substrate and introduces a certain number of dangling bonds into the structure which is no more ideal. The fit of the resulting histogram with their own experimental RDF recorded on an as-deposited sputtered a-Ge film is shown in Figure 2.36. It has all the already discussed features of Polk's histogram, but goes as far as 10 Å. To a first order the rotation of neighbouring tetrahedra about the common bonds had to be random in order to achieve the best fit with experiment. The

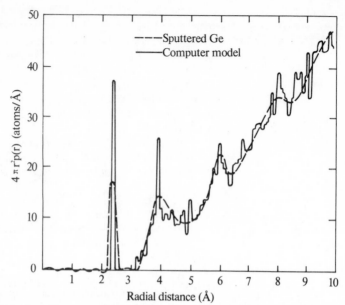

Fig. 2.36. The histogram of a non-ideal random continuous network computer model
of a-Ge and a-Si. Broken curve – experimental RDF of sputtered a-Ge film
(after Shevchik and Paul (1972)).

authors stress however that differently prepared a-Ge films showed characteristic differences in their RDFs. The highest density was found in electrolytic films (6% below the crystal), while sputtered and electron-beam
evaporated films were not only 11 and 10% less dense than the crystal,
respectively, but showed strong small angle scattering, as found also by Moss
and Graczyk (1969) in sputtered a-Si films. This is interpreted as a sign of
the presence of voids which are estimated to contribute about half of the
density deficit. The rest has to be attributed to a uniformly distributed
deficit.

The role of voids in the structure of a-Ge and a-Si was discussed in some
detail by Ehrenreich and Turnbull (1970) which propose a tentative 'swiss
cheese' model in which voids of different sizes are included into a random
network. The presence and coalescence of such voids by the annealing of
a-Ge films has been detected by electron microscopy by Barna, Barna, and
Pocza (1972).

The data of Polk's model for a-Si and a-Ge and Shevchik's results on
a-$A^{III}B^{V}$ compounds were used by Connell and Paul (1972) to calculate the
heat of crystallization of a-Si, a-Ge, a-GaP, and a-GaAs on the basis of the
expression of the strain energy given by Keating (1966). In particular the
value of 0.16 eV/atom for a-Ge agrees well with the experimental value of
0.13 eV/atom found by Chen and Turnbull (1969). A similarly good agreement was obtained also with the proximity cell model by Grigorovici and
Mănăilă (1970). It must therefore be supposed that both models have about
the same total configurational entropy, but differ by the way it is distributed over bond lengths, bond angles and dihedral angles.

The second modelling procedure by computer used by Henderson (1971)
and Henderson and Herman (1972), consisting in interconnecting the atoms

of a previously completely disordered system in the right first- and second-neighbours configurations, yielded results very similar to those obtained by simulating the building up of a continuous network atom by atom.

—Amorphous *selenium* and *tellurium.* Supposing that a-Se and a-Te consist of 'ideal' networks of atoms linked together statistically in one of the two three-atoms configurations characteristic to either the trigonal chain (adjacent dihedral angles of the same sign) or to the 8-fold ring (adjacent dihedral angles of opposite sign), Grigorovici (1969b) proposed a simplified Voronoi polyhedron (Figure 2.37) in which only the A and B faces corresponding to the $d^2 sp^3$ bonds are maintained and the edges are rounded off in order to simulate the weakness of second- and third-neighbours bonds. No attempt has yet been made of actually using this proximity cell for the modelling of a-Se and a-Te.

Merits and short-comings

The modelling of amorphous semiconductors based on the continuous network approach yields essentially idealized, chemically satisfied structures. The contradiction between the maintainance of the short-range order and the destruction of the long-range order is solved by mixing different first-, second- and third-neighbours configurations and this introduces the compulsory presence of fluctuating bond lengths, bond angles and dihedral angles. The explanation given by Myuller (1966) to the easing of vitrification of alloys by the presence of relatively small quantities of certain impurities by the formation around the impurity atoms of microdomains which present geometrical incompatibilities with the lattice of the host substance fits well into the continuous network approach. Finally the blurring of the band edges in the electronic structure of amorphous semiconductors might be supposed to be linked to the above-mentioned fluctuations.

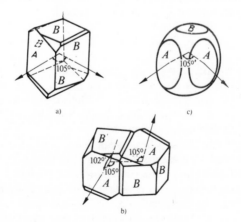

Fig. 2.37.   Voronoi polyhedron of trigonal Se.
     a) exact polyhedron;
     b) two polyhedra in staggered configuration;
     c) simplified 'polyhedron' in order to fit the monoclinic lattices.

Unsatisfied bonds introduced accidentally during the formation of the amorphous phase might form the analogue to point defects like vacancies and interstitials in a crystal lattice. Such defects might be linked to the relatively few unpaired electrons or acceptor and, perhaps, donor levels found in amorphous semiconductors.

The above picture also easily explains the unsensitiveness of the electrical properties of amorphous semiconductors to the addition of impurities which introduce shallow levels into the related semiconducting crystal. To satisfy all its bonds the host atoms have only to arrange themselves conveniently around a tri- or pentavalent impurity atom introduced, for instance, into a-Ge in order to annihilate the acceptor or donor properties we usually ascribe them (Mott, 1969).

Finally, the quantitative agreement reached between model and experiment in such matters as the shape of the RDF, the density and the heat of crystallization, as well as the demonstration of the thermodynamic stability of small non-crystalline clusters of Ge atoms can also be listed among the merits of the continuous network approach.

As far as can be judged on the basis of the existing attempts the continuous network approach seems at least to present no major contradictions with the existing experimental data and thus to be the most satisfactory of the three approaches analysed in this review.

2.7   CONCLUSIONS

Recent progress in our knowledge about the structure of amorphous semiconductors was achieved by refining the classical methods of investigation by X-ray and electron diffraction, and by X-ray absorption, but even more by using new approaches in modelling. There is little hope of finding new experimental techniques which would yield directly a three-dimensional picture of the short-range order in amorphous materials. Improvements in modelling, by hand or by computer, consisting in a better simulation of reality seem more promising for instance, a better understanding of the physics involved in the process of growth by atoms arriving on the surface of a cluster (Shevchik (1973)) or a better simulation of the minimization of free energy by springs (Grigorovici and Belu (1972)) or by appropriate protentials in computer arrays may improve the agreement between models and experiment. Information provided by other experimental methods and by the theory of the chemical bond might also contribute to a better choice between different proposed models.

For the time being the situation is still fluid. New ideas are pouring in, others are eliminated. A convenient formalism of describing statistically the three-dimensional short-range order in amorphous semiconductors seems to emerge from the continuous network approach and will be highly appreciated by both the experimentalist and the theoretician.

At the present rate of publication of papers devoted to this subject it can not only be hoped that progress in knowledge will be rapid, but also that many conflicting views will find themselves less contradictory than they seem to be now.

REFERENCES

Andreyevskii, A. I., Nabitovitch, I. D. and Voloshchuk, Ya. V., (1960), *Kristallografiya*, **5**, 369; (1961), *Kristallografiya*, **6**, 662; (1962), *Kristallografiya*, **7**, 865.
Azaroff, L. V., (1963), *Rev. Mod. Phys.*, **35**, 1012.
Badzian, A. R., (1967), *Mater. Res. Bull.*, **2**, 987.
Barna, A., Barna, P. B. and Pocza, J. F., (1972), *J. Non-Cryst. Solids*, **8–10**, 36.
Bates, C. H., (1966), Ph.D. Thesis, Pennsylv. State Univ.
Baumann, F., (1966), *Verhandl. Deutsch. Phys. Ges.*, **3**, 214.
Berry, M. V. and Doyle, P. A., (1973), *J. Phys. Chem.*, **6**, 16.
Betts, F., Bienenstock, A., Keating, D. T. and de Neufville, J. P., (1972), *J. Non-Cryst. Solids*, **7**, 417.
Betts, F., Bienenstock, A. J., and Bates, C. W., (1972), *J. Non-Cryst. Solids*, **8–10**, 364.
Bienenstock, A., (1973), *J. Non-Cryst. Solids*, **11**, 447.
Bienenstock, A., Betts, F. and Ovshinsky, S. R., (1970), *J. Non-Cryst. Solids*, **2**, 347.
Borshchevskii, A. S., Goriunova, N. A., Kesamanly, F. P. and Nasledov, D. N., (1967), *Phys. Stat. Solidi*, **21**, 9.
Brady, G. W., (1959), *J. Phys. Chem.*, **63**, 1119.
Breitling, G., (1972), *J. Non-Cryst. Solids*, **8–10**, 395.
Brodsky, M. H. and Title, R. S., (1969), *Phys. Rev. Letters*, **23**, 581.
Brown, R. W., Miller, A. R. and Allgaier, R. S., (1970), *Thin Solid Films*, **5**, 157.
Červinka, L., Hosemann, R. and Vogel, W., (1970), *J. Non-Cryst. Solids*, **3**, 294.
Červinka, L., Hrubý, A., Matyáš, M., Šimeček, T., Skácha, J., Štourač, L., Tauc, J., Vorlíček, V. and Höschl, P., (1970), *J. Non-Cryst. Solids*, **4**, 258.
Chaudhari, P., Graczyk, J. F. and Herd, S. R., (1972), *Phys. Stat. Solidi*, (b), **51**, 801.
Chaudhari, P., Graczyk, J. F. and Charbnau, H. P., (1972), *Phys. Rev. Letters*, **29**, 425.
Chen, H. S. and Turnbull, D., (1969), *J. Appl. Phys.*, **40**, 4214.
Chopra, K. L. and Bahl, S. K., (1969), *J. Appl. Phys.* **40**, 4171.
Coleman, M. V. and Thomas, D. J. D., (1967a), *Phys. Stat. Solidi*, **22**, 593. (1967b), *Phys. Stat. Solidi*, **24**, K 111; (1968), *Phys. Stat. Solidi*, **25**, 241.
Connell, G. A. N. and Paul, W., (1972), *J. Non-Cryst. Solids*, **8–10**, 215.
Di Salvo, F. J., Bagley, B. G., Tauc, J. and Waszczak, J. V., (1973), Proc. 5th Int. Conf. Amorphous and Liquid Semiconductors in Garmisch-Partenkirchen, Taylor and Francis, London.
Ditina, Z. Z., Strahov, L. P. and Helms, H., (1968), *Fiz. Tehn. Poluprov.*, **2**, 1199.
Donovan, T. M., Spicer, W. E., Bennett, J. M. and Ashley, E. J., (1970), *Phys. Rev.*, **2B**, 397.
Dove, D. B., Chang, J. and Molnar, B., (1972), *J. Non-Cryst. Solids*, **8–10**, 376.
Dove, D. B., Heritage, M. B., Chopra, K. L. and Bahl, S. K., (1970), *Appl. Phys. Letters*, **16**, 138.
Duwez, P. and Tsuei, C. C., (1970), *J. Non-Cryst. Solids*, **4**, 345.
Eckstein, B., (1968), *Mater. Res. Bull.*, **3**, 199.
Ehrenreich, H. and Turnbull, D., (1970), *Comments Solid State Phys.*, **3**, 75.
Fawcett, R. W., Wagner, C. N. J. and Cargill, G. S. III, (1972), *J. Non-Cryst. Solids*, **8–10**, 369.
Feinleib, J., de-Neufville, J. and Moss. S. C., (1970), *Bull. Amer. Phys. Soc.*, **15**, 245.
Feltz, A., Buttner, H. J., Lippmann, F. J. and Maul, W., (1972), *J. Non-Cryst. Solids*, **8–10**, 64.
Ferrier, R. P. and Herrell, D. J., (1970), *J. Non-Cryst. Solids*, **4** 338.
Ferrier, R. P., Prado, J. and Anseau, M., (1972), *J. Non-Cryst. Solids*, **8–10**, 798.
Filipovitch, V. N., (1955), *J. Tehn. Fiz.*, **25**, 1604.
Franklin, R., (1950), *Acta, Cryst.*, **3**, 107.
Furukawa, K., (1962), in Reports on Progress in Physics (A. C. Stickland editor), Vol. 25, p. 395. Institute of Physics, London.
Furukawa, K., (1964), *Nihon Kessho Gakkaishi*, **6**, 101.
Goriunova, N. A., (1965), *Izv. Akad. Nauk SSSR, Seriya Neorg. Mater.*, **1**, 885.
Graczyk, J. F. and Moss, S. C., (1970), in Proceed. Internat. Conf. Phys. Semicond. (S. P. Keller, J. C. Hensel and F. Stern, editors), p. 658. Cambridge, Mass.

Grigorovici, R., (1969a), *J. Non-Cryst. Solids*, 1, 303; (1969b), Abstracts Internat. Conf. Amorph. Liq. Semicond., Cambridge, p. 60; (1970), in Compt. Rend. Congr. Internat. Couches Minces (M. R. Gerbier, editor), p. 481. Cannes. (1973), Lectures of the 1972 Scottish Universities Summer School of Physics, Aberdeen (W. Spear and P. Le Comber, editors), Academic Press, London, p. 191.

Grigorovici, R. and Belu, A.. (1972), Proc 11th Internat. Conf. Phys. Semicond. (M. Miasek. editor), Polish Scientific Publishers, Warsaw, p. 453.

Grigorovici, R., Croitoru, N., Dévényi, A. and Teleman, E., (1964), in Proceed. Internat. Conf. Phys. Semicond., (M. Balkanski, editor), p. 423. Dunod, Paris.

Grigorovici, R., Croitoru, N., Marina, M. and Năstase, L., (1968), *Rev. Roum. Phys.*, 13, 317.

Grigorovici, R. and Mănăilă, R., (1968), Thin Solid Films, 1, 343; (1969), *J. Non-Cryst. Solids*, 1, 371; (1970), *Nature, (London)*, 226, 143; (1971), in Proceed. Internat Conf. Heterojunct. and Layer Struct. (G. Szigeti, editor), Vol. III, p. 37. Akadémiái Kiadó, Budapest.

Grigorovici, R., Mănăilă, R. and Vaipolin, A. A., (1968), *Acta Cryst.*, B 24, 535.

Henderson, D., (1971), *Bull. Amer. Phys. Soc.*, 16, 348.

Henderson, D. and Herman, F., (1972), *J. Non-Cryst. Solids*, 8–10, 359.

Henninger, E. H. and Buschert, R. C., (1967), *J. Phys. Chem. Solids*, 28, 423.

Henninger, E. H., Buschert, R. C. and Heaton, L., (1967), *J. Chem. Phys.*, 46, 586.

Hilton, A. R., Jones, C. E. and Brau, M., (1966), *Phys. Chem. Glasses*, 7, 105.

Hilton, A. R., Jones, C. E., Dobrott, R. D., Klein, H. M., Bryant, A. M. and George, T. D., (1966), *Phys. Chem. Glasses*, 7, 116.

Howie, A., Krivanek, O. and Rudee, M. L., (1973), *Phil. Mag.*, 27, 235.

Hrubý, A. and Honserová, J., (1972), *Czech. J. Phys.*, B, 22, 89.

Hrubý, A. and Štourač, L., (1969), *Mat. Res. Bull.*, 4, 745; (1971), *Mat. Res. Bull.*, 6, 247.

Ichikawa, T., (1973), *Phys. Stat. Solidi*, (b), 56, 707.

Jensen, A., (1964), Patentschrift 1 213 075 BRD.

Joffe, A. F., (1947), *Dokl. Akad. Nauk SSSR*, Vol. I, p. 305. Anniversary Collection.

Joffe, A. F. and Regel, A. R., (1960), in Progress in Semiconductors (A. F. Gibson, editor), Vol. 4, p. 237. Heywood, London.

Kakinoki, J., (1965), *Acta Cryst.*, 18, 578.

Kakinoki, J., Katada, K., Hanawa, T. and Ino, T., (1960), *Acta Cryst.*, 13, 171.

Kaplow, R., Rowe, T. A. and Averbach, B. L., (1968), *Phys. Rev.*, 168, 1068.

Kaplow, R., Strong,S. L. and Averbach, B. L., (1965), *Phys. Rev.*, 138 A, 1336.

Keating, P. N., (1966), *Phys. Rev.*, 145, 637.

Kinser, D. L., Wilson, L. K., Sanders, H. R. and Hill, D. J., (1972), *J. Non-Cryst. Solids*, 8–10, 823.

Klug, H. P. and Alexander, L. E., (1954), X-Ray Diffraction Procedures for Poly-crystalline and Amorphous Materials, John Wiley, New York.

Kolomiets, B. T., (1964), *Phys. Stat. Solidi*, 7, 359, 713.

Kolomiets, B. T., Pajasová, L. and Štourač, L., (1965). *Fiz. Tverd. Tela*, 7, 1588.

Krebs, H., (1952), in Semiconducting Materials (H.K. Henisch, editor), Butterworths, London; (1966), *Angew. Chemie* 78, 577; (1968), Grundzüge der Anorganischen Kristallchemie, F. Enke, Stuttgart; (1969), *J. Non-Cryst. Solids*, 1, 455.

Krebs, H. and Schultze-Gebhardt, F., (1955), *Acta Cryst.*, 8, 412.

Krebs, H., Schultze-Gebhardt, F. and Thees, R., (1955), *Z. anorg. Chemie*, 282, 177.

Krebs, H. and Steffen, R., (1964), *Z. anorg, Chemie*, 327, 224.

Light, T. B. and Wagner, C. N. J., (1968), *J. Appl. Cryst.*, 1, 199.

Lin, S. C. H. and Joshi, M., (1969), *J. Electrochem. Soc.*, 116, 1740.

Lucovsky, G., Mooradian, A., Taylor, W., Wright, G. B. and Keezer, R. C., (1967), *Solid State Commun.*, 5, 113.

Luo, H. L., (1964), Ph.D. Thesis, California Institute of Technology.

Luo, H. L. and Duwez, P., (1963), *Appl. Phys. Letters*, 2, 21.

Mader, S., (1971), *J. Vacuum Sci. Tech.*, 8, 247.

Mănăilă, R. and Popescu, M., (1971), private communication. (1972), Communication at Conference on Amorphous, Liquid and Vitreous Semiconductors, Sofia.

Mikolaitchuk, A. G. and Dutchak, Ya. J., (1964), *Kristallografiya*, 9, 106.

Moss, S. C., de Neufville, J. P. and Ovshinsky, S. R., (1973), *Bull. Amer. Phys. Soc.*, 18, 389.

Moss, S. C., Flynn, P. and Bauer, L.-O., (1973), *Phyl. Mag.*, 27, 441.

Moss, T. S., (1959), Optical Properties of Semiconductors, p. 179. Butterworth, London.

Moss, S. C. and Graczyk, J. F., (1969), *Phys. Rev. Letters*, **23**, 1167.
Mott, N. F., (1969), *Phil. Mag.*, **19**, 835.
Myuller, R. L., (1966), in Solid State Chemistry (Z. V. Borisova, editor), p. 1. Consultants Bureau, New York.
Nicotera, E., Corchia, M., De Giorgi, G., Villa, F. and Antonini, M., (1972), *J. Non-Cryst. Solids*, **11**, 417.
Noda, T. and Inagaki, M., (1964), *Bull. Chem. Soc. Japan*, **37**, 1534.
Ovshinsky, S. R., (1966), USA Patent 3 271 591 (filed Sept. 1963); (1968), *Phys. Rev. Letters*, **21**, 1450.
Paalman, H. H. and Pings, C. J., (1963), *Rev. Mod. Phys.*, **35**, 389.
Pauling, L., (1960), The Nature of the Chemical Bond, 3rd ed., Cornell Univ. Press, Ithaca.
Phillip, S. V., Booth, R. E. and McMillan, P. W., (1970), *J. Non-Cryst. Solids*, **4**, 510.
Polk, D. E., (1971), *J. Non-Cryst. Solids*, **5**, 365.
Porai-Koshits, E. A. and Vaipolin, A. A., (1963), *Fiz. Tverd. Tela*, **5**, 246.
Richter, H., (1972), *J. Non-Cryst. Solids*, **8–10**, 388.
Richter, H. and Breitling, G., (1958), *Z. Naturforsch.*, **13a**, 988; (1966), *Z. Naturforsch.*, **21a**, 1710.
Richter, H., Breitling, G. and Herre, F., (1956). *Z. angew. Physik*, **8**, 433.
Richter, H. and Gommel, G., (1957), *Z. Naturforsch.*, **12a**, 996.
Richter, H. and Herre, F., (1958), *Z. Naturforsch.*, **13a**, 874.
Rudee, M. L., (1971), *Phys. Stat. Solidi*, (b) **46**, K 1; (1972), *Thin Solid Films*, **12**, 207.
Rudee, M. L. and Howie, A., (1972), *Phil. Mag.*, **25**, 1001.
Sayers, D. E., Lytle, F. W. and Stern, E. A., (1972), *J. Non-Cryst. Solids*, **8–10**, 401.
Sayers, D. E., Lytle, F. W. and Stern, E. A., (1970), in Adv. in X-Ray Analysis (W. N. Mueller, editor), Vol. 13, p. 248. Plenum Press, New York.
Sayers, D. E., Stern, E. A. and Lytle, F. W., (1971), *Phys. Rev. Letters*, **27**, 1204.
Seeger, A. and Chik K-P., (1969), Abstracts Internat. Conf. Amorph. Liq. Semicond., p. 96. Cambridge.
Shevchik, N. J., (1971a), Ph.D. Thesis, Harvard University; (1971b), *Phys. Stat. Solids* (b), **52**, K 121.
Shevchik, N. L., (1973), Private communication.
Shevchik, N. L., Lannin, J. S. and Tejeda, J., (1973), *Phys. Rev., B*, **7**, 3987.
Shevchik, N. J. and Paul, W., (1972), *J. Non-Cryst. Solids*, **8–10**, 381.
Sie, C., Dugan, P. and Moss, S. C., (1972), *J. Non-Cryst. Solids*, **8–10**, 377.
Spear, W. E., (1960), *Proc. Phys. Soc. (London)*, **B76**, 826.
Şteeb, S. and Hezel, R., (1966), *Z. Physik*, **191**, 398.
Stouraĉ, L., Kolomiets, B. T. and Shilo, V. P., (1968), *Czech. J. Phys.* B 18, 92.
Tatarinova, J. K. and Kazmazovskaya, T. S., (1961), *Kristallografiya*, **6**, 668.
Temkin, J., Connell, G. A. N. and Paul, W., (1972), *Solid State Commun.*, **11**, 1591.
Tauc, J., Di Salvo, F. J., Peterson, G. E. and Wood, D. L., (1972), Amorphous Magnetism (H. O. Hooper and A. M. de Graaf, editors) Plenum Press, New York and London, p. 119.
Title, R. S., Brodsky, M. H. and Crowder, B. L., (1970), in Proceed. Internat. Conf. Phys. Semicond., (S. P. Keller, J. C. Hensel and F. Stern, editors), p. 794. Cambridge, Mass.
Tsuei, C. C. and Kankeleit, E., (1967), *Phys. Rev.*, **162**, 312.
Tsushihashi, S. and Kawamoto, Y., (1971), *J. Non-Cryst. Solids*, **5**, 286.
Turnbull, D. and Polk, D. E., (1972), *J. Non-Cryst. Solids*, **8–10**, 19.
Ugai, Ya. A., Ziubina, T. A. and Aleinikova, K. B., (1968), *Neorgan. Materialy*, **4**, 17.
Vaipolin, A. A., (1972), in Troinyie Poluprovodniki (S. I. Radautsan, editor), Shtiintsa, Kishinev, p. 26.
Vaipolin, A. A., Kuzmenko, G. S. and Osmanov, E. O., (1972), in Troinyie Poluprovodniki (S. I. Radautsan, editor), Shtiintsa, Kishinev, p. 90.
Vaipolin, A. A., Osmanov, E. O. and Rud, Yu. V., (1965), *Fiz. Tverd. Tela*, **7**, 2266.
Warren, B. E., (1937), *J. Appl. Phys.*, **8**, 645.
Wells, A. F., (1962), Structural Inorganic Chemistry, 3rd ed., Oxford University Press, London.
Wihl, M., Cardona, M. and Tauc, J., (1972), *J. Non-Cryst. Solids*, **8–10**, 172.
Wyckoff, R. W. G., (1964). Crystal Structures, 2nd ed., J. Wiley, New York.
Yasaitis, J. A., and Kaplow, R., (1972), *J. Appl. Phys.*, **43**, 995.
Young, P. A. and Thege, W. G., (1971), *Thin Solid Films*, **7**, 41.

Chapter 3

Electronic Structure of Disordered Materials

E. N. Economou

Department of Physics and Center for Advanced Studies, University of Virginia, Charlottesville, Virginia 22901

Morrel H. Cohen, Karl F. Freed,† and E. S. Kirkpatrick

James Franck Institute, The University of Chicago, Chicago, Illinois 60637

*This work was supported by the ARO(D), NASA and NSF and has benefited from general support of Materials Sciences by ARPA at The University of Chicago. It was started when one of the authors (E.N.E.) was at The University of Chicago.
†Alfred P. Sloan, Foundation Fellow.

3.1   INTRODUCTION

It has been known since the work of Bloch (1928) that there are universal features in the electronic structures of crystals. The most important of these are energy bands separated by gaps, crystal momentum as a good quantum number, and the form of the wave function; they provide the conceptual foundations of much of solid state physics.

The existence of such universal features is an immediate consequence of the regularity of arrangement of the atoms in crystals. Disordered materials lack such regularity. Are there corresponding universal features in the electronic structures of disordered materials? If so, what are they?

The theory of the electronic structures of disordered materials is in its infancy. The first hint at the answers to these basic questions emerged about two decades after the theory of crystals was well started, (Fröhlich (1947)) the next only twelve years ago, (Anderson (1958)) and it is only in the last three years that definitive answers have begun to appear. In this paper, we shall review the answers as they have developed to date.

Proofs of the existence of universal features have been based on highly simplified model Hamiltonians. Discussion of real disordered materials has been based on highly simplified model band structures possessing these features. Accurate models for real materials are presently beyond our grasp, so we confine ourselves in this paper to simple models.

We begin with a discussion of perfect and imperfect crystals, observing the effects of increasing disorder on the electronic structure. We then discuss the nature of disordered materials and present a model which, we believe, incorporates or illustrates the main characteristics of their electronic structure. The various assumptions underlying the model are proved or made plausible through formal or physical analysis of several concrete, simple systems. We review these analyses and thereby introduce the reader to the theoretical techniques which are proving effective in such problems. In particular, we discuss the fluctuation approach, band tailing, the proof of the existence of mobility edges, and Anderson's transition. We make application of the general theoretical results to binary alloys and one-dimensional systems. Edwards' path integral formulation is reviewed in some detail. In conclusion, we summarize the present state of the theory and indicate the most immediate unsolved problems.

3.2   THE UNIVERSAL FEATURES OF THE ELECTRONIC STRUCTURES OF CRYSTALS

3.2.1   Perfect Crystals

The universal structural feature of perfect crystals is their periodicity. An immediate consequence of this periodicity is the Bloch-Floquet theorem stating that the one-electron eigenfunctions are of the form

$$\psi(r) = \psi_{nk}(r) = e^{ik \cdot r} u_{nk}(r), \tag{3.1}$$

where the wave vector k is in the first Brillouin zone, $u_{nk}(r)$ has the periodicity of the crystal structure and $n = 1, 2, 3, \ldots$ is the so-called band index. Thus, all eigenfunctions in Eq. (3.1) are *extended:* an electron described by Eq. (3.1) goes everywhere in the crystal with equal probability (apart from a periodic modulating factor $|U|_{nk}^{u}(r)$) just as in the free electron case.

Perfect periodicity of the atomic structure implies perfect ordering of the atoms, as regards position and composition both on an atomic, or short, distance scale and on a macroscopic, or long, scale. In other words, crystals possess perfect long- and short-range structural order. This structural order is reflected in the wave function in Eq. (3.1) as well. Its amplitude shows long-range order because of its extended nature, as in the free-electron case. There is another property of the free electron wave function which is retained in the periodic case: perfect *phase coherence,* i.e., long-range order in the phase. Given the phase at one point, one can determine the phase at *any* other point in the crystal provided only that one knows the wave vector k.

Periodicity has a profound and characteristic influence on the energy spectrum, which consists of continuous bands at allowed levels separated by forbidden gaps. The eigenenergies $E_n(k)$ are continuous functions of k within each band n, analytic except at certain symmetry points and lines. There are well defined band edges which correspond to the absolute minimum or maximum of E as a function of k for each band. The density of states n(E) approaches zero at the edges as the square root of the energy difference. Topological considerations show that other critical points, saddle points, must occur within the band, also giving rise to square root singularities in n(E) (Figure 3.1a).

All these features, and particularly the extended nature of the states, their perfect phase coherence, and the sharp band edges are universal properties of the electronic structures of perfect crystals stemming from the periodicity of the lattice.

3.2.2   Single Localized Imperfection in a Perfect Crystal

Let us consider a crystal containing a single localized imperfection such as an impurity. The problem can then be thought of as one of scattering by the imperfection, where the Bloch states of the perfect crystal play the role of the incident waves. If the localized change ΔV introduced by the imperfection in the crystal potential V is weak one can use the Born approximation to obtain the solution of the scattering problem. The result is then that the states remain extended and retain their phase coherence. There are only minor changes in the density of states and in particular the sharp band edges remain. As ΔV increases, a resonance structure develops in the density of states near the bottom of the band for ΔV attractive, or near the top of the band for ΔV repulsive. The amplitudes of the states associated with the resonance are no longer spatially uniform although they still extend throughout the crystal. Finally, when the strength of the potential change exceeds a certain critical value, the resonance splits off the band as a δ-function in the band gap. It can be shown that the state corresponding to this δ-function is bound to and consequently localized around the imperfection. Accordingly, both attractive and repulsive localized potential changes can bind an electron in a crystal, in contrast to the ordinary scattering problem in which one cannot produce bound states out of repulsive potentials. The reason is that the energy spectrum in a crystal is comprised of continua which possess both lower and upper bounds. A repulsive potential, if strong enough, can push a localized state above the upper bound of the continuum in the same way that an attractive potential can push a localized state below the lower bound of the continuum. These features are summarized in Figure 3.1b.

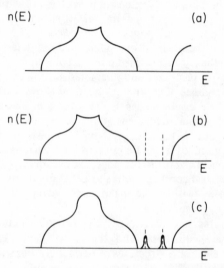

Fig. 3.1.   Density of states n(E) as a function of energy E for a.) a perfect crystal, b.) a crystal containing only one localized imperfection, and c.) a crystal containing a low concentration of localized imperfections. In all three cases there are continuous bands of energy levels separated by gaps with square root singularities at the band edges. The square root singularities associated with saddle points within the band in a.) and b.) are eliminated by scattering in c.). If the potential change ΔV introduced by the imperfection is strong enough in b.), localized levels split off from the bands (off the bottom for attractive ΔV or the top for repulsive) which broaden into bands in c.).

3.2.3   Imperfect Crystals

The next step is to consider an imperfect crystal such as would result from a low but finite concentration of randomly distributed individual imperfections of the kind considered above. For weak scattering potential and for energies well inside the band, we expect that the wave functions remain extended the same way that the states for the free particle, multiply-scattered by weak potentials, remain extended. However, the perfect phase coherence is lost. Because of the multiple scattering by many *randomly* distributed scatterers, the phase at a certain point $\mathbf{r}$ becomes uncorrelated with the phase at another point $\mathbf{r}'$ if the distance $|\mathbf{r} - \mathbf{r}'|$ is much larger than a characteristic length ξ, called the phase coherence length. The mean free path l is another measure' of the linear dimension over which phase coherence is retained and as such differs from the phase coherence length by a factor of order unity. The wave function ψ can be expanded in the original Bloch functions,

$$\psi = \Sigma_k \, a_k \, \psi_k, \tag{3.2}$$

where

$$a_k = (\psi_k^*, \psi). \tag{3.2a}$$

The integral in Eq. (3.2a) can be broken up into a sum of integrals each over regions of linear dimension of order ξ. The phase of ψ in different regions is uncorrelated. Thus the contributions of each region to the integral have random phases, and so must their sum, a_k. Further analysis leads to the conclusion that the a_k are roughly constant in magnitude on a constant energy surface in k space, have a maximum magnitude on some particular energy surface, and have a width in k space of order $1/\xi$ (or $1/l$). The random multiple scattering introduces an extra factor $e^{-R/\xi}$ in the average autocorrelation function

$$c(\mathbf{R}) = <\psi^*(\mathbf{r} + \mathbf{R}) \, \psi(\mathbf{r})>, \tag{3.2b}$$

where the average is carried out over an ensemble of imperfect crystals with a random distribution of imperfections. The factor $e^{-R/\xi}$ means simply that phase coherence has been lost over distances greater than ξ. There is no long-range order in the phase (which would require $\xi = \infty$), but only short-range order.

Since the concentration N_i of imperfections is much smaller than the concentration N of regular atoms, we expect the change in the density of states to be small, of the order N_i/N. In particular, uncertainty in k leads to rounding off of the square root singularities caused by the saddle points. As the strength of the imperfection scattering potential ΔV is increased and resonances develop, there can be a profound change in the density of states in the vicinity of the resonance. Finally, when each $|\Delta V|$ is large enough to produce localized states at each imperfection, there will be a non-zero density of states inside the gap with a peak expected around the position of the bound state corresponding to a single impurity. If the imperfections were regularly arranged in a superlattice there would be a narrow impurity band possessing the typical features stemming from the periodicity. However, they are randomly arranged in this example and consequently, in order to describe the properties of these states in the gap, a theory of fully-disordered systems is needed.

We return to the states lying well within the original band for which the characteristic motion is propagation with occasional scattering. It is easy to show that as the amount of imperfection or the strength of the scattering potential increases the mean free path l decreases. A limit is reached when l becomes so short, smaller than interatomic separation, that scattering and propagation no longer can be clearly distinguished. This limiting case of an extremely small effective mean free path will be discussed later in connection with the nature of the extended states.

The results reviewed in this subsection are summarized in Figure 3.1c. The discussion by which they were obtained, however, is incomplete, and, as a consequence, only trivial and readily anticipated modifications of the features of crystals were found. What is ignored is the effect of fluctuations in the scatterer concentration, which was treated, in effect, as everywhere uniform. A band model including the consequences of such fluctuations is presented in Section 3.3. A detailed physical discussion is given in Section 3.4.

3.3   MODELS OF THE ELECTRONIC STRUCTURES
OF DISORDERED MATERIALS

3.3.1   The Basic Model

We present here without derivation what we regard as the simplest possible model of electronic structure containing the essential features common to all disordered materials. Any other more detailed or sophisticated model of specific materials should incorporate the universal features contained in this basic model. It has been developed over the last decade largely through the efforts of Mott, (1967a–1968d), Cutler (1969), Austin (1969) who synthesized Anderson's (1958) work with work on the nature of the eigenstates in the tails of the bands (Fröhlich (1947) Gubanov (1963) Banyai (1964) Lifshitz (1964a) Bonch-Bruevich (1964) Halperin (1966) Zittartz (1966)) and with results on one-dimensional systems (Mott (1961) Landauer (1954) Lax (1958a) Frisch (1960) Borland (1961, 1963, 1964)). Important contributions were made by Edwards (1961, 1962, 1970a, b) Ziman (1961, 1968, 1969) and more recently by Cohen (1969, 1970) and collaborators, (Economou (1970a, b)).

The model (Mott, 1967a) can be considered as a natural extrapolation of the results presented earlier for perfect and imperfect crystals and is displayed in Figure 3.2 via a sketch of the density of states. A perfect crystal possessing a single isolated band has been taken as the starting point, and randomness has been introduced continuously through some disordering process. As the randomness increases, the band becomes broader and the nature of the wave functions changes. For $E_c < E < E_c'$, the states remain extended with a finite phase coherence length. At the energies E_c, E_c' the

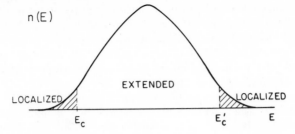

Fig. 3.2. The simplest model for the density of states of a single isolated energy band
in a disordered material. There is a band of extended states inside the
energies E_c and E'_c, with tails of localized states outside (shaded areas).

character of the states is assumed to change abruptly from extended to
localized so that for $E < E_c$ and $E'_c < E$ there are tails of localized states.
According to the present model the loss of long-range order in the potential
has a dual effect. For the states in the middle of the band, the long-range
phase order is lost, but the states remain extended.

A few words are in order to clarify what we mean by extended states. In
the free particle case the eigenstates are extended in the sense that the
particle can be found anywhere in space with equal probability. In the
periodic case the particle can also be found in any unit cell with equal
probability. Finally, in the more general case of disordered systems we call a
wave function extended if there are paths extending to infinity in both
directions lying entirely in regions where the wave function is not negligible.
Thus, in an extended state, an electron can travel through the system from
one end to the other and consequently make a non-zero contribution to the
conductivity of the system (Ziman (1968)).

On the other hand, for the states in the tails of the band, the effect of
the randomness is more drastic, forcing the amplitude of the wave function
to be different from zero only in a finite region and thus changing the
character of the states from extended to localized. Localized and extended
states cannot belong to the same energy except accidentally; any infini-
tesimally small perturbation would mix the two states, transforming both of
them to extended. There should therefore be two characteristic energies E_c,
E'_c separating the regions of localized states from that of extended states.
Such energies E'_c, E_c are called mobility edges and are insensitive to variations in the microscopic state of the material for reasons which will be
made apparent later.

The localized states are not in general associated with single imperfections. They appear however small the degree of randomness or strength of
scattering, being associated instead with fluctuations present in any disordered material.

As the randomness increases, more and more localized states are created,
and the mobility edges E'_c, E_c move inward into the band. At the same time,
the mean free paths of the extended states are reduced. When the randomness reaches a certain critical value, the two mobility edges merge, and, for

any randomness greater than this critical one, all the states in the band are localized. This is the Anderson (1958) transition.

It should be pointed out that the model presented is appropriate for three dimensions. In the one-dimensional case all states are localized by disorder because a particular interrelation between the phase and the amplitude of the wave function implies the localization of the wave function within the phase coherence length. We shall return to this point in Section 3.7.

The rest of this section is devoted to a discussion of amorphous semiconductors, which play a special role within the field of the electronic structures of disordered materials for two reasons. First, as discussed below, the transport properties of amorphous semiconductors are dominated by carriers within kT of the transition energy E_c where the states are uniquely characteristic of disordered materials. Secondly, the amorphous semiconductors are all covalent, and it is the electronic structures of the covalent materials which should be most sensitive to disorder. Simple metals, where the electrons interact weakly with the atoms via small pseudopotentials are free-electron-like near the Fermi surface both as solids and liquids. Insulating materials with large band gaps but narrow bands again have electronic structures relatively insensitive to order. The covalent semiconductors correspond to intermediate cases of maximal sensitivity of electronic structure to atomic structure and composition.

3.3.2 Amorphous Semiconductors: phenomenology

The most important elemental constituents of amorphous semiconductors are Si and Ge in Group IV; P, As, Sb and Bi in Group V; and the chalcogenides (S, Se, and Te) in Group VI. Elements, compounds, or multicomponent alloys varying widely in composition can be prepared in amorphous form variously by cooling a melt or condensing a vapor.

Diffraction experiments (Moss (1970)) indicate that their structure is characterized by a good short-range order closely similar to that of the corresponding crystal, when the latter exists. Small-angle scattering (Moss (1970)) spin resonance, (Brodsky (1969)) and density measurements (Brodsky (1971)) on unannealed films of variously, Ge, Si and SiC suggest that dangling bonds occur only on the internal surfaces of voids which are accidents of preparation. These can apparently be largely eliminated by annealing or by high temperature evaporation (Donovan (1970a, b)).

Taken together, these facts argue for structures close to that of an ideal covalent glass, (Mott (1967a)) which we define as a covalent network in which the valence requirement of each atom is locally satisfied so that there are no dangling bonds. Ideal structural models have been proposed for Ge (Grigorovici (1968), Polk (1971)) and Si (Moss (1970)). Real covalent amorphous materials differ from the ideal through the presence of chain ends, vacancies, dangling bonds, microscopic voids, etc., in varying degrees.

Our present state of understanding of these materials is so primitive that even the d.c. electrical conductivity σ presents a challenge. It appears to be intrinsic in its temperature dependence even for alloys with components of differing valence over wide ranges of composition.

Thus,

$$\sigma = \sigma_o \, e^{-\Delta E/kT}, \tag{3.3}$$

holds over a wide range of temperature, with σ_o and ΔE slowly varying functions of composition. Such intrinsic behavior of σ for the alloys is readily explained by supposing them to be nearly ideal covalent glasses (Mott (1967a)). With few dangling bonds present carriers are produced primarily by the breaking of bonds. The whole picture of the transport properties can be summarized by saying that amorphous semiconductors somewhat resemble low-mobility, intrinsic, crystalline semiconductors.

Simple chemical considerations, the activated temperature dependence Eq. (3.3) of σ, and the optical absorption data (Donovan (1970a, b)) all suggest a band model with valence and conduction bands separated by a gap. In the next subsection we present such models, but ones based on the ideas of the previous subsection and not on crystalline energy band theory.

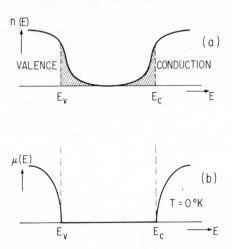

Fig. 3.3. The simplest band model for an amorphous or liquid semiconductor (a). Localized states (shaded areas) appear as tails of the conduction and valence band, leaving a well defined mobility gap (b).

3.3.3 Amorphous Semiconductors: band models

The simplest model that can be used for the discussion of amorphous (or liquid) semiconductors and which incorporates the basic properties of disordered materials is shown in Figure 3.3. There is a valence band v with a tail of localized states (shaded area) above E_v and a conduction band c with one below E_c. What are the implications of this model for E? The conductivity of any semiconductor can be expressed (using the results of Lax (1958b)) as

$$\sigma = \sum_{b=c,v} \int dE \, n_b(E) \, e \, \mu_b(E) \, f_b(E), \tag{3.4}$$

where b is the band index, n_b (E) is the density of states, μ_b (E) the average mobility for all carriers of energy E and f_b (E) is the probability of occupation for a state of energy E by electrons for b = c and holes for b = v. In the crystalline case, it is the sharp cut off of n_b (E) which leads to the well-defined activation energy observed in σ. In the disordered case, there is no sharp cut off of n_b (E) as can be seen from Figure 3.3a and the observed activation energy must have another origin.

We have supposed that the states are localized from E_v to E_c and extended outside it. Anderson (1958) has shown that electrons in localized states cannot diffuse at T = $0°$K. At finite temperatures they presumably can contribute to the conductivity only by phonon assisted hopping. We therefore expect a fall in μ_b (E) near E_v or E_c of at least several orders of magnitude, as shown in Figure 3.3b. The energies E_v and E_c are therefore mobility edges (Cohen (1969)). Evaluation of Eq. (3.4) with n_b (E) and μ_b (E) given by Figures (3.3a) and (3.3b) leads to an activation energy in σ which relates to the mobility gap $E_c - E_v$ instead of a gap in the density of states. The electrons and holes in extended states within about kT of the mobility edges carry the current. This model of a band with tails of localized states and sharp energies of transition to extended states which are mobility edges we term the Mott-CFO model after Mott (1967a) and Cohen, Fritzsche, and Ovshinsky (1969).

Actual specimens of the elements and compounds do not in general have an ideal glassy structure. Because the local connectivity of the network is well defined, it can have well defined defects: chain ends, vacancies, dangling bonds, etc. This might lead to nonmonotonicity in the density of states, which could become as shown in Figure 3.4.

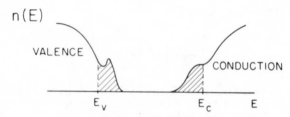

Fig. 3.4. The density of states for a non-ideal glassy structure. Non-monotonic behavior is expected.

In the semiconducting alloys, on the other hand, ideality of structure, i.e., local satisfaction of valence requirements, implies that the connectivity of the network varies locally as the valence of the constituents vary. Thus, in addition to the translational disorder of the elements and compounds, there is compositional disorder at the nodes of the network and additional translational disorder associated with the randomness of connectivity. This increased disorder could broaden the tails until they overlap, (Cohen (1969)) as shown in Figure 3.5. Such overlapping may also be the case for some liquid semiconductors because of their reduced short-range order. The important consequences that a model like the one shown in Figure 3.5 has for transport and other properties and related problems will not be discussed here (Fritzsche (1971), considers these questions in detail)). Rather, we shall concentrate our attention on a presentation of the theoretical foundation of these models and of the relevant theoretical techniques for attacking the problem of the electronic motion in disordered systems.

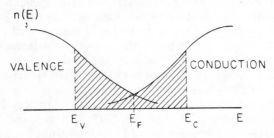

Fig. 3.5.   The CFO model of the density of states for a covalent amorphous alloy.

3.4   FLUCTUATIONS AND THE NATURE OF THE WAVE FUNCTIONS

Until recently, the standard treatments of the electronic structures of disordered materials employed on 'effective medium' approach, which treated the system as translationally invariant (Lax (1951) Foldy (1945) Edwards (1961)). The basic idea underlying the calculations and arguments reviewed in this section, however, is that in disordered materials localized states are associated with fluctuations in atomic configuration always present in such materials. Within a region in which the configuration differs from an average or typical configuration, so does the potential acting on the electron differ from its average or typical set of values. When the fluctuation is strong enough, or large enough in spatial extent, the associated potential change can bind, or localizes, one or more states. The associated energy levels depend on the details of the fluctuation, and the density of states can be determined from the probability distribution of fluctuations. For example, the localized states furthest from the center of the band are associated with the widest, or deepest, and hence least likely, fluctuations, explaining the existence of band tails.

In the last decade many quantum mechanical techniques have been employed to relate gross properties of the potential fluctuations to average characteristics of the energy spectrum (Lifshitz (1964a) Bonch-Bruevich (1964) Halperin (1966) Zittartz (1966) Kane (1963)). On the other hand, Ziman (1968) made a detailed study of classical motion within potential fluctuations and was the first to point out the relevance of percolation theory to the motion of electrons in disordered materials.

Recently Cohen (1970) has examined the case of a random binary alloy and has demonstrated through a simple probabilistic analysis that fluctuations able to bind an electron always exist. The same elementary analysis leads to a rough estimate of the density of states in the tails. Better estimates have been obtained without explicit reference to the character of the wave functions (Lifshitz (1964a) Bonch-Bruevich (1964) Halperin (1966) Zittartz (1966) Landauer (1954) Kirkpatrick (1970) Thouless (1970)).

Eggarter and Cohen (1970) combined percolation theory (Ziman (1968) Frisch (1963)) with a simple semiclassical calculation of the density and character of states (Kane (1963)) and were able to obtain not only the tails of localized states but also the position of the mobility edge. Extension of their arguments yields insight into the nature of the extended states. Their derivation of the relation between the potential fluctuations and the properties of the eigenfunctions, being a semiclassical one, lacks rigor in comparison with earlier quantum mechanical calculations (Lifshitz (1964a) Bonch-Bruevich (1964) Halperin (1966) Zittartz (1966)). Nevertheless their result for the conductivity agrees with the sophisticated quantum mechanical calculations of Coopersmith (1965, 1967) and Neustadter (1969). In this section we present in some detail the method of Eggarter and Cohen because it illustrates clearly and simply the essential physical concepts and ideas of electronic motion in a disordered system.

Consider the Coopersmith-Neustadter model consisting of a set of hard-core scatterers with scattering length a, randomly distributed with average density ρ. Suppose that the system is sufficiently dilute so that

$$\rho \, a^3 \ll 1. \tag{3.5}$$

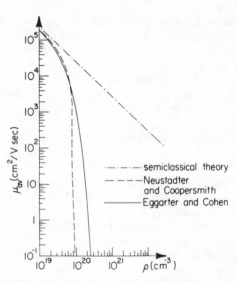

Fig. 3.6. Average electron mobility μ_{av} vs. density ρ. The semiclassical results, the results of Neustadter and Coopersmith, and the results of Eggarter and Cohen are indicated.

Suppose also that the temperature is sufficiently low that the thermal wavelength is long both compared to the interparticle spacing and the scattering length. Under these circumstances, Coopersmith (1965) showed that it was possible to systematize multiple scattering theory, collecting and summing the leading terms in every order of the multiple scattering expansion. This procedure was utilized by Neustadter and Coopersmith (1969) to obtain the average mobility μ_{av} as a function of density for fixed temperature and scattering length. Their results are shown in Figure 3.6 for $T = 3.96°K$ and $a = .62$ Å; they find a drop of 5 orders of magnitude in μ_{av} for a factor of 2 change in ρ for ρ in the range $10^{19} - 10^{20}/cm^3$.

This dramatic drop in μ_{av} can be explained in terms of the Mott-CFO model, (Eggarter (1970)). Briefly, a tail of localized states forms which is associated with density fluctuations. The width of the tail increases with density and becomes comparable to kT at a certain density. At that point the mobility drops precipitously because the electron spends most of its time in localized states which do not contribute to transport.

If the scatterers were distributed uniformly, the energy of an electron in the system would be given by

$$E(k) = V_o + \frac{h^2 k^2}{2m} , \qquad (3.6)$$

where

$$V_o = \frac{h^2}{2m} 4\pi \rho a, \qquad (3.7)$$

is the optical potential (Foldy (1945)) which shifts the bottom of the energy spectrum from zero to a positive value. Eq. (3.5) is the condition for the applicability of the optical model. Eq. (3.6) corresponds to the usual square-root density of states

$$n(E) \sqrt{E - V_o}, E > V_o,$$
$$= 0, \qquad E < V_o. \qquad (3.8)$$

Eq. (3.6) is definitely not valid, since the density ρ is not constant but possesses local fluctuations. However, their influence is important only for low energies, or long wavelengths and in this case a semiclassical approach like that of Kane (1963) can be used.

The space is divided into cubic cells of edge L. Each cell i has a number N_i of scatterers which fluctuates around the average value

$$<N> = \rho L^3. \qquad (3.9)$$

This in turn implies that the local optical potential

$$V_i = V_o N_i/<N>, \qquad (3.10)$$

fluctuates around the average value V_o. Provided the cell is large enough, it is possible to treat the electron semiclassically and take the density of states as a sum of contributions from each cell

$$n(E) \propto L^3 \sum_i (E - V_i)^{1/2} \ U(E - V_i),$$ (3.11)

where

$$U(x) = 1, x > 0$$
$$= 0, x < 0 .$$ (3.12)

The cell size is as yet unspecified. If we take it too large, density fluctuations would be smeared out. If we take it too small, we violate the uncertainty principle because we are simultaneously specifying the position and the energy of the electron. We therefore take L as the smallest length consistent with the uncertainty principle

$$\Delta x \sim L \sim \frac{h}{P_x} \sim h \, (3/2 \, m \, E)^{1/2},$$ (3.13)

and since we are interested in energies around V_0 we can make the unnecessary but simpler assumption

$$L \sim h \left(\frac{3}{2m \, V_0} \right)^{1/2} = \left(\frac{3}{4\pi \rho a} \right)^{1/2},$$ (3.14)

$$<N> \sim \left(\left(\frac{3}{4\pi \rho a} \right)^3 \frac{1}{\rho} \right)^{1/2}$$ (3.15)

Eqs. (3.5) and (3.15) imply that $<N> \gg 1$, which permits the conversion of the sum in Eq. (3.11) to an integral (Eggarter (1970)) of the form

$$n(E) = 2 \frac{(2m)^{3/2}}{4\pi^2 \, \hbar^3} \frac{V_0^{1/2}}{<N>^{1/4}} \ F(\epsilon),$$ (3.16a)

where

$$F(\epsilon) = \frac{1}{(2\pi)^{1/2}} \int_0^\infty X^{1/2} \exp \left[-\frac{(\epsilon - x)^2}{2} \right] \ dx,$$ (3.16b)

and

$$\epsilon = \frac{E - V_0}{V_0} <N>^{1/2} .$$ (3.16c)

The density of states given by Eq. (3.16) is shown in Figure 3.7a.

The density of states now shows a tail and deviates appreciably from the optical model over an energy range of order $V_0/<N>^{1/2}$ on either side of V_0. The tail comes from occasional cells in which there is a low density of scatterers surrounded by cells in which the density of scatterers is near the average level. Thus an electron within that particular cell cannot escape from it when its energy lies below the potential in the surrounding cells. Within the semiclassical picture then, the states in the tail are localized.

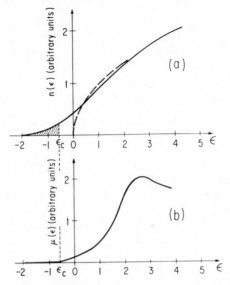

Fig. 3.7. a) The density of states as given by Eqs. (3.16). The dotted line is the
density of states for the optical model. The states below E_c (shaded area) are
localized.
b) The mobility as given by Eq. (3.18)

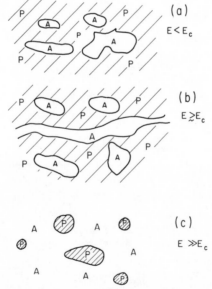

Fig. 3.8. Semiclassical states of motion in a random potential $V(r)$ for various
energies. P designates a forbidden region of space where $V(r) > E$ and A an
allowed region where $V(r) \leqslant E$. E_c is the mobility edge in this semiclassical
approach.

Fig. 3.9. Allowed (white) and prohibited (tiny P's) regions in a computer-generated random two-dimensional potential. The fraction of cells allowed in each picture is:

a) 0.396 ($\ll P_c$, the critical concentration for percolation);

b) 0.594 ($\approx P_c$);

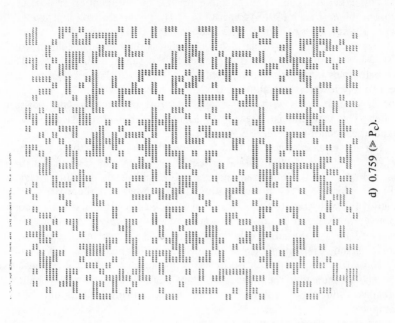

d) 0.759 ($\geqslant P_c$).

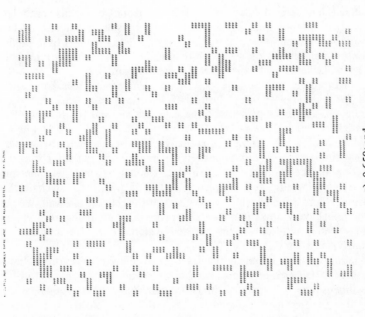

c) 0.650; and

This argument can be made more precise by formulation as a percolation problem (Frisch (1963) Eggarter (1970)). Put an electron in a given cell with a given energy. What is the probability that we can draw a closed surface around that particular cell which passes only through cells in which the value of the potential is higher than the value of the energy? That probability is unity below a value of ϵ of -0.52 and decreases continuously and rapidly towards zero above -0.52. Thus all states with $\epsilon < \epsilon_c = 0.52$ are localized. The states with $\epsilon > \epsilon_c$ are a mixture of extended and localized states. The situation is sketched in Figure 3.8 then demonstrated by an explicit calculation in Figure 3.9.

For any given energy E we find the allowed region (A) of space for which $V_o(\mathbf{r}) < E$ ($V_o(\mathbf{r})$ is the optical potential in the present case; in the most general case it is the potential energy) and the classically prohibited regions (P) for which $V_o(\mathbf{r}) > E$. Percolation theory shows that there is a critical energy E_c such that for $E < E_c$ all the allowed regions in a macroscopic sample are localized and disjoint (Figure 3.8a) so that every A region can be enclosed by a finite surface lying entirely in P regions. As E exceeds E_c nonlocalized A regions appear as channels, as in Figure 3.8b for E just above E_c most of the space is still occupied by localized A and by P regions. As the energy increases, the localized regions coalesce into channels, channels broaden, and the prohibited regions shrink, Figure 3.8c, and finally disappear.

All three regions of behavior can be studied on the computer-generated pictures shown in Figures 3.9 a-d. In this two-dimensional system the area considered has been divided into 44×50 rectangular cells, each large enough that the potential in each cell can be considered statistically independent. A constant potential has been assigned to each cell by a pseudo-random number generator. For each of four energies E, the forbidden cells have been blackened, with the results shown in Figure 3.9.

In Figure 3.9a, $E \ll E_c$, and only 40 percent of the cells are allowed. These form isolated (white) clusters of moderate size. When $E \approx E_c$, in Figure 3.9b, about 59 percent of the cells are allowed (p_c for the planar square lattice $= 0.592$) (Frisch (1961) Sykes (1964)) and a channel may be observed crossing the sample horizontally. This first channel to appear is multiply-connected and very sponge-like in appearance, but still covers only a fraction of the sample. The remaining 'resonances' have become quite large. A slight increase in E permits these to join with the first channel, as shown in Figure 3.9c, in which 65 percent of the cells are allowed. We note that although very few 'resonances' remain isolated in Figure 3.9c, and those few are quite small, the allowed regions of Figure 3.9c are still extremely convoluted in shape.

It should be noted that as the fraction of allowed cells approaches p_c from below the average size of an isolated region grows more slowly in three-dimensional systems than in two dimensions (Sykes (1964)), so it is likely that the 'resonant' isolated regions shrink more slowly as E is raised above E_c in three than two dimensions. Thus figure 3.9c probably underestimates the importance of resonant states in a real three-dimensional system.

In the final Figure, 3.9d, with $E \gg E_c$, 76 percent of the sites are allowed and only small separated forbidden regions remain. In such a situation it becomes possible to talk of an electron propagating freely in the allowed region with occasional scattering.

What are the modifications that quantum mechanics brings to this picture? First of all, the semiclassical energies within each A region are shifted upward as, e.g., for the zero point energy. Second, there is no longer a sharp distinction between P and A regions. The electron wave function penetrates into the P regions, in effect increasing the effective volume of the A regions. These two effects tend to cancel and produce at most quantitative changes in our picture; they may be ignored. Third, and much more important, the penetration into the P regions leads to tunneling between separated A regions so that these can no longer be described as disjoint. For $E < E_c$, tunneling from one localized A region to another raises the possibility of quantum mechanical delocalization of the classically localized states. However, if we start by ignoring tunneling and find the eigenstates independently for each A region, we can generate a quantum mechanical representation for expressing the matrix elements of the Hamiltonian coupling the A regions. The problem becomes identical to the one considered by Anderson (1958) and by Mott (1970) who show, in effect, that tunneling will not in general destroy localization. More explicitly, the energies of the levels within each A region are different because of their random shape and internal potential. Resonant tunneling does not occur with appreciable probability but only nonresonant tunneling with loss of amplitude. Thus tunneling leads to an exponential decrease of amplitude from A region to A region and does not destroy localization for $E < E_c$. On the other hand, if we ignore tunneling above E_c, we have a coexistence of localized and extended states. The extended states are of an entirely novel character and are peculiar to disordered materials. They do not cover all of the volume of the material but are restricted to the neighborhood of the classical percolation channels. Within them the energy levels are dense because the states are extended. Consequently, tunneling from localized A to a channel can be resonant, and a single tunneling event from a localized state to the nearest channel serves to delocalize that state. Above E_c, there are no localized states, but only channel states and states resonant with these. Just at E_c the resonant states are indistinguishable from bound states. As E increases the localized states become resonances, channel states open up, and the ratio of channel to resonant states increases. The resonances gradually disappear, and the channel states gradually spread out into ordinary extended states. The proportion of channel states near E_c, where the distinction is meaningful, is given roughly by the fraction of the classically allowed volume encompassed by channels in the potential. This in turn is given by the classical percolation probability, p (E). We have found arguments which strongly suggest the following energy dependence for p (E):

$$p(E) = 0, \quad E < E_c \qquad\qquad (3.17a)$$

$$p(E) \propto (E - E_c)^s, \quad 0 < s < \frac{2}{3}, \quad E \geqslant E_c ,$$

with $\frac{1}{3}$ perhaps a better lower limit (Shante (1971a)). The density of channel states is given therefore by

$$n_c(E) = p(E) n(E).$$ (3.17b)

The channel states contribute appreciably to the conductivity, the resonances contribute little, and the localized states not at all. Estimates for the mobility edge can therefore be made from classical percolation theory, an important advance in itself.

The calculation of the mobility within the framework of the Eggarter-Cohen model will be examined next. The mean free path is limited by two scattering processes. The first is single scattering, $l = 1/\rho\sigma$, which together with the optical model gives the semiclassical mobility indicated in Figure 3.6. The second is reflection off the walls of cells which are prohibited to it semiclassically, i.e., off the long-wavelength density fluctuations. As $< N >$ is about 80 for ρ about 10^{19}, this second process involves very high-order multiple scattering. It is an easy matter to work out the probability of traversing n cells without reflection and have reflection occur at the (n + 1) cell such that the contribution to the mean free path is (n + 1)L and thence to carry out the average for all n. The result is an energy-dependent mean free path $\lambda(\epsilon)$. The corresponding mean free time $\tau(\epsilon)$ is computed from

$$\tau(\epsilon) = \lambda(\epsilon) \left\langle \frac{1}{v} \right\rangle_\epsilon,$$ (3.18)

where $< 1/v >_\epsilon$ is a conditional average of the inverse velocity taken only over allowed cells and for fixed energy ϵ. The energy dependent mobility is then computed from

$$\mu(\epsilon) = \frac{2}{3} \frac{e}{m} \tau(\epsilon) p(\epsilon),$$ (3.19)

where the factor of $\frac{2}{3}$ is required to get back the semiclassical mobility at low densities and $p(\epsilon)$ is the percolation probability for an electron of energy ϵ, i.e., the probability that an electron is not in a classically localized state and is instead in an allowed channel which extends throughout the entire medium. Since $p(\epsilon)$ vanishes below $\epsilon_c = -0.52$, ϵ_c is a true mobility edge within the limits of our semiclassical treatment. The results of Eggarter and Cohen for $\mu(\epsilon)$ are shown in Fig. 3.7b. The fall off for larger ϵ occurs when μ approximately equals the semiclassical value which is proportional to $\frac{1}{v}$ or $\epsilon^{-1/2}$.

To make a comparison with the results of Neustadter and Coopersmith (1969) it is necessary to average $\mu(E)$ over a Boltzman distribution

$$\mu_{av} = \frac{\int \mu(E) n(E) e^{-\beta E} dE}{\int n(E) e^{-\beta E} dE}$$ (3.20)

The result is shown in Figure 3.6. There is good agreement with the elaborate quantum mechanical approach of Neustadter and Coopersmith.

This simple analysis confirms the Mott-CFO model in every detail. There is a Gaussian tail of localized states associated with density fluctuations, a mobility edge at $\epsilon_c = -0.52$, channel and resonant extended states just above ϵ_c, and ordinary extended states further above ϵ_c. The mobility is of course zero below ϵ_c and positive above it. A final comment is in order regarding the behavior of μ (E) for E just above E_c. Figure 3.7b shows a linear increase which comes from an assumed linear increase in the percolation probability. However, we believe it more likely that the percolation probability, and consequently the classical mobility, would have the critical index behavior of Eq. (3.17a).

$$\mu \propto (E - E_c)^s, \ 0 \leqslant s \leqslant \frac{2}{3}, \ E > E_c,$$

$$= 0 \quad E < E_c .$$

(3.21)

The way that the quantum-mechanical mobility behaves just above E_c for $T = 0°K$ is an open problem of considerable interest. The semiclassical arguments developed in this section together with the tunneling arguments for including quantum effects support a continuous drop, i.e., $s > 0$ in Eq. (3.21). Davis and Mott (1970) on the other hand, argue by an analysis of the Kubo formula that the mobility drops discontinuously to zero at E_c. Mott assumed that above E_c all states extend everywhere throughout the material. If one asserts that the resonances make little contribution to σ, one has only the contribution of the channel states to the Kubo formula. Repeating Mott's analysis, but for the channel states now, one finds that the energy dependence of the mobility is dominated by the percolation probability because of Eq. (3.17b). The energy dependence of the mobility near E_c is the same as classically, Eq. (3.21). The conclusion is not modified by including the resonances.

3.5 MOBILITY EDGES AND ANDERSON'S TRANSITION

In this chapter we shall review some recent work (Economou (1970b) (1972a)) aiming at a first principle derivation of the Mott-CEO model. The basis of this approach in Anderson's (1958) classic paper on 'The Absence of Diffusion in Certain Random Lattices'. Anderson considers a tight binding model in which a band is formed from s-like atomic orbitals $|l>$ with energies ϵ_l, for the orbital centered around the site l. The bandwidth is $B = 2VZ$ where Z is the coordination number and V is an overlap energy integral. Randomness is introduced into the system by assuming that the quantities ϵ_l are random variables with a distribution having a width W. He finds that there is a critical value $W_c \sim BlnZ$ of W such that for $W > W_c$ the states at the middle of the band (and by inference all the states) are localized and transport ceases to exist.

The matrix elements of the Hamiltonian of the system under consideration are

$$<l \mid H \mid m> = \epsilon_l \delta_{lm} + V_{lm},$$

(3.22)

where V_{lm} is a constant V for nearest neighbors and zero otherwise.

We consider a particle initially localized in a region of space, e.g. such that $| \psi(0) > = | 0 >$. If there are localized eigenstates overlapping with the state $| 0 >$ the particle will have finite probability of being initially in each one of these eigenstates. Since this probability is time independent there will be a finite probability p_{oo} of rediscovering the particle at the initial state as $t \to \infty$. On the other hand, if no localized eigenstates overlapping with the state $| 0 >$ exist, the particle will diffuse away and the probability of rediscovering it at the initial state will approach zero as $t \to \infty$. Thus by determining whether p_{oo} is different or equal to zero, one can discover whether localized states exist in the neighborhood of the site 0. One can show (Anderson (1958)) that

$$p_{oo} = \lim_{s \to 0^+} \frac{s}{\pi} \int_{-\infty}^{\infty} dE \, G_0 (E + is) \, G_0 (E - is), \qquad (3.23)$$

where $G_0 (z) = <0| \dfrac{1}{z - H} |0>$. Eq. (3.23) can also be written as $p_{oo} = \displaystyle\int_{-\infty}^{\infty}$

$dE \, f_0 (E)$ where $f_0 (E)$ is a positive quantity given by

$$f_0(E) = n_0 (E) \lim_{s \to 0^+} \frac{1}{1 - \dfrac{\Delta_0 (E + is) - \Delta_0 (E - is)}{2 \, is}}, \qquad (3.24)$$

where $\Delta_0(E)$ is the 'self energy' defined as $\Delta_0(z) = z - \epsilon_0 - G_0^{-1} (z)$ and $n_0(E)$ is the contribution to the total density of states from the state $|0>$. A non-zero value of $f_0(E)$ would require that $n_0(E) \neq 0$ (which means that eigenstates with eigenenergy E overlapping the state $| 0 >$ exist) and that $\lim_{s \to 0+} [1 - (\Delta_0(E + is) - \Delta_0(E - is))/2 \, is]^{-1} \neq 0$ (which means that these eigenstates are localized). Since all the singularities of $\Delta_0(z)$ and $G_0(z)$ lie on the real axis, being either branch cuts or simple poles, and since their branch cuts coincide while their poles do not, one can conclude that extended states correspond to the simple poles of $G_0(E)$.

The analytic properties of $\Delta_0(z)$ can be studied conveniently by using the renormalized perturbation series (RPS) expansion (Anderson (1958) Watson (1957)) for this quantity,

$$\Delta_0 = \sum_{n \neq o} V_{on} G_n^o V_{no} + \sum_{\substack{n \neq o \\ \ell \neq o,n}} V_{on} G_n^o V_{nl} G_l^{o,n} V_{lo} + \dots , \qquad (3.25)$$

where $G_m^{o,n,l,\dots} (z)$ is the m, m matrix element of the operator $(z - H^{o,n,l},)^{-1}$ and $H^{o,n,l}\dots$ differs from H in that $\epsilon_0 = \epsilon_n = \epsilon_l = \dots = \infty$. The usefulness of Eq. (3.25) lies in the fact that it converges everywhere except on the branch cuts (Economou (1972a)). Thus if one can find the regions of the real E-axis for which the RPS (3.25) diverges, one has determined the regions of extended states. On the other hand, the regions for which the RPS converges correspond either to localized states or to no states at all.

As Anderson has pointed out, the problem of the convergence of a series of random terms (in this case the RPS for Δ_0) should be approached in a probabilistic way: for each energy E one should find the probability that the series (3.25) converges. Following a lengthy and complicated analysis (Economou (1972a)) of the structure of the Nth order term $\Delta_0^{(N)}$ of the RPS, (Watson (1957)) one can show that when there is no long-range statistical correlation among the variables ϵ_m, there exists a non negative function L(E) such that the random variable $|\Delta_0^{(N)}(E)|$ is sharply distributed around the quantity $L(E)^N$ in the sense that

$$\text{Prob.} \left[L(E)^{N-N^q} < |\Delta_0^{(N)}(E+is)| < L(E)^{N+N^q} \right]_{N \to \infty} 1,$$

where q satisfies the relation $\frac{1}{2} < q < 1$, with the following consequences. Hence, there is a localization function L(E) for the disordered system under consideration such that the regions of the energy spectrum for which L(E) < 1 consist entirely of localized states and those for which L(E) > 1 consist entirely of extended states. The solutions of the Eq. $L(E_c) = 1$ give the positions of the mobility edges E_c. When E belongs to an energy gap, L(E) < 1.

Because of a strong statistical correlation among the terms that contribute to $\Delta_0^{(N)}$, for N large, one can obtain an explicit formula for the function L(E),

$$L(E)^N = \Sigma' V_{on_1} \bar{G}_{n_1}^o V_{n_1 n_2} \bar{G}_{n_2}^{o,n_1} \ldots V_{n_N o}, \tag{3.26}$$

where

$$\log \bar{G}_{n_i}^{o,n_1 \cdots n_i - 1} = < \log | \frac{1}{E - \epsilon_{n_i} - \Delta_{n_i}^{o,\ldots,n_i-1}} | >, \tag{3.27}$$

and Σ' indicates summation over all the indices $n_1, \ldots n_N$ with the restrictions $n_1 \neq 0, n_2 \neq n_1, 0, \ldots n_N \neq n_{N-1}, \ldots, 0$.

It should be pointed out that the function L(E) one can infer from Anderson's paper differs from the one presented here. The reason is that Anderson's approach to the statistical analysis of $\Delta_0^{(N)}$ requires for its applicability that all the terms which contribute to $\Delta_0(N)$ are statistically independent, (Thouless (1970)) whereas we have shown that strong statistical correlation of these terms occurs (Thouless (1970) Economou (1970b)). In any case the estimates one obtains from Eqs. (3.26) and (3.27) differ from those obtained from Anderson's approach only by a factor of about 2.

The results stated above constitute a first principle derivation of the Mott-CFO model at least for the system we consider here. Let Γ be a measure of the spread of the random variables $\{\epsilon_n\}$ around an average value ϵ_0 that will be chosen as zero. Consider first the case where $\Gamma = 0$. From Eqs. (3.26) and (3.27) and the properties of periodic systems one can show that L(E) $\geq$ 1 for E inside the band with the equality obtaining at the band edges. This is in agreement with the fact that all the eigenstates are extended for a periodic system. On the other hand when $\Gamma \to \infty$ one can see that a particle initially localized at $|0>$ cannot escape since with probability unity it is surrounded by infinite potential barriers. Thus $\Delta_0(E) \to 0$ everywhere and so does $L(E)$. Assuming a continuous behavior of $L(E)$ as a function of Γ for each E we can conclude that for each E there is a critical value of Γ, $\Gamma_{c,E}$ such that for $\Gamma \geqslant \Gamma_{c,E}$, $L(E) \leqslant 1$. If we call Γ_c = max $\Gamma_{c,E}$, it follows that for $\Gamma \geqslant \Gamma_c$, $L(E) < 1$ for every E. Thus when $\Gamma \geqslant \Gamma_c$ all the eigenstates are localized and all the transport coefficients vanish. This is Anderson's transition. For $\Gamma < \Gamma_c$ there will be regions of extended states only ($L(E) > 1$) separated from regions of localized states only ($L(E) < 1$) by mobility edges E_c ($L(E_c) = 1$), in complete agreement with the Mott-CFO model. In the framework of this analysis, Anderson's transition can be defined more generally as the disappearance of a region of extended states through the merging of two adjacent mobility edges. This definition coincides with that used by Anderson for the case of only one region of extended states. If more than one region of extended states is present, then the disappearance of all the extended states would take the form of successive Anderson's transition, each one eliminating one region of extended states.

An order of magnitude estimate for the localization function $L(E)$ can be obtained by neglecting the quantities $\Delta_{n_1}^0, \cdots, {n_i-1}$ in Eq. (3.27.) The resulting expression for $L(E)$ is

$$L(E) \approx \alpha \, KV \, e^{-<\log |E - \epsilon_i|>}, \tag{3.28}$$

where K is the connective constant defined in percolation theory (Frisch (1963)) as the Lth root of the number of all possible self-avoiding paths of length L on the lattice as $L \to \infty$. K is generally (Frisch (1963)) of the order $Z-2$ where Z is the number of nearest neighbors. The quantity α has been introduced in formula (3.28) as a correction factor determined by the relation $L(E_b)_{\Gamma=0} = 1$ where E_b is the position of the band edge for the periodic case $\epsilon_n = 0$ for every n. For the case $\Gamma = 0$ Eq. (3.28) can be written as

$$L_{\Gamma=0} = \frac{\alpha \, KV}{|E|}$$

from which one concludes that $\alpha = \dfrac{Z}{K}$ and consequently

$$L(E) \approx ZV \, e^{-<\log |E - \epsilon_i|>}, \tag{3.28a}$$

Formula (3.28a) is identical to the estimate used somewhat arbitrarily by Ziman (1969). In the case of a rectangular distribution function for ϵ_i of total width W explicit results have been obtained (Ziman (1969)). The critical value of the width W is

$$W_c \approx 2.7B, \tag{3.29}$$

where B = 2ZV is the bandwidth for the periodic case of zero randomness.

It is worthwhile to notice that Anderson's statistical analysis as clarified and corrected by Thouless (1970) would give $W_c \approx 5B$ for the critical width W_c.

For the case where the ϵ_i's are distributed according to a Lorentzian of half width Γ, Eq. (3.28a) gives

$$L(E) \approx \frac{ZV}{(E^2 + \Gamma^2)^{1/2}} \tag{3.30}$$

The critical value of Γ for Anderson's transition to occur is

$$\Gamma_c \approx B/2. \tag{3.31}$$

Anderson's approach would give $\Gamma_c \approx 0.9B$. Using the correspondence W/4 $\leftrightarrow \Gamma$ (pr. $[|\epsilon_i| < \frac{W}{4}] = \frac{1}{2}$ and pr. $[|\epsilon_i| < \Gamma] = \frac{1}{2}$ for the rectangular and Lorentzian distribution respectively) one can see that the estimates for the critical randomness in the Lorentzian case are about 25% less than the estimates for the rectangular one. This is probably due to the long tails of the Lorentzian distribution.

Eq. (3.27a) is the simplest estimate one can obtain for L(E). We shall summarize here more accurate estimates pertinent to a special case of significant practical importance.

Suppose that

$$\bar{G}_{n_i}^{o,n_1,\cdots}(E) = |\mathscr{G}_{n_i}^{o,n_1,\cdots}(E - \Sigma(E))|, \tag{3.32}$$

where $\bar{G}_{n_i}^{o,n_1,\cdots}$ is the n_i,n_i matrix element of the Green's function corresponding to the Hamiltonian (3.22) with $\epsilon_o = \epsilon_{n_i} = \ldots = \infty$ and $\epsilon_m = 0$ for all remaining sites m, and $\Sigma(E)$ is a complex function of E. Then one can show that

$$L(E) < 1 \tag{3.33a}$$

if

$$F(E) = \frac{ZV}{|E - \Sigma(E)|} \leqslant 1. \tag{3.33b}$$

Moreover, for small randomness Γ, when F approaches unity L approaches a value which is less than unity by an amount of order $[\Gamma/ZV]^{1/2}$.

The physical meaning of Eq. (3.33a) is that all eigenstates with eigenenergies E satisfying Eq. (3.33b) are localized. In addition, the mobility edges E_c lie always in the region where $F(E) > 1$. As the randomness becomes smaller the mobility edges E_c approach the positions E_c^* where $F(E_c^*) = 1$. In other words, the function F(E) not only can be used as a reasonable estimate of the localization function L(E) for small randomness, but can serve also as an exact upper limit for it, for deriving the positions of the mobility edges.

We conclude this Section by examining again the case of the Lorentzian distribution for $\{\epsilon_i\}$. This particular case is important because it permits an exact evaluation of $\tilde{G}^{o,n_1},...$ by the same technique used in Lloyd's work. Lloyd (1969) was able to calculate exactly the quantity $\langle G_n \rangle$, and Brouers (1970) did the same thing for the quantity $\langle \Delta_n \rangle$. Both of them inferred from their results that localized states do not exist, assuming incorrectly that the non-existence of localized states is equivalent to a non zero lm $\langle G_n \rangle$ or lm $\langle \Delta_n \rangle$. The result for $\tilde{G}^{o,n_1}_{n_i},\cdots$ is of the form (3.32) with $\Sigma(E) = - is(E)\Gamma$ where $s(E) = 1$ if lm $E > 0$ and -1 if lm $E < 0$. Thus

$$F(E) = \frac{ZV}{(E^2 + \Gamma^2)^{1/2}}. \tag{3.34}$$

Using Eqs. (3.33) and (3.34) we see that all the eigenstates with eigenenergy E outside the interval $[-E_c^*, E_c^*]$, where $F(\pm E_c^*) = 1$ or $E_c^* = (Z^2V^2 - \Gamma^2)^{1/2}$, are localized. The mobility edges E_c lie always inside this interval. Anderson's transition occurs for Γ_c such that the two mobility edges merge, i.e., $E_c = 0$. However, at Γ_c, $E_c^* > 0$ so that $\Gamma_c < ZV$. In other words for $\Gamma > ZV$, all the states are localized, although the transition occurs always at a lower value of Γ.

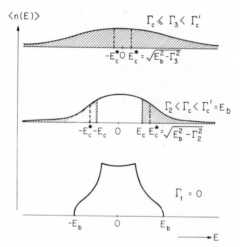

Fig. 3.10. Sketches of the (average) density of states per atom n(E) for three different values of the halfwidth Γ of the Lorentzian distribution of single site energies. The mobility edges $\pm E_c$ separate regions of localized states (shaded) from those of extended states, and always lie within the interval $[-E_c^*, E_c^*]$. Γ_c is Anderson's critical value of the randomness.

This discussion is summarized in Figure 3.10 where the density of states is plotted schematically for three different values of Γ and the positions of $\pm E_c$ and $\pm E_c^*$ are given.

In the next section we shall apply the procedure based on Eqs. (3.33) to the case of random binary alloys, where Eq. (3.32) is approximately true.

A detailed review of the material of this section, its history, and of the mathematical and physical analysis leading to the proof of the existence and properties of L(E) and F(E) can be found in a paper by Economou and Cohen (1971a).

3.6   APPLICATIONS TO BINARY ALLOYS

The techniques of the preceding Section prove exceedingly fruitful when applied to random, substitutional binary alloy systems of the form $A_x B_{1-x}$. These may be described with the Hamiltonian (3.22) by letting $\epsilon_n = \epsilon^A$ or ϵ^B with probability x and $(1-x)$, respectively. By appropriate choices of x and the dimensionless parameter $\delta = (\epsilon^A - \epsilon^B)/E_b$, where E_b is the half bandwidth in the periodic case $\epsilon^A = \epsilon^B = 0$, one may obtain concrete examples of all the general phenomena of disordered systems which were summarized in the opening sections. Mobility edges, a mobility gap between two regions of extended states, and an Anderson transition are obtained and discussed below.

In what follows we shall take A to be the minority constituent ($x \leqslant 0.5$), and choose $\delta > 0$. Except when explicitly noted, we restrict V_{lm} to have a constant value V, independent of x, and connect nearest neighbors only. Although in this section we consider electrons in an alloy, a very similar Hamiltonian is used to describe lattice vibrations in the presence of mass disorder. The results of this section may be extended by a simple transformation to describe such phonon systems (Economou, 1971b).

There are several exact results available in this model to serve as checks on approximate calculations of $<G_n(E)>$ and L(E). In the limit of small x or δ, all averaged quantities may be expanded in powers of a small parameter. For arbitrary x and δ, a number of moments of the density of states can be calculated exactly (Velicky, 1968). When $\delta > 2$, the density of states is split into two separated sub-bands, centered about ϵ^A and ϵ^B, each of width B. Thus in the limit $\delta \to \infty$, a site containing a B atom is forbidden to an electron with energy near ϵ^A, and percolation theory (Frisch, 1963) may be used to determine the probability that such an electron is trapped or free to move across the crystal. When x is greater than x_c, the critical value for the onset of percolation, there will be extended states in the A sub-band. Since x_c is less than $1/2$ for all three-dimensional lattices, we observe that at least one of the strongly split sub-bands will always contain extended states, in contrast to the complete localization observed for $\Gamma > \Gamma_c$ in the Lorentzian model of the preceding section.

3.6.1   Single Site Approximations

The approximations needed to calculate F(E) and from it, by Eqs. (3.26) and (3.27) L(E), for the binary alloy may be conveniently discussed in terms of an effective self-energy defined by

$$\left\langle \frac{1}{Z-H} \right\rangle = \frac{1}{Z-W-\Sigma} , \qquad (3.35)$$

where W is the periodic part of H, defined in Eq. (3.23.) Since Σ, an averaged property, is also periodic, its matrix elements in the orbital basis of Eq. (3.32) have the simple form

$$\langle n \mid \Sigma \mid m \rangle = \delta_{nm} \, \Sigma^{(0)} \, (E) + \Sigma^{(1)} \, (n-m, E). \tag{3.36}$$

Useful and accurate approximation schemes may be generated by neglecting $\Sigma^{(1)}$ in Eq. (3.36), and are termed single site approximations (SSA's). Use of any SSA to calculate $\langle G \rangle$ in this model Eq. (3.32) yields.

$$\langle G(E) \rangle = g(E - \Sigma^{(0)}(E)), \tag{3.37}$$

i.e., a single complex self-energy, $\Sigma^{(0)}$, replaces the random diagonal element ϵ_i, on each site. Soven (1967) has shown, by expanding in powers of the atomic scattering matrix t_n, that one may solve Eq. (3.37) self-consistently and obtain $\langle G \rangle$ through order $\langle t \rangle^3$ by choosing $\Sigma^{(0)}$ such that

$$\langle t_n (E - \Sigma^{(0)}) \rangle_{SSA} = 0. \tag{3.38}$$

The single site average $\langle \; \rangle_{SSA}$ in Eq. (3.37) is taken by treating the site n exactly, and replacing the ϵ_i on all other sites by $\Sigma^{(0)}$. The solution to Eq. (3.38) he termed the coherent potential approximation, Σ_{CPA}.

It is important to note that Σ_{CPA} is the unique self-consistent SSA, in the sense that the average of any analytic function f(G) could have been used instead of Eq. (3.36) to introduce Σ. As a consequence of Eq. (3.38) one can show that

$$\langle f(G) \rangle = f(g(E - \Sigma_{CPA})), \tag{3.39}$$

through at least order $\langle t \rangle$, for any analytic f(G). One such function is the partially excluded Green's function G_0^n defined in Eq. (3.25); another is log (G). We shall therefore use Σ_{CPA} in $g(E - \Sigma^{(0)})$, calculating F(E), the approximate localization function, from Eq. (3.33b).

It can also be shown (Velicky (1968)) that the CPA is valid outside the regions in which expansion in t_n is appropriate, for it interpolates correctly between the low concentration and weak scattering limits, and extrapolates more accurately from either limit than do the respective limiting expressions for the self-energy. The second statement is tested by comparing moments of the calculated density of states with the exact results mentioned above. It must be noted, however, that the CPA provides a good description of extended states only. Because fluctuations about the average properties of the medium are neglected in solving Eq. (3.38) the calculated densities of states have sharp band edges, and lack the tails associated with localized states. Extensions of the CPA have been suggested (Schwartz (1970) Freed (1971a)) in which the average Eq. (3.38) is generalized to treat scattering from more than one site, and off-diagonal terms from $\Sigma^{(1)}$ appear in the self-consistent solution. For the study of localization phenomena, however, we have seen in Sections 3.4 and 3.5 that fluctuations involving very large numbers of sites are essential. Treatment of scattering from a finite number of sites therefore adds no significant new information to the SSA, and will not be pursued here.

Unlike the simplest approximation (3.28a) which we have seen to be unrelated to $L(E)$ when δ is large, $F(E)$, defined by Eq. (3.33b), is strictly greater than $L(E)$ whenever Eqs. (3.32) and (3.33a) hold. Use of $F(E)$ will not overestimate the amount of localization present in the system. Finally Eq. (3.32) is, as noted above, correct within the framework of the CPA. Thus we have calculated $F(E)$, as defined in Eq. (3.33b).

Before making use of the CPA, we note that the simplest approximation to $L(E)$, Eq. (3.28a), in which Δ_0^n is neglected entirely, has been applied by Ziman (1969) to this system. The approximation predicts regions of localized states inside both band edges for small values of δ, but there are no extended states in the limit $\delta \to \infty$, in contradiction with the percolation theory limit. For large, finite δ and any x, Ziman always finds some extended states in both bands, which, as is seen below, is also incorrect. The SSA used in that work to obtain an averaged density of states is in error at large δ, for it gives one of the sub-bands a width greater than B, in contradiction with the exact limits of the spectrum (Kirkpatrick (1971a)). However, this has no effect on the calculation, via Eq. (3.28a) of $L(E)$, using the CPA self-energy,

$$F(E) = E_b / | E - \Sigma_{CPA}(E) |, \qquad (3.40)$$

to provide an estimate of the true localization function, $L(E)$.

Figure 3.11 exhibits the types of behavior obtained in this way. The four sections show the density of states $n(E) = \langle n_0(E) \rangle$ and the function $F(E)$ for x = 0.1 and four values of δ, increasing from bottom to top. Also plotted is the fractional A parentage $n^A(E)$, of states with energy E, calculated in the CPA as $n^A(E) = x \, \rho^A(E)/n^A(E) = x \, \rho^A(E)/n(E)$, where $\rho^A(E)$ is the conditionally averaged density of states $\langle n_0(E) \rangle_{0=A}$, subject to the condition that site 0 contain an A atom. This quantity is used below to aid in analyzing the character of the states at given energies. For convenience, the calculations described in this figure, as well as in Figures 3.12 and 3.13, were performed using the analytic 'semicircular' model density of states, (Velicky (1968)) resulting from a Hamiltonian in which V is not restricted to nearest neighbors only. The conclusions to be drawn are not qualitatively affected by this.

Mobility edges, indicated in Figure 3.11 by arrows labelled E_c, are found inside the band edges for all values of δ, moving further in as δ is increased. In Figure 3.11b there appears a mobility gap in the interior of the band, a region of localized states separating two regions of extended states, a feature first hypothesized by Mott (1967a). As δ is increased (Figure 3.11c), an impurity sub-band is drawn off. Mobility edges are observed at the top and bottom of both sub-bands. For still larger δ, shown in Figure 3.11d, the states in the impurity sub-band are all localized, while the mobility edges in the majority sub-band remain practically unshifted in Figures 3.11b–d. This

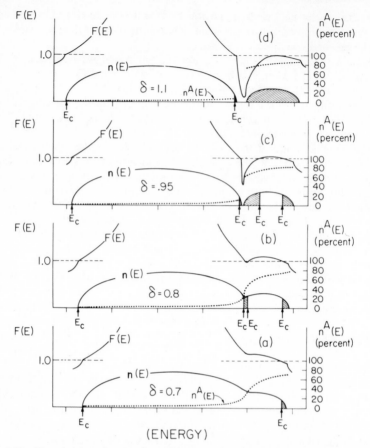

Fig. 3.11. Alloy density of states (in arbitrary units) and properties related to localization, calculated by the CPA with x = 0.1, and δ = (a) 0.7, (b) 0.8, (c) 0.95,
and (d) 1.1. Regions of localized states have been shaded. The heavy line
above each density of states indicates the 'localization' function F(E)
defined by Eq. (3.32b) the dotted line the A parentage $n^A(E)$ defined in the
text (Economou 1970a).

is consistent with the percolation theory result that there must remain
extended states in the majority sub-band as $\delta \to \infty$.

The sequence of events seen in Figures 3.11 defines, for each concentration, several critical values of δ which we shall use to characterize the
general behavior of the model. For a given x we define δ_M, the value of δ at
which the mobility gap first appears (Figure 3.11b), δ_G, at which two
sub-bands separate (Figure 3.11c), and δ_A, at which the Anderson transition
occurs in the minority sub-band (Figure 3.11d). These critical values of δ,
plotted as functions of x in Figure 3.12, provide a 'phase diagram' for the
model. We note that for concentrations greater than a critical value x_c, no
Anderson transition occurs, and there are extended states in both sub-bands
as $\delta \to \infty$.

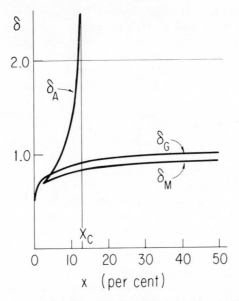

Fig. 3.12. Phase diagram of the alloy model, as obtained from the CPA self-energy. δ_M denotes the critical value of δ for the appearance of a mobility gap at a given concentration, δ_G the opening of a gap in the density of states, and δ_A the Anderson transition for the impurity states. The concentration above which the Anderson transition no longer takes place is denoted x_c.

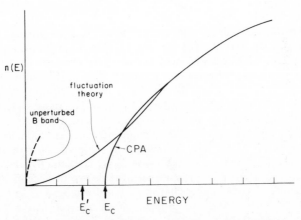

Fig. 3.13. Tail of the density of states, as calculated from the fluctuation theory, compared with the CPA result for $x = 0.1$, $\delta = 0.5$. E_c denotes the mobility edge predicted from the CPA and F(E), and E'_c the semiclassical result. E'_c and E_c differ by .004 times the unperturbed B bandwidth.

There are two sources of error in the calculations shown in Figure 3.12: we have used F(E) instead of L(E) to predict localization; and F(E) was calculated with the approximate CPA self-energy. Still we expect that Figure 3.12 contains all the features of the 'phase diagram' corresponding to an exact treatment, and differs only in quantitative details, principally in that it underestimates the amount of localization present at any concentration.

The quantity δ_G in Figure 3.12 is affected only by the errors inherent in the CPA, specifically the neglect of the tails of localized states. As Lifshitz (1964b) first observed, these must fill the region between two sub-bands until $\delta = 2$, i.e., until $\epsilon^A - \epsilon^B$ exceeds the full bandwidth B. This exact result is indicated in Figure 3.12 by a light line.

Both sources of error affect the determination of mobility edges, and hence δ_M, but to some extent these errors appear to counteract each other. This conclusion is suggested by an independent calculation, using the treatment of fluctuations outlined in Section 3.4, of the density of states and mobility edge near the lower band edge. The optical potential in this case is simply t_n^A, for the scattering of an isolated A atom in a B host, and the results are valid semiclassically for all δ if the concentration is sufficiently low. Figure 3.13 compares the CPA results with the fluctuation theory for a typical case. The calculated density of states has a low energy tail, but joins smoothly with the CPA density at higher energies. The two mobility edges are in rough agreement.

The agreement is not too surprising, since the CPA is known to provide a good description of the extended states inside the mobility edge, and $(Im\Sigma)^{1/2}$, which was found in V to characterize the difference between F(E) and the true L(E), is small near the band edges. Results obtained with the CPA by extrapolating from a region in which the states are quite extended, in particular the predicted δ_M of Figure 3.12, should be quite reliable.

The errors made when δ is large are more difficult to estimate, since both the CPA and the use of F(E) may be seriously in error. Some qualitative features of the CPA description of a majority sub-band in this limit may be tested numerically by direct numerical estimation of the bandwidth and the rough magnitude of the tails (Kirkpatrick (1971c)). The CPA majority band, of course, lacks tails, but its width is found to be consistently within the calculated bounds, and in close agreement. Thus use of the CPA self-energy to describe a majority sub-band should not introduce serious error.

A comparison of the large δ results with the predictions of percolation theory shows that the use of F(E) with Σ_{CPA} consistently underestimates the amount of localization present, with the degree of underestimate slight in the majority regime, and greater in a minority sub-band. As noted at the beginning of this section, the problem of an electron in the A sub-band becomes equivalent to a classical percolation problem in the limit $\delta \to \infty$, since the electron can neither hop to nor tunnel coherently through the B sites. The number of electronic states lying in extended channels in the material is then given by Np(x), where p(x) is the classical percolation probability for the lattice considered, and vanishes at a critical concentration x_c, which depends somewhat on the choice of lattice. x_c defined in this way should of course correspond to the x_c defined in Figure 3.12 as the asymptote to the δ_A curve. Using the methods of this section we have estimated the number of extended states for a simple cubic nearest neighbor model alloy in three dimensions, and compare this quantity with the percolation theory prediction (Broadbent (1957) Frisch (1962)) in Figure 3.14. Although the predicted x_c is for too low (using F(E) gives $x_c \approx 12\%$, while the exact result is $\approx 30\%$), the two curves are quite similar, a very encouraging sign. F(E) can be seen in Figure 3.14 to underestimate consistently the amount of localization present. Thus the exact 'phase boundary' δ_A in Figure 3.12 will lie everywhere below the plotted δ_A, with its asymptote x_c shifted to the left of the plotted result.

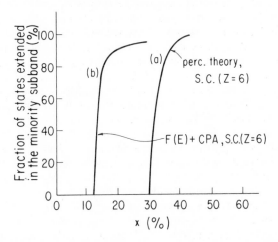

Fig. 3.14. The fraction of the extended states in the minority subband (%) as a function of the minority element concentration for $\delta = \infty$. Curve (a) has been calculated from percolation theory for a simple cubic lattice. Curve (b) is calculated from the approximate localization function F(E) computed within the CPA for simple cubic lattice.

We can now extend the analysis of the character of the states which was made in connection with the fluctuation theory of Section 3.4 to describe this system. The localized states at the bottom of the majority sub-band in, e.g. Figure 3.11c, occupy large isolated regions of pure B material. This interpretation is in agreement with both the low values of the fractional A parentage, n^A, and the low total number of localized states in this region. As we cross the lowest mobility edge, the electron can percolate from one island of pure B material to another opening up extended channels. A considerable portion of the total volume is still inaccessible to these eigenfunctions, as was discussed in Section 3.4. As we proceed towards the middle of the majority sub-band the extended eigenfunctions become more spatially uniform. For still larger energies they again narrow to channels, and at the mobility edge near the top of the majority sub-band, are interrupted by fluctuations as higher A content, isolating localized states. A similar description holds for the minority sub-band as one moves from its top to its bottom; the role of B is played instead by A. This description is in agreement with the behavior, in the minority sub-band of the function $n^A(E)$, which is from 5 to 7 times as great as the concentration of A atoms, and with the total number of localized states in each region.

3.7   ONE-DIMENSIONAL DISORDERED SYSTEMS

3.7.1   Review of Earlier Work

Mott and Twose (1961) were the first to suggest that all the eigensolutions of the Schrödinger equation describing the electronic motion in a one-dimensional disordered system are localized, although the existence of localized states in such systems had been discussed before (Landauer (1954)). These authors using partly intuitive arguments have demonstrated the correctness of their statement for the particular case of a disordered Kronig-Penney lattice. Makinson and Roberts (1962) have found that the eigenstates in the band tails of a one-dimensional liquid are localized. Borland (1961, 1963, 1964) was the first to show in some generality that *all* the eigenstates of disordered one-dimensional systems are localized. Halperin (1967) and Hori (1966) in review articles have critically discussed Borland's proof and have clarified some of the concepts involved. Although Borland's proof is rigorously valid at sufficiently high energies only, it can be easily generalized (Economou (1971c)) for every energy.

In this section we shall discuss briefly Borland's proof, emphasizing the physical aspects of his approach. We then present some new results (Economou (1971c)) on the localization of eigenstates in the one-dimensional case described by a Hamiltonian of the form Eq. (3.22). We conclude by discussing the existence of pseudomobility edges in one dimension, where the localization length changes rather abruptly from small values in the band tails to large values inside the band.

The basic property that distinguishes the one-dimensional case from those of higher dimensionality and the one that is responsible for having all eigenstates localized in one dimension is the unique relation between the ratio of the amplitude of the wave function at two different points and the corresponding phases at these two points, i.e.,

$$\frac{A^2(x_1)}{A^2(x_2)} = \frac{d\phi(x_2)}{d\phi(x_1)} \tag{3.41}$$

where A is the amplitude and ϕ is the phase (Borland (1963)) of the wave function. The derivative $d\phi(x_2)/d\phi(x_1)$ is obtained by comparing the results, at x_2, of integrating the Schrödinger equation from x_1 to x_2 for two different initial conditions $\psi(x_1)$, each with $A(x_1)$, but differing in phase by $d\phi(x_1)$. In general, we expect that the randomness between x_1 and x_2 will cause a loss of correlation in the phases $\phi(x_1)$ and $\phi(x_2)$, as described by the appearance of a finite correlation length in the autocorrelation function. Thus $\phi(x_2)$ becomes quite insensitive to the choice of $\phi(x_1)$ as $|x_1 - x_2| \to \infty$. Hence the RHS of Eq. (3.41) tends to zero as x_2 approaches infinity and consequently $A(x_2) \xrightarrow[A(x_2) \to \infty \text{ as } x_2 \to \infty]{\to \infty}$. This result is a consequence of the fact that the solutions of Schrödinger's equation for almost every energy have an amplitude that approaches infinity as $|x| \to \infty$ in one direction. In this case it is always possible (Halperin (1967)) to find two independent solutions ψ_1, ψ_2 for almost every energy such that

$$|\psi_1(x)| \to \infty, x \to -\infty, \qquad (3.42a)$$
$$\to 0, \quad x \to +\infty,$$

$$|\psi_2(x)| \to 0, \quad x \to -\infty, \qquad (3.42b)$$
$$\to \infty, x \to +\infty.$$

However, for certain energies E_i (which form a dense set when the length of the system considered is infinity) there is a unique solution $\psi_{E_i}(x)$ such that

$$|\psi_{E_i}(x)| \xrightarrow[\text{as } |x| \to \infty]{} 0. \qquad (3.43)$$

Thus E_i are the eigenvalues of the energy and $\psi_{E_i}(x)$ are the eigenstates which are localized according to Eq. (3.43).

Borland's paper is devoted to rigorous mathematical formulation of the physical description presented here. In particular the problem of the disappearance of long-range order in the phase is examined in detail. The formulation is the following: The distribution $p(\phi,x)$ of the phase ϕ at a certain point x can be expressed in terms of the distribution $p(\phi_0,x_0)$ of the phase ϕ_0 at an initial point x_0, i.e.,

$$p(\phi, x) = \int_{-\pi/2}^{\pi/2} K(\phi, \phi_0, x) \, p(\phi_0, x_0) \, d\phi_0, \qquad (3.44)$$

where the kernel K depends on the distributions determining the random potential between x_0 and x. Borland has shown that provided certain conditions are satisfied, as $|x - x_0| \to \infty$, $p(\phi,x)$ converges to a limiting function which is *independent* of the distribution $p(\phi_0,x_0)$ of the initial phase ϕ_0 at the point x_0. This statistical statement is equivalent to asserting that for a given disordered potential the phase ϕ at a point x becomes independent of the phase ϕ_0 at the initial point x_0 as the distance $|x - x_0|$ becomes very large. Thus no matter what the phase is at the point x_0 the phase at the point x will have a value which depends only on the particular form of the disordered potential between x and x_0. Some apparent contradiction arising when one considers the point x as the initial one and the point x_0 as the final one have been discussed and resolved in the literature (Borland (1963) Halperin (1967)). One can easily explain the behavior of the phase by taking into account the fact that the general solution $\psi(x)$ is a linear combination of the functions ψ_1, ψ_2 defined in Eq.(3.42). No matter what the linear combination is at a given point, as $x \to \infty (-\infty)$, $\psi(x) \to \psi_2 (\psi_1)$, and consequently the phase converges towards a unique point.

Borland has also discussed in detail how the eigenfunctions satisfying Eq. (3.43) and the corresponding eigenvalues E_i arise from his analysis by considering a finite system of length L and examing the limit as $L \to \infty$. A slightly different discussion of this point has been given by Halperin (1967).

3.7.2 Some Recent Developments (Economou (1971c))

We shall examine first the localization of an electron in a one-dimensional Anderson model. The basic set of states $|1>$ would then be localized around each site 1 of the periodic one-dimensional lattice. The Hamiltonian is

$$< l |H| m> = \epsilon_l \delta_{lm} + V_{lm}, \tag{3.45}$$

where we assume that ϵ_l are independent random variables and $V_{lm} = V$ if $l = m \pm 1$ and zero otherwise.

As has been stated in section V the renormalized perturbation expression (RPE) for the self-energy $\Delta_o(E)$ converges everywhere except on those portions of the real axis which correspond to extended states. Eq. (3.25) can be written in the present case as

$$\Delta_o(E) = V_{o1} G_1^o V_{1o} + V_{0-1} G_{-1}^o V_{-1o} = V^2 \frac{1}{E - \epsilon_1 - \Delta_1^o}$$

$$V^2 \frac{1}{E - \epsilon_{-1} - \Delta_{-1}^o} \tag{3.46}$$

The quantities Δ_1^o and Δ_{-1}^o can be expanded again using Eq.(3.46) in terms of $\Delta_2^{o,1}$ and $\Delta_2^{o,-1}$ and the same is true for $\Delta_2^{o,1}$, etc. Thus the final expression for $\Delta_o(E)$ has the form

$$\frac{\Delta_o(E)}{V} = \cfrac{1}{\cfrac{E}{V} - \cfrac{\epsilon_1}{V} - \cfrac{1}{\cfrac{E}{V} - \cfrac{\epsilon_2}{V} - \cfrac{1}{\cfrac{E}{V} - \cfrac{\epsilon_3}{V} - \cfrac{1}{\cdots}}}} + \cfrac{1}{\cfrac{E}{V} - \cfrac{\epsilon_{-1}}{V} - \cfrac{1}{\cfrac{E}{V} - \cfrac{\epsilon_{-2}}{V} - \cfrac{1}{\cfrac{E}{V} - \cfrac{\epsilon_{-3}}{V} - \cfrac{1}{\cdots}}}} \tag{3.47}$$

Thus in one dimension the RPE for $\Delta_o(E)$ is the sum of two terms and not an infinite series as in three dimensions. This is an immediate consequence of the fact that there are only two self-avoiding paths starting from site 0, linking nearest neighbors and returning to site 0, while in higher dimensionality there is an infinite number of such paths. This has important consequences for the convergence properties of the RPE for $\Delta_o(E)$. In higher dimensionality than one, the RPE for $\Delta_o(E)$ might diverge because of the divergence of the series or because of the divergence of the iteration procedure implicit in the definition of the RPE for $\Delta_o(E)$. In one dimension in which the series is reduced to a sum of two terms, the only possibility for divergence is through divergence of the iteration procedure. As we shall see, any nonzero degree of randomness is enough to assure the convergence of the iteration procedure everywhere and consequently the localizability of all eigenfunctions. Thus, in the framework of the present approach, the localizability of all eigenfunctions, no matter how small the degree of randomness is, is an immediate consequence of the low connectivity of any one-dimensional lattice, i.e., of the fact that there is only one self-avoiding path linking two different points. This relation between the existence of localized states only and the lack of alternative ways of propagation from one point to another is an extremely desirable feature of the present method, since it provides a quite clear physical explanation for the unique localization properties of the random one-dimensional lattices.

For our purposes the best way to study the convergence of the continued fractions in Eq. (3.47) is by reducing them to a succession of Möbius transformations (Wall (1948)). Let t be the first continued fraction in Eq. (3.46) Then

$$t = W_1(W_2(W_3(\ldots))),$$ (3.48)

where $W_i(z)$ denotes the transformation

$$W_i(z) = \cfrac{1}{\cfrac{E - \epsilon_i}{V} - z}$$ (3.49)

If we terminate the continued fraction at the nth step we obtain

$$t_n = W_1(W_2(\ldots W_n(0)\ldots)).$$ (3.50)

The continued fraction t converges by definition if the $\lim_{n \to \infty} t_n$ exists. A study of the periodic case, where all the ϵ_i have a common value ϵ_o, using the properties of the Möbius transformation (3.49), shows (Economou (1971c)) that the RPE (3.47) diverges - - and consequently the eigenstates are extended - - for $\epsilon_o - 2V < E < \epsilon_o + 2V$. Thus one retrieves the standard result of a band of extended states centered around the common value ϵ_o with a bandwidth equal to 4V.

The case where the quantities ϵ_i are random variables can be treated in a statistical way: The distribution function $P_n(t_n)$ of the quantity t_n defined by Eq. (3.50) is found in terms of the distribution function for the quantities ϵ_i. If the function $P_n(x)$ converges to a limiting function $P(x)$ as $n \to \infty$ then one can show that for almost every set of values $\{\epsilon_i\}$ (except of some of zero total measure) the continued fraction t converges. The problem of convergence of the function $P_n(t_n)$ is mathematically equivalent to that considered by Borland. Using a slight generalization of Borland's proof it can be shown (Economou (1971c)) that the continued fraction t converges everywhere (in the statistical sense explained above) for every set of independent random variables $\{\epsilon_i\}$ provided only that the distribution function $P(\epsilon_i)$ is bounded. This last restriction excludes cases like the random alloy considered in Section 3.6 where $P(\epsilon_i)$ contains δ-functions. However, on physical grounds one expects that the localizability of all eigenfunctions will remain even for unbounded distributions, since the latter can result from bounded distributions by a limiting process during which no discontinuous changes in the observable properties of the system can occur.

It should be noted that the model Hamiltonian (3.45) only approximately describes the motion of an electron in a random potential, and consequently the proof of localizability of the eigenfunctions should not be considered as an alternative proof to that given by Borland. Rather one should view this work as a proof that localization in one-dimensional random systems extends to systems that are not described by a simple potential. E.g., the Hamiltonian (3.45) considered here describes accurately, the atomic vibrations in the case of nearest-neighbor coupling only. When second-neighbor interactions are included, the RPE was found not to terminate, and no proof of complete localization was found. These observations suggest that the localizability of all eigenstates in random one-dimensional lattices is a universal property of wave propagation in such media, when the propagation must occur by a unique path, the same way that the Bloch theorem is a universal property of wave propagation in periodic media (Brillouin 1953).

We conclude this section with a brief discussion of recent work (Williams (1969) Economou (1970c)) indicating that there are energies within the band at which the localization length changes rapidly from small values in the tails to large values in the interior of the band. These energies are the one-dimensional analogues of the mobility edges in three dimensions. In the latter case as we cross the mobility edges the character of the states changes from localized to extended with a *finite* mean free path. In one dimension a finite mean free path always implies localization, as can be seen from Eq. (3.41) with a localization length of the same order of magnitude as the mean free path or the phase coherence length. Thus all the states are localized in one dimension but as a reminder of what would happen if we had a higher dimensionality, the localization length changes rather abruptly at certain energies from values characteristic of the tails to values of the order of magnitude of the mean free path in the interior of the band.

A rather abrupt change in the localization length at certain energies is shown in almost all numerical calculations (Borland (1963) Williams (1969) Makinson (1962)) in which quantities depending on the localization length or characterizing the transport properties of the eigenstates have been plotted as a function of the energy. The abruptness of the change depends of course on the calculated quantity, the degree of randomness, etc. For the cases considered by Williams and Matthews it seems that the localization length changes by two orders of magnitude for energy changes of less than 1% of the bandwidth at energies in the band tails. Preliminary direct calculations (Bush (1971)) of the localization length support the estimations inferred from Williams' and Matthews' work.

It should be noted that Williams' and Matthews' numerical analysis is based upon the introduction of a maximized average momentum, called the characteristic momentum, which determines the transport properties of the eigenstates and depends critically on the localization length. The mobility and the conductivity can be expressed directly in terms of the characteristic momentum, (Economou (1970c)) and using numerical results (Williams (1969)) one sees that the mobility drops by many order of magnitudes within a range of less than 1% of the bandwidth around the critical energies. This property justifies the name pseudomobility edges for these critical energies.

3.8   EDWARDS' PATH INTEGRAL FORMULATION

3.8.1   Introduction and Model

Edwards (1971a, b) has discussed the phenomenon of the localization of electrons in disordered systems for the case of a simple model which is conveniently expressed in terms of Feynman-like path integrals (Feynman (1965)). The model is that of an electron in a system of very dense, random, weak scatterers. If ρ is the density of the scatterers and $v(r)$ is the scattering potential, the model is employed in the limit

$$\lim (\rho \to \infty, v \to 0, \rho v_2 \to \text{finite}). \tag{3.51}$$

Edwards has chosen this limit because in it an individual scatterer cannot bind an electron. Thus all bound states arise solely from the randomness. However, this restriction is unnecessary. Localized states in a dense system arise only from the randomness, irrespective of the strength of the potential. A periodic arrangement of scatterers gives rise to extended states even when the potential of a scatterer is strong enough to bind the electron.

Given such a model, we can ask if there are localized states at energies below a certain critical value, E_c, and extended states at energies above this value. It is clear that the localization phenomenon bears some resemblence to phase changes in statistical mechanics. For example, below the Curie point a ferromagnet has finite magnetization, while for higher temperatures the magnetization vanishes. Except in the region near the critical point, the ferromagnet can be described in terms of the Curie-Weiss mean field. By analogy, we can ask if there could be a similar 'mean field' which localizes electrons below E_c in disordered systems. Edwards' (1962, 1970a) calculation in the high density, weak-scattering limit is of just this mean field variety and indeed gives rise to electron localization as well as directly providing the size of the region of localization. As Edwards' presentation is quite brief and as some of the methods employed are not widely used (Feynman (1965) Gelfand (1960)) we give here an expanded and improved version of his theory with the important details made explicit. First we consider the one electron Green's function $\mathscr{G}(\mathbf{r}\mathbf{r}', t \mid \{\mathbf{R}_j\})$ for an electron in the presence of N scatterers at the fixed positions $\{\mathbf{R}_j \mid j = 1, \dots, N\}$. It is obvious that this one electron Green's function must depend explicitly on the positions of the scatterers. Since the scatterers are randomly distributed, such properties of the system as the density of states are obtained from the average Green's function $G(\mathbf{rr}'; t)$, the average of $\mathscr{G}$ over the random scatterer positions. In the next subsection we consider the path integral representation of G in the limit Eq. (3.51).

For the model of N scatterers, the Hamiltonian for the electron is

$$H\left(\{\mathbf{R}_j\}\right) = -\frac{h^2}{2m}\nabla_{\mathbf{r}}^2 + \sum_{j=1}^{N} v(\mathbf{r} - \mathbf{R}_j), \tag{3.52}$$

where m is the electron mass and v is the scattering potential. The positions $\mathbf{R}_j$ of the scatterers are taken to be random; the probability distribution for the scattering centers is therefore

$$P(\{\mathbf{R}_j\}) = \Omega^{-N} \tag{3.53}$$

where Ω is the volume of the system.

The Schrödinger equation for the one electron causal Green's function is

$$[i\hbar\frac{\partial}{\partial t} - H(\{\mathbf{R}_j\})]\,\mathscr{G}(\mathbf{rr}'; t \mid \{\mathbf{R}_j\}) = \delta(\mathbf{r}-\mathbf{r}')\,\delta(t), \tag{3.54}$$

and depends explicitly on the scatterer positions. The average Green's function is then

$$G(\mathbf{rr}'; t) = \Omega^{-N} \int \ldots \int \prod_{j=1}^{N} d\mathbf{R}_j\,\mathscr{G}(\mathbf{rr}'; t \mid \{\mathbf{R}_j\}) \tag{3.55}$$
$$= <\mathscr{G}>.$$

Since Eq. (3.55) implies that $G(\mathbf{rr}'; t)$ has the general properties of usual Green's functions, e.g., $\lim_{t \to 0} G(\mathbf{rr}'; t) - \delta(\mathbf{r} - \mathbf{r}')$, etc., G can be thought to describe the propagation of a particle, even though it does not correspond to a physical electron in a specific configuration. Approximation techniques which are useful in calculating G for physical particles will also be valuable for this averaged G. In this section, therefore, we speak of G as describing the motion of a fictitious 'average electron' in the averaged system. The density of states is obtained from the Fourier transform of Eq.(3.55) by analytically continuing G for negative times, omitting the step function $\Theta(t)$ which is present in the causal Green's function, and then taking the trace. Thus, we have

$$n(E) = \frac{1}{2\pi\hbar}\int d\mathbf{r}\int_{-\infty}^{\infty} dt\,\exp(i\,E\,t/h)\,G(\mathbf{rr}; t). \tag{3.56}$$

3.8.2 The Path Integral Representation
The Schrödinger Eq. (3.54) can be solved for very short times ϵ to give

$$\mathscr{G}(\mathbf{rr}'; \epsilon \mid \{\mathbf{R}_i\}) \simeq \left(\frac{m}{2i\hbar\epsilon\pi}\right)^{3/2}\exp\left\{\frac{im}{2\hbar\epsilon}(\mathbf{r}-\mathbf{r}')^2 - \frac{i\epsilon}{\hbar}v\left[\frac{1}{2}(\mathbf{r}+\mathbf{r}') - \mathbf{R}_j\right]\right\}$$

$$\tag{3.57}$$

For longer times the propagator can be obtained from Eq. (3.56) by successive propagations for times ϵ_k,

$$\mathscr{G}(\mathbf{rr'};t.|\{\mathbf{R}_i\}) \simeq \prod_{k=1}^{n} \left\{ \int d\mathbf{r}_k\, \mathscr{G}(\mathbf{r}_k\mathbf{r}_k;\epsilon_k\,|\{\mathbf{R}\}) \right\} \int \delta(\mathbf{r}_n - \mathbf{r}) \quad (3.58)$$
$$\mathbf{r}_0 = \mathbf{r'}$$

where

$$t = \sum_{k=1}^{n} \epsilon_k , \qquad (3.59)$$

and $\mathbf{r}_k$ denotes the intermediate position of the electron at $t_k = \Sigma \epsilon_k$. Eq.
(3.58) for small enough $\max_{1 \leqslant k \leqslant n} \epsilon_k k$ represents an accurate 'finite-difference'
approximation to the solution of Eq. (3.54). This approximation becomes
exact in the limit that $\max_{1 \leqslant k \leqslant n} \epsilon_k \to 0$, $n \to \infty$, such that Eq. (3.59) is satisfied.
The resultant solution is the Feynman path-integral representation of the
one electron Green's function (Feynman (1965) Gelfand (1960)).

$$\mathscr{G}(\mathbf{rr'};t\,|\{\mathbf{R}_j\}) = \int_{\mathbf{r}(0)=\mathbf{r'}}^{\mathbf{r}(t)=\mathbf{r}} \mathscr{D}[\mathbf{r}(\tau)] \exp\left\{ \frac{i}{\hbar} \frac{m}{2} \int_0^t d\tau\, \dot{\mathbf{r}}(\tau)^2 \right.$$
$$\left. - \frac{i}{\hbar} \int_0^t d\tau \sum_{j=1}^{N} v[\mathbf{r}(\tau) - \mathbf{R}_j] \right\}. \qquad (3.60)$$

In (3.60) $\mathbf{r}(\tau)$ is the continuous curve which is formed as the limit of the set
of points $\mathbf{r}(t_k) \equiv \mathbf{r}_k$, $\dot{\mathbf{r}}(\tau) = \dfrac{d\mathbf{r}(\tau)}{d\tau}$, and $\mathscr{D}[\mathbf{r}(\tau)]$ is the Feynman measure in
configuration space. The latter is formally infinite, but just as for the Dirac
delta functions, mathematically acceptable transcriptions of the formalism
are available (for a readable account see Mortensen (1969) Schilder (1970)).

As Edwards and Gulyaev (1964) note, the average (3.53) of (3.60) can be
explicitly obtained because the $\{\mathbf{R}_j\}$ and therefore the $v[\mathbf{r}(\tau) - \mathbf{R}_j]$ are
independent random variables, (Jones 1969, Kubo 1962)

$$G(\mathbf{rr'};t) = \int_{\mathbf{r}(0)=\mathbf{r'}}^{\mathbf{r}(t)=\mathbf{r}} \mathscr{D}[\mathbf{r}(\tau)] \exp\left\{ \frac{i}{\hbar} \frac{m}{2} \int_0^t d\tau\, \dot{\mathbf{r}}^2(\tau) \right.$$
$$\left. + \rho \int d\mathbf{R} \left[\exp\left(-\frac{i}{\hbar} \int_0^t d\tau v[\mathbf{r}(\tau) - \mathbf{R}] \right) - 1 \right] \right\}, \qquad (3.61)$$

where $\rho = N/\Omega$ is the density. In the limit Eq. (3.51), we can expand the
exponential of v and only the linear and quadratic terms survive. The
average potential

$$\bar{V} = \rho \int d\mathbf{R}\, v[\mathbf{r}(\tau) - \mathbf{R}] = \rho \int d\mathbf{R}\, v(\mathbf{R}), \qquad (3.62)$$

is infinite in the limit Eq. (3.51). However, we are free to choose our energy origin as the average of energy, removing the infinity in Eq. (3.62). Thus, we are left with only the quadratic term

$$-\frac{\rho}{2\hbar^2} \int d\mathbf{R} \int_0^t d\tau \int_0^t d\tau' \, v \, [\mathbf{r}(\tau) - \mathbf{R}] \, v \, [\mathbf{r}(\tau') - \mathbf{R}]$$

$$= -\frac{\rho}{2\hbar^2} \int_0^t d\tau \int_0^t d\tau' \, W \, [\mathbf{r}(\tau) - \mathbf{r}(\tau')] , \qquad (3.63)$$

where

$$W(\mathbf{r}) = \int \frac{d^3k}{(2\pi)^3} \, \exp \, (i\mathbf{k}\cdot\mathbf{r}) \, |v_k|^2 , \qquad (3.64)$$

and v_k is the Fourier transform of $v(r)$. Thus, in this limit (Edwards (1970a, b) (1964))

$$G(\mathbf{rr}';t) = \int_{\mathbf{r}(0)=\mathbf{r}'}^{\mathbf{r}(t)=\mathbf{r}} \mathscr{D} \, [\mathbf{r}(\tau)] \, \exp \left\{ \frac{i}{\hbar} \, \frac{m}{2} \int_0^t d\tau \, \dot{\mathbf{r}}^2 \, (\tau) \right.$$

$$\left. -\frac{\rho}{2\hbar^2} \int_0^t d\tau \int_0^t d\tau' \, W \, [\mathbf{r}(\tau) - \mathbf{r}(\tau')] \right\} \qquad (3.65)$$

Note that for imaginary time Eq. (3.65) becomes the distribution function for the end vectors of a polymer chain with the segment-segment interaction $-\rho W$; it gives the dimensions of a polymer (Edwards (1965)). For a sufficiently attractive potential, i.e., $-\rho W$ sufficiently negative, the polymer can collapse, while for a repulsive interaction the polymer spreads out in space more rapidly than by a random walk (Edwards (1966)).

The density of states is then given by

$$n(E) = \int_{-\infty}^{\infty} \frac{dt}{2\pi\hbar} \oint \mathscr{D} \, [\mathbf{r}(\tau)] \, \exp \left\{ \frac{i}{\hbar} \, \frac{m}{2} \int_0^t d\tau \, \dot{\mathbf{r}}^2 \, (\tau) - \frac{\rho}{2\hbar^2} \int_0^t d\tau \int_0^t d\tau' \right.$$

$$\left. \times W[\mathbf{r}(\tau) - \mathbf{r}(\tau')] + \frac{i}{\hbar} \, Et \right\} , \qquad (3.66)$$

where the contour integral notation indicates that $\mathbf{r}(0) = \mathbf{r}(t) = \mathbf{r}$ and we take the trace $\int d\mathbf{r}$. We note that

$$G \, (\mathbf{r},\mathbf{r}';t) = G \, (\mathbf{r} - \mathbf{r}';t) \qquad (3.67)$$

because of the translational invariance of the averaged system: there is no preferred origin or direction. In problems of magnetism, such as the Ising model, a vector 'mean field' which is spatially uniform but has a preferred orientation may break the symmetry of the averaged system. The electrons, however, are subject only to scalar potential fields. Thus the simplest symmetry-breaking 'mean field' for this problem is one which specifies an origin, and may result in localization. Such a field can be introduced just as has been done in the Ising model of a ferromagnet by the use of the method of random fields (Gelfand (1960) Siegert (1963)).

3.8.3 The Method of Random Fields

The well known integral

$$w = \int_{-\infty}^{\infty} dx \exp\left(-w^{-1} x^2/2\right) x^2 \Big/ \int_{-\infty}^{\infty} dx \exp\left(-w^{-1} x^2/2\right), \quad (3.68a)$$

can be written in the case of many variables as

$$w_{ij} = \left[\prod_m \left(\int_{-\infty}^{\infty} dx_m\right) x_i x_j \exp\left(-\sum_{k,\ell} x_k w_{k\ell}^{-1} x_\ell\right)\Big/\left[\prod_m \left(\int dx_m\right)\right]\right.$$

$$\exp\left(-\sum_{k,\ell} x_k w_{k\ell}^{-1} x_\ell\right), \quad (3.68b)$$

where w_{ij} is the matrix inverse of w_{ij}^{-1}, $\Sigma_j w_{ij} w_{jk}^{-1} = \delta_{ik}$. Letting the indices, i,k, etc., refer to points r_i, r_j, etc., in space and defining $\phi(r_i) = x_i$, and then passing to the limit of continuous variables, we get (Edwards (1966) (1970c))

$$w(r-r') = \mathcal{N} \int \phi(r)\, \phi(r') \exp\left[-\frac{1}{2}\int dr \int dr'\, \phi(r)\, w^{-1}(r-r')\, \phi(r)\right] \delta\phi, \quad (3.68c)$$

where

$$\mathcal{N}^{-1} = \int \exp\left[-\frac{1}{2}\iint \phi w^{-1} \phi\right]\delta\phi, \quad (3.68d)$$

$$\delta\phi = \prod_r d\,\phi(r), \quad (3.68e)$$

and w and w^{-1} are inverses

$$\int w\,(r-r'')\, w^{-1}\,(r''-r')\, dr'' = \delta\,(r-r'). \quad (3.68f)$$

The $\phi(r)$ occurring in the integrand of Eq. (3.68c) is a Gaussian random function. The identity

$$\exp\left\{-\frac{\rho}{\hbar^2}\int_0^t d\tau \int_0^t d\tau'\, W[r(\tau)-r(\tau')]\right\} =$$

$$= \mathcal{N}\int \delta\phi \exp\left\{-\frac{i}{\hbar}\int_0^t \phi\,[r(\tau)]\, d\tau - \right.$$

$$\left. -\frac{1}{2}\int dr \int dr'\, \phi(r)\frac{1}{\rho}W^{-1}\,[r(\tau)-r(\tau')]\, \phi(r')\right\} \quad (3.69a)$$

with

$$\mathcal{N}^{-1} = \int \delta\phi \exp\left\{-\frac{1}{2\rho}\iint \phi\, W^{-1}\, \phi\right\}, \quad (3.69b)$$

can easily be verified by using the transformation

$$\phi(\mathbf{r}) = \phi'(\mathbf{r}) - i \int_o^t d\tau \frac{\rho}{\hbar} W [\mathbf{r} - \mathbf{r}(\tau)], \delta\phi = \delta\phi' \tag{3.69c}$$

in the right hand side of Eq. (3.69a) and then changing the dummy integration variable ϕ' to ϕ, (Edwards (1965) (1966)).

Introducing Eq. (3.69a) into Eq. (3.65) and changing the orders of the $\mathscr{D} [\mathbf{r}(\tau)]$ and $\delta\phi$ integrations, gives

$$G(\mathbf{rr}'; t) = \mathscr{N} \int \delta\phi G (\mathbf{rr}'; t; [\phi]) \exp\left[-\frac{1}{2\rho} \iint \phi W^{-1} \phi \right], \tag{3.70}$$

$$= < G (\mathbf{rr}'; t; [\phi]) >_\phi$$

where

$$G(\mathbf{rr}'; t; [\phi]) = \int_{\mathbf{r}(0)=\mathbf{r}'}^{\mathbf{r}(t)=\mathbf{r}} \mathscr{D} [\mathbf{r}(\tau)] \exp \left\{ \frac{i}{\hbar} \frac{m}{2} \int_o^t d\tau \, \dot{\mathbf{r}}^2 (\tau) \right.$$

$$\left. -\frac{i}{\hbar} \int_o^t \phi [\mathbf{r}(\tau)] \, d\tau \right\}. \tag{3.71}$$

Comparing Eqs. (3.55) and (3.60) with Eqs. (3.70) and (3.71) reveals that the limit Eq. (3.51) just replaces an average over the positions of random scattering centers by an average over a Gaussian random potential. The latter is easily seen by writing $G([\phi])$ of Eq. (3.71) alternatively as the solution to the Schrödinger equation (Edwards (1970a) Feynman (1965) Gelfand (1960))

$$\left[i\hbar\frac{\partial}{\partial t} + \frac{\hbar^2}{2m}\nabla_\mathbf{r}^2 - \phi(\mathbf{r}) \right] G (\mathbf{rr}'; t; [\phi]) = \delta (\mathbf{r} - \mathbf{r}')\delta (t), \tag{3.72}$$

which describes the motion of an electron in the presence of the external random field $\phi(\mathbf{r})$.

The statistical mechanics of the Curie Weiss mean field or the van der Waals mean field can likewise be discussed by the method of random fields (Siegert (1963) Jalickee (1969)). In these cases the mean field analogous to $\phi(\mathbf{r})$ is a position-independent vector. The existence of this mean field, however, implies the destruction of the isotropy of space, i.e., the breaking of a symmetry. As Edwards (1970a, b) notes, therefore, there must also be a breaking of symmetry in order to obtain electron localization in the translationally invariant averaged system.

In order to see how symmetry is broken in the electron localization problem, (for a discussion of the analogous polymer problem see Freed (1971b)), consider an averaged electron which is created at $\mathbf{r}'$ at time zero and then propagates to $\mathbf{r}$ at t where it is annihilated. In the averaged system, the electron sets up a charge density in space which is obtained from $G(\mathbf{rr}'; f)$. Because the presence of $\int_o^t d\tau \int_o^t d\tau' W[\mathbf{r}(\tau) - \mathbf{r}(\tau')]$ in Eq. (3.65) implies that the effective electron self interaction is nonlocal in time, we could obtain a reasonable approximation to the self-field seen by the electron by taking some functional of W and G. The resulting approximation leads to closed equations if the latter are to be solved self-consistently. It is thus reasonable to expect that there exists a self-consistent field (SCF) theory of electron localization. The symmetry breaking aspect of this SCF arises because we specify in G_{SCF} $(\mathbf{rr}'; t)$ that the electron is initially at $\mathbf{r}'$ and finally at $\mathbf{r}$, thereby introducing two spatial origins [or one origin in the case that $\mathbf{r} = \mathbf{r}'$ in the evaluation of n(E)].

Edwards (1965, 1966) has discussed the SCF theory for the polymer problem, and the details for the case of electron localization are essentially the same. However, Edwards' SCF treatment of the polymer problem was criticized on the grounds both of principle and of the mathematical approximations employed. Recently, Edwards' approach has been shown to follow from very general, sound, physical principles (Freed (1971b)). We therefore can present his SCF theory for the case of electron localization. The development of the SCF theory, although in the spirit of Edwards' polymer approach, avoids the introduction of extraneous mathematical approximations. The resultant equations are sufficiently complicated that the simplifying power of Edwards' approximations can well be appreciated. Nevertheless, it should be possible to obtain better approximations.

We now note that if an SCF theory exists, the SCF Green's function, G_{SCF}, must satisfy the equation of motion

$$\left[i\hbar \frac{\partial}{\partial t'} + \frac{\hbar^2}{2m} \nabla^2_{\mathbf{r}''} - V_{SCF} (\mathbf{r}'' \,|\, \mathbf{rr}'; t) \right] G_{SCF} (\mathbf{r}'' \, \mathbf{r}'''; t' \, t'' \,|\, \mathbf{rr}'; t)$$

$$= \delta (\mathbf{r}'' - \mathbf{r}''') \delta (t' - t''), \tag{3.73}$$

where the explicit parametric dependence of the SCF V_{SCF} and G_{SCF} on the origins $\mathbf{rr}'$; t have been indicated. The V_{SCF} is some, as yet unspecified, functional of W and G_{SCF}. Comparing Eqs. (3.73) with (3.71) and Eq. (3.70) it is clear that V_{SCF} must be that field, call it ϕ_0, which makes the dominant contribution to the averaging in Eq. (3.70). We therefore proceed to find this dominant field.

3.8.4   Derivation of the SCF Theory
Eq. (3.71) can be rewritten as

$$G (\mathbf{rr}'; t) = \mathcal{N} \int \delta \phi \exp \left\{ - B(\mathbf{rr}'; t; [\phi]) \right\} \tag{3.74}$$

where

$$B(\mathbf{rr}'; t; [\phi]) = - \ln G(\mathbf{rr}'; t; [\phi]) + \frac{1}{2\rho} \int d\mathbf{R} \int d\mathbf{R}' \, \phi(\mathbf{R}) \, W^{-1} \, (\mathbf{R} - \mathbf{R}')\phi(\mathbf{R}').$$

$$\tag{37.4a}$$

To calculate the density of states we only need G($\mathbf{rr}$;t). The dominant field ϕ_0 is therefore obtained as the solution to

$$\frac{\delta\, B\,(\mathbf{rr};t[\phi])}{\delta\,\phi(\mathbf{r}'')}\bigg|\,\phi = \phi_{o} = 0.\tag{3.75}$$

Note that since B depends parametrically on $\mathbf{rt}$, ϕ_{o} must also depend on these variables, so we write

$$\phi_{o}(\mathbf{r}'') \rightarrow \phi_{o}(\mathbf{r}''|\mathbf{rt}).\tag{3.76}$$

More explicitly Eq. (3.75) can be written

$$-\frac{1}{G(\mathbf{rr};t;[\phi_{o}])}\,\frac{\delta\,G\,(\mathbf{rr};t;[\phi])}{\delta\phi(\mathbf{r}'')}\bigg|\,\phi = \phi_{o} = \frac{1}{\rho}\int d\mathbf{R}\,W^{-1}\,(\mathbf{r}'' - \mathbf{R})\,\phi_{o}(\mathbf{R}) = 0\tag{3.77}$$

Formally writing

$$G([\phi])\,G([\phi])^{-1} = 1,\tag{3.78a}$$

gives

$$\frac{\delta G([\phi])}{\delta\phi}\,G([\phi])^{-1} + G([\phi])\,\frac{\delta G([\phi])^{-1}}{\delta\phi} = 0,\tag{3.78b}$$

so that

$$\frac{\delta G([\phi])}{\delta\phi} = -\,G([\phi])\,\frac{\delta G([\phi])^{-1}}{\delta\phi}\,G([\phi]).\tag{3.78c}$$

From the equation of motion (3.73)

$$G^{-1}(\mathbf{rr}';tt';[\phi]) = \left[\,i\hbar\frac{\partial}{\partial t} + \frac{\hbar^{2}}{2m}\nabla_{\mathbf{r}}^{2} - \phi(\mathbf{r})\,\right]\delta(\mathbf{r} - \mathbf{r}')\,\delta(t - t'),\tag{3.79}$$

so that

$$\frac{\delta\,G^{-1}\,(\mathbf{rr}';tt';[\phi])}{\delta\,\phi\,(\mathbf{r}'')} = -\,\delta(\mathbf{r} - \mathbf{r}'')\,\delta\,(\mathbf{r} - \mathbf{r}')\,\delta\,(t - t'),\tag{3.80}$$

and therefore

$$\frac{\delta\,G\,(\mathbf{rr};t;[\phi])}{\delta\,\phi(\mathbf{r}'')}\bigg|\,\phi = \phi_{o} = \int_{o}^{t} d\tau\,G\,(\mathbf{rr}'';t\tau,[\phi_{o}])\,G\,(\mathbf{r}''\mathbf{r};\tau 0;[\phi_{o}]).\tag{3.81}$$

Substituting Eq. (3.81) into (3.77) and applying $\rho\int d\mathbf{r}''\,W(\mathbf{r}' - \mathbf{r}'')$ on the left gives finally

$$\phi_{o}(\mathbf{r}'\,|\mathbf{r};t) = \rho\int d\mathbf{r}''\,W\,(\mathbf{r}' - \mathbf{r}'')\int_{o}^{t} d\tau\,G(\mathbf{rr}'';t\tau;[\phi_{o}])$$

$$\times\,G\,(\mathbf{r}''\mathbf{r};\tau 0;[\phi_{o}])\,/\,G(\mathbf{rr};t;[\phi_{o}])\tag{3.82}$$

$$\equiv V_{SCF}\,(\mathbf{r}'\,|\mathbf{r};t).$$

The auxiliary propagators entering Eq. (3.82) for only a portion of the time interval t must obey the equation of motion (3.72) *for the particular value of the random field* ϕ_0!

Thus, defining e.g.

$$G_{SCF} (r''r; \tau 0|r;t) = G(r''r; \tau 0; [\phi_0])$$ (3.83)

The closed pair of Eqs. (3.72) for $\phi \to \phi_0$ and (3.82) leads to the SCF theory (3.73), (3.82), and (3.83).

Eq. (3.72) with (3.82) represents a rather complicated self-consistency problem. It is, of course, quite evident that $V_{SCF}(r'|r;t)$ or its Fourier transform $V_{SCF}(r'|r;E)$ could be primarily attractive or repulsive, leading to localized or extended states respectively. This potential is in general expected to be complex for reasons which are given in more detail below. Note that in this context 'states' are obtained from the residues of the averaged Green's function at poles or along cuts. It is convenient to consider these states as the physical states of the self-interacting electron in the averaged environment.

Before discussing Edwards' asymptotic solution to Eq. (3.72) with (3.82) - - obtained for the polymer case - - it is convenient to discuss his simpler mean field theory, in which the mean-field is constant in certain regions of space (Edwards (1970a, b)).

3.8.5 Edwards' Mean-Field Theory

Consider a situation in which the SCF is attractive and leads to bound states. In this case the potential would have a well in some region of space surrounded by effectively inpenetrable barriers. The simplest such potential is the square well potential. However, in order to include the possibility of the existence of localized states as well as extended states, we must allow the size of this cubic square well potential to be either finite or infinite. As an approximation to the functional integral Eq. (3.70) − (3.71) we therefore restrict the random field $\phi(r)$ to the class of cubic square well potentials of magnitude ϕ_0 inside the cube of side L and infinite outside. By analogy with the choice of the dominant field in Eq. (3.75), we can pick the value of L which gives the dominant contribution to Eq. (3.70). Thus, we take

$$\phi(r) \xrightarrow[\text{field}]{\text{mean}} \begin{cases} \phi_0 & r \in L^3 \\ \infty & r \notin L^3 \end{cases}.$$ (3.84)

Neglecting all other possible fields $\phi(r)$ and any contributions from fluctuations in L about its dominant value, the approximation implies

$$\delta\phi \xrightarrow[\text{field}]{\text{mean}} d\phi_0.$$ (3.85)

Rigorously, in using the restriction Eq. (3.84) on the random variable Eqs. (3.84), (3.85) should be

$$\delta\phi \xrightarrow[\text{field}]{\text{mean}} d\phi_0 \, df(L),$$ (3.85a)

where df(L) is the measure for a particular L. In the spirit of the SCF theory we can assume this measure df(L) to be slowly varying and just take the dominant contribution L_o. Since the potential Eq. (3.84) admits the symmetry breaking which leads to localized states, L_o can be chosen to give the dominant contribution to the density of states n(E,L). This is to be contrasted with the SCF theory in which it was necessary to choose the dominant ϕ for G(**rr**;t) in order to include the appropriate symmetry breaking.

We now require the following integrals:

Inside L^3,

$$\int_o^t \phi\,[\mathbf{r}(\tau)]\,d\tau \xrightarrow[\text{field}]{\text{mean}} \int_o^t \phi_o\,d\tau = \phi_o^t, \qquad (3.86)$$

while outside L^3,

$$\int_o^t \phi\,[\mathbf{r}(\tau)]\,d\tau \xrightarrow[\text{field}]{\text{mean}} \infty \qquad (3.87)$$

Therefore defining

$$\cup [\mathbf{r}(\tau)] = \begin{cases} o & \mathbf{r}\epsilon L^3 \\ \infty & \mathbf{r}\notin L^3 \end{cases} \qquad (3.88)$$

as the usual cubic box square well potential,

$$\int_o^t \phi\,[\mathbf{r}(\tau)]\,d\tau \xrightarrow[\text{field}]{\text{mean}} \phi_o t + \int_o^t \cup [r(\tau)]\,d\tau. \qquad (3.89)$$

Inside L^3

$$\int_{L^3} d\mathbf{r} \int_{L^3} d\mathbf{r}'\,\phi\,(\mathbf{r})\,W^{-1}\,(\mathbf{r}-\mathbf{r}')\phi(\mathbf{r}') \xrightarrow[\text{field}]{\text{mean}} \int_{L^3} d\mathbf{r} \int_{L^3}\,d\mathbf{r}'\,\phi_o^2\,W^{-1}\,(\mathbf{r}-\mathbf{r}')$$

$$\sim L^3\,\phi_o^2\,w^{-1}, \qquad (3.90)$$

where

$$w^{-1} = \int_{L^3} d\mathbf{r}\,W^{-1}\,(\mathbf{r}).$$

The infinite contributions in $\iint \phi W^{-1}\phi$ from outside L^3 exactly cancel in both numerator and denominator of Eq. (3.70). Substituting Eq. (3.84) through (3.90) into (3.70) and (3.71) leads to

$$\mathcal{N} \int \delta\phi \exp\left\{ -\frac{i}{\hbar} \int_0^t \phi\, [\mathbf{r}(\tau)]\ d\tau - \frac{1}{2\rho} \iint \phi\, W^{-1}\, \phi \xrightarrow{\text{mean field}} \right.$$

$$\frac{\int d\phi_0 \exp\left\{ -\frac{i}{\hbar} t\, \phi_0 - \frac{1}{2\rho} \phi_0^2\, L^3\, w^{-1} - \frac{i}{\hbar} \int_0^t \cup [\mathbf{r}(\tau)]\ d\tau \right\}}{\int d\phi_0 \exp\left\{ -\frac{1}{2\rho} \phi_0^2\, L^3\, w^{-1} \right\}}$$

$$= \exp\left\{ -\frac{i}{\hbar} \int_0^t \cup [\mathbf{r}(\tau)]\ d\tau - \frac{t^2\, \rho\, w}{2\,\hbar^2\, L^3} \right\} \tag{3.91}$$

The latter follows by elementary integration. Substituting Eq. (3.91) into (3.70) and (3.71) gives

$$G_{MF}(\mathbf{r}\mathbf{r}';t) = \int_{\mathbf{r}(0)=\mathbf{r}'}^{\mathbf{r}(t)=\mathbf{r}} \mathscr{D}\, [\mathbf{r}(\tau)] \exp\left\{ \frac{i}{\hbar} \int_0^t\ d\tau \left(\frac{m}{2}\, \dot{\mathbf{r}}^2(\tau) - \cup [\mathbf{r}(\tau)] \right) \right.$$

$$\left. \frac{t^2\rho w}{2\hbar^2 L^3} \right\}. \tag{3.92}$$

Note that apart from the $\mathbf{r},\mathbf{r}'$ independent exponential factor, Eq. (3.92) is just the Green's function for a particle in a cubic box of side L. (Edwards (1970b) (1969)). Thus, ignoring the step function θ (t) of the causal Green's function, Eq. (3.92) is given exactly by

$$G_{MF}(\mathbf{r}\mathbf{r}';t) = \sum_n \psi_n(\mathbf{r})\, \psi_n(\mathbf{r}') \exp\left\{ -\frac{i}{\hbar} \frac{\hbar^2\, \pi^2\, t}{2m\, L^2}\, n^2 - \frac{t^2\rho w}{2\hbar^2\, L^3} \right\}, \tag{3.93}$$

where ψ_n are the usual box eigenfunctions which are properly normalized and $n^2 = n_x^2 + n_y^2 + n_z^2$. For low enough energies it is only necessary to consider the lowest eigenvalue $n^2 = 3$. Substituting Eq. (3.93) into (3.55) yields

$$n(E) = \frac{1}{2\pi\hbar} \int_{-\infty}^{\infty} dt \exp\left\{ -\frac{it}{\hbar} \left[\frac{3\hbar^2\, \pi^2}{2m\, L^2} - E \right] - \frac{t^2\rho w}{2\hbar^2\, L^3} \right\}$$

$$= \left[\frac{L^3}{2\pi\rho\, w} \right]^{1/2} \exp\left\{ -\frac{\left[\frac{3\hbar^2\, \pi^2}{2m\, L^2} - E \right]^2 L^3}{2\rho w} \right\} \tag{3.94}$$

The dominant L is obtained as that which maximizes the exponential term. This procedure is analogous to the minimization of the 'free energy' — In n(E) in thermodynamic systems; however, it does not rest upon such fundamental principles. The result is

$$L_o = \left\{ \frac{2m(-E)}{\pi^2\hbar^2} \right\}^{-1/2}, \tag{3.95}$$

so that finally

$$n(E) = \pi \hbar \left[\frac{\hbar}{2\rho_w \, [-2m \, E]^{3/2}} \right]^{1/2} \exp \left[-\frac{8 \, (-E)^{1/2}}{\rho \, w} \left(\frac{\pi^2 \hbar^2}{2m} \right)^{3/2} \right]. \quad (3.96)$$

As Edwards (1970a, b) notes Eq. (3.83) is only valid for $E < 0$, while for $E = 0$, $L_o = \infty$. Since for $E > 0$ L_o is complex, the usual methods based upon extended states should then be used. Eq. (3.96) is the well known exponential tail of the density of states in a disordered system which arises from the localized states.

Having shown in a simple approximation how localized states may appear, we now return to the more general SCF theory.

3.8.6 Asymptotic Solution to the SCF Theory

For the polymer problem, Edwards (1965, 1966) obtained the asymptotic SCF solution by first solving the analog of Eq. (3.72) asymptotically (effectively for large t) by WKB approximations (assuming a spherically symmetric field). Using the approximate $G([\phi])$ so obtained in Eqs. (3.75) − (3.76), Edwards (1965, 1966) obtained an approximate SCF solution. We avoid extraneous mathematical approximations by taking $W = w\delta \, (\mathbf{r} - \mathbf{r}')$ for simplicity. In order to make the correspondence between the polymer solution and the present problem of electron localization it is convenient to quote the Green's function for a self-interacting polymer chain of contour length L and average bond length l; (Edwards (1965) (1966) Freed (1971b))

$$G \, (\mathbf{r}\mathbf{r}';L) = \int_{\mathbf{r}(0) = \mathbf{r}'}^{\mathbf{r}(L) = \mathbf{r}} \mathscr{D} \, [\mathbf{r}(s)] \, \exp \left\{ -\frac{3}{2l} \int_o^L ds \, \dot{\mathbf{r}}^2 \, (s) \right.$$

$$\left. -\frac{1}{2l^2} \int_o^L ds \int_o^L ds' \, v \, [\mathbf{r}(s) - \mathbf{r}(s')] \right\}. \quad (3.97)$$

The correspondence between Eqs. (3.97) and (3.64) is then

$$\frac{3}{2l} \longleftrightarrow \frac{m}{2\hbar^2} \, ; \, t \longleftrightarrow - \, i\hbar L; \, \rho \, w \longleftrightarrow - \, v/l^2. \quad (3.98)$$

Finally noting that the SCF polymer Green's function is of the form of a Markov process,

$$G_{SCF}(\mathbf{r}\mathbf{r}';L) = \int_{\mathbf{r}(0) = \mathbf{r}'}^{\mathbf{r}(L) = \mathbf{r}} \mathscr{D} \, [\mathbf{r}(s)] \, \exp \left\{ -\int_o^L ds \left(\frac{3}{2} \dot{\mathbf{r}}^2 \, (s) + V_{SCF} \, [\mathbf{r}(s)|\ldots] \right) \right\},$$

the final correspondence is   (3.99)

$$V_{SCF} \text{ (polymer)} \longleftrightarrow V_{SCF} \text{ (electron)}. \quad (3.100)$$

Edwards' (1965, 1966) asymptotic solution for V_{SCF} (polymer) is

$$V_{SCF}\text{ (polymer)} = \left(\frac{l}{6}\right)^{1/3} \left(\frac{v}{4\pi l^2}\right)^{2/3} \frac{1}{r^{4/3}} \quad (L \to \infty) \tag{3.101}$$

in the region where V_{SCF} (polymer) is large (we ignore the tails here). Making the correspondence Eqs. (3.98) and (3.100) we have the electron SCF (the origin of V_{SCF} is the initial and final electron's position; r measures the distance from this origin).

$$V_{SCF}(\mathbf{r}) = \left(\frac{\hbar^2}{2m}\right)^{1/3} \left[\frac{-\rho w}{4\pi}\right]^{2/3} r^{-4/3}$$

$$= -(1 - i\sqrt{3})\frac{1}{2}\left(\frac{\hbar^2}{2m}\right)^{1/3}\left[\frac{\rho w}{4}\right]^{2/3} r^{-4/3} \quad (t \to \infty \text{ or } E \to 0).$$

$$\tag{3.102}$$

Note that $V_{SCF}(\mathbf{r})$ is attractive and hence gives rise to complex localized 'states' for $E < 0$ and extended 'states' for $E > 0$. V_{SCF} and its eigenstates are complex as a consequence of the averaging over a distribution of configuration-dependent energies. As a simple example of this effect, one could consider a single bound state with energy E. If one were to average its wave function e^{iEt} over a Lorentzian distribution of energies of width Γ, centered at E_o, such as might result in an ensemble of different configurations, the result, $e^{iE_ot - \Gamma t}$, has a 'width' Γ due solely to randomness. V_{SCF} is strongly attractive for small r and therefore it is expected that there be 'resonance-like states' for small $E > 0$. The occurrence of these resonances is easily seen by considering the s-wave states of Eq. (3.102).

$$\left[-\frac{\hbar^2}{2m}\frac{1}{r^2}\frac{d}{dr}r^2\frac{d}{dr} + C/r^{4/3} - E\right]\psi(r;E) = 0, \tag{3.103a}$$

where

$$C = -(1 - i\sqrt{3})\frac{1}{2}\left(\frac{\hbar^2}{2m}\right)^{1/3}\left[\frac{\rho w}{4\pi}\right]^{2/3}. \tag{3.103b}$$

As the $r^{-4/3}$ potential is not commonly encountered in quantum mechanics, the transformation $\chi(r;E) = \psi(r;E)/r$, $r = y^3$, $\chi(r;E) = y\, I(y;E)$ gives

$$\left[-\frac{\hbar^2}{18m}\frac{d^2}{dy^2} + C - E y^4 - \frac{\hbar^2}{18my^2}\right] I(y;E) = 0. \tag{3.104}$$

Thus, the solutions are analogous to the case of a quartic oscillator of strength $-E$, *which can be either positive or negative,* with a small attractive

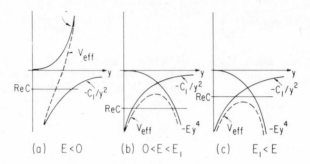

Fig. 3.15. The effective potential ($C_1 = h^2/18m$:
  a) for $E < 0$ it is attractive and the states are localized.
  b) For $0 < E < E_1$ it is attractive around the origin and repulsive for large y and the states are 'resonance' extended states.
  c) For $E_1 < E$ the potential is negative and the states are extended.

centrifugal 'barrier' and complex 'energy' C. There are three cases $E < 0$, $0 < E \dfrac{3}{8\pi^2} \left[\dfrac{m}{2h^2} \right] [\rho w]^2 = E_1$ and $E > E_1$, as shown in Figure 3.15.

Thus, the nature of the 'effective wave functions', i.e., residues of the average Green's function, change from localized to extended - - as we pass from $E < 0$ to $E > 0$, with a transition region of resonance states for $0 < E < E_1$. The existence of these resonance states might be suggestive of a smooth change in the conductivity through the transition region $0 < E < E_1$ from zero conductivity for the bound states $E < 0$ to the high conductivity of the extended states for $E > E_1$. However, it is completely hazardous to make inferences about the conductivity from the residues of the average Green's function, since the latter could differ substantially from the residues of the average of the tensor product of two Green's functions in the above transition region. It is therefore of considerable interest to consider the conductivity within the SCF theory and also to investigate improved solutions for V_{SCF} and G_{SCF}.

As we have noted above, Edwards' asymptotic solution for the SCF Eq. (3.101) is only approximate and improved solutions should be sought. However, the $-4/3$ power law in Eq. (3.101) does give the 'critical exponent' 6/5 for the mean square end-to-end distance of a self-avoiding random walk in agreement with the extrapolations obtained from lattice calculations (Domb (1963)). Thus, the use of Eq. (3.102) is justified in order to obtain the qualitative results of this subsection.

3.8.7 The Fluctuations (Edwards 1965, 1966)

The SCF theory of electron localization only considers the dominant contribution to the averaging in Eq. (3.70). However, the full use of Eq. (3.70) can give information as to the range of validity of such an SCF theory.

Expanding $B(\mathbf{rr};t;[\phi])$ in Eq. (3.74) in a functional Taylor series in $\bar{\phi} = \phi - \phi_0$ gives upon substitution into Eq. (3.70)

$$G(\mathbf{rr};t) = \mathscr{N} \int \delta\bar{\phi} \exp \left\{ - B(\mathbf{rr};t;[\phi_0]) \right.$$

$$-\frac{1}{2} \int d\mathbf{r}' \int d\mathbf{r}'' \, \bar{\phi}(\mathbf{r}') \left[\frac{\delta^2 B(\mathbf{rr};t;[\phi_0])}{\delta\phi(\mathbf{r}')\,\delta\phi(\mathbf{r}'')} \right]_{\phi=\phi_0} \bar{\phi}(\mathbf{r}'')$$

$$\left. + \mathscr{O}([\bar{\phi}]^3) \right\}$$

$$\cong G(\mathbf{rr};t;[\phi_0]) \left[\det \left\{ \rho W \frac{\delta^2 B(\mathbf{rr};t\,[\phi_0])}{\delta\phi_0^2} \right\}^{-1/2} \right. \qquad (3.105)$$

where cubic and higher terms have been neglected. The det in Eq. (3.105) implies the Fredholm determinant of the operator in brackets which has parametric dependence upon r,t. The evaluation of this determinant requires, of course, a suitable solution for $G_{SCF} = G([\phi_0])$.

3.8.8 Conductivity in the Simple Mean Field Theory

As noted above, the SCF theory of the conductivity deserves further attention. Nevertheless in the case of the simple mean field theory of Section 3.8.5, Edwards (1970b) notes that the conductivity is identically zero for the localized states of $E < 0$. The details proceed rather simply as before. In the Kubo-Greenwood-Peierls (Kubo (1956) (1957)) expression for the conductivity, it is necessary to obtain the average of the product of two Green's functions

$$< \mathscr{G}(\mathbf{xx}';t_1 | \{\mathbf{R}_j\}) \, \mathscr{G}(\mathbf{yy}';t_2 | \{\mathbf{R}_j\}) > \qquad (3.106)$$

In the simple limit of Eq. (3.50) this can be shown to yield the path integral (Edwards (1970b)).

$$\int_{x(0)=x'}^{x(t_1)=x} \mathscr{D}[x(\tau)] \int_{y(0)=y'}^{y(t_2)=y} \mathscr{D}[y(\tau)] \exp\left\{ \frac{i}{\hbar} \int_0^{t_1} \frac{m}{2} \dot{x}^2(\tau)\,d\tau \right.$$

$$+\frac{i}{\hbar} \int_0^{t_2} \frac{m}{2} \dot{y}^2(\tau)\,d\tau \, \frac{-\rho}{2\hbar^2} \int_0^{t_1} \int_0^{t_1} W[x(\tau) - x(\tau')]\,d\tau\,d\tau'$$

$$-\frac{\rho}{\hbar^2} \int_0^{t_1} \int_0^{t_2} W[x(\tau) - y(\tau')]\,d\tau\,d\tau'$$

$$\left. -\frac{\rho}{2\hbar^2} \int_0^{t_2} \int_0^{t_2} W[y(\tau) - y(\tau')] \qquad d\tau\,d\tau' \right\} \qquad (3.107)$$

Using the method of random fields Eq. (3.107) can be written as

$$\mathscr{N} \int \delta\phi \int_{x(0)=x'}^{x(t_1)=x} \mathscr{D}\,[x(\tau)]\,\exp\left\{\frac{i}{\hbar}\int_0^{t_1}\frac{m}{2}\,\dot{x}^2\,(\tau)\,d\tau - \frac{i}{\hbar}\int_0^{t_1}\phi\,[x(\tau)]\,d\tau\right.$$

$$\times \int_{y(0)=y'}^{y(t_2)=y} \mathscr{D}\,[y(\tau)]\,\exp\left\{\frac{i}{\hbar}\int_0^{t_2}\frac{m}{2}\,\dot{y}^2\,(\tau)\,d\tau - \frac{i}{\hbar}\int_0^{t_2}\phi\,[y(\tau)]\,d\tau\right\}$$

$$\times \exp\left\{-\frac{1}{2\rho}\int d\mathbf{r}\int d\mathbf{r}'\,\phi\,(\mathbf{r})\,W^{-1}\,(\mathbf{r}-\mathbf{r}')\,\phi\,(\mathbf{r}')\right\}$$

$$= <G(xx';t_1;[\phi])\,G(yy';t_2;[\phi])>_\phi. \tag{3.108}$$

Eq. (3.108) can be verified by using the transformation

$$\phi(\mathbf{r}) = \phi'(\mathbf{r}) - i\int_0^{t_1} d\tau\frac{\rho}{\hbar}\,W[\mathbf{r}-x(\tau)] - i\int_0^{t_2}\frac{\rho}{\hbar}\,W[\mathbf{r}-y(\tau)]\,d\tau, \tag{3.109}$$

with $\delta\phi = \delta\phi'$, and then changing the dummy integration variable from ϕ' to ϕ.

If we now introduce the mean field approximation Eq. (3.84) and (3.85), the approximation to Eq. (3.108) becomes

$$G_W\,(xx';t_1)\,G_W\,(yy';t_2)\,\exp\left\{-\frac{(t_1+t_2)^2\,\rho w}{2\hbar^2\,L^3}\right\}, \tag{3.110}$$

where G_W denotes the one electron Green's function for the square well potential (3.88). The conductivity is obtained from the Fourier transform of the current-current matrix elements of Eq. (3.10). For the dominant L finite (i.e., $E < 0$), the spectrum of G_W is discrete and the wave functions are real, so the current matrix elements must vanish (Edwards (1970b)). Thus, for $E < 0$, the conductivity is identically zero in this mean field approximation (Edwards (1970b)). We require, however, a more sophisticated theory than this simple mean field version to obtain information about the detailed behavior of the mobility above the mobility edge.

3.9 CONCLUSIONS AND PROBLEMS OUTSTANDING

The basic thesis of this review is that the disordered materials show certain universal features in their electronic structure. They are best summarized in the Mott-CFO model: there are bands of allowed states; within the bands critical energies — mobility edges — occur at which the nature of the states changes abruptly from localized in the tails of the band to extended in the interior of the band; bands may overlap; although all the extended states have the common characteristic that they extend to infinity, they differ vastly among themselves regarding the volume they fill; the mobility is zero in the region of localized states and increases to positive values in the region of extended states — probably without any discontinuity.

This point of view was substantiated in several ways: By partly intuitive arguments it was shown that localized states are associated with potential fluctuations and that they give rise to tails in the density of states. Anderson's tight binding model was considered next and within its framework the existence of mobility edges was derived from first principles. Approximate quantitative expressions were obtained for the positions of the mobility edges. The case of random binary alloys was considered next as an application of the general theory, and the results were in striking agreement with the expectations of the Mott-CFO model as well as with independent calculations in the limiting cases where the latter were available.

The electronic structure in a one-dimensional disordered potential, a very special, but pedagogically interesting case, was also considered. Two different proofs were given for the basic theorem that all the eigenstates in a random one-dimensional system are localized. Rather abrupt changes of the localization length seem to occur at critical energies termed pseudomobility edges, which divide the energy spectrum into regions of short localization length (tails of the bands) and long localization length (interior of the band).

Edwards' beautiful and promising approach, utilizing functional integral techniques, was reviewed in detail, emphasizing his derivation of the existence of localized states through the introduction of a symmetry-breaking local self-consistent field.

Although the work reviewed in this article represents considerable progress in characterizing the existence of the proposed, universal features of disordered systems, more challenging problems of calculating experimental quantities dependent on the details of the eigenstates remain. Of these, the experimental quantity of most immediate importance appears to be the mobility of the resonant and channel extended states at energies just above a mobility edge.

Two theoretical programs appear to promise useful information on these difficult questions. The first is the extension of Edwards' techniques for direct evaluation of the necessary two particle correlation functions, as sketched in the preceding section. The second approach is to exploit the connection to percolation theory, established in Sections 3.4 and 3.6 using classical statistical methods to characterize the allowed regions and thus the eigenstates occupying them. The drawback to such a calculation, at present, is its semiclassical nature, and Edwards' methods remain the only fully quantum mechanical treatment of highly disordered systems.

Once the general picture of disordered systems has been filled in in these respects it should be straightforward to consider such transport properties at non-zero temperature and frequency as the ac conductivity and optical properties, and the more complicated classical transport coefficients, such as the Hall coefficient and thermopower. It will then become possible to deal more accurately with the properties which distinguish the different classes of disordered systems: the nature of the regions which trap localized states, effects of charges and external fields, and the energetics of the observed structures. Finally, it remains to incorporate the effects of electron-electron correlations explicitly into this model. Description of all these properties, whenever possible from a microscopic theory, should pose a significant challenge to solid state physics for some time to come.

REFERENCES

Anderson, P. W., (1958), *Phys. Rev.*, **109**, 1492.
Austin, I. G. and Mott, N. F., (1969), *Adv. Phys.*, **18**, 41.
Banyai, L., (1964), Physique de Semi-Conducteurs, p. 417. Dunod, Paris.
Bloch, F., (1928), *Z. Physik*, **52**, 555.
Bonch-Bruevich, V. L., (1964), *Fizika tuerd. Tela*, **5**, 1852. (Translation: (1964) *Soviet Phys. Solid St.*, **5**, 1353).
Borland, R. E., (1961), *Proc. Phys. Soc.*, **78**, 926.
Borland, R. E., (1963), *Proc. R. Soc.*, **A274**, 529.
Borland, R. E. and Bird, N. F., (1964), *Proc. Phys. Soc.*, **83**, 23.
Brillouin, L., (1953), Wave Propagation in Periodic Structures, 2nd ed. Dover, New York.
Broadbent, S. R. and Hammersley, J. M., (1957), *Proc. Cambridge Phil. Soc.*, **53**, 629.
Brodsky, M. H. and Title, R. S., (1969), *Phys. Rev. Letters*, **23**, 581.
Brodsky, M. H., (1971), *J. Vac. Science and Technology*, **8**, 125.
Brouers, F., (1970), *J. Non-Cryst. Solids*, **4**, 428.
Cohen, M. H., Fritzsche, H. and Ovshinsky, S. R., (1969), *Phys. Rev. Letters*, **22**, 1065.
Cohen, M. H., (1970), *J. Non-Cryst. Solids*, **4**, 391.
Coopersmith, M. H., (1965), *Phys. Rev.*, **139**, 1359.
Coopersmith, M. H. and Neustadter, H. E., (1967), *Phys. Rev.*, **161**, 168.
Cutler, M. and Mott, N. F., (1969), *Phys. Rev.*, **181**, 1336.
Davis, E. A. and Mott, N. F., (1970), *Phil. Mag.*, **22**, 903.
Domb, C., (1963), *J. Chem. Phys.*, **38**, 2957.
Donovan, T. M., Spicer, W. E., Bennett, J. M. and Ashley, E. J., (1970a), *Phys. Rev.*, **B2**, 397.
Donovan, T. M., Ashley, E. J. and Spicer, W. E., (1970b), *Physics Letters*, **32A**, 85.
Economou, E. N., Kirkpatrick, S., Cohen, M. H. and Eggarter, T. P., (1970a), *Phys. Rev. Letters*, **25**, 520.
Economou, E. N. and Cohen, M. H., (1970b), *Phys. Rev. Letters*, **25**, 1445.
Economou, E. N. and Cohen, M. H., (1972), *Phys. Rev.*, **5B**, 2931.
Economou, E. N., (1971a), *Solid State Comm.*, **9**, 1317.
Economou, E. N. and Cohen, M. H., (1971b), *Phys. Rev.*, **4B**, 396.
Economou, E. N. and Cohen, M. H., (1970c), *Phys. Rev. Letters*, **24**, 218.
Economou, E. N. and Papatriantafillou, C., (1972b) *Solid State Comm.*, **11**, 197.
Edwards, S. F., (1961), *Phil. Mag.*, **6**, 617.
Edwards, S. F., (1962), *Proc. R. Soc.*, **A267**, 518.
Edwards, S. F., (1970a), *J. Phys.*, **C3**, L30.
Edwards, S. F., (1970b), *J. Non-Cryst. Solids*, **4**, 417.
Edwards, S. F. and Gulyaev, Y. B., (1964), *Proc. Phys. Soc. (London)*, **83**, 495.
Edwards, S. F., (1965), *Proc. Phys. Soc. (London)*, **85**, 613.
Edwards, S. F., (1966), *Natl. Bur. Std. (U.S.) Misc. Publ.*, **273**, 225.
Edwards, S. F. and Freed, K. F., (1970c), *J. Phys.*, **C3**, 739.
Edwards, S. F. and Freed, K. F., (1969), *J. Phys.*, **A2**, 145.
Eggarter, T. P. and Cohen, M. H., (1970), *Phys. Rev. Letters*, **25**, 807.

Feynman, R. P. and Hibbs, A. R., (1965), Quantum Mechanics and Path Integrals, McGraw-Hill, New York.

Fröhlich, H., (1947), *Proc. Roy. Soc. (London)*, **A188**, 521.

Foldy, L. L., (1945), *Phys. Rev.*, **67**, 107.

Freed, K. and Cohen, M. H., (1971a), *Phys. Rev.*, 3 3, 3400.

Freed, K. F., (1971b), *J. Chem. Phys*, **55**, 3910.

Fritzsche, H., (1971), present volume.

Frisch, H. L. and Lloyd, S. P., (1960), *Phys. Rev.*, **120**, 1175.

Frisch, H. L. and Hammersley, J. M., (1963), *S.I.A.M.*, **11**, 894.

Frisch, H. L., Sonnenblick, E., Vyssotsky, V. A. and Hammersley, J. M., (1961), *Phys. Rev.*, **124**, 1021.

Gelfand, I. M. and Taglow, A. M., (1960), *J. Math. Phys.*, **1**, 48.

Grigorivici, R., (1968), *Materials Res. Bulletin*, **3**, 13.

Gubanov, A. I., (1963), Quantum Electron Theory of Amorphous Conductors, Academy of Sciences Press, Moscow (Translation: Consultants Bureau, New York, 1965).

Halperin, B. I. and Lax M., (1966), *Phys. Rev.*, **148**, 722.

Halperin, B. I., (1967), *Adv. Chem. Phys.*, **13**, 123.

Hori, J., (1966), Progr. Theor. Phys., Suppl. No. 36, 3.

Jalickee, J. B., Siegert, A. J. F. and Vezzetti, D. J., (1969), *J. Math. Phys.*, **10**, 1442.

Jones, R. and Lukes, T., (1969), *Proc. Roy. Soc.*, **A309**, 457.

Kane, E. O., (1963), *Phys. Rev.*, **131**, 79.

Kirkpatrick, S., Velicky, B. and Ehrenreich, H., (1970), Phys. Rev., **B1**, 3250.

Kirkpatrick, S., Economou, E. N., Cohen, M. H. and Papatriantafillou, C., (1971) unpublished.

Kubo, R., (1962), *J. Phys. Soc. (Japan)*, **17**, 1100.

Kubo, R., (1956), *Can. J. Physics*, **34**, 1274.

Kubo, R., (1957), *J. Phys. Soc. (Japan)*, **12**, 570.

Landauer, R. and Helland, J. C., (1954), *J. Chem. Phys.*, **22**, 1655.

Lax, M. and Phillips, J. C., (1958a), *Phys. Rev.*, **110**, 41.

Lax, M., (1958b), *Phys. Rev.*, **109**, 1921.

Lax, M., (1951), *Reviews Mod. Phys.*, **23**, 287.

Lifshitz, I. M., (1964a), *Adv. Phys.*, **13**, 483.

Lifshitz, I. M., (1964b), *Usp. Fiz. Nauk*, **83**, 617 (Translation: *Soviet Phys. Usp.* **7**, 549 (1965)).

Llyod, P., (1969), *J. Phys.*, **C2**, 1717.

Makinson, R. E. B. and Roberts, A. P., (1962), *Proc. Phys. Soc.*, **79**, 630.

Mortensen, R. E., (1969), *J. Stat. Phys.*, **1**, 271.

Moss, S. C. and Graczyk, J. F., (1970), Proceedings, Tenth International Conference on the Physics of Semiconductors, Cambridge, Mass.

Mott, N. F., (1967a), *Adv. Phys.*, **16**, 49.

Mott, N. F. and Twose, W. D., (1961), *Adv. Phys.*, **10**, 107.

Mott, N. F., (1966), *Phil. Mag.*, **13**, 989.

Mott, N. F., (1968a), *J. Non-Cryst. Solids*, **1**, 1.

Mott, N. F. and Allgaier, R. S., (1967b), *Phys. Stat. Sol.*, **21**, 343.

Mott, N. F., (1968b), *Phil. Mag.*, **17**, 1259.

Mott, N. F. and Davis, E. A., (1968c), *Phil. Mag.*, **17**, 1269.

Mott, N. F., (1969a), *Phil. Mag.*, **19**, 835.

Mott, N. F., (1968d), *Revs. Mod. Phys.*, **40**, 677.

Mott, N. F., (1969b), *Contemp. Phys.*, **10**, 125.

Mott, N. F., (1968c), *Festkorperprobleme*, **9**, 22.

Mott, N. F., (1970), *Phil. Mag.*, **22**, 7.

Neustadter, H. E. and Coopersmith, M. H., (1969), *Phys. Rev. Letters*, **23**, 585.

Polk, D. E., (1971), *J. Non-Cryst. Solids*, **5**, 365.

Schilder, M., (1970), *J. Stat. Phys.*, **1**, 475.

Schwartz, L., (1970), Thesis, Harvard; Schwartz, L. and Ehrenreich, H., (1971), *Ann. Phys. (N.Y.)*, **64**, 100.

Shante, V. K. S. and Kirkpatrick, S., (1971a) *Adv. in Physics,* **30**, 325.
Siegert, A. J. F., (1963), statistical Physics Vol. III of Brandeis Lectures W. A. Benjamin, New York.
Soven, P., (1967), *Phys. Rev.,* **156**, 809.
Sykes, M. F. and Essam, J. W., (1964), *Phys. Rev.,* **133**, A310.
Thouless, D., (1970), *J. Phys.,* **C3**, 1559.
Velicky, B., Kirkpatrick, S. and Ehrenreich, H., (1968), *Phys. Rev.,* **175**, 747.
Wall, H. S., (1948), Analytic Theory of Continued Fractions, D. VanNostrand, New York.
Watson, K. M., (1957), *Phys. Rev.,* **105**, 1388.
Williams, F. W. and Matthews, N. F. J., (1969), *Phys. Rev.,* **180**, 864.
Ziman, J. M., (1961), *Phil. Mag.,* **6**, 1013.
Ziman, J. M., (1968), *J. Phys.,* **C1**, 1532.
Ziman, J. M., (1969), *J. Phys.,* **C2**, 1230.
Zittartz, J. and Langer, J. S., (1966), *Phys. Rev.,* **148**, 741.

Chapter 4

Optical Properties of Amorphous Semiconductors

J. Tauc

Division of Engineering and Department of Physics, Brown University, Providence, Rhode Island*.

4.1 INTRODUCTION

The sharp structure observed in the fundamental optical spectra of crystals, both vibrational and electronic, can be classified and interpreted by symmetry arguments based explicitly on the existence of long-range order. Indeed, this is one of the few properties of crystals which cannot be accounted for on the basis of short-range order alone: If the long-range order is destroyed, the sharp structural detail, which is typical for crystals, disappears. However, the broad features of the spectra are similar if the short-range order is similar.

*The first version of this paper was written during a stay at Bell Laboratories, Inc. Murray Hill, New Jersey.

In their review paper, Ioffe and Regel (1960) quote the following simple rule (cf. Section 2.4.1): the basic nature of a solid depends on the short-range order, and more explicitly, on the number of the first neighbours of an atom (the first coordination number). By basic nature they meant whether a crystal is a metal, a semiconductor or an insulator. A striking example of this rule is germanium and germanium-like semiconductors which become metallic after melting. In Figure 4.1 the fundamental reflectivity bands of crystalline, amorphous and molten Ge are compared. The spectrum of molten Ge is typically metallic as it is dominated by the optical properties of free carriers. This change, from semiconducting to metallic behavior, can be interpreted on the basis of Ioffe's rule, because during melting the first coordination number changes from 4 to 8.

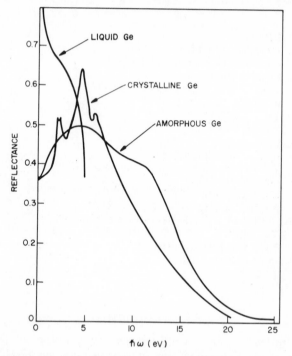

Fig. 4.1. Fundamental reflection spectra due to electronic transitions in crystalline, amorphous and liquid Ge (after Philip and Ehrenreich (1963), Donovan *et al.* (1970b) and Abrahám *et al.* (1963) respectively).

Therefore, if we want to study those changes in the optical spectra which are caused by the loss of long-range order, we must compare amorphous materials with crystalline materials with the same short-range order; which means that at least the first coordination numbers must be the same, e.g. we must compare the properties of crystalline Ge with those of semiconducting amorphous Ge composed of approximately the same tetrahedra.

The higher coordination spheres vary differently in different materials (Chapter 2); whereas some features of the change caused by the loss of long-range order are similar for all materials, others differ from one class of materials to another.

Let us mention the following general rule: optical transitions between states which may be described by wavefunctions localized over distances of the order of the lattice constant are changed relatively little by disorder. We shall see several examples of this rule. This may also explain why the optical properties of amorphous insulators often differ little from those of crystal-line insulators. On the other hand, if the bonds are very delocalized (i.e. valence electron wavefunctions spread over long distances) the changes are also small. Indeed, the optical properties of liquid metals differ little from those of crystalline metals. Semiconductors with covalent bonding are the materials most sensitive to disorder. Their transport properties are more drastically changed than their optical properties.

Amorphous materials are, of course, optically isotropic. However, it is not easy to prepare them in a macroscopically homogeneous form (discussed in Section 4.2.).

4.2   VIBRATIONAL SPECTRA

4.2.1   Fundamental Vibrational Spectra
The fundamental vibrational spectra of crystals are properly interpreted on the basis of normal vibrations of the crystal as a whole. They are usually described as one- or multi-phonon transitions. These are governed by the laws of conservation of energy and wavevector in the photon-phonon system, e.g., for one phonon transition this latter rule allows only transitions creating phonons with the wavevector $q = 0$. In addition to these fundamental rules, there are selection rules based on symmetry considerations. For example, in elemental crystals which contain only 2 atoms per unit cell (such as Ge) the one optical phonon transition is forbidden (see Zallen (1968)).

It is not possible to transfer these considerations to noncrystalline solids because here the concepts used in the crystal are not meaningful. A rigorous general theory applicable to highly disordered structures, such as amorphous semiconductors, is not available. Taylor's (1967) work on the vibrational spectra of crystals with large defect concentration may be a step towards such a theory. Under these circumstances two approximate methods may be useful: distorted lattice model and localized model.

Distorted Lattice Model
One can consider the amorphous solid as a disordered crystal. Let us suppose that the conservation of the wavevector and the symmetry selection rules cease to be effective and only energy is conserved during the transitions. Then the one-phonon absorption will dominate the spectrum and the absorption curve will reflect the one-phonon density of states. Of course, it

may not be possible to describe amorphous solids well enough by such crude assumptions. In addition, the density of states of phonons, which corresponds in non-crystalline solids to the frequency distribution of vibrations, may be significantly changed by disorder. Nevertheless, we would still expect absorption bands at approximately the maxima of the crystalline phonon-density curves, even if these bands do not exist in the crystal because of the selection rules. In such a case we will speak about disorder induced absorption bands.

Spitzer and Fan (1958) and Fan and Ramdas (1959) observed an absorption band in silicon at 490 cm^{-1} induced by irradiation. They ascribed it to the fundamental optical mode of Si made infrared active by the introduction of lattice defects. Tauc *et al.* (1970a) observed a band in amorphous Ge at 270 cm^{-1}, and interpreted it as disorder induced one-phonon absorption (Figure 4.2). Lurio and Brodsky (1972) studied the infrared spectra of a-Ge and Si and their results were interpreted by Alben *et al.* (1973). They used the continuous random network model and showed that for the understanding of the relative strengths of the infrared bands the transition matrix elements must be taken into account.

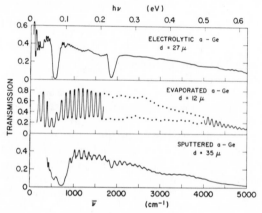

Fig. 4.2.   Transmission of a-Ge films prepared by various methods; d is the film thickness (after Tauc *et al.* (1970a)).

Hadni *et al.* (1965), Amrhein (1969), and Amreheim and Müller (1969) presented experimental evidence for the one-phonon absorption (forbidden in pure and perfect crystals) introduced by impurities and disorder in solids in the far infrared region. Armheim interprets in this way the broad absorption band which is observed in oxide glasses and other disordered solids in the frequency range between 1 and 300 cm^{-1}.

The absorption bands observed both in the crystalline and amorphous state, are often much broader in the latter case; the broadening may be so large that some bands effectively disappear (cf. Goryunova *et al.* (1970), Abrahám *et al.* (1970).

Analogous considerations apply also to the Raman spectra of amorphous semiconductors. In the absence of the q-vector conservation, the intensity of the scattered radiation is approximately proportional to the band state densities of vibrations. For the Stokes components, Shuker and Gammon (1970) deduced the following expression

$$I(\omega) = \sum_b C_b (1/\omega) [1 + n(\omega,T)] \ g_b (\omega) \qquad (4.1)$$

where $I(\omega)$ is the intensity of the scattered radiation, ω is the frequency of vibration, $n(\omega,T) = [\exp (\hbar\omega/kT) -1]^{-1}$ is the thermal population of vibrational states, $g_b (\omega)$ is their density in the band b. The resultant intensity is the sum over all bands. The constants C_b depend on the polarization of the incident and scattered radiations. It is assumed that inside the bands C_b is constant but has different values for different bands b. Under these assumptions, the maxima in the Raman spectra of amorphous semiconductors correspond approximately to the maxima of the state densities of vibrations.

Raman spectra of amorphous films of group IV and III–IV compounds were studied by Smith *et al.* (1971) and Wihl *et al.* (1971). In Figure 4.3 the Raman spectrum of a-Ge multiplied by $\omega/[1 + n(\omega)]$ is shown and compared with the band state densities of phonons in c-Ge as calculated by Dolling and Cowley (1966). We see that the general features of both curves

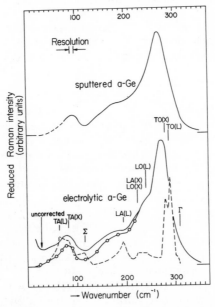

Fig. 4.3. Raman spectrum of a-Ge at room temperature (after Wihl *et al.* (1971)). The measured intensity was reduced by multiplication with the factor $\omega[n(\omega,T) + 1]^{-1}$ (cf. Eq. (4.1)). The reduced spectrum is compared with the density of phonon states in c-Ge calculated by Dolling and Cawley (1966) T = 90°K. The temperature correction brings the two curves into a closer agreement.

are similar and the agreement can be made better if the calculated state densities of c-Ge are smeared out by Gaussian broadening with properly chosen parameters (cf. Smith (1971)). This shows that the band state densities of vibrations in these films are not much changed by the loss of the long-range order. The more sophisticated calculations of Alben *et al.* (1973) mentioned above reproduced very well the observed Raman spectra of a-Ge and Si and their differences relative to the infrared spectra.

The Raman spectrum of c-Ge is governed by the q-vector conservation and is therefore very different from the Raman spectrum of a-Ge (Figure 4.3). The one-phonon spectrum consists of a rather sharp band situated at the frequency of the optical phonon at the center of the zone. This frequency is higher than the frequencies of the maxima of the state densities of optical phonons which occur at the boundary of the zone.

Prettl *et al.* (1973) studied the far infrared spectra of a-Ge, GaP, GaAs and InAs and compared them with the Raman spectra. They showed that the spectra are similar in the optical phonon frequency regions but the absorption is much smaller at low frequencies. They argue that this effect is due to the different dependence of the coupling constant for Raman scattering and infrared absorption on the wavevector q.

Stimets *et al.* (1973a, b) studied the infrared absorption spectra of a-Ge, GaAs and $Ge_x Sn_{1-x}$ system. Lannin (1973) studied Raman spectra of a-$Ge_x Si_{1-x}$ system.

Wihl *et al.* (1972) observed a Raman band in a $-$ SnTe which is Raman inactive in the crystalline form. They studied also the Raman spectra of a $-$ Sb and their evolution during the crystallization of the film.

Localized (Molecular) Model

Another point of view utilized for an approximate interpretation of the vibration spectra is the one based on a localized model (Simon (1960), Borelli and Su (1968)). One can attempt to isolate some structural elements in the glass which can be assumed to vibrate as an independent unit (a 'molecule'). Of course, the unit is not free and the boundary conditions may change the character of some vibrations relative to the free molecule and shift the frequencies of all vibrations. Nevertheless, if such structural elements are present in both the crystalline and amorphous solids, their vibrations are usually found in both cases. For example, if a restrahlen band can be ascribed to a vibration of two neighbouring ions (or all ions of one kind relative to the other) then one would also expect to observe it in the corresponding amorphous form. In this way Felty *et al.* (1967) interpreted their measurements of the reststrahlenband in the mixed amorphous quasi-binary system $As_2 S_x Se_{3-x}$ (Figure 4.4). They observed two peaks at frequencies approximately 300 and 215 cm^{-1} which did not vary in position but only in strength. The frequencies corresponded to the reststrahlen bands in pure $As_2 S_3$ and $As_2 Se_3$ respectively, and the authors ascribed them to the vibrations of As-S and As-Se bonds. No structure, which would corres-

pond to Se-S, Se-Se or S-S bond, was observed. We note that Taylor *et al.*
(1973) observed a qualitatively different behavior in the
$Tl_2 Se_x Te_{1-x} As_2 Se_y Te_{3-y}$ system. The intensity of the two observed infrared bands did not change much with composition but their frequencies
shifted. Both types of behavior have been observed also in crystalline solid
solutions (Lucovsky *et al.* (1968)).

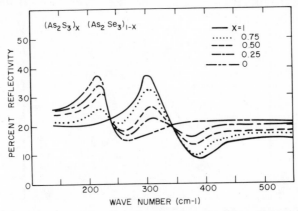

Fig. 4.4.   The reststrahlen bands of the quasi-binary amorphous system $(As_2 S_3)_x$
$(As_2 Se_3)_{1-x}$ (after Felty *et al.* (1967)).

The infrared and Raman spectra of crystalline $As_2 S_3$ and $As_2 Se_3$ have
been investigated by Zallen *et al.* (1970), and these authors discuss a comparison of the lattice optical properties of the crystals to those of the
corresponding glasses. The strongest reststrahlen bands they observe for the
crystals are at 299 and 311 cm^{-1} for $As_2 S_3$ and at 219 and 226 cm^{-1} for
$As_2 Se_3$ (the two frequencies for each crystal refer to two principal axes of
polarization). They interpret the corresponding vibrations as consisting
largely of 'rigid-sublattice' motions in which the arsenic and the chalcogenide sublattices undergo oppositely-directed rigid translations. Because
the broad reststrahlen bands seen for the glasses span the same frequencies,
they conclude that the dominant infrared-active vibrations in the glasses are
also predominantly of this character, with each arsenic atom vibrating
oppositely to its three neighboring sulfur-or-selenium atoms, and each
sulfur-or-selenium atom vibrating oppositely to its two neighboring arsenic
atoms. This idea is attractive from the viewpoint that just such vibrations
might be expected to retain their character in the glasses because of the
survival of the short-range order, while other infrared-active crystalline vibrations more dependent on unit-cell structure and symmetry would not
persist.
     More recently, the vibrational spectra of a-$As_2 S_3$ and $As_2 Se_3$ were
analyzed with a molecular model by Lucovsky and Martin (1972) and

Lucovsky (1972). Their basic unit is the As S_3 molecule weakly coupled
through a As-S-As bond to the next AsS_3 molecule (Figure 4.5a). Applying
the selection rules for the AsS_3 molecule they were able to explain the
observed difference between the effective phonon densities determined
from the i.r. and Raman spectra (Figure 4.5b). Taylor *et al.* (1973) interpret
their extensive infrared and NMR studies as evidence for the remnants of the
layers in the vitreous As_2S_3 and As_2Se_3 which persists into the liquid state
(Taylor *et al.* (1971)).

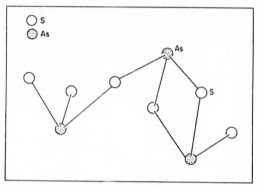

Fig. 4.5a. Schematic representation of the local molecular order in a-As_2Se_3
(Lucovsky (1972)).

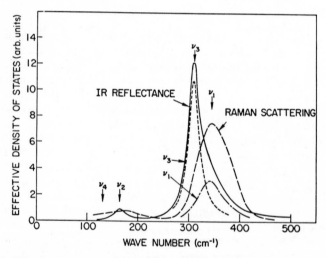

Fig. 4.5b. Effective densities of states determined from the i.r. spectra ($\sim \omega^2 \epsilon_2$) and
Raman spectra (cf. Eq. (4.1)) in a-As_2S_3. Dotted lines show the decomposi-
tion of the observed spectra into bands (Lucovsky (1972)).

Infrared spectra of a-As_2S_3 and As_2Se_3 were studied also by Onomichi
et al. (1971). Raman spectra of a-As_2S_3 were measured by Ward (1968),
Kobliska and Solin (1971, 1973) and Markov and Reshetnyak (1972).
Finkman *et al.* (1973) studied the Raman spectra of liquid As_2S_3 and
interpreted the observed temperature dependences by changes in structural
configurations which relate the 'crystal-like' and 'molecule-like' models.
Resonant Raman scattering from a-As_2S_3 obtained by temperature tuning
the band edge was studied by Kobliska and Solin (1972).
Another example of the similarity of the vibrational spectra of some

complexes in the amorphous and crystalline states is the work of Lucovsky *et al.* (1967) on Se. Using infrared and Raman spectra of trigonal and monoclinic Se they made an assignement of the fundamental vibrational mode frequencies. In the spectra of a-Se, they found bands which were broadened but situated at energies for the one or the other crystalline structures.

The bands at 95,120 and 254 cm^{-1} are common to the α-monoclinic modification, and the band at 135 cm^{-1} to the trigonal modification. Some bands observed in the crystalline form were missing. The authors conclude from these results that a-Se contains Se_8 rings which are the structural elements responsible for the observed bands in the α-monoclinic modification, and helical chains playing a similar role in trigonal Se. The conclusion that a-Se contains Se_8 rings was previously made in an indirect way, by Srb and Vaško (1963) from their comparison of the spectra of a-Se and a-S with orthorhombic S. The observation of the combination bands in a-Se and their temperature dependence by Siemsen and Riccius (1969) is consistent with the above interpretation of the vibrational spectra.

Brodsky *et al.* (1972a) found that the Raman spectrum of a-Te consists of two broad maxima at 90 and 157 cm^{-1}. The latter frequency is higher than the frequency of the corresponding vibration in c-Te. This increase of frequency was ascribed to the weakening of the interaction between the chains which the authors assume are conserved in a-Te. In the amorphous $GeTe_{1-x}$ system structural studies have had difficulty distinguishing between various models for the bonding (Betts *et al.* (1972)) especially at the 50% composition (Bienenstock (1973)). Fisher *et al.* (1973b) have found that the Raman and infrared spectra are similar throughout the composition range. This supports models which do not require changes in the bonding throughout the $Ge_x Te_{1-x}$ system, such as a covalent-random network of two-fold coordinated Te atoms and four-fold coordinated Ge atoms (Betts *et al.* (1970)).

Ing *et al.* (1969) used the Raman spectroscopy to study the structural elements present in evaporated As-S films of various compositions. They found evidence for the presence of $As_2 S_3$, S_8 rings, $As_4 S_4$ monomers and perhaps short segments of sulfur chains; the $As_4 S_4$ monomers were found even in bulk $As_2 S_3$ glasses.

The above examples serve to illustrate the value of infrared and Raman spectroscopy for the investigation of the structure of amorphous semiconductors. The methods can, in favourable cases, give more detailed information than X-ray or electron diffraction.

The situation is much more complicated in glasses with a complex network structure, for example, as in the oxide glasses, the simplest of which is amorphous SiO_2. There is much literature on the interpretation of these spectra (Simon (1960), Zarzycki and Naudin (1960), Su *et al.* (1962) and others). Bell *et al.* (1968) constructed three dimensional models of SiO_2, GeO_2 and BeF_2 glasses with typical atomic arrangements, then made plausible assumptions about the force constants, and calculated the frequency distributions of the vibrations. A similarity of observed infrared and Raman spectra with computed results was shown to exist. However, they did not calculate the optical and Raman matrix elements.

4.2.2 Infrared Optical Glasses

If we disregard impurities, defects and carriers, which will be dealt with in Section 4.2.3, the infrared transmission of solids is limited intrinsically at high energy by the electronic absorption edge E_g and at low energy by lattice vibrations. The first transparency range is between E_g and the first reststrahlenband at energy E_R. Further transparency ranges may occur between and below the reststrahlenbands, but strongly temperature dependent combination bands (multiphonon transitions in crystals) may contribute significantly to the absorption. The minimum of the absorption constant α can be very low in some extremely pure glasses; e.g. in a-SiO$_2$ α can be below 10^{-6} cm^{-1} in the range 0.7 to 1.1 μm (Keck et al., (1973)). The existence of a transmission window between the electronic and vibrational transitions with a very low absorption level is essential for the application of glass fibers in the optical long-distance communication lines.

The usefulness of non-oxide chalcogenide glasses as optical materials is due to their transparency in the infrared region. They are composed of heavier atoms than oxide glasses, and therefore, the vibration frequencies are shifted further into the infrared region. At the same time the edge E_g shifts towards smaller energies so that the first transparency range is deeper in the infrared compared to oxide glasses.

In crystalline semiconductors free carriers cause an increase in the absorption constant proportional to λ^2. This effect tends to make the crystals opaque at high temperatures. In glasses, no absorption of this kind has been observed. However, Edmond (1966) and Mitchell et al. (1971) observed in a-As$_2$Se$_3$ and TlSeAs$_2$Te$_3$ at high temperatures an additional absorption in the infrared region close to the gap which looks like an enhanced tail C discussed in Section 4.4.5 in which the absorption decreases with λ. The absorption constant is correlated with electrical conductivity, and it appears therefore likely that the effect is due to free carriers. The mechanism, however, must be different from that in crystalline semiconductors. Mott and Davis (1971) ascribed the increasing optical absorption to the increasing density of available final states. They did not obtain the observed shapes of the absorption curves, and this problem is open to further studies.

In spite of this effect, glasses are in general more transparent at high temperatures than crystals with the same value of E_G. On the other hand in glasses, at very low frequencies, there is a broad absorption band probably due to one phonon absorption mentioned in Section 4.2.1.

An important advantage of glasses compared to crystals is their optical isotropy and, in principle, easier technological processing, i.e., casting in various shapes and sizes. However, there are serious difficulties with the technology of glasses because they tend to be macroscopically inhomogeneous. This difficulty is well known in oxide glasses and has been studied in Se and compound chalcogenide glasses by the Schlieren method by Vaško (1969). Even if one is able to reduce significantly the presence of segregated impurities, gross defects such as voids and crude phase separation, one often still faces small variations of the index of refraction through the sample, which may make it turbid. These inhomogeneities have been reduced below tolerable limits in commercially produced glasses. The properties of infrared glasses available commercially in the United States are summarized in Table 4.1. Other commercial suppliers of infrared glasses are mentioned by Hilton (1970).

Non-oxide chalcogenide glasses are mechanically weaker than oxide glasses; hardness and softening points are lower. For some applications glasses transparent enough in at least one of the two atmospheric windows

(3 to 5, 8 to 14 μ) are required; often a relatively high softening temperature is necessary (Hilton (1966), Hilton and Brau (1963), Hilton et al. (1964, 1966 a, b, c), Hilton and Jones (1966)); cf. also Edmond et al. (1962). These problems have been successfully solved for the glasses No. 20 and No. 1173 (See Table 4.1).

TABLE 4.1. *Properties of Commercially Produced Chalcogenide Glasses*

	Arsenic Sulphide*	Glass No 20**	Glass No 1173**
Composition	As_2S_3	$Ge_{33}As_{12}Se_{55}$	$Ge_{28}Sb_{12}Se_{60}$
Transmission range (μ)	0.7 to 11	1 to 15	1 to 15
Refractive index at 10μ	2.3726†	2.4919	2.6002
Δ index/$°C$	-8.6×10^{-6}††	–	79.0×10^{-6}
Density (g/cm^3)	3.20	4.40	4.67
Expansion coefficient (per $°C$)	26×10^{-6}	13.3×10^{-6}	15.0×10^{-6}
Thermal conductivity cal/s cm $°C$	4×10^4	6.1×10^4	7.2×10^4
Strain point $10^{14.6}$	$160°C$	$354°C$	$240°C$
Anneal point $10^{13.4}$ Poise	$177°C$	$364°C$	$259°C$
Softening point $10^{7.6}$ Poise	$267°C$	$474°C$	$370°C$

* Produced by Servo Corporation of America under the name Servofrax, and by American optical Company.
** Produced by Texas Instruments
† After Rodney et al. (1958)
†† At 5μ (after Hilton and Jones (1967))

The index of refraction of chalcogenide glasses is considerably higher than of oxide glasses, and therefore, it is usually necessary to coat the optical elements with antireflecting coatings (Hilton (1966)). The temperature dependence of the index of refraction was studied by Hilton and Jones (1967).

Important applications of chalcogenide glasses are acousto-optic modulators in the infrared region and ultrasonic delay lines. Glasses for these applications must have high optical quality and very low acoustical absorption losses, comparable to the loss in fused SiO_2 (about 10^{-2} decibels/cm at 20 MHz at room temperature for the longitudinal mode). Krause, Sigety and Kurkjian (1970) have discovered chalcogenide glasses (such as glasses No. 20 and 1173 in Table 4.1) which fulfill this condition and have an acoustic-optic figure of merit higher than any available crystalline material. They observed a large decrease in acoustic loss when the structure of the glass changed from a two-dimensional polymeric structure in the V – VI binary systems to a completely cross-linked network in certain composition ranges of the ternary systems IV – V – VI.

Also, films of amorphous semiconductors have found optical applications; for example, in multilayer infrared filters or as antireflective coatings.

4.2.3 Vibrations of Impurities and Defects

The transparency of semiconducting glasses in the infrared region is often severely reduced by contamination, primarily due to oxygen. The bands caused by vibrations of oxygen bonds (and other impurity bonds) in various chalcogenide glasses have been reported by Vaško (1965), Savage and Nielsen (1965) Hilton and Jones (1966), Tausend (1970), Vaško *et al.* (1970) and others.

In amorphous solids there may exist atomic configurations which may be described as defects. Various kinds of network defects introduced by irradiation into glasses were discussed by Stevels (1960). A simple defect of this kind can be visualized in amorphous Ge. Structural fluctuations may produce places where effectively one atom is missing. These atomic vacancies are believed to act as acceptors similar to the atomic vacancies introduced into a-Ge by irradiation. However, it appears probable from the work by Brodsky and Title (1969) and Galeener (1971) that all or a part of these acceptors in a-Ge films is associated with atoms at the internal surfaces of voids. Tauc *et al.* (1970a) observed an absorption band at 560 cm^{-1} in a-Ge which was present in amorphous films prepared by several different techniques, and was too strong to be explained by impurities (Figure 4.2). They tentatively ascribed it to the vibration of a defect intrinsic to a-Ge of the kind mentioned above. This assignment is supported by the observation of Chittick (1969) who found the same band in a-Ge films prepared by yet another technique (the glow-discharge decomposition of germane gases). This band disappeared when the films crystallized by annealing.

Another strong band at 720 cm^{-1} was observed in sputtered a-Ge films (Figure 4.2). The samples were known to contain a large concentration of oxygen (between 3 to 7%), and the band is probably due to its presence. However, it is not situated at the frequency (860 cm^{-1}) of the Ge-O-Ge vibration in c-Ge, which is close to the dominant reststrahlen frequency in the trigonal, quartz like structure of GeO_2 and in amorphous GeO_2. It is known, however, that this band disappears when c-Ge containing oxygen is subjected to irradiation. It is replaced by bands centered between 700 and 750 cm^{-1} which have been attributed to oxygen-defect complexes (Whan and Stein (1963)). The reststrahlen frequency in the tetragonal rutile structure of GeO_2 observed at 720 cm^{-1} falls into this interval. In trigonal GeO_2, oxygen is interstitial and two-fold coordinated, while in tetragonal GeO_2, it is three-fold coordinated, as it would be when located substitutionally at a Ge site adjacent to a vacancy or, possibly at a surface state. Therefore the band at 720 cm^{-1} in sputtered a-Ge was ascribed to the vibration of an oxygen-vacancy complex. Again, this kind of absorption is typical for a disordered structure.

4.3   ELECTRONIC TRANSITIONS IN IMPURITIES AND DEFECTS

From the point of view of electronic properties, we may consider two kinds of impurities which may be present in an amorphous solid.

As discussed in Chapter 5, shallow impurities such as elements of column 3 or 5 of the periodic table, which so significantly influence the transport properties of c-Ge and c-Si, appear unimportant in a-Ge and a-Si. An explanation of this observation was suggested by Mott (1969). In the amorphous state the structure can always adjust itself so that all the valence bonds are satisfied. Such impurities are not expected to form distinct absorption centers but would only contribute to the fluctuations of the internal potential. In the optical spectra they would increase the absorption near the absorption edge (cf. Section 4.4.4).

However, if the change in potential due to a defect or an impurity atom is more profound (exceeding the fluctuations of the internal potential), localized optical transitions can take place. In this case, the influence of the fluctuating internal potential is merely a broadening of the lines. Impurities of this kind have been extensively studied in classical oxide glasses. The energy levels of an ion introduced into the glass can be often described by ligand field theory, as reviewed by Bates (1962) for the transition-metal ions. Recently, much attention has been paid to the rare-earth ions because of their application in laser materials (cf. reviews by Snitzer (1966), Rindone (1966)). The advantage of glasses for lasers is easier fabrication but there are cases in which the broader lines are useful. It is also well known that glasses, like crystals, can be colored by doping or by irradiation similarly as crystals (cf. Weyl (1951), Lell et al. (1966), Bishay (1970)).

No comparable work has been done on semiconducting glasses. Tauc et al. (1969) and,Chittick (1969) reported an observation of a band at 0.23 eV in a-Ge films which disappeared upon crystallization by annealing simultaneously with the band observed at 0.07 eV (560 cm^{-1}) which was ascribed to vibrations of a defect characteristic of a-Ge (Section 4.2.3). The band at 0.23 eV was tentatively ascribed to an electronic transition on the same defect.

As mentioned in Section 4.2.2, no free carrier absorption is usually observed in amorphous semiconductors. Another effect of free carriers is transitions between various branches of the valence band split by spin-orbit interaction. This absorption which is observable in p-type crystalline semiconductors (p-bands) was reported also in a-Se (Kessler and Sutter (1963)) and in a-Ge (Tauc et al. (1966)). However, in the case of a-Se the energy difference between the corresponding bands deduced from experiment is much higher than the spin-orbit splitting observed in the emission spectra of c-Se and predicted theoretically. In the case of a-Ge, the results of more accurate experiments on thicker films could not be interpreted as p-bands (Tauc et al. (1970a)). The suggested explanation of the observed infrared bands in a-Ge was discussed in connection with Figure 4.2 in this Section and in Sections 4.2.1 and 4.2.3. There is therefore no evidence of p-bands in amorphous semiconductors.

4.4   ABSORPTION EDGE

4.4.1   Introduction

Electronic transitions between the valence and conduction bands in the crystal start at the absorption edge which corresponds to the minimum energy difference E_g between the lowest minimum of the conduction band and the highest maximum of the valence band. If these extrema lie at the same point of the k-space, the transitions are called direct. If this is not the case, the transitions are possible only when phonon-assisted and are called indirect. The rule governing these transitions is the conservation of quasimomentum during the transitions, either of the electron alone in direct transitions, or the sum of the electron and phonon quasimomenta in indirect transitions.

The value of the gap E_g depends in a rather subtle way on the structure and actual values of the pseudopotential in the crystal. It is to be distinguished from the gap E_G, which is characteristic of the whole absorption band and is connected with the basic chemical properties of the material (cf. Sections 4.5 and 4.6).

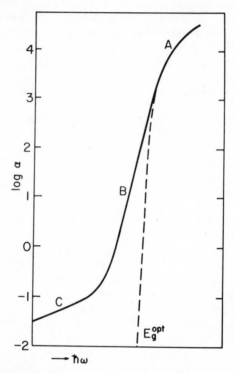

Fig. 4.6. Parts A, B, C of the absorption edge.

When the semiconductor becomes amorphous, one observes a shift of the absorption edge either towards lower or higher energies. No simple general rule governing these changes has been suggested. In a group of similar materials certain dependencies are observed; e.g. Rockstad and De Neufville (1973) reported on a relationship between the optical gap and the glass transition temperature in chalcogenide glasses. The shape of the absorption curve appears to be similar for many amorphous semiconductors.

In many amorphous compound semiconductors the absorption edge has the shape shown in Figure 4.6. One can distinguish the high absorption region A ($\alpha > 10^4$ cm^{-1}), the exponential part B which extends over 4 orders of magnitude of α and the weak absorption tail C. These regions will be discussed in Sections 4.4.2 to 4.4.5. The absorption edge of As_2S_3 is shown in Figure 4.7.

Edmond (1966) observed similar edges in As_2Se_3 and As_2Te_3 glasses. Parts A and B were observed by Bahl and Chopra (1969) and Tsu *et al.*

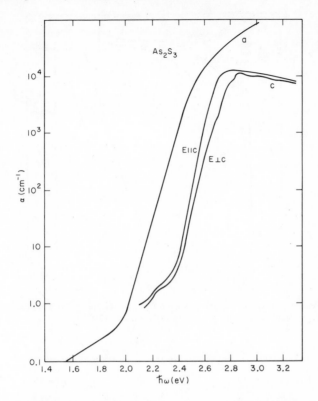

Fig. 4.7. Absorption edge of c-As$_2$S$_3$ for 2 direction of polarization relative to the axis c (after Evans and Young (1967)) compared with the absorption edge of a-As$_2$S$_3$ (after Kosek and Tauc (1970)). Optical properties of a-As$_2$S$_3$ were also studied by Young (1971).

(1970) in a-GeTe, by Fagen and Fritzsche (1970a) in complicated chalcogenide glasses, by Tauc *et al.* (1968) and Červinka *et al.* (1970) in semiconducting glasses based on CdAs$_2$. Weiser and Brodsky (1969) studied part A in a-As$_2$Te$_3$ films.

Absorption edges of crystalline and amorphous Se and Te are shown in Figure 4.8. We see that the shifts of absorption are in the opposite direction compared with Figure 4.7.

Absorption edges of c-Ge and a-Ge are compared in Figure 4.9. The sharpness of the edge in a-Ge depends on its preparation and thermal history. Ge and Si are discussed separately in Section 4.4.6.

Absorption edges of a-InSb and c-InSb are shown in Figure 4.10.

4.4.2 High Absorption Region

It is often observed in semiconducting glasses that, at high enough absorption levels ($\alpha \geqslant 10^4$ cm^{-1}) the absorption constant α has the following frequency dependence:

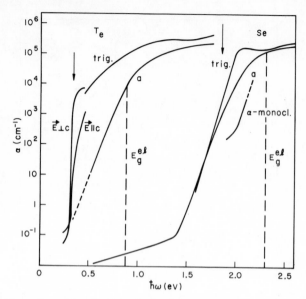

Fig. 4.8. Absorption edges of c-Te and c-Se (for two directions of polarization) compared with those of a-Te and a-Se (after Stuke (1970)).

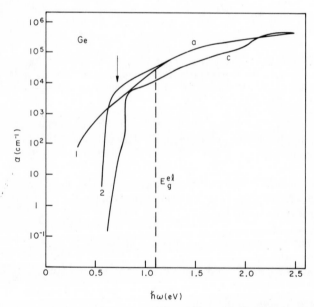

Fig. 4.9. Absorption edges of c-Ge and a-Ge (after Stuke (1970)). The curve 1 was measured by Clark (1967) and Abrahám (private communication), the curve 2 by Donovan *et al.* (1969).

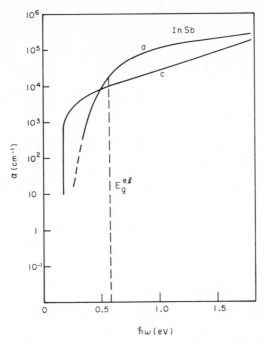

Fig. 4.10. Fundamental absorption bands in III–V compounds (after Stuke (1970)).

$$\hbar\omega\alpha(\omega) \sim \left(\hbar\omega - E_g^{opt}\right)^r \qquad (4.2)$$

where r is a constant of the order 1. An example is shown in Figure 4.11. We see that for a-As_2S_3 r = 2. Eq. (4.2) has been used to define the optical gap E_g^{opt} as distinguished from the electrical gap E_g^{el} determined from the temperature dependence of electrical conductivity. The range in which the dependence (4.2) is observed is too small to be sure of the exact value of the exponent r.

We shall now discuss a model we shall use for the description of optical transitions in amorphous semiconductors. The basic difference compared to crystals is the non-conservation of the k-vector. This is due to the change of the character of the wave-functions, some of which become localized over a certain volume V(E) rather than extended over the whole volume of the sample as in crystals. When the wavefunctions are localized, the transition probabilities between states localized at different sites are reduced by a factor depending on the overlap of the wavefunctions of the initial and final states. The effects of correlation were considered by Stern (1971), Dow and Hopfield (1971) and Sak (1972). We shall neglect the correlation and consider only transitions between states of which one is localized and the other extended, or of which both are weakly localized so that the over-

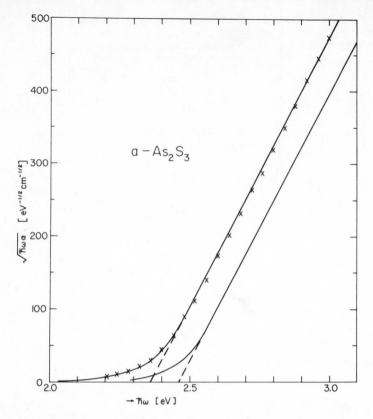

Fig. 4.11. The dependence of $\sqrt{\hbar\omega\alpha}$ on photon energy in a-As_2S_3 (after Kosek and Tauc (1970)) from which the optical gap E_g^{opt} is determined.

lapping of the wavefunctions is large enough. Under these circumstances the absorption constant

$$\alpha = \frac{2\pi^2 e^2}{m^2 c\omega n} \cdot p^2 \cdot \int V(E) g_i(E) g_f(\hbar\omega + \dot{E}) \, dE \qquad (4.3)$$

n is the index of refraction, g_i, g_f are the densities of the initial and final states. The optical matrix element p was assumed constant. If the localized wavefunctions can be constructed as linear combinations of the extended (Bloch) functions in the same band of the corresponding crystal then p has the same value as in the crystal ($\approx$ h/a, where a is the lattice constant) (Tauc et al. (1965)). If the initial and final wavefunctions are weakly localized we can consider transitions between wavefunctions which are not too much different from the Bloch functions, for which, however, the k-vector is not conserved ('non-direct transitions' as introduced by W. E. Spicer). In this case V(E) is equal to the volume of the primitive crystal cell V_{cell}.

For transitions between a state localized over V(E) and an extended state the absorption is enhanced by $V(E)/V_{cell}$. This enhancement of absorption by localization is actually observed in crystals. E.g. Eagles (1960) could describe the transitions from hydrogen-like impurities into the band states of crystalline GaAs with an equation basically similar to Eq. (4.3).

However, Davis and Mott (1970) argued that this enhancement by localization $V(E)/V_{cell}$ does not take place in amorphous solids. When we calculate p in the crystal for the wavefunction extending over V we sum over $V(E)/V_{cell}$ cells because in each cell p has the same sign, and therefore p ~ $V(E)/V_{cell}$. Davis and Mott suggested that in an amorphous solid the signs of p in different cells are random, and therefore p ~ $(V(E)/V_{cell})^{1/2}$. It follows that $p^2_{amorphous}/p^2_{crystal} = V_{cell}/V(E)$. This factor cancels out V(E) in Eq. (4.3) and replaces it by V_{cell} = const. The absorption under these assumptions is independent of the volume of localization.

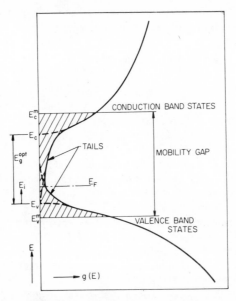

Fig. 4.12. Density of states g(E) as function of energy E in amorphous semiconductors, according to the Mott-CFO model (cf. Chapter 3). E_g^{opt} as used in the text is determined by extrapolation of the delocalized states. E_v^m and E_c^m are mobility edges.

Davis and Mott's hypothesis simplifies considerably the following discussions and we shall adopt it. However, its validity is subject to justification in any particular theoretical model of the electronic structure of an amorphous solid. Hindley (1970) has shown that it follows from his random phase model for the wavefunctions in amorphous semiconductors.

The density of states g_i, g_f we shall consider is shown in Figure 4.12. This is the model of state densities usually denoted as Mott-CFO model.

Part A of the absorption curve (Figure 4.6 and Eq. (4.3)) is very probably associated with transitions from the localized valence band states below E_v into the delocalized conduction band states above E_c^m, or vice-versa. If we assume that the densities of states not too close to E_v or E_c are

$$g_v \sim (E_v - E)^{r_1}, \; g_c \sim (E - E_c)^{r_2} \qquad (4.4)$$

then it follows from Eq. (4.3)

$$\omega \alpha \sim (\hbar\omega - E_g^{opt})^{r_1 + r_2 + 1} \qquad (4.5)$$

where $E_g^{opt} = E_c - E_v$.

In the simplest case we would expect that $r_1 = r_2 = 1/2$ as in crystals, and therefore $r = 2$. This has been actually observed in many crystals such as As_2S_3 (Figure 4.11), $As_2S_2Se_2$, $As_4S_3Se_3$, As_2Se_3, As_2Te_3, GeTe, $CdGeAs_2$ and Ge. In the empirical relation $\alpha\hbar\omega = A \, (\hbar\omega - E_g^{opt})^2$ the constant A has been found to be between 10^5 to 10^6 cm^{-1} eV^{-1} (Davis and Mott (1970)). We can calculate its value from Eq. (4.3):

$$A = \left(e^2 mp^2 V_{cell} / 2\pi c\hbar^5 n \right) \left(m_v m_c / m^2 \right)^{3/2} \qquad (4.6)$$

m_v, m_c are the density effective masses. Taking $V_{cell} \approx 100$ Å^3, $p \approx h/a \approx 10^{-19}$ gcm s^{-1}, $m_v = m_c = m$ we obtain $A \approx 3 \times 10^5$ cm^{-1} eV^{-1} in a reasonable agreement with the experimental data.

In more complicated glasses such as $Si_5Ge_5As_{25}Te_{55}$ Fagen (1971) observed $r = 3$ in Eq. (4.2). Cohen (1971) suggested that in these highly disordered systems the tails are important, and the linear parts $r_1 = r_2 = 1$ in the density curves close to the inflexion points (cf. Figure 4.12) extend over appreciable energy ranges and are responsible for the observed $r = 3$.

In our interpretation we determine E_g^{opt} from an extrapolation of the densities of states deeper in the bands. Davis and Mott (1970) interpreted the experimentally observed relation Eq. (4.2) in a different way, namely, as transitions from the localized states at the top of the valence band into the delocalized states in the conduction band (or vice versa). Their reasoning is based on calculations of Mott (1969, 1970) showing that under certain assumptions the density of delocalized states inside the band near its maximum is a linear function of energy. In this interpretation E_g^{opt} has the meaning of the smaller energy difference between the boundaries of the localized states in the valence band and delocalized states in the conduction band or vice versa (approximately the smaller one of $E_c^m - E_v$ or $E_c - E_v^m$ in Figure 4.12).

Davis (1970) found $r = 1$ in a-Se.

4.4.3 The Exponential Region of the Absorption Edge

We shall discuss the exponential region of the absorption edge (part B in Figure 4.6). It has the following typical properties:

a) In the absorption constant range from 1 cm^{-1} (or less) to about 10^4 cm^{-1}

$\alpha(\omega)$, the absorption constant, is described by the formula

$$\alpha(\omega) \sim \exp(\hbar\omega/E_e) \tag{4.7}$$

b) The energy E_e characterizing the slope is almost temperature independent at low temperatures (usually below room temperature) and has, in many semiconducting glasses, the value between 0.05 eV and 0.08 eV.

c) At high temperatures, the slope decreases with temperature (at high enough temperatures $E_c \sim T$) (cf. Cervinka et al. (1970)).

d) In many amorphous semiconductors parts A and B move as a whole*. The small but non-zero temperature dependence of the slope of the exponential part makes the temperature shift, measured at low absorption levels, somewhat larger than the temperature shift of E_g^{opt} (cf. Kosek and Tauc (1970)).

We shall discuss various attempts to explain this exponential region.

Theoretical work has led to the conclusion that disorder may introduce energy levels in the forbidden energy gap of a semiconductor. As seen in Figure 4.12, tails of band state densities extending into the gap appear; the electron states in these tails are localized. The exponential part of the absorption edge was suggested (Tauc et al. (1968)) to be evidence for such states. Using Eq. (4.3) and Davis-Mott assumption $V(E) = V_{cell}$ we obtain for the absorption constant due to transitions between the tail states adjacent to the valence bands and extended states in the conduction band (above E_c^m in Figure 4.12, $\hbar\omega < E_g^{opt}$)

$$\alpha = \frac{\pi^2 e^2 \hbar f V_{cell}}{mcn} \int_{E_g^{opt} - \hbar\omega}^{\infty} g_{tail}(E_i)\, g_c\,(\hbar\omega - E_g^{opt} + E_i)\, dE_i \tag{4.8}$$

where the energy E_i is measured as shown in Figure 4.12, $g_{tail}(E)$ is the density of states in the valence band tail, the oscillator strength $f = 2p^2/m\hbar\omega$. A similar equation holds for transitions from the extended states of the valence band into the tail of the conduction band. If we assume

$$g_{tail}(E_i) = E_t^{-1} N \exp[-E_i/E_t] \tag{4.9}$$

where E_t is constant, N is the total concentration of states between $E_i = 0$ and ∞, then

$$\alpha = \frac{\pi^{5/2} e^2 h}{2mc} \frac{f}{n} g_c(E_t) V_{cell} N \exp[\hbar\omega/E_t]. \tag{4.10}$$

We can associate Part B and Part C of the absorption edge with this kind of absorption. For $g_c(E)$ we may take the free electron density. In this way Wood and Tauc (1972) analyzed tentatively Part B and C in a-As_2S_3, $Ge_{28}Sb_{12}Se_{60}$ and $Ge_{35}As_{12}Se_{55}$. They found the total concentration of states associated with Part B and C to be of the order 10^{20} cm^{-3} and $10^{16}-10^{17}$ cm^{-3} respectively. This interpretation appears to be acceptable

---

*Siemsen and Fenton (1967) reported a different behavior in a-Se which will be discussed later.

for Part C as we shall discuss in Section 4.4.5. For Part B, however, although it cannot be ruled out one meets with several difficulties. The value of E_t in a large variety of amorphous materials is close to 0.05 eV. It appears unlikely that the density tails in different materials would have the same E_t. Similar exponential tails have been observed in crystalline semiconductors with E_t of the same order of magnitude. These tails are usually referred to as Urbach edges (or Urbach tails) (Urbach (1953)). We shall now review briefly the basic facts about the Urbach edges and their possible origin in crystalline materials, and then discuss the application of the same considerations to the explanation of Part B of the absorption edge of amorphous semiconductors.

Urbach edges

The frequency and temperature dependence of Urbach edges is empirically described by the equation

$$\alpha(\omega) \sim \exp\left[\sigma(\hbar\omega - E_g(T))/kT^*\right], \quad \hbar\omega < E_g(T) \tag{4.11}$$

where σ is a constant of the order of unity, $E_g(T)$ the temperature dependent gap and T^* an effective temperature (Mahr (1963)). The effective temperature T^* is approximately constant at low temperatures and proportional to T at high temperatures.

The theories of the Urbach edge are based on the idea that a sharp absorption edge is broadened by some mechanism. In ionic crystals there is little doubt that optical phonons are responsible for the Urbach edges. If their frequency is ω_0 then by a general argument given below

$$T^* = (\hbar\omega_0/2k)\text{ctnh}(\hbar\omega_0/2kT) \tag{4.12}$$

It is seen that $T^* = \hbar\omega_0/2k$ for $T \ll \hbar\omega_0/2k$ and $T^* = T$ for $T \gg \hbar\omega_0/2k$.

Eq. (4.12) can be deduced from the assumption that the broadening of the edge is proportional to the mean square displacement of ions $\langle u^2 \rangle$ which in turn is proportional to the mean potential energy of their oscillations, and therefore to their total energy. In the classical case $\langle u^2 \rangle_{clas} \sim kT$; in the quantum-mechanical case,

$$\langle u^2 \rangle_{q.m.} \sim kT^* = \hbar\omega_0 \left(\langle n \rangle + \tfrac{1}{2}\right) \tag{4.13}$$

where $\langle n \rangle$ is the average number of phonons in the state considered

$$\langle n \rangle = \left[\exp(\hbar\omega_0/kT) - 1\right]^{-1} \tag{4.14}$$

Putting this into Eq. (4.13) we obtain Eq. (4.12).

If the Urbach tail is interpreted in this way, we may say that at low temperatures the broadening is determined by the zero-point vibrations; at higher temperatures the contribution of thermal vibrations becomes more and more important.

There have been many attempts to theoretically explain the exact nature of interactions leading to this rule in crystals. The basic difficulties were briefly discussed by Hopfield (1968).

Siemsen and Fenton (1969) applied the theory developed for the crystals by Toyozawa (1958 and 1959) (cf. also Sumi and Toyozawa (1971)) to the temperature dependence of the absorption edge of a-Se shown in Figure 4.13. Toyozawa explains the exponential edge of ionic crystals by broadening of the exciton line by phonon interaction and his results are in accord with the phenomenological formula (4.11). Siemsen and Fenton obtained good agreement with experiment by using this equation with $\hbar\omega_o = 4.6 \times 10^{-2}$ eV, which they considered as an average of the strong vibration modes of the short range atomic order. In a-Se this approach appears to be justifiable because a-Se is composed of structural elements which have small interactions with the rest of the material. The excitonic line in question was associated with the Se_8 — molecular component of a-Se (Lucovsky (1970)). The interpretation is in accord with the measurement of photoconductivity by Hartke and Regensburger (1965).

Toyozawa's and related theories are based on a two-mode configuration-coordinate model and do not appear applicable to amorphous semiconductors in general. Another approach is associated with the work of Segall (1966) on the Urbach tails observed in II—VI compounds. The Urbach tail is a high temperature effect. At low temperatures a structure appears in the absorption edge, well understood as a quantum effect due to electron (or exciton)-phonon interaction. According to the theory of Segall and others (Dunn (1968), (1969)) these side bands are broadened at high temperatures forming an exponential edge.

We shall now consider theories which are based on interactions with internal electric fields. These theories may have a general applicability. They explain the exponential edges in pure crystals as due to electric fields produced by longitudinal optical phonons and the exponential edges in compensated semiconductors (Redfield and Afromovitz (1967)) as due to the electric fields produced by charged impurities. For the electric fields which are responsible for this effect Dow and Redfield (1971) suggested the term 'microfields' used in the same sense as in plasma physics. The spatial variation of microfields occurs over distances large compared with atomic sizes but small compared with macroscopic dimensions (i.e., $1\text{Å} < |F|/|.\nabla F| \ll 10^4$ Å). There is some experimental evidence (see below) that such fields may be important in amorphous semiconductors, and therefore may be responsible for the exponential edges.

Internal electric fields as a source of absorption-edge broadening were suggested by Redfield (1963). Later, he and his associates found experimental evidence that internal fields produced by charged impurities did in fact produce broadening (Redfield and Afromovitz (1967), Afromomovitz and Redfield (1968)). Prakash (1967) showed that the Franz-Keldysh effect (the broadening of the absorption edge by an electric field) with a Gaussian

---

*The data of Siemsen and Fenton appear also to indicate that the energy of the exciton line changes little with temperature, so that all their $\alpha(\omega)$ curves at various temperatures converge at high-absorption levels (Figure 4.13). The excitonic line is assumed to be also Gaussian-broadened. However, this broadening is seen only close to the excitonic line; at low absorption levels the exponential tail prevails.

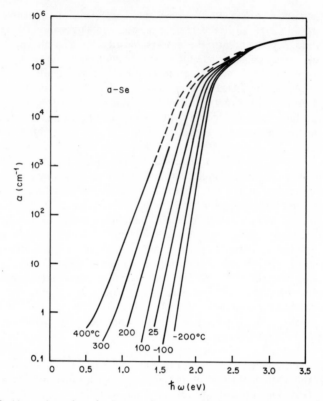

Fig. 4.13. Absorption edge of a-Se at various temperatures (after Siemsen and Fenton (1967)).

distribution of electric fields gives an exponential edge; however, he obtained an incorrect temperature dependence of the slope.

A simple model

Before we discuss the more sophisticated theories of Dexter, and Dow and Redfield, we shall briefly discuss a simple model describing intuitively optical transitions in the presence of electric fields (Tauc (1970b)).

One assumes that an amorphous semiconductor is penetrated by internal electric fields the possible origins of which we shall discuss in the Section 4.4.4. The internal potential as shown in Figure 4.14 has maxima at which the hole states in the valence band are localized and minima at which the electron states in the conduction band are localized. The transitions from the maxima of the valence band to the minima of the conduction band have a smaller energy than E_g^{opt} but their probability is decreased by the spatial separation of both states. Assuming that the amplitudes of the wave-functions decrease as exp (− r/L) at distances r sufficiently far from the center (L is a constant) the transition probability decreases as exp (− 2R/L)

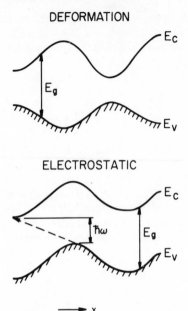

Fig. 4.14. Deformation and electrostatic parts of internal potential fluctuations.

where R is the distance between the centers (wavefunctions decaying exponentially with distance appear in Stern's (1971) theory). The energy of the transition differs from E_g^{opt} by ΔE where

$$|\Delta E| = eFR \tag{4.15}$$

where F is the average electric field between the centers. From energy conservation $\Delta E = E_g^{opt} - \hbar\omega$. Therefore the factor governing the absorption just below the energy E_g^{opt}

$$\alpha \sim \exp[-2R/L] = \exp[-2|\Delta E|/eFL] = \exp[2(\hbar\omega - E_g^{opt})/eFL] \tag{4.16}$$

If the distribution of F decreases fast enough in the high field range one can justify an approximation $F \approx F_{av} = const$. If L_D is the distance between potential barriers, then their average height φ is

$$\varphi \approx FL_b/2. \tag{4.17}$$

The absorption edge will be dominated by the factor

$$\alpha(\omega) \sim \exp(\hbar\omega L_b/e\varphi L) \tag{4.18}$$

Comparing this expression with the experimental data on amorphous semiconductors at low temperatures mentioned above, we see that $e\varphi L/L_b \approx 0.05$ eV. Assuming further that the wavefunctions are localized over one-half of the distance between the barriers ($L \approx L_b/2$) we obtain $e\varphi$ of the order 0.1 eV. The same order of magnitude of internal potentials was assumed by Cohen *et al.* (1969) and Boer and Haislip (1970) to explain some transport properties of amorphous semiconductors. (Cf. also Bagley (1970) and Chapter 5.4).

The temperature dependence of the slope $d\log \alpha / d\hbar\omega$ at higher temperatures must be ascribed to an increase of the internal electric fields due to thermal vibrations.

Excitonic theories

We shall now discuss theories based on the Coulombic interaction between the electron and the hole, that is the excitonic effects. The theory of Dexter is based on the shift of the excitonic line by electric fields, the theory of Dow and Redfield on the exciton broadening by electric fields.

Dexter (1967) showed that internal electric fields with a Gaussian distribution of field intensities F causing the Stark shift of an excitonic line give an exponential edge. Let the probability that F lies between F and F + dF be

$$P(F) = (2\pi F_{av}^2/3)^{-3/2} \, 4\pi F^2 \exp(-3F^2/2F_{av}^2) \tag{4.19}$$

where $F_{av} = \sqrt{\langle F^2 \rangle}$. The Stark shift of a line situated at E_g is $- aF^2$ (a is a constant) and the shape of the edge is

$$\alpha(\omega) \sim \int_{-\infty}^{+\infty} F^2 \exp(-F^2/2F_{av}^2) \, \delta \, (\hbar\omega - E_g + aF^2)) \, dF$$

$$\sim (E_g - \hbar\omega) \exp[(\hbar\omega - E_g)/2aF_{av}^2] \tag{4.20}$$

In this way Dexter showed how the quadratic (normal) Stark effect can produce an exponential edge at $\hbar\omega \ll E_g$ if the field distribution is Gaussian. Dexter considers the ionic motion to be the source of the fields. The field intensity is proportional to the relative ion displacement q. Their distribution in thermal equilibrium is Gaussian. $\langle u^2 \rangle$ is determined by Eq. (4.13). It follows that $F_{av}^2 \sim \langle u^2 \rangle \sim T^*$. By comparison of Eq. (4.20) with Eq. (4.11) we have

$$2aF_{av}^2 = kT^*/\sigma \tag{4.21}$$

For a-Se at low temperatures $kT^*/\sigma \approx 0.05$ eV (Siemsen and Fenton (1968)). Taking a value for a $\approx 6 \times 10^{-15}$ eVcm^2/V^2 determined from the shift of the absorption edge by an external d.c. electric field (cf. Section 4.7) (Stuke and Weiser (1966)), we obtain an estimate of the average electronic field intensity needed to explain the low temperature slope of the absorption edge, $F_{av} \sim 2 \times 10^6$ V/cm.

An important objection to Dexter's model is that at such high fields the exciton Stark shift does not remain quadratic and, furthermore, the exciton line broadening becomes larger than the shift (Dow and Redfield (1971b)). In other words, the effect of the electric field on the motion of the center of mass of an exciton becomes less important than the influence of the electric field on the relative motion of the electron and the hole (that is the internal motion of an exciton). Dow and Redfield's theory of the Urbach tail is based on the exciton broadening by electric fields with a Gaussian distribution. The problem has not been solved exactly. Dow and Redfield (1970) found a solution for a uniform electric field and applied the results to non-uniform fields by making some plausible assumptions.

In the effective mass approximation (Elliott theory of excitons) the internal motion of an exciton is described by the wavefunction $U_\nu(r)$

$$\left[-\frac{\hbar^2}{2m_r} \nabla^2 - \frac{e^2}{\epsilon_0 r} + V_0(r) \right] U_\nu(r) = E_\nu U_\nu(r) \qquad (4.22)$$

where r is the position vector of the electron relative to the hole, $V_0(r)$ is the potential energy associated with the microfields in the solid, ϵ_0 is the static dielectric constant, m_r is the reduced mass. Knowing $U_\nu(r)$ we can calculate the optical absorption constant

$$\alpha(\omega) \sim \sum_\nu |U_\nu(0)|^2 \delta(\hbar\omega - E_\nu) \qquad (4.23)$$

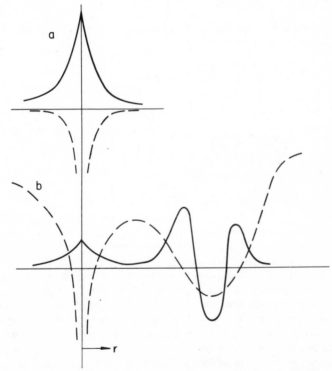

Fig. 4.15. Potential − − − and wavefunction $\phi(r)$——— for the relative electron-hole motion in the effective mass approximation. a: field-free case (enhancement of α by the Elliott factor $|\phi(0)|^2$). b: the potential fluctuations decrease $|\phi(0)|^2$ by giving the electron or hole the possibility to tunnel into a nearby potential minimum.

The summation extends over all electron and hole wave vectors and states of relative motion ν.

In Figure 4.15 the potential $V = V_o(r) - e^2/\epsilon_o r$ in which the electron moves in the presence of microfields is sketched. In such a potential, the electron has a certain probability to escape by tunneling from the bound state with the hole. The situation in the figure corresponds to a very high tunneling probability. A large part of the wavefunction $U(r)$ shown also in the figure is located in the outer potential trough. The amplitude of the wavefunction at the center-of-mass $U(0)$ is very small.

The case of a uniform field $V(r) = - eFr$ is similar to the actual situation, with one important exception: in a uniform field the electron tunneling from the hole has available a continuum of final states, in non-uniform fields the final states are the states associated with the potential trough.

Dow and Redfield (1970) calculated α from Eq. (4.23) in a uniform field F and found for energies E sufficiently below the exciton energy E_o

$$\alpha(E) \sim \exp\left[- \text{const.} \frac{E-E_o}{F} \right] \tag{4.24}$$

They use this result for the calculation of α in electric fields with a distribution $P(F)$ taking

$$\alpha(E) = \int_0^\infty P(F)\alpha(E,F)\,dF \tag{4.25}$$

If we assume that $P(F)$ is a delta function centered around F, it follows that $\alpha \sim \exp[- \text{const. } E/F_o]$; this is the energy dependence observed in the Urbach rule. By numerical calculations, Dow and Redfield (1972) showed that this is true also in the relevant part in the spectrum in the case that the electric fields have a Gaussian distribution with reasonable parameters; F_o in this case is $\sqrt{\langle F^2 \rangle} = F_{av}$ as in Eq. (4.19).

It is more difficult to account for the temperature dependence of Urbach tails in ionic crystals. As it was shown above (Eq. 4.21) $F_o \sim (T^*)^{1/2}$ and therefore this simple model does not give the proper temperature dependence, and one may hope a more correct consideration of the influence of non-uniform fields on exciton broadening may give the correct answer. Dow and Redfield (1972) point out some plausible reasons that it may be indeed so. Their arguments are based on the consideration that F_{av} is proportional not only to $(T^*)^{1/2}$ as shown above but also to the magnitude of the polaron cut-off vector q_c (cf. Section 4.4.4) which may be proportional to $(T^*)^{1/2}$.

Bonch-Bruevich (1970 a, b, 1972) calculated the absorption constant of amorphous semiconductors using his general theory of the optical properties of solids, and obtained under certain conditions exponential edges.

4.4.4 Internal Potentials
The long-wavelength component of internal potential fluctuations can be divided into 2 parts: the electrostatic part and the deformation or elastic

part* as defined in Figure 4.14. We shall first show that the deformation part produces a Gaussian shape of the edge (Dexter (1958), Tauc (1970 b)).

The deformation part produces fluctuations of the gap-width E_g^{opt} around its mean value E_{go}^{opt}. In the simplest case they are due to density fluctuations $\delta\rho/\rho$:

$$\delta E_g = E_g - E_{go} = (E_{lc} - E_{lv})\delta\rho/\rho \qquad (4.26)$$

where E_{lc} and E_{lv} are the deformation potentials of the conduction and valence band respectively. At thermal equilibrium, $\delta\rho$ has a Gaussian distribution $P(\rho)$ around $\delta\rho = 0$, and so does δE_g. We calculate α at $\hbar\omega$

$$\alpha(\hbar\omega) = \int P(E_g)\alpha_o(\hbar\omega - E_g)dE_g \qquad (4.27)$$

where $\alpha_o(\hbar\omega)$ is the absorption constant for $E_g^{opt} = E_{go}^{opt}$. For $\hbar\omega$ sufficiently below E_{go} we find

$$\alpha(\omega) \sim \exp\left[-(\hbar\omega - E_{go})^2/2(\delta E_g)^2\right] \qquad (4.28)$$

This Gaussian shape is not observed, and we may conclude that the density fluctuations are not responsible for Part B of the absorption edge (Tauc (1970b)). The fluctuations of E_g^{opt} may also be due to the non-homogeneity of the sample associated with the changing composition, as it has been considered by Inglis and Williams (1970). This effect may contribute to the edge broadening, however, no experimental evidence about it has been reported.

It is probable, therefore, that Part B is associated with the electrostatic part of potential fluctuations as discussed in the previous section. There are several possible sources of internal electric fields. One of them is density fluctuations via the piezoelectric effect in semiconductors with a piezoelectric constant different from zero. It has been shown (Tauc (1970)) that this effect is probably small.

The oscillating density fluctuations in crystalline solids are associated with longitudinal acoustic phonons. In an amorphous solid, the fluctuations do not change with time, and we may think about them as due to frozen-in phonons.

It has been suggested (Tauc (1970)) that in analogy with longitudinal acoustic phonons one may consider also frozen-in longitudinal optical phonons as a way to describe a kind of disorder in amorphous solids. It is plausible to assume that the 'frozen-in' phonons have a Gaussian distribution as the actual phonons have (Dow and Redfield (1972)). Electric fields produced by frozen-in phonons will be subject to screening effects, which may be taken approximately into account by limiting the extention of potential barriers to the screening length. The optical phonons can create electric fields large enough to explain Part B of the absorption edge with plausible assumptions (Tauc (1970)). With this interpretation of the absorption edge we would expect that the exponential broadening of the edge (Part B) should be present in compound semiconductors in which electric

---

*This terminology may be confusing as a deformation produces in general both symmetrical and antisymmetrical components.

dipole moment is non-zero. This seems to be in agreement with the observation of sharp edges in a-Ge and Si (to be discussed in Section 4.4.6). If 'frozen-in' phonons are responsible for Part B one would expect that E_e should become larger (the edge broader) with increasing ionicity of the compound.

Another source of internal electric fields are charged impurities. Redfield (1965) suggested to describe the fields F due to them by Haltsmark (1919) distribution. This distribution has a tail which decreases rather slowly towards $F \to \infty$. This circumstance makes it difficult to obtain exponential absorption tails without introducing, rather arbitrarily, a cut-off field (Krieg (1967), Afromowitz and Redfield (1968)). From the more recent work by Morgan (1965) it seems plausible to assume that the distribution of electric fields created by charged compensated impurities is Gaussian, and therefore the deduction of the exponential tail is essentially the same as for phonons (Dow and Redfield (1972)).

Another source of internal fields may be the fluctuations of chemical composition which may contribute both to the deformation and electrostatic parts of the internal potential fluctuations.

We have seen that there are many possible and plausible sources of potential fluctuations in amorphous solids which may be responsible for the broadening of the edges. It is rather surprising that with such a large variety of possible sources Parts B of many amorphous semiconductors are closely similar. This observation suggests that there may be some internal mechanism which adjust the average value of internal fields into relatively narrow limits.

Menth et al. (1971) have produced a change of the slope of Part B of As_2S_3 by doping with noble metals. Olley and Yoffe (1971) observed a similar effect induced by ionic bombardment in a-As_2Se_3.

The electrostatic part of internal potential fluctuations is seen indirectly in the optical spectra as a broadening of the edge. However, the fluctuations of the edge of the valence band (Figure 4.14) should be seen in the photoemission experiments in the analysis of the energies of emitted electrons. Nielsen (1971) has checked this point in As_2S_3, and did not find any evidence for the undulation of the valence band edge. From this it follows that the average value of the potential fluctuations is below the resolution of his experiments which he claims to be 0.15 eV. This is in agreement with our previous estimate of these fluctuations in Section 4.4.3. It seems therefore likely that the concentration of potential fluctuations extending deep into the gap is small. Most of these fluctuations have amplitudes below 0.1 to 0.2 eV.

4.4.5 Weak Absorption Tail

Below the exponential part of the absorption edge an absorption tail is observed (Part C in Figure 4.6). Its strength and shape were found to depend on the preparation, purity and thermal history of the material (Edmund (1966), Vaško (1969), Vaško et al. (1970)) even if the material is in the bulk form rather than as a film. It is, of course, difficult to study this

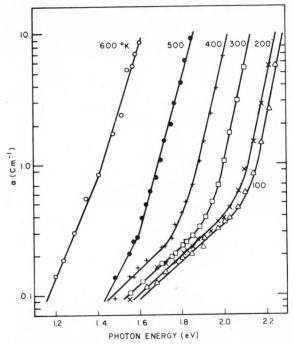

Fig. 4.16. Absorption edge of a-As$_2$S$_3$ at low absorption levels (regions B and C shown in Figure 4.6) at various temperatures (after Tauc *et al.* (1970b)).

absorption in thin films because of the low absorption levels. Recently, Wood and Tauc (1972) did a detailed study of this tail and its temperature dependence in very pure a-As$_2$S$_3$ and some other chalcogenide glasses. The absorption constant was below 0.5 cm^{-1}.

Absorption at such low levels may be only apparent, as the attenuation of the beam passing through the material may be due to light scattering. If the light scattering centers are spherical, their diameter small compared with the wavelength of light, and their concentration sufficiently low so that their interference effects can be neglected, the attenuation α_{scat} (cm^{-1}) (cf. e.g. Kerker (1969)) is

$$\alpha_{scat} = 745 \frac{\epsilon''^2}{\lambda^4} \left(\frac{\epsilon' - \epsilon''}{\epsilon' + 2\epsilon''} \right)^2 V_s^2 N \qquad (4.29)$$

where ϵ' and ϵ'' are the dielectric constants (relative to the dielectric constant of vacuum) of the medium and of the scatterers respectively, V_s the volume of one scatterer, N their concentration.

Light scattering was experimentally tested in a-As$_2$S$_3$ and the glasses shown in Figure 4.16 and 4.17 by Wood and Tauc (1971). The vitreous samples must be free of large macroscopic inhomogeneities which appear as striations. Two methods developed by Tynes and Bisbee (1970) were used.

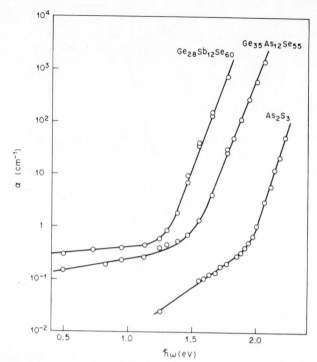

Fig. 4.17. Weak absorption tails in some chalcogenide glasses (after Wood and Tauc (1972)).

Either the light scattered perpendicularly to the transmitted beam was measured with a microscope and compared with some scattering standard, or the total scattered radiation was measured by putting the sample into a box with walls made of photocells. Samples of a-As_2S_3-Servofrax (cf. Table 4.1) scattered light from particles clearly visible in the microscope. The methods mentioned above yielded a value of $\alpha_{scat} \approx 0.04$ cm^{-1}, that is about an order of magnitude lower than that of the weak absorption tail shown in Figure 4.16.

Because of these experimental observations it seems legitimate to ascribe the weak absorption tail in the glasses shown in Figure 4.17 to optical transitions rather than scattering. This does not imply that this is always the case, and the scattering may be important in this part of the spectrum, especially for poorly prepared samples.

The optical transitions may be of the following kinds:
a) both the initial and final states are localized at one center
b) the initial and final states are localized at different centers
c) the initial state is localized and the final state extended, or vice versa.

It is difficult to imagine what centers would give the kind (a) transitions and produce part C of the absorption edge in As_2S_3 and As_2Se_3. Such transitions would rather form separate bands (cf. Section 4.3).

Transitions (b) would correspond to electron transfer from an impurity ion to an ion of the host lattice, or vice versa. In such cases, one has observed formulations of similar tails in crystals (for example impurities in garnets (Wood and Remeika (1966 and 1967)). Fe ions in a-As_2S_3 were studied by Tauc et $al.$ (1972).

Transitions (c) have been discussed as a possible interpretation of part C (Tauc et $al.$ (1970 b), Tauc and Menth (1972), Wood and Tauc (1972)). Transitions from the band tail states (localized) into the opposite band states (extended) were considered and Eq. (4.10) was applied. From the analysis of the data it invariably appeared that the oscillator strength f was considerably smaller than unity (10^{-2} to 10^{-1}). This was interpreted as evidence of the correlation effects (Dow and Hopfield (1972)) which reduce the effective optical cross-sections for transitions from localized states into delocalized states in glasses relative to the corresponding case in crystals. (cf. also Bonch-Bruevich (1972)). We may say in an intuitive way that in glasses, because of the internal potential fluctuations, the extended states close to the band gap exist only in a part of the sample and are not necessarily at every site of the centers. (Chapter 5). It was suggested that because of this effect glasses are more transparent below the absorption edge than crystals with the same content of impurities.

Actually, there is not sufficient evidence for this conclusion. First, part C of the edge has not been studied quantitatively from the point of view of transitions (b). Second, the oscillator strength f can be estimated from Eq. (4.10) only if we know the concentration of centers N which is not easily measurable. Earlier estimates of the concentration of localized states introduced by the loss of the long-range order were rather high (N ≈ 10^{18} to 10^{20} cm^{-3}) but later work suggests that the concentration of the states in the gap induced by the loss of the long-range order is much smaller. For example, from the measurements of the magnetic susceptibility Bagley et $al.$ (1972) and DiSalvo et $al.$ (1972) found that the concentration of singly occupied states in very pure vitreous S, Se, As_2S_3 and As_2Se_3 is below 3 × 10^{16} cm^{-3}. As one may expect a relatively large part of the states in the gap to be singly occupied (Pollak (1970), Kaplan et $al.$ (1972)) the total concentration of states in the gap (singly and doubly occupied) may be also small. Anderson (1972), however, suggested that the interaction of electrons with lattice may produce pairs of electrons located at centers which would not produce paramagnetic terms. Of course, additional states are introduced by defects (e.g. in non-annealed films (cf. Section 4.4.6)) and impurities which increase absorption in part C.

In connection with the optical observation of localized states in amorphous solids let us mention the work by Kolomiets et $al.$ (1970) on radiative recombination in vitreous As_2S_3 and As_2Se_3. They observed a luminescence band at an energy of about half the gap and attributed it to transitions involving levels at about the middle of the gap. However, they observed this band, somewhat shifted and sharpened with a much higher intensity also in the corresponding crystals. Fischer et $al.$ (1971) observed strong photoluminescence in a-$(As_2Te_3)_2 As_2Se_3$ films at low temperatures.

They interpret the observed luminescence peak and its temperature dependence as caused by the competition of the recombination and relaxation mechanisms, rather than evidence for a peak in the density of states. Bishop and Guenzer (1973) observed enhancement bands in the photoluminescence of chalcogenide glasses (cf. Bishop (1973)). Engemann and Fischer (1973) observed photoluminescence in a-Si.

4.4.6 Absorption Edge of Amorphous Ge and Si

The shape and energy of the absorption edge of a-Ge depends on its preparation and thermal history. At high absorption levels ($\alpha > 10^4$ cm^{-1}) Tauc *et al.* (1966) found the usual relation $\hbar\omega\alpha \sim (\hbar\omega - E_g^{opt})^2$ which was discussed in Section 4.4.2. Bahl *et al.* (1973a) found this relation also for a-Ge films deposited at 4.2 K and measured in situ. For as deposited films $E_g^{opt} = 0.35$ eV but by annealing at 300 K E_g^{opt} changed to 0.75, 0.83 and 0.85 eV at 300, 77 and 4.2 K, respectively.

Below $\alpha \approx 10^4$ cm^{-1} two different kinds of edges were observed in evaporated a-Ge films (cf. Figure 4.9). Some authors (Clark (1967) and others) observed a more or less exponential edge with a slope of about 0.15 eV at room temperature (curve 1 in Figure 4.7). Závětová and Vorlíček (1971) found that the edge shifted with temperature, with the rate of -3.8×10^{-4} eV/K at $\alpha = 10^3$ cm^{-1} between 80 and 300 K.

On the other hand, Donovan *et al.* (1969, 1970) found a sharp edge and no low energy tails (curve 2 in Figure 4.9). A sharpening of the edge at lower absorption levels was also reported by Chopra and Bahl (1970) and Theye (1970).

The results of Donovan *et al.* (1970) are in accord with the observation of the photoemission (Donovan and Spicer (1968), Spicer and Donovan (1970). On the basis of both experiments they concluded that there is no evidence for states in the gap the concentration of which would exceed 2×10^{17} cm^{-3}; they had to use, of course, some rather arbitrary estimates about the optical matrix elements.

A sharp edge in a-Ge is well understood on the basis of the considerations in Section 4.4.3. Contrary to compound semiconductors or Se, optical phonons in crystalline Ge have no dipole moment, because there are only 2 atoms in the elementary cell (cf. Zallen (1968)). Therefore, broadening by internal electric fields due to frozen-in optical phonons is not expected.

From this point of view, the broad edge observed by many authors (Clark (1967), Tauc (1968), Tauc *et al.* (1969). Závětová and Vorlíček (1971), Davey and Pankey (1969)) should be due to some other effect. It seems plausible to assume that the additional absorption observed in non-annealed samples is due to states in the gap. Brodsky and Title (1969) measured the ESR on thin films of a-Ge, Si and SiC. In non-annealed samples they found free spin concentrations of the order 10^{20} cm^{-3} which decreased drastically by annealing. From the measured g value of these states they concluded that they are associated with the dangling bonds on internal surfaces of voids.

From the work of Theye (1971 a, b) (Figure 4.18) and of Connell and Paul (1971) it appears likely that the same states are responsible for the additional optical absorption (broad absorption edges) observed in non-annealed a-Ge films. Donovan and Heinemann (1971) confirmed the existence of internal voids by direct electron-microscopic observation on very thin (100 Å) a-Ge films. Galeener (1971 a, b) predicted a band in the UV region due to the voids of linear dimensions of the order 10 Å which has been actually observed in non-annealed a-Ge films (cf. Section 4.5). Camphausen *et al.* (1971) showed that the results of their transport measurements are compatible with the existence of the states postulated by Brodsky and Title. This interpretation of the shape of the absorption edge of a-Ge was discussed by Paul (1972) and Connell *et al.* (1973).

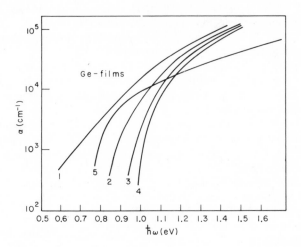

Fig. 4.18. Dependence of the absorption edge of a-Ge on the temperature of annealing. Curve 1: deposited at 20°C, nonannealed. Curves 2, 3, 4, 5: annealed at 200, 300, 400, 500°C respectively. During the annealing at 500°C the sample crystallized (after Theye (1971b)).

Some recent results appear to contradict the model which ascribes the broad edges to states in the gap which disappear on annealing. Knotek and Donovan (1973) deposited and measured in situ a-Ge films in ultra-high vacuum and found that the absorption edge in as deposited films was at 0.6 eV. This was corroborated by the photoconductivity measurements for which a threshhold near 0.56 eV was obtained. When these films were exposed to air and annealed at 400°C at 10^{-6} Torr a large decrease of absorption was observed between 0.6 and 1.2 eV. The reaction with oxygen was suggested as a possible reason for this behavior. Previously, Koc *et al.* (1972) and Závětová *et al.* (1972) reported some observations about the influence of oxygen on the absorption edges of a-Ge.

It is obvious that the absorption edges of a-Ge films depend very much on the conditions during the preparation and annealing. However, at the present state of research a model which would correlate the variety of observed dependencies is not available. The situation is similar for a-Si.

Grigorovici and Vancu (1968), Brodsky *et al.* (1970) and more recently by Donovan and Fischer (1971) studied the absorption edge in evaporated a-Si films. At high absorption levels ($\alpha > 10^4$ cm^{-1}) they found the quadratic dependence of $\alpha\hbar\omega$ on energy as shown in Figure 4.11 and discussed in Section 4.4.2. At lower absorption levels, Brodsky et al. (1970) observed a tail with an approximately exponential shape, and a sharpening at low energies; the edge moves towards higher energies and sharpens upon annealing, as was also observed in evaporated a-Ge films. Fischer and Donovan (1972) found considerably sharper edges in their samples of a-Si. There is a disagreement also in the photoemission experiments. Peterson et al. (1970) and Fisher and Erbudak (1971) found in a-Si evidence for an exponential tail in the density of states extending from the valence band on the basis of their photoemission measurements. Pierce and Spicer (1971, 1972) conclude from their measurements that there is no evidence for tailing.

Bahl et al. (1973) measured the dependence of optical absorption in evaporated a-Si films on the rate of deposition.

Stern (1971) calculated the density of states and optical absorption in a-Si assuming that the internal potential is the crystalline potential perturbed by a fluctuating potential with the root-mean-square amplitude 0.89 eV and the correlation length 6A. With these parameters, this model describes well the optical absorption edge as observed by Brodsky et al. (1970) and Beaglehole and Závětová (1970), and also some transport data. He does not consider any long-wave-length fluctuations such as those considered in Section 4.4.3.

A different approach for analyzing the absorption edge of a-Si films was suggested by Brodsky et al. (1972b). These authors used the spin concentration measured by ESR as a measure of the film perfection, and extrapolated the optical spectra to zero spin concentration. For a-Si they found a sharp edge at 1.8 eV.

The results of Donovan, Spicer, Theye and others suggest that it is possible to prepare films of a-Ge and Si with low concentrations of states in the gap. D. Weaire (1971) and V. Heine (1971) consider a simple tightbinding model for the electronic structure of a-Ge and obtain sharp band edges if they neglect the fluctuations of the parameters. It is probably impossible to construct a non-crystalline covalent structure without a spread of bondangles and lengths which produces a tailing of states into the gap. Phillips (1971a, b) and Van Vechten (1972) suggests that this broadening of the edges should be small because of the following general principle:

In carefully prepared amorphous films the atoms arrange themselves selfconsistently by maximizing bond energies. A large concentration of states in the gap is energetically unfavorable, and therefore during the self-consistent rearrangement these states are ejected above the band edge. It appears therefore that a purely geometrical disorder does not introduce large concentrations of states in the gap, in accordance with the measurements of sharp edges in carefully prepared ('defect-free') a-Ge and a-Si.

The existence of energy gaps in amorphous covalent semiconductors was also discussed by Klíma and McGill (1970).

Beaglehole and Závětová (1970) studied absorption in amorphous Ge-Si alloys at high absorption levels and their results are discussed in Section 4.5.

4.5   FUNDAMENTAL ABSORPTION BAND

A typical property of the fundamental absorption band of crystalline semi-conductors is a relatively sharp structure (cf. Figures 4.1, 4.19 to 4.24).* This is due to the singularities of the joint density of states (cf. e.g. Tauc (1965), Phillips (1966)) to which ϵ_2 is proportional in the crystals. These singularities depend on the existence of the k-vector as a good quantum number and its conservation during the transition. It is therefore a typical long-range effect. Actually, the sharp structure of the fundamental absorption band is one of the few fundamental properties of semiconductors which is basically impossible to interpret without invoking the long-range order.

This point of view is fully confirmed by the changes in the fundamental absorption spectrum occurring when long-range order disappears. In Figures 4.19, 4.20 and 4.21 we see that the fundamental absorption bands are often smooth single bands; for some materials (Figures 4.22, 4.23, 4.24) there are 2 smooth bands.

In the following, we compare the spectra of the corresponding amorphous and crystalline materials. In this connection another feature is significant, the shift of absorption which may occur in either direction. Going from the crystal to the amorphous material, this shift is towards smaller energies in Ge, Si and III—V compounds, and in the opposite direction in Se and Te.

The fundamental absorption bands were determined from the measurements of the reflectivity at normal incidence and the subsequent use of Kramers-Kronig analysis (cf. e.g. Tauc (1965)). To be able to perform this analysis it is necessary to make some assumptions about the extrapolation of the reflectivity curve above the maximum measured frequency to infinity. The shape and position of the sharp structure observed in crystal spectra is little affected by various choices of this extension. However, the position of a smooth band as observed in amorphous semiconductors may be shifted differently with different extrapolations. The data might be also affected by the surface condition. To be sure about the absorption shift, Tauc and Abrahám (1970) determined the absorption constant in a-Ge by a different method, the measurement of the transmission of thin a-Ge films, and confirmed the results obtained by Kramers-Kronig analysis.

The observed shift makes the interpretation of the fundamental absorption band non-trivial. Tauc et al. (1965) suggested that the absorption band

---

*At the absorption edge, we plot the frequency dependence of the absorption constant α, which is proportional to the energy absorbed in the material if, in the frequency range considered, the index of refraction n is constant. If, in addition, the oscillator strength f can be assumed constant, then α is proportional to the properly combined densities of initial and final states. However, in a broad frequency range n cannot be considered constant and it is necessary to use the correct formula for the absorbed energy. It is proportional to the imaginary part of the dielectric constant $\epsilon_2 = 2nk$ where $k = c\alpha/2\omega$ (cf. e.g. Tauc (1965). It is ϵ_2 which we plot for the fundamental absorption bands.

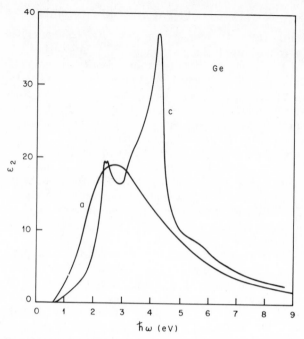

Fig. 4.19. Fundamental absorption bands of a-Ge and c-Ge (after Spicer and Donovan (1970a)).

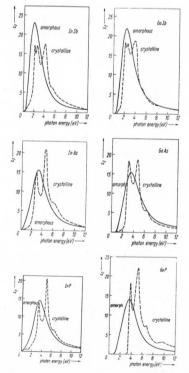

Fig. 4.20. Fundamental absorption bands of III-V compounds (after Stuke and Zimmerer (1972)).

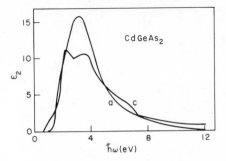

Fig. 4.21. Fundamental absorption bands of a-CdGeAs$_2$ and c-CdGeAs$_2$ (after Tauc *et al.* (1968); cf. also Gorjunova *et al.* (1970)).

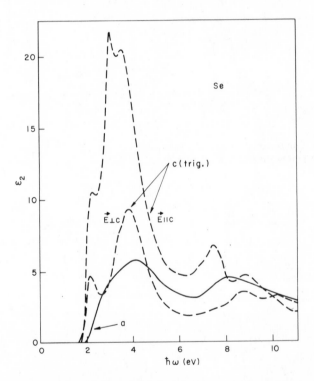

Fig. 4.22. Fundamental absorption band of a-Se, and c-Se for 2 directions of light polarization (after Stuke (1970)).

of amorphous Ge can be understood by assuming that only energy is conserved during the optical transitions and the optical matrix element is constant in the whole energy range. In this case $\omega^2 \epsilon_2$ is proportional to the energy conserving convolution of the density of states of the conduction and valence bands

$$\omega^2 \epsilon_2 \sim \int g_v(E) g_c \, (\hbar\omega + E) \, dE \qquad (4.30)$$

As a first approximation it was assumed that the band state densities may change little by the transition from the crystalline to the amorphous state.

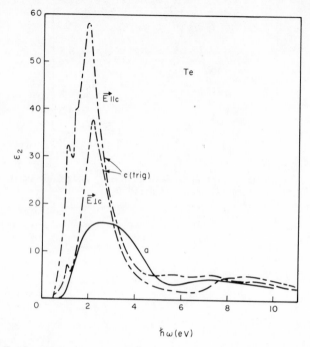

Fig. 4.23. Fundamental absorption band of a-Te, and c-Te for 2 directions of light polarization (after Stuke (1970)).

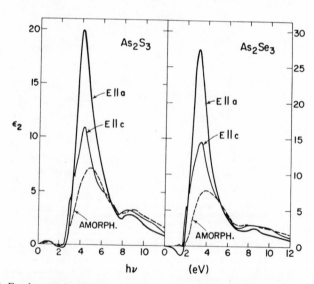

Fig. 4.24. Fundamental absorption bands of crystalline and amorphous As_2S_3 and As_2Se_3 (after Drews *et al.* (1972)).

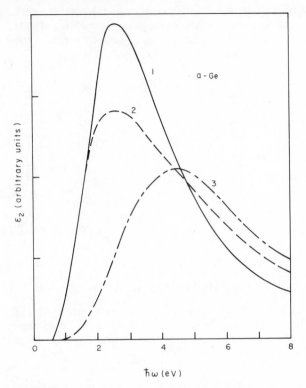

Fig. 4.25. Imaginary part of the dielectric constant ϵ_2 of a-Ge. Curve 1: measured values. Curve 2: ϵ_2 calculated from Eq. (4.30) using the band state densities g_v, g_c determined from photoemission measurements and shown in Figure 4.27. Curve 3: ϵ_2 calculated from Eq. (4.30) with the densities g_v, g_c calculated for c-Ge (after Donovan and Spicer (1968)).

However, actual computation by Donovan and Spicer (1968) of the integral in Eq. (4.25) using g_v, g_c calculated for c-Ge by Herman et al. (1967) gave an absorption without structure but peaked at about the same energy as in c-Ge (4.4 eV), in disagreement with experiment (Figure 4.25). It was clear, therefore, that if the 'non-direct, constant matrix model' (Berglund and Spicer (1964)) is applicable, it was necessary to assume important changes in the band state densities.

Herman and Van Dyke (1968) suggested a possible origin of the difference: a-Ge has a lower density, and it is therefore necessary to use g_v, g_c for a dilated Ge crystal. In fact, the densities shift in the desired direction; unfortunately, to obtain complete agreement with experiment it is necessary to assume that the density of a-Ge is as much as 28% smaller than that of c-Ge. This value was actually reported by Clark (1967), however, later determinations of the densities (Donovan et al. (1970)) have shown that a difference of the 'true' density larger than 10% is improbable. Larger differences are very likely due to the presence of macroscopic voids. Herman's suggestion can therefore explain only a part of the observed shift.

A different simple explanation was suggested by Brust (1969 a, b). He starts with the formula for absorption in crystals

$$\epsilon_2(\omega) = \frac{4\pi^2 e^2}{m^2 \omega^2} \frac{2}{8\pi^3} \sum_{s,n} \int_{BZ} d^3k |p_{n\underline{s}}(k)|^2 \, \delta(\hbar\omega - E_n(k) + E_s(k)) =$$

$$= \frac{4\pi^2 e^2}{m^2 \omega^2} \frac{2}{8\pi^3} \sum_{s,n} \int_0^{\hbar\omega} dE_f |p_{ns}(k)|^2 \, \delta(E_f - E_n(k)) \, \delta(E_f - \hbar\omega - E_s(k))$$

$$(4.31)$$

where the summation $\sum_{s,n}$ is over all unfilled (n) and filled (s) bands, p(k) is the optical matrix element between these bands, E_f is the energy of the final state. In amorphous semiconductors the δ-functions are replaced by broader spectral weight functions, the first one by

$$A_n(k, E_f) = \frac{\mathrm{Im}\,\Sigma_n(k,E_f)/\pi}{[E_f - \mathrm{Re}\,\Sigma_n(k,E_f) - E_n(k)]^2 + [\mathrm{Im}\,\Sigma_n(k,E_f)]^2} \qquad (4.32)$$

and the second one by $A_s(k,E_i)$ of the same form with $E_i = E_f - \hbar\omega$. $\Sigma_n(k,E_f)$ is the complex self-energy function. We shall neglect the real part of the self-energy and its k-dependence. For the imaginary part we have

$$\mathrm{Im}\Sigma(E_f) = \pi |M_{scat}|^2 \rho(E_f) \qquad (4.33)$$

and similarly for $\mathrm{Im}\Sigma(E_i).\rho(E_f)$, $\rho(E_i)$ refer to the densities of final and initial (hole) states. The scattering matrix element between the crystalline states ψ_{nk} and amorphous states $\psi_{n'k'}$

$$M_{scat} = \langle \psi_{nk} | H_{scat} | \psi_{n'k'} \rangle \qquad (4.34)$$

is assumed to be a constant adjustable parameter. This simple relation has been justified for electron-phonon-scattering (Brust and Kane (1963)). Brust suggested its use for the electron scattering due to disorder.

The calculations have shown that the band-state densities

$$g(E_f) = \int g_c(E_k) A_n(E_k, E_f) dE_k \qquad (4.35)$$

are little changed compared with the densities in the crystal but that the transition probabilities significantly decrease with photon energy. Therefore the decrease of absorption at high energies (disappearance of the peak at 4.4 eV) is ascribed to this effect. Brust obtained good agreement between his calculations and the experimental $\epsilon_2(\omega)$.

A similar but more general method for the calculation of the broadening of the crystalline band structure by disorder was developed by Kramer (1970) ('complex band structure' calculations). With a few adjustable parameters he obtained a good agreement with experiment for the optical absorption band of a-Se (Kramer et al. (1970)) and also for the electron photoemission from a-Se (Laude *et al.* (1971). Also the absorption band of a-Ge, Si and III—V compounds can be well described by this theory (Kramer (1971, 1972)). In Brust's and Kramer's methods the wavefunctions in the disordered state will remember the k-vector of the crystalline wavefunctions from which they originate and therefore the k-vector plays a role even for the transition in the amorphous solid. On the contrary, in the constant-matrix, nondirect transition model only energy conservation is considered.

Spicer and Donovan (1970b) argue that the photoemission data on a-Ge cannot be understood on the basis of a theory which interprets the difference between the crystalline and amorphous spectra by a change of the transition probabilities only as the simple Brust's theory does. The change of state densities is essential. We shall discuss their argument.

In an external photoemission experiment, one measures the energy distribution of electrons emitted from a solid upon irradiation with photons of energy $\hbar\omega$. A photon of this energy excites an electron from an initial state E_i in the valence band into a final state $E_f = E_i + \hbar\omega$ in one of the conduction bands of the solid. If E_f exceeds the minimum energy which the electron must have to be able to leave the solid, it can escape. One measures the energy distribution curve of emitted electrons $N(E,\hbar\omega)$ at various photon frequencies $\hbar\omega$. The observed $N(E,\hbar\omega)$ is affected by the scattering processes which the excited electrons undergo before leaving the solids. The usefulness of the photoemission method is based on the observation that often the scattering process does not wipe out from the emitted electrons all the information about their origin. Electron-phonon interactions often change the energies of excited electrons insignificantly. Auger interaction (electron-electron) changes their energies so drastically that the escaping scattered electrons have much lower energies than the nonscattered ones and can be easily distinguished from them (cf. Spicer and Eden (1968)). Under such conditions, E differs from E_f by a constant energy. The structure of the energy distribution function can give us more information about the electronic states in the solid than the optical data, because we essentially measure the average transition probabilities only to a limited assembly of states (the states in the conduction bands with a fixed energy E_f) rather than to the whole assembly of conduction band states.

A point of interest here (cf. Spicer and Eden (1968)), is that photoemission studies allow us to distinguish between direct (k-vector conserving) and non-direct transitions. For direct transitions, the structure of $N(E,\hbar\omega)$ is determined by the joint density of states which has peaks at particular values of $\hbar\omega$. Therefore structure appears and disappears as $\hbar\omega$ is varied (cf. Spicer and Eden (1968)). For nondirect transitions only energy is conserved, and

$$N(E,\hbar\omega) \sim g_c(E)g_v(E-\hbar\omega) \tag{4.36}$$

Therefore, if we observe a peak in the energy distribution curve at the position E_o which changes with $\hbar\omega$ so that $\Delta E_o = \Delta\hbar\omega$, we may ascribe it to a peak of the density of states in the valence band and consider it as an evidence for non-direct transition. Donovan and Spicer (1968) interpreted

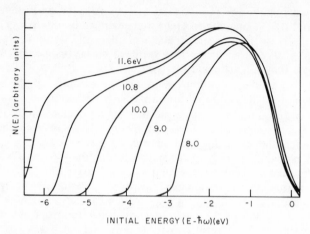

Fig. 4.26. Electron distribution curve in a-Ge at various photon energies in a-Ge (after Spicer and Donovan (1970b)).

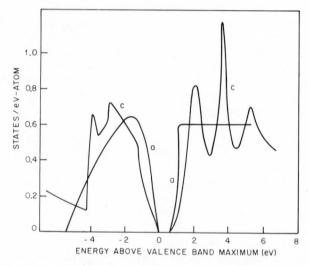

Fig. 4.27. The density of states $g_v(E)$, $g_c(E)$ in the valence and conduction band of a-Ge as determined from the photoemission data compared with the values calculated for c-Ge (after Donovan and Spicer (1968)).

their measurements on a-Ge (Figure 4.26) in this way. The high-energy part of the energy distribution curve is always due to transitions from the top of the valence band. We see that the leading edges, plotted as a function of $E-\hbar\omega$, superpose well; therefore Eq. (4.36) applies. The transitions are non-direct and one can use Eq. (4.36) to estimate the density of the valence band states. Donovan and Spicer (1968) found $g_c(E)$ to be a flat function and obtained for $g_v(E)$ the result shown in Figure 4.27. Its important feature is the shift of the maximum of $g_v(E)$ towards the top of the band. With g_v and g_c determined in this way they calculated ϵ_2 from Eq. (4.30) and obtained good agreement with experiment (Figure 4.25).

Brust's calculations do not give correctly the experimental curves shown in Figure 4.26 and Spicer and Donovan (1970b) show that the disagreement is due to the partial conservation of the k-vector in Brust's calculation. They conclude that the photoemission data can be understood only on the basis of a non-direct model with the important consequence that the band state densities are appreciably different in the amorphous and crystalline states.

An approach in which the amorphous solid is not considered as a perturbed crystal was suggested by Phillips (1971a). He proposed a model which is isotropic and independent of the crystalline structure, except for the assumption that the local symmetry corresponding to the covalent bonds of the basic unit is conserved but distorted.

Phillip's starting point is the similarity of the observed absorption band of a-Ge and the absorption band calculated on the basis of Penn's model (Penn (1962)). This is a nearly free electron model isotropic in three dimensions. It has only one energy gap E_G appearing at the Fermi surface. Its value in the crystal can be estimated from the formula suggested by Heine and Jones (1969)

$$E_G = 2V_s(220) + 2|V_s(111)|^2/\Delta T \qquad (4.37)$$

where $V_s(220)$, $V_s(111)$ are Fourier components of the pseudopotential at $k = (220)$ or (111) respectively,

$$\Delta T = (\hbar^2/2m)[(110)^2 - (001)^2].$$

In c-Ge, $\Delta T = 4.6$ eV and $V_2(220) = 0.1$, $V_s(111) = 3.1$ eV. From Eq. (4.37) we obtain $E_G = 4.4$ eV.

The gap E_G is not the minimum gap E_g which determines the absorption edge and temperature dependence of carrier concentration, but the average gap corresponding to the separation of the bonding and antibonding states in the covalent bond picture. It has fundamental importance for many basic properties of the solid and appears in the theories of the relation between the chemical and optical properties of crystals (Phillips (1970)).

Bardasis and Hone (1967) calculated the imaginary part of the dielectric constant $\epsilon_2(\omega)$ using Penn's model. For $\hbar\omega > E_G$ it is approximately equal

$$\epsilon_2(\omega) = \frac{\text{const.}}{(\hbar\omega - E_G)^{1/2}} \qquad (4.38)$$

This formula is applicable for describing the high energy part of the fundamental absorption band in crystals.

The typical features of the fundamental absorption band of a-Ge (Figure 4.19) — a single band which is much steeper at low energies than at high energies — is qualitatively well reproduced by Eq. (4.38) if we allow for some broadening. Indeed, Penn's model may be useful for amorphous semiconductors because it is isotropic and does not contain more gaps than just the fundamental gap E_G which is responsible for the one broad peak; both optical and photoemission data do not give any evidence about additional gaps present in the crystal.

Cardona and Pollak (1971) suggested a Gaussian broadening of the band described by Eq. (4.38) and improve so the agreement with experiment (cf. also Stuke and Zimmerer (1972)).

There is, however, a difficulty in explaining the shift of the peak of $\epsilon_2(\omega)$ from 4.4 eV in c-Ge (which agrees with the Penn gap E_G) to 2.5 eV observed in a-Ge. According to Eq. (4.38) the gap is predominantly determined by the second term on the right hand side of Eq. (4.37), which corresponds to the second-order perturbation, rather than by the first term corresponding to the first-order perturbation. The reason is that the Fourier components of the pseudopotential $V_s(111)$ are much larger than $V_s(220)$. In other words E_G is determined by the component related to the covalent bond. Phillips (1971a) suggested some possible explanations of the decrease of E_G.

Stuke (1970) noticed that the intensity of the E_2 peak in the spectra of ZnS depends on the crystallographic modification (zincblende or wurzite) much more than the E_1 peak. He suggested that the E_2 peak which dominates the spectra of crystalline tetrahedral semiconductors is much more sensitive to the long-range order than the E_1 peak. In his interpretation, this effect is the reason of the shift of the maximum of the absorption band towards the E_1 peak observed in the amorphous forms. Ortenburger *et al.* (1972) calculated the band state densities and ϵ_2 for a hypothetical hexagonal structure of Ge, and observed a shift of E_2 towards smaller energies.

Joannopoulos and Cohen (1972) considered several polytypes of Ge (of which some exist as high-pressure phases) and found that the band state density in Ge III (which contains 12 atoms per unit cell with 2 interatomic distances and a considerable range of angles) is similar to that of a-Ge as determined by photoemission measurements. The peak dominating the spectrums of GeIII is due to transitions along the [001] direction in the Brillouin zone and should be classified as E_2 peak (rather than the E_1 peak associated in c-Ge with transitions along the [111] direction). Joannopoulos and Cohen (1972) could also distinguish between the features of the band state density curves which depend only on the tetrahedral coordination and features which are sensitive to the more distant order, and draw some conclusions about the amorphous state. They atribute the similarity between Ge III and a-Ge to the presence of pentagonal rings in both forms. They conclude that it is the difference in the short-range order (extending, of course, the first coordination sphere) which is responsible for the difference in the band state densities and the optical absorption between the crystalline and amorphous sorms of Ge and Si. In the previously discussed calculations the short-range order was assumed to be the same in both cases, and only the loss of the long-range order was considered. Joannopoulos and Cohen ((1972, 1973)) argue that this approach is not adequate for Ge, Si and similar semiconductors and, indeed, their results explain remarkably well the observed facts which the previous theories were unable to explain.

Tight binding methods were developed for the calculation of the electronic states in amorphous solids by Weaire (1971), Weaire and Thorpe (1971), Thorpe and Weaire (1971a, b), Heine (1971), Thorpe and Weaire (1971a, b) and L. Roth (1972). Their calculations of the band state densities are explicitly based on the consideration of the short-range. Chen

(1973a, b) did molecular-orbital calculations of a-Se, As_2S_3 and As_2Se_3.

Wiech and Zopf (1972) measured the soft X-ray emission spectra of a-Si, Eastman (1972) and Ley et al. (1972) UV and X-ray photoemission from a-Si and a-Ge and obtained information about the state densities in the valence band. They observed two bands as predicted by Thorpe and Weaire (1971) instead of the three bands observed in crystals (Figure 4.28). This experimental result agrees with the above mentioned considerations of Joannopoulos and Cohen (1972).

Brown and Rustgi (1972) measured extreme ultraviolet transmissions of c-Si and a-Si and obtained information about the electronic structure of the conduction bands. The interpretation of the results is complicated by the Coulombic potential of the core hole. Cardona et al. (1970) measured absorption in Se between 50 and 70 eV and did not find any significant differences between a-Se and c-Se. Vaško (1970) collected the experimental data and computed the optical constants of a-Se in the range 10^{-2} to 10^5 eV.

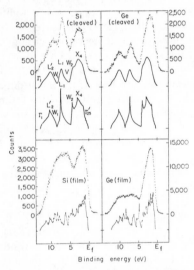

Fig. 4.28. X-ray photoelectron emission spectra for crystalline and amorphous Ge and Si as measured by Ley et al. (1972). The solid curves were calculated by Joannopoulos and Cohen (1972) for the diamond structures (the middle curve in the upper graph is a broadened version of the theoretical curve below it), and for the Ge III structure (lower graph) (After Ley et al. (1972)).

Photoemission studies of amorphous materials with ultraviolet or x-ray radiation or both have been performed on amorphous Si (Peterson et al. (1970), Fischer and Erbudak (1971), Erbudak and Fischer (1972), Pierce and Spicer (1971, 1972)), Ge (Donovan and Spicer (1968), Spicer and Donovan (1970a), Ribbing et al. (1971)), Ge doped with As (Ribbing and Spicer (1971)), Si and Ge (Pierce et al. (1972), Ley et al. (1972)), GeTe (Fisher and Spicer (1972, 1973), Shevchik et al. (1973a, 1973b)), Ge_xTe_{1-x} (Fisher et al. (1973a)), S and Se (Nielsen (1973)), Se (Nielsen (1972b), Laude et al. (1971)), Se and Te (Laude and Fitton (1972), Shevchik et al. (1973c)), Se and As_2Se_3 (Nielsen (1971)), As_2S_3 and As_2Se_3 (Nielsen (1972a)), Sb_2Se_3 (Wood et al. (1973)), As, Sb, and Bi (Ley et al. (1973)).

The electron energy loss function in c-Ge and a-Ge was measured by Zeppenfield and Raether (1966). They observed a small shift of the peak from 15.6 to 15.5 eV. The shift is in the expected direction (the peak energy is proportional to the square root of density) and corresponds to a small change of density. The shape of the electron energy loss function is similar in both cases. The same result was found from optical measurements. The electron energy loss function is proportional to $-\operatorname{Im}\epsilon^{-1}$ ($\epsilon = \epsilon_1 + i\epsilon_2$) is here the complex dielectric constant, cf. Tauc (1965). This quantity was determined for a-Ge by Donovan *et al.* (1970b). At high energies ($\hbar\omega \gg E_G$) there appears to be little difference between the optical constants of amorphous and crystalline solids; of course, the optical properties in this region are very sensitive to surface conditions, and it is difficult to obtain reliable results on the bulk optical constants. Didden (1972) measured the electron energy loss functions of a-Se and a-Te.

An interesting effect in the UV region was predicted by Galeener (1971a) and found in agreement with the UV data on a-Ge (Galeener (1971b)). Galeener showed that the submiscropic voids postulated in a-Ge to explain ESR and other experimental results (cf. Section 4.4.6) should diminish the value of the maximum of $\epsilon_2(\omega)$ spectrum, and in addition produce a peak due to the void resonance. Jung (1971) measured ϵ_2 of a-Ge in the range 2-4eV as a function of substrate temperature during the deposition. The observed changes of the maximum of $\epsilon_2(\omega)$ are in accord with Galeener's theory. The void resonance can explain a peak observed by Donovan *et al.* (1970b) in $\epsilon_2(\omega)$ of a-Ge at about 8eV. If the linear dimensions of the voids are much smaller than the wavelength of light, the resonance peak depends only on the optical constants of the host material, the volume fraction δ of the voids and their depolarization factor L. Analyzing the experimental data, Galeener found the film considered $\delta \approx 0.05$ and $L \approx 0.08$. The large depolarization factor suggest that the voids are disks with the normals randomly oriented in the plane of the film, corresponding to very fine cracks perpendicular to the plane of deposition (cf. Section 1.4.2). Such cracks may exist between the islands which started to grow simultaneously at various places of the substrate. A more detailed experimental work (Bauer *et al.* (1971)) appears to be consistent with Galeener's theory.

Beaglehole and Závětová (1970) found that the fundamental absorption band of a-Si is similar to that of a-Ge; the maximum is at 3.1 eV and the band is narrower (at $\epsilon_2 = 10$ its width is 2.1 eV as compared to 3 eV in a-Ge). They also observed a smooth shift of the band in Ge-Si alloys with the change in composition from Ge to Si without any additional broadening due to the compositional disorder.

As seen in Figure 4.22 and 4.23 the absorption in a-Se and Te shifts significantly to higher energies relative to the crystalline forms. This has been ascribed to the decrease of interaction between the chains in the disordered structures (Stuke (1970)).

4.6   INDEX OF REFRACTION, SUM RULES

Polarizability of solids is proportional to the real part of the dielectric constant $\epsilon_1 = n^2 - k^2$, where n is the index of refraction and $k = c\alpha/2\omega$. In the region where the material is transparent , the index of refraction $n = \sqrt{\epsilon_1}$.

The frequency dependence of ϵ_1 in the fundamental absorption bands of c-Ge and a-Ge are compared in Figure 4.29.

Polarization and absorption are interconnected by the Kramers-Kronig dispersion relation (cf. e.g. Tauc (1965))

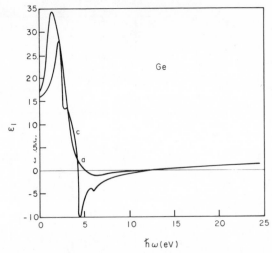

Fig. 4.29. The real parts of the dielectric constant of a-Ge and c-Ge in the fundamental absorption band (after Tauc and Abrahám (1969)).

$$\epsilon_1(\omega) = 1 + \frac{2}{\pi} \int\limits_0^\infty \frac{\xi \epsilon_2(\xi)}{\xi^2 - \omega^2} d\xi \tag{4.39}$$

which was deduced under very general conditions and is valid for amorphous materials and liquids as well as for crystals.

Let us now consider the electronic part of the index of refraction at zero frequency. It is identical to the index of refraction measured at very low frequencies if there is negligible absorption due to lattice vibrations (as in a-Ge, a-Si), impurities or defects. If they are not negligible, we must determine n at low enough frequencies, but still above the onset of these absorption mechanisms. Or, alternatively, we may define it from relation (4.39) putting $\omega = 0$.

$$n_0^2 = n^2(0) = \epsilon_1(0) = 1 + \frac{2}{\pi} \int\limits_0^\infty \omega \epsilon_2(\omega) d\omega \tag{4.40}$$

where $\epsilon_2(\omega)$ is due to electronic transitions only.

In Ge, Si and III–V compounds and CdGeAs$_2$ n_0 is equal or somewhat larger in the amorphous state than in the crystalline state; in Se and Te the change is in the opposite direction and is large (Table 4.2). The large decrease of n_0 in a-Se and Te is associated with the large shift of the absorption spectrum towards higher energies (Figure 4.22 and 4.23). Poly-crystalline films of GeTe have $n_0 = 6$ while in a-GeTe films $n_0 = 3.76$ (Fischer and Spicer (1973); such large changes of n_0 are observed also in other compounds containing Te.

The large change of the index of refraction, and consequently of the reflectivity, of the system Te$_{81}$Ge$_{15}$X$_4$ where X consists of Group V and VI additives were studied by Feinleib et al. (1970) in an arrangement which may be useful for technical applications. They exposed amorphous films to short intense laser pulses. The crystallized spots on the films had different transmission and reflectivity in appropriate frequency ranges. The spots could be erased by an application of laser pulses of different integrated intensity (cf. Chapters 1 and 6). Crystallization was believed to be partially

J. Tauc

TABLE 4.2 Optical Properties of Simple Amorphous Materials

	density (g/cm³)		n_O		E (α = 100 cm⁻¹) (eV)		E_g (eV)	E_e (meV)	$dn_O/n_O dp$ (10⁻³ kbar⁻¹)		dE_g^{opt}/dp (meV/kbar)		$\dfrac{n_\parallel - n_\perp}{stress}$ (kbar⁻¹)
	c	a	c	a	c	a	a	a	c	a	c	a	a
Ge	5.35[a]	from 4.54[b] to 5.4[b]	4.01[e]	4.05[e]	0.76[f]	0.59[f] 0.99[e]	1.0[e]		− 1.0[e]	− 0.8[e]	5.0[e]	3.5[e]	
Si	2.33[a]	from 1.63[c] to 2.26[c]	3.42[e]	3.5[e]	1.23[g]	0.69[g] 1.33[e]	1.4[e]		− 0.3[e]	− 0.05[e]	− 1.5[e]	0.25[e]	
InSb	5.77[d]	5.3 ± 0.5[d]	3.94[d]	4.5[d]	0.17[h]	0.3[h]							
InAs	5.66[d]	4.9 ± 0.5[d]	3.44[d]	3.6 to 4[d]									
InP	4.79[d]	4.8 ± 0.5[d]	3.09[d]	3.5[d]									
GaSb	5.61[d]	5.5 ± 0.5[d]	3.81[d]	4.3[d]									
GaAs	5.31[d]	5.0 ± 0.5[d]	3.28[d]	3.6[d]	1.38[i]	0.59[e]	0.61[e]		− 0.7[e]	0.7[e]	11.3[e]	0.7[e]	
GaP	4.13[d]	4.0 ± 0.5[d]	2.93[d]	3.1 to 3.2[d]	2.19[j]	0.40[e]	0.42[e]		− 0.3[e]	− 0.25[e]	− 1.1[e]	0.2[e]	
CdGeAs₂	5.58[k]	5.72[k]	3.40[l]	3.83[l]	0.55[l]	0.82[l]		52[l]					

	density g/cm³ (a)	density g/cm³ (c)	n_o (a)	n_o (c)	$E(\alpha = 100\,cm^{-1})$ eV (a)	$E(\alpha = 100\,cm^{-1})$ eV (c)	E_g eV (a)	E_e meV (a)	$dn_o/n_o dp$ 10^{-3}kbar^{-1} (a)	$dn_o/n_o dp$ (c)	dE_g^{opt}/dp meV/kbar (a)	dE_g^{opt}/dp (c)	$\frac{n_\parallel - n_\perp}{stress}$ kbar^{-1} (a)
Se	4.28[a]	4.81 trigonal[a] 4.50 monocl.[a]	2.50[n]	∥ 3.41[m] ⊥ 2.64	1.75[o]	1.74[o]		59[o]	11.0[zi]				8.4 × 10^{-4}zj
Te		6.25[a]	3.4[p]	5.3[p]	0.62[o]	∥ 0.37[o] ⊥ 0.32[o]		55[o]					
As$_2$S$_3$	3.20[a]	3.43[a]	2.41[s]	2.98 (2.04) 2.64[r]	2.26[t]	2.50[t] 2.54	2.36[t]	59[u]	7.3[zk]				5.4 × 10^{-4}zk
As$_2$Se$_3$	4.58[v]	4.75[v]	3.5[z]		1.62[w]	1.78[z]	1.76[x]	50[w]			-7.6[zl]	-14[zl]	
As$_2$Te$_3$					0.70[y]		0.82[y]	53[y]					
GeS	3.49[za]	4.24[za]	2.3[zb]	3.5[zb]	1.41[za]			98[za]					
GeTe	5.6 ± 0.5[zc]	6.17[zc]	3.8[ze]	6[zb]	0.52[zh]	zg	0.70[zh]	68[zd]					

Caption to Table 4.2

c . . . crystalline, a . . . amorphous
n_0 is the index of refraction at $\omega = 0$ corresponding to the electronic transitions.
E is photon energy for $\alpha = 100$ cm.$^{-1}$
E_g is the optical gap defined in Eq. (4.2) (r = 2).
E_e determines the slope of the Urbach tail (Eq. 4.7).
$dn_0/n_0 dp$ and dE^{opt}/dp were measured with hydrostatic stress.
E^{opt} in amorphous materials was E_g in Ref. e, but is defined differently in Ref. zl.
$(n_\parallel - n_\perp)$/stress is the piezobirefringence constant
The parameters of amorphous films vary considerably with the conditions of preparation and annealing as discussed in the text.
The data were taken at room temperature.

a   Handbook of Chemistry and Physics (Chemical Rubber Co), 52nd edition, 1972.
b   Donovan *et al.* (1970a). The upper value is for films containing voids, the lower
    value for samples with a considerably reduced concentration of voids.
c   Brodsky *et al.* (1972c). Same comment as in b applies.
d   Stuke and Zimmerer (1972). Density and n_0 of amorphous films depend on sample
    preparation but this point has not been studied in III–V compounds in so much
    detail as in Ge and Si films (cf. Connell and Paul (1972)).
e   Connell and Paul (1972). The values were measured on annealed amorphous films.
    The dependence of n_0 on sample preparation is discussed in the text.
f   Figure 4.9
g   Donovan and Fischer (1972).
h   Figure 4.10
i   Hill, D. E., (1964), *Phys. Rev.,* **133**, A 866
j   Tietjen, J. J. and Amick, J. A., (1966), *J. Electrochem. Soc.,* **113**, 724.
k   Červinka *et al.* (1970b).
l   Tauc *et al.* (1968).
m   Gampel, L. and Johnson, F. M., (trigonal modification 1969), *J. Opt. Soc. Amer.*
    59, 72.
n   Koehler, W. F., Odencrantz, F. K. and White, W. C., (1959), *J. Opt. Soc. Amer.,* **49**,
    109.
o   Figures 4.8 and 4.13.
p   Stuke, (1970) (polycrystalline and amorphous films).
r   Evans and Young (1967). The three values of n_0 correspond to E-vector of radi-
    ation parallel to a, b, c axis respectively. The value of n_0 for E∥b was only esti-
    mated.
s   Rodney *et al.* (1958).
t   Figures 4.7, 4.11, 4.16
u   Wood and Tauc (1972).
v   Kolomiets (1964).
w   Edmond (1966).
x   Davis and Mott (1971).
y   Rockstad H. K., (1970), *J. Non-Crystalline Solids,* **2**, 192.
z   Shaw, R. F., Liang, W. Y., Yoffe, A. D., (1970), *J. Non-Crystalline Solids,* **4**, 29.
    Shaw, R., Thesis, (1969), University of Cambridge.

za Trousil *et al.* (1971).
zb derived from Pajasova (1971)
zc Bahl and Chopra (1969a)
zd Brodsky and Stiles (1970).
ze Bahl and Chopra (1969).
zf Fischer and Spicer (1973)
zg Free carrier absorption makes $\alpha > 100$ cm^{-1}
zh Howard and Tsu (1970)
zi Schneider and Vedam (1970).
zj Yu and Cardona (1971).
zk Galkiewicz and Tauc (1972).
zl Grant and Yoffe (1970).

due to direct photon effects (photocrystallization) which was previously studied in a-Se films by Dresner and Stringfellow (1968). In more complicated materials the crystallized material has another composition than the host glass (e.g. Te crystallizes out from Te-rich alloys). The reverse process is due to melting of the spot by high temperature and subsequent quenching.

Optical switching in thin film amorphous semiconductors was reviewed by Adler and Feinleib (1971) and by Fritzsche (1973). Its physical nature is discussed in Section 1.3.3.

Large changes of the index of refraction are an indication of some differences in the bonding and thus the polarizability of the material. For example, c-GeTe has a coordination of 6 and a-GeTe an average coordination of 3. A model involving 2- and 4-fold coordinated Te and Ge, respectively, appears most consistent with the infrared (Fisher *et al.* (1973b)) and photoemission data (Fisher *et al.* (1973a), Shevchik *et al.* (1973b)). Bonding differences between the amorphous and crystalline Ge chalcogenides and other materials are discussed by Lucovsky and White (1973). The differences in the structures of a-CdGeAs$_2$ and c-CdGeAs$_2$ are also important (cf. Chapter 2). It appears that the loss of long-range order alone changes n_o only a little, while changes in bonding in the first and second nearest neighbor region may result in large shifts in n_o.

The changes of n_o due to the changes in the short range order can be estimated by the same methods as for crystalline materials. Brodsky and Stiles (1970) successfully used Phillip's theory developed for crystals to calculate the difference of n_o in a-GeTe and c-GeTe. Also the phenomenological theory of Wemple and DiDomenico (1969) appears to be applicable to amorphous materials. It is based on the one-oscillator expression

$$n^2 (\omega) = 1 + E_o E_d / [E_o^2 - (\hbar\omega)^2] \qquad (4.41)$$

which describes the frequency dependence of the index of refraction of semiconductors and insulators close to $\omega = 0$ with 2 adjustable parameters E_o and E_d. Wemple (1972) argues that E_d (corrected for differences in densities) depends on the short range order only and is the same in the crystalline and amorphous forms if the short-range order (the first coordination number) is the same.

One can obtain good reproducibility of the value of n_o in bulk amorphous materials, but in films n_o depends considerably on their preparation and thermal history. This was shown for a-Ge films by Wales *et al.* (1967) and Theye (1970). For the edges shown in Figure 4.18 and numbered 1, 2, 3, 4, 5 she found n_o = 4.7, 4.6, 4.5, 4.5, 4.1 respectively. A similar effect was reported for Si by Brodsky *et al.* Their results are shown in Figure 4.30. One notices that even after crystallization n_o keeps changing with thermal treatment.

Brodsky and Stiles (1970) suggested a formula according to which such changes are due to changes in density, the nearest neighbour distance d, and the fraction s of dangling bonds:

$$\frac{\epsilon_1{}^{film} (0) - 1}{\epsilon_1{}^{cryst} (0) - 1} = \frac{1}{(1-s)^2} \frac{\rho^{film}}{\rho^{cryst}} \left(\frac{d^{film}}{d^{cryst}} \right)^5 \qquad (4.42)$$

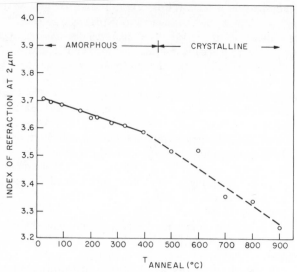

Fig. 4.30. The dependence of the index of refraction of Si films on the annealing temperature (after Brodsky *et al.* (1970)).

They could explain the changes of n_o in a-Si and a-Ge films with annealing using plausible parameters.

As we mentioned, in bulk amorphous materials the reproducibility of n_o is much better than in films. Rodney *et al.* (1958) gave a table of $n(\omega)$ for vitreous $As_2 S_3$. An advantage of glasses is that solid solutions exist in a much broader range than in crystals, and it is often possible to adjust the composition so that the desired value of n is attained with a high precision. Hilton *et al.* (1966) found a linear dependence of n on x for the system $Ge_{15} As_{15} Te_{70-x} Se_x$; Kosek and Čermák (1969) observed the same relation in the system $As_2 S_{3+x}$ $(0 < x < 2)$. The high index of refraction and optical isotropy make glasses convenient materials for such applications as antireflective coatings.

Besides Eq. (4.40) there is another very general relation connecting an integral of $\epsilon_2(\omega)$ over the fundamental absorption band with the concentration of the valence band electrons N:

$$\int \omega \epsilon_2(\omega)d\omega = 2\pi^2 e^2 N/m \tag{4.43}$$

This sum rule must also hold for amorphous materials. Relations (4.40) and (4.43) provide a good test on the overall accuracy with which $\epsilon_2(\omega)$ was determined.

4.7   EFFECT OF EXTERNAL FIELDS

The effect of external fields on the optical properties of glasses is usually much smaller than in crystals; in an intuitive way this is understandable because of the much larger effective masses (smaller mobilities) in glasses.

No magnetoabsorption or magnetoreflection effect has been reported; the Faraday rotation, although small, has been observed (see below).

An applied external electric field shifts the absorption edge as has been found in a-Se (Stuke and Weiser (1966), Drews (1966), Aoko and Okano (1966)) and some chalcogenide glasses (Fagen and Fritzsche (1970)). According to this work the shift is proportional to the square of the external field; the proportionality factor is very small, of the order 10^{-15} $eVcm^2/V^2$. These observations can be interpreted in two different ways. One interpretation is based on Franz' (1958) theory for the effect of electric fields on exponential absorption tail which corresponds to interband transitions (Franz-Keldysh effect). The exponential edge shifts as a whole according to the relation

$$\Delta E = \frac{\hbar^2 e^2}{24 E_e^2 m^*} \; F^2 \tag{4.44}$$

where E_e is determined by the slope of the exponential part of the absorption curve $\alpha(\omega) \sim \exp(\hbar\omega/E_e)$ (Section 4.4.3.), $m^* = m_v m_c/(m_v + m_c)$. Drews (1966) found $m^* \approx (4.5 \pm 0.5)m$ in a-Se.

Another interpretation is based on the normal Stark effect of localized or excitonic states. The energy shift can be estimated to be of the order

$$\Delta E \approx \frac{\hbar^2 e^2}{2 m E_g^2} F^2 \tag{4.44}$$

The proportionality factor in this case is about 3×10^{-15} $eVcm^2/V^2$, of the same order as observed. It is difficult to decide between the two possible interpretations.

Weiser and Stuke (1969) observed a relatively broad band in a-Se near 2.1 eV (that is below the electrical band gap and the photoconduction edge) in electroreflectance. They interpreted it as additional evidence for the excitonic nature of the absorption edge in a-Se (excitons in Se_8 rings, cf. Section 4.4.3). This is a strong point for this interpretation, because electro-reflectance is known to be particularly sensitive to excitonic effects.

Piller et al. (1969) observed changes in the electroreflectance of Ge-films with increasing disorder in these films. They found that when the film is fully disordered (substrate temperature during evaporation below 200°C) all structure disappears except for the peak corresponding to the transitions from the top of the valence bands (both the highest band and the split-off band) into the conduction band. The peaks move towards smaller energies as disorder increases. This result was interpreted as evidence that the split-off valence band exists in a-Ge (cf. Section 4.3), however, some objections have been raised against this interpretation (Fischer (1971)).

Kolomiets and Raspopova (1970) studied the influence of hydrostatic pressure on the gap in a-As_2Se_3 and c-As_2Se_3 determined from the measurement of the spectral dependence of photoconductivity. At room temperature, they found $dE_g/dp \approx -16 \times 10^{-6}$ eV/bar both in the crystalline and amorphous materials. The temperature shift of the absorption edge dE_g/dT is negative ($\approx -12 \times 10^{-4}$ eV/K and much larger than the shift which corresponds to the above pressure coefficient and the compressibility. It was concluded that the temperature shift of the gap is due to the electron-phonon interaction. This effect appears to depend little on disorder. The measurements of the pressure dependence of the optical gap by Grant and Yoffe (1970) gave results in qualitative agreement with the above work. Similar conclusions were reached by Fagen *et al.* (1970) from studies of the pressure and temperature dependence of the absorption edge and electrical conductivity of more complicated chalcogenide glasses.

Connell and Paul (1971) studied shifts of the absorption edges and indices of refraction of a-Ge, Si and several III—V compounds by hydrostatic pressure. (Table 4.2) They observed no hysteresis on pressure release. The pressure coefficients are very different from those observed in the corresponding crystals indicating a profound change of the wavefunctions. The pressure dependence of the index of refraction is similar in amorphous and crystalline films. Kastner (1972) studied pressure dependence of n for several chalcogenide glasses.

Schneider and Vedam (1970), Yu and Cardona (1971) and Galkiewicz and Tauc (1971) studied stress-induced birefringence of a-Se and a-As$_2$S$_3$ respectively (Table 4.2). a-Se exhibits hysteresis effects.

Gobrecht *et al.* (1962), Garben and Selinger (1968) and Mort and Scher (1971) studied Faraday rotation in a-Se. Mort and Scher suggested a theoretical interpretation of the results. Vorlíček and Zvára (1971) studied Faraday rotation in a-CdGeAs$_2$.

ACKNOWLEDGEMENTS

The first version of the manuscript was written during my stay at Bell Laboratories, at Murray Hill, N. J. and I owe very much to their hospitality and the general stimulating atmosphere. In particular, I profited there from discussions with P. W. Anderson B. Bagley, F. J. DiSalvo A. Menth, J. C. Phillips, K. K. Thornber, J. H. Wernick, D. L. Wood and others.

I thank also my colleagues from the Institute of Solid State Physics in Prague, in particular, A. Abrahám, V. Vorlíček and M. Závětová, with whom some of the work summarized in the paper was started. I thank further J. D. Dow, G. B. Fisher, H. Fritzsche, B. O. Seraphin, W. E. Spicer and R. Zallen for helpful discussions and comments on the manuscript.

REFERENCES

Abrahám, A., Tauc, J. and Velický, B. (1963). *Phys. Stat. Sol.*, 3, 767.
Adler, D. and Feinkib, J., (1971), The Physics of Opto-Electronic Materials, (editor W. A. Albers, Jr.), Plenum Press, New York and London, p. 233.
Afromowitz, M. A. and Redfield, D. (1968), Proc. Int. Conference Physics of Semiconductors, p. 98. Nauka, Moscow.
Alben, R., Smith, J. E. Jr., Brodsky, M. H. and Weaire, D., (1973), *Phys. Rev. Lett.*, 30, 1141.
Amrheim, E. M., (1969), *Glasstechnische Berichte*, 42, 52.
Amrheim, E. M. and Muller, F. H. (1969), *Kolloid Zeitschrift und Zeitschrift für Polymere*, 234, 1078.
Anderson, P. W., (1972), *Nature Physical Science*, 235, 163.
Aoki, M. and Okano, S. (1966), *J. Appl. Phys. Japan*, 5, 980.
Bagley, B. G., (1970), *Solid State Communications*, 8, 345.

Bagley, B. G., DiSalvo, F. J. and Waszczak, J. V., (1972), *Solid State Communications*, 11, 89.
Bahl, S. K. and Chopra, K. L., (1969a), *J. Appl. Phys.*, 40, 4171, (1969b), *J. Appl. Phys.*, 40, 4940.
Bahl, S. K., Bluzer, N., and Glover, R. E., (1973a), *Bull. APS*. 18, 132.
Bahl, S. K., Bhagat, S. M., and Glover, R., (1973b), *Sol. State Comm.*, in press.
Bardasis, A. and Hone, D., (1967), *Phys. Rev.*, 153, 849.
Bates, T., (1962), Modern Aspects of the Vitreous State (J. D. Mackenzie, editor), p. 195. Butterworth, London.
Bauer, R. S., Galeener, F. L. and Spicer, W. E., (1972), *J. Non-Crystalline Solids*, 8/10, 196.
Beaglehole, D. and Závětová, M., (1970), *J. Non-Crystalline Solids*, 4, 272.
Bell, R. J., Bird, N. F. and Dean, P., (1968), *Physics*, 1, 299.
Berglund, C. N. and Spicer, W. E., (1964), *Phys. Rev.*, 136A, 1030 and 1044.
Betts, F., Bienenstock, A. and Ovshinsky, S. R., (1970), *J. Non-Crystalline Solids*, 4, 554.
Bishay, A., (1970), *J. Non-Crystalline Solids*, 3, 54.
Bishop, S. G., (1973), to be published.
Bishop, S. G. and Guenzer, C. S., (1973), Proc. 5th Int. Conf. Amorphous and Liquid Semiconductors in Garmisch-Partenkirchen, Taylor and Francis, London.
Bonch-Bruevich, V. L., (1970a), *J. Non-Crystalline Solids*, 4 410: (1970b), *Phys. Stat. Sol.*, 42, 35; (1972), Proc. 11th Int. Conf. Physics of Semiconductors, Polish Scientific Publishers, Warsaw, p. 502.
Böer, K. W., (1969), *Phys. Stat. Sol.*, 34, 721, 733.
Böer, K. W. and Haislip, R., (1970), *Phys. Rev. Lett.*, 24, 230.
Borrelli, N. F. and Su, G. J., (1968), *Mat. Res. Bull.*, 3, 181.
Brodsky, M. H. and Stiles, P. J., (1970), *Phys. Rev. Lett.*, 25, 798.
Brodsky, M. H. and Title, R. S., (1969), *Phys. Rev. Lett.*, 23, 581.
Brodsky, M. H., Title, R. S., Weiser, K. and Pettit, G. D., (1970), *Phys. Rev.*, B1, 2632.
Brodsky, M. H., Gambino, R. J., Smith, J. E. Jr. and Yacoby, Y., (1972a), *Phys. Stat. Sol.* (b), 52, 609.
Brodsky, M. H., Kaplan, D. and Ziegler, J. F., (1972b), Proc. 11th Int. Conf. Phys. Semiconductors, p. 529, Polish Scientific Publishers, Warsaw.
Brown, F. C. and Rustgi, O. P., (1972), *Phys. Rev. Lett.*, 28, 497.
Brust, D., (1969a), *Phys. Rev. Lett.*, 23, 1232; (1969b), *Phys. Rev.* 186, 768.
Brust, D., and Kane, E. O., (1963), *Phys. Rev.*, 176, 894.
Burley, R. A., (1968), *Phys. Stat. Sol.*, 29, 551.
Camphausen, D. L., Connell, G. A. N. and Paul, W., (1972), *J. Non-Crystalline Solids*, 8/10, 223.
Cardona, M. and Pollak, F. H., (1971), in Physics of Opto-Electronic Materials (W. A. Alberts, Jr. editor), p. 81. Plenum Press, New York.
Cardona, M., Gudat, W, Sonntag, B. and Yu, P. Y., (1970), Proc. 10th Int. Conf. Physics of Semiconductors, p. 209 Cambridge, Mas., U.S. Atomic Energy Commission.
Červinka, L., Hosemann, R. and Vogel, W., (1970a), *J. Non-Crystalline Solids*, 3, 294.
Červinka, L., Hrubý, A., Matyáš, M., Simeček, T., Škácha, J., Štourač, L., Tauc, J., Vorlíček, V. and Höschl, P., (1970b), *J. Non-Crystalline Solids*, 4, 258.
Chen, I., (1973a, b) to be published.
Chittick, R. C., (1970), *J. Non-Crystalline Solids*, 3, 255.
Chopra, K. L. and Bahl, S. K., (1970), *Phys. Rev.* B1, 2545.
Clark, A. H., (1967), *Phys. Rev.*, 154, 750.
Cohen, M. H., (1971), private communication.
Cohen, M. H., Fritzsche, H. and Ovshinsky, S. R., (1969), *Phys. Rev. Lett.*, 22, 1065.
Connell, G. A. N. and Paul, W., (1972), *J. Non-Crystalline Solids*, 8/10, 215.
Connell, G. A. N., Temkin, R. J. and Paul, W., (1973), to be published.
Davis, E. A., (1970), *J. Non-Crystalline Solids*, 4, 107.
Davis, E. A., and Mott, N. F., (1970), *Phil. Mag.*, 22, 903
Davey, J. E. and Pankey, T., (1969), private communication.
Dexter, D. L., (1958), Nuovo Cimento 7, Series X, Suppl. p. 245.
Dexter, D. L., (1967), *Phys. Rev. Lett.*, 19, 1383.
Didden, N., (1972), *Z. Physik*, 257, 310.

Di Salvo, F. J., Menth, A., Waczszak, J. V. and Tauc, J., (1972) to be published.
Dolling, G. and Cowley, R. A., (1966), *Proc. Phys. Soc.*, **88**, 463.
Donovan, T. M. and Spicer, W. E., (1968), *Phys. Rev. Lett.*, **21**, 1572.
Donovan, T. M., Ashley, E. J. and Spicer, W. E., (1970a), *Phys. Lett.*, **32A**, 85.
Donovan, T. M., Spicer, W. E. and Bennett, J. M., (1969), *Phys. Rev. Lett.*, **22**, 1058.
Donovan, T. M., Spicer, W. E., Bennett, J. M. and Ashley, E. J., (1970b), *Phys. Rev.*, **B2**, 397.
Dresner, J. and Stringfellow, G. B., (1968), *J. Phys. Chem. Solids*, **29**, 303.
Dow, J. D. and Hopfield, J. J., (1972), *J. Non-Crystalline Solids*, **8/10**, 664.
Dow, J. D. and Redfield, D., (1970), *Phys. Rev.*, **B1**, 3358; (1971), *Phys. Rev. Letters*, **26**, 762; (1972), *Phys. Rev.*, **5**, 594.
Drews, R. E., (1966), *Appl. Phys. Lett.*, **9**, 347.
Drews, R. E., Emerald, R. L., Slade, M. L. and Zallen, R., (1972), *Solid State Communications*, **10**, 293.
Dunn, D., (1968), *Phys. Rev.*, **174**, 855.
Dunn, D., (1969), *Can. J. Phys.*, **47**, 1703.
Eagles, D. M., (1960), *J. Phys. Chem. Solids*, **16**, 76.
Eastman, D. E. and Grobman, W. D., (1972), Proc. 11th Int. Conf. Phys. Semiconductors, p 889, Polish Scientific Publishers, Warsaw.
Edmond, J. T., (1966), *Brit. J. Appl. Phys.*, **17**, 979.
Edmond, J. T., Anderson, A. and Gebbie, H. A., (1963), *Proc. Phys. Soc. (London)*, **81**, 378.
Engemann, D. and Fischer, R., (1973), Proc. 5th Int. Conf. Amorphous and Liquid Semiconductors in Garmisch-Partenkirchen, Taylor and Francis, London.
Erbudak, M. and Fischer, T. E., (1972), *J. Non-Crystalline Solids*, **8/10**, 965.
Evans, B. L. and Young, P. A., (1967), *Proc. Roy. Soc.*, **A297**, 230.
Fagen, E. A., (1970), private communication.
Fagen, E. A. and Fritzsche, H., (1970a), *J. Non-Crystalline Solids*, **2**, 180.
Fagen, E. A. and Fritzsche, H., (1970b), *J. Non-Crystalline Solids*, **4**, 480.
Fagen, E. A., Holmberg, S. H., Seguin, R. W., Thompson, J. C. and Fritzsche, H., (1970), Proc. Int. Conf. Semiconductor Physics p. 672, Cambridge, Mass., US Atomic Energy Commission.
Fan, H. Y. and Ramdas, K. A., (1959), *J. Appl. Phys.*, **30**, 1127.
Feinlieb, J., DeNeufville, J., Moss, S. C. and Ovshinsky, S. R., (1971), *Appl. Phys. Lett.*, **18**, 254.
Felty, E. J., Lucovsky, G. and Myers, M. B., (1967), *Solid State Communications*, **5**, 555.
Finkman, E., DeFonzo, A. P. and Tauc, J., (1973), Proc. 5th Int. Conf. Amorphous and Liquid Semiconductors in Garmisch-Partenkirchen, Taylor and Francis, London.
Fischer, J. E., (1971), *Phys. Rev.*, **27**, 1131.
Fischer, J. E. and Donovan, T. M., (1972), *J. Non-Crystalline Solids*, **8/10**, 202.
Fischer, R., Heim, U., Stern, F. and Weiser, K., (1971), *Phys. Rev. Lett.*, **26**, 1182.
Fischer, T. E. and Erbudak, M., (1971), *Phys. Rev. Lett.*, **27**, 1220.
Fisher, G. B. and Spicer, W. E., (1972), *J. Non-Crystalline Solids*, **8/10**, 978.
Fisher, G. B., Lindau, I., Spicer, W. E., Verhalle, Y., and Weaver, H. E., (1973a), Proc. 5th Int. Conf. Amorphous and Liquid Semiconductors in Garmisch-Partenkirchen, Taylor and Francis, London.
Fisher, G. B., Tauc, J. and Verhalle, Y., (1973b), Proc. 5th Int. Conf. Amorphous and Liquid Semiconductors in Garmisch-Partenkirchen, Taylor and Francis, London.
Franz, W., (1958), *Z. Naturforsch.*, **13a**, 484.
Fritzsche, H., (1969), *Bull. Am. Phys. Soc.*, **14**, 342; (1971), *J. Non-Crystalline Solids*, **6**, 49; Lectures of the 1972 Scottish Summer School of Physics, (W. Spear and P. Le Comber, editors), Academic Press, London and New York.
Galeener, F. L., (1971a), *Phys. Rev. Lett.*, **27**, 421; (1971b), ibid. **27**, 1716.
Galkiewicz, R. and Tauc, J., (1972), *Solid State Communications*, **10**, 1261.

Garben, B. and Seliger, H., (1968), *Phys. Stat. Sol.,* 29, K27.
Gobrecht, H., Tausend, A. and Bach, J., (1962), *Z. Phys.,* 166, 76.
Goryunova, N. A., (1968), Proc. Int. Conference Physics of Semiconductors, p. 1198 Nauka, Moscow.
Goryunova, N. A., Gross, E. F., Zlatkin, L. B. and Ivanov, E. K., (1970), *J. Non-Crystalline Solids,* 4, 57.
Grant, A. J. and Yoffe, A. D., (1970), *Solid State Communications,* 8, 1919.
Grigorvici, R., (1968), *Mat. Res. Bull,* 3, 13.
Grigorovici, R. and Vancu, A., (1968), *Thin Solid Films,* 2, 105.
Hadni, A., Morlot, G., Gervaux, X., Chanal, D., Brehat, F. and Strimer, P., (1965), *Comptes Rendus, Acad. Sc. Paris,* 260, 4973.
Hartke, J. L. and Regensburger, P. J., (1965), *Phys. Rev.,* 139, A970.
Hass, M., (1966), Semiconductors and Semimetals (Willardson, R. K. and Beer, A. C. editors), Academic Press, New York.
Heine, V., (1971), *J. Phys.,* C4, L221.
Heine, V. and Jones, R. O., (1969), *J. Phys.,* C2, 719.
Herman, F. and Van Dyke, J. P., (1968), *Phys. Rev. Lett.,* 21, 1575.
Herman, F., Kortum, R. L., Glin, C. D. K. and Shay, J. L., (1967), Proc. Int. Conference on II–VI Semiconducting Compounds, p. 271. Benjamin, New York.
Hilton, A. R., (1966), *Appl. Optics,* 5, 1877; (1970), *J. Non-Crystalline Solid,* 2, 28.
Hilton, A. R. and Brau, M., (1963), *Infrared Physics,* 3, 69.
Hilton, A. R. and Jones, C. E., (1966), *Phys. Chem. Glasses,* 7, 112; (1967), *Appl. Optics,* 6, 1513.
Hilton, A. R., Jones, C. E. and Brau, M., (1964), *Infrared Physics,* 4, 213; (1966a), *Physics and Chemistry of Glasses,* 7, 105; (1966b), *Infrared Physics,* 6, 183.
Hilton, A. R., Jones, C. E., Dovrott, R. D., Klein, H. M., Bryant, A. M. and George, T. D., (1966c), *Phys. Chem. Glasses,* 7, 116.
Hindley, N. K., (1970), *J. Non-Crystalline Solids,* 5, 17 and 31.
Holtsmark, J., (1919), *Ann. Physik,* 58, 577.
Hopfield, J. J., (1968), *Comments on Solid State Physics,* 1, 16.
Howard, W. E. and Tsu, R., (1970), *Phys. Rev.,* B1, 4709.
Hutson, A. R., (1961), *J. Appl. Phys.,* 32, 2287.
Ing. Jr., S. R., Chiang, Y. S. and Ward, A., (1969), Communication at the American Ceramic Society Meeting (May 1969).
Ioffe, A. F. and Regel, A. R., (1960), *Progr. Semiconductors,* 4, 239.
Inglis, G. B. and Williams, F., (1970), *Phys. Rev. Lett.,* 25, 1275.
Jung, G., (1971), *Phys. Stat. Sol.,* (b) 44,239.
Joannopoulos, J. D. and Cohen, M. L., (1972), *Solid State Communications,* 11, 549.
Kaplan, T. A., Mahanti, S. D. and Hartmann, W. M., (1971) *Phys. Rev. Lett.,* 27, 1796.
Kastner, M., (1972), *Phys. Rev.,* B6, 2273.
Keck, D. B., Maurer, R. D. and Schultz, P. C., (1973), *Appl. Phys. Lett.,* 22, 307.
Kerker, M., (1969), The Scattering of Light, Academic Press, New York.
Kessler, F. R. and Sutter, E., (1963), *Z. Physik,* 173, 54.
Knotek, M. L., and Donovan, T. M., (1973), *Phys. Rev. Lett.,* 30, 652.
Kobliska, R. J. and Solin, S. A., (1971), *J. Non-Crystalline Solids,* 8/10, 191; (1972), *Solid State Comm.,* 10, 231; (1973), *Phys. Rev.,* B8, 756.
Koc, S. Renner, O., Závětová, M. and Zemek, J., (1972), *Czech. J. Phys.,* (B), 22.
Kolomiets, B. T., (1964), *Phys. Stat. Sol.,* 7, 359, 713.
Kolomiets, B. T. and Raspopova, E. M., (1970), *Fizika i Technika Poluprovodnikov,* 4, 157.
Kolomiets, B. T., Mamontova, T. N. and Babaev, A. A., (1970), *J. Non-Crystalline Solids,* 4, 289.
Kosek, F. and Čermák, J., (1969), *Čs. čas. fyz.,* A19,271.
Kosek, F. and Tauc, J., (1970), *Czech. J. Phys.,* B20, 94.
Kramer, B., (1970), *Phys. Stat. Sol.,* (b) 41,649; (1971), ibid 47, 501.

Kramer, B., Maschke, K., Thomas, P. and Treusch, J., (1970), *Phys. Rev. Lett.*, **25**, 1020.
Kramer, B., Maschke, K. and Thomas, P., (1972), *Phys. Stat. Sol.*, (b)**49**,525.
Krause, J. T., Kurkjian, C. R., Pinnow, D. A. and Sigety, E. A., *Appl. Phys. Lett.*, **17**, 367.
Krieg, J. G., (1967), *Z.f. Physik*, **205**, 425.
Lannin, J. S., (1973), Proc. 5th Int. Conf. Amorphous and Liquid Semiconductors in Garmisch-Partenkirchen, Taylor and Francis, London.
Laude, L. D. and Fitton, B., (1972), *J. Non-Crystalline Solids*, 8/10, 971.
Laude, L. D., Willis, R. F., Fitton, B., (1972), *Phys. Rev. Lett.*, **29**, 472.
Lell, E., Kreid, N. J. and Hensler, J. R., (1966), Progress in Ceramic Science (J. Burke, editor), vol. 4. Pergamon, London.
Ley, L., Kowalczyk, S., Pollak, R. and Shirley, A. D., (1972), *Phys. Rev. Lett.*, **29**, 1088.
Ley, L., Pollak, R., Kowalczyk, S., McKeely, F. R. and Shirely, D. A., (1973), *Phys. Rev.*, in press.
Lucovsky, G., Brodsky, M. H. and Burstein, E., (1968), in Localized Excitations in Solids (R. F. Wallis, editor), Plenum Press, New York and London, p. 592.
Lucovsky, G., (1969), The Physics of Selenium and Tellurium (W. C. Cooper, editor), p. 255; Pergamon Press, New York. (1970), Proc. Int. Conf. Semiconductor Physics, p. 799; Cambridge, Mass. U.S. Atomic Energy Commission; (1972), *Phys. Rev.*, **B6**, 1480.
Lucovsky, G. and Martin, R. M., (1972), *J. Non-Crystalline Solids*, 8/10, 185.
Lucovsky, G., Mooradian, A., Taylor, W., Wright, G. B. and Keezer, R. C., (1967), *Solid State Communications*, **5**, 113.
Lurio, A. and Brodsky, M. H., (1972), *Bull. APS*, **17**, 322.
McGill, T. C. and Klíma, J., (1972), *Phys. Rev.*, **B5**, 1517.
Mahr, H., (1963), *Phys. Rev.*, **132**, 1880.
Markov, Yu. F., Reshetnyak, N. B., (1972), *Fiz. Tverdovo Tels.*, **14**, 1242 (Sov. Phys. – Sol. State 14, 1063).
Mitchell, D. L., Bishop, S. G. and Taylor, P. C., (1971), *Solid State Communications*, **9**, 1833.
Morgan, T. N., (1965), *Phys. Rev.*, **139**, A343.
Mort, J. and Scher, H., (1971), *Phys. Rev.*, **B3**, 334.
Mott, N. F., (1969), *Phil. Mag.*, **19**, 835; (1970) *Phil. Mag.*, **22**, 1.
Mott, N. F. and Davis, E. A., (1971), Electronic Processes in Non-Crystalline Materials, Clarendon Press, Oxford.
Nielsen, P., (1971), *Solid State Communications*, **9**, 1745; (1972a), *Bull. Am. Phys. Soc.*, **17**, 113; (1972b), *Phys. Rev.*, **B6**, 3739; (1973), Proc. 5th Int. Conf. Amorphous and Liquid Semiconductors, Garmisch-Partenkirchen, Taylor and Francis, London.
Olley, J. A. and Yoffe, A. D., (1972), *J. Non-Crystalline Solids*, 8/10, 850.
Onomichi, M., Arai, T. and Kuds, K., (1971), *J. Non-Crystalline Solids*, **6**, 362.
Ortenbruger, I. B., Rudge, W. E. and Herman, F., (1972), *J. Non-Crystalline Solids*, 8/10, 653.
Pajasova, L., (1971), Third Int. Conf. on Vacuum Ultraviolet Radiation Physics, Tokyo, Japan.
Paul, W., (1972), Proc. 11th Int. Conf. Phys. Semiconductors, p. 38, Polish Scientific Publishers, Warsaw.
Penn, D. R., (1962), *Phys. Rev.*, **128**, 2093.
Peterson, C. W., Dinah, J. H. and Fischer, T. E., (1970), *Phys. Rev. Lett.*, **25**, 861.
Philipp, H. R. and Ehrenreich, H., (1963), *Phys. Rev.*, **129**, 1550.
Phillips, J. C., (1966), *Solid State Physics*, **18**, 55; (1968), *Phys. Rev. Lett.*, **20**, 550; (1970), *Rev. Mod. Phys.*, **42**, 317; (1971a), *Phys. Stat. Sol.*, (b) **44**,Kl; (1971b), *Comments on Solid State Physics*, **4**, 9.
Pierce, D. T., Ribbing, C. G. and Spicer, W. E., (1972), *J. Non-Crystalline Solids*, 8/10, 959.
Pierce, D. T. and Spicer, W. E., (1971), *Phys. Rev. Lett.*, **27** 1217; (1972), *Phys. Rev.*, **B5**, 3017.

Piler, H., Seraphine, B. O., Markel, K. and Fischer, J. E., (1969), *Phys. Rev. Lett.*, **23**, 775.

Pollak, M., (1970), *Discussions Faraday Soc.*, **50**, 13.

Powell, R. A. and Spicer, W. E., (1973), *Bull. Am. Phys. Soc.*, **18**, 390.

Prakash, V., (1967), Techn. Rep. HP–13, Div. of Eng. and Appl. Physics, Harvard University, Cambridge, Mass.

Prettle, W., Shevchik, N. J. and Cardona, M., (1973), to be published.

Redfield, D., (1963), *Phys. Rev.*, **130**, 914, 916.

Redfield, D. and Afromowitz, M. A., (1967), *Appl. Phys. Lett.*, **11**, 138.

Ribbing, C. G., Pierce, D. T. and Spicer, W. E., (1971), *Phys. Rev.*, **B4**, 4417.

Ribbing, C. G. and Spicer, W. E., (1971), *Physics Letter*, **37A**, 85.

Rindone, G. E., (1966), Luminescence in Inorganic Solids, (P. Goldberg, editor), p. 419. Academic Press, New York.

Rockstad, H. K. and DeNeufville, J. P., (1973), Proc. 5th Int. Conf. Amorphous and Liquid Semiconductors in Garmisch-Partenkirchen, Taylor and Francis, London.

Rodney, W. S., Malitson, I. H. and King, T. A., (1958), *J. Opt. Soc. Am.*, **48**, 633.

Sak, J., (1972), *J. Phys.*, C 5.

Savage, J. A. and Nielsen, S., (1965), *Phys. Chem. Glasses*, **6**, 90.

Schneider, W. C. and Vedam, K., (1970), *J. Opt. Soc. Amer.*, **60**, 800.

Shevchik, N. J., Tejeda, J., Langer, D. W. and Cardona, M., (1973a), *Phys. Rev. Lett.* **30**, 659; (1973b), *Phys. Stat. Sol.*, (b) **57**, 245.

Shuker, R. and Gammon, R. W., (1970), *Phys. Rev. Lett.*, **25**, 222.

Siemsen, K. J. and Fenton, E. W., (1967), *Phys. Rev.*, **161**, 632.

Siemsen, K. J. and Riccius, H. D., (1969), *J. Phys. Chem. Solids*, **30**, 1897.

Simon, I., (1960), Modern Aspects of the Vitreous State (J. D. Mackenzie, editor), vol. 1 p. 120. Butterworths, London.

Slater, J. C., (1956), Handb.d.Physik (S. Flügge, editor) XIX, part 1, 67.

Smith, J. E., Jr. Brodsky, M. H., Crowder, B. L., Nathan, M. I. and Pinczuk, A., (1971), *Phys. Rev. Lett.*, **26**, 642.

Snitzer, E., (1966), *Proc. of the IEEE*, **54**, 1249.

Spicer, W. E. and Donovan, T. M., (1970a), *J. Non-Crystalline Solids*, **2**, 66; (1970b), *Phys. Rev. Lett.*, **24**, 595.

Spicer, W. E. and Eden, R. C., (1968), International Conference on the Physics of Semiconductors p. 65. Moscow, Nauka, Leningrad.

Spitzer, W. G. and Fan, H. Y., (1958), *Phys. Rev.*, **109**, 1011.

Srb, I. and Vaško, A., (1963), *Czech. J. Phys.*, **B13**, 827.

Stern, F., (1971), *Phys. Rev.*, **B3**, 2636.

Stevels, J. M., (1960), Non-Crystalline Solids, (V. D. Fréchette, editor), p. 412. J. Wiley, New York.

Stimets, R. W., Waldman, J., Lin, J., Chang, T. S., Temkin, R. J. and Connell, G. A. N., (1973a), *Sol. State Communications*, in press.

Stimets, R. W., Waldman, J., Lin, J., Chang, F., Temkin, R. J., Connell, G. A. V. and Felterman, H. R., (1973b), Proc. 5th Int. Conf. Amorphous and Liquid Semiconductors in Garmisch-Partenkirchen, Taylor and Francis, London.

Stuke, J., (1970), *J. Non-Crystalline Solids*, **4**, 1.

Stuke, J. and Weiser, G., (1966), *Phys. Stat. Sol.*, **17**, 343.

Stuke, J. and Zimmerer, G., (1972), *Phys. Stat. Sol.*, (b) **49**, 513.

Su, G. J., Borrelli, N. F. and Miller, A. R., (1962), *Phys. Chem. Glasses*, **3**, 167.

Sumi, H. and Toyozawa, Y., (1971), *J. Phys. Soc. (Japan)*, **31**, 342.

Tauc, J., (1965), *Prog. Semiconductors*, **9**, 87; (1968), *Mat. Res. Bull.*, **3**, 37; (1970a), The Optical Properties of Solids (F. Abelès, editor), p. 277. North Holland, Amsterdam. (1970b), *Mat. Res. Bull.*, **5**, 721.

Tauc, J. and Abrahám, A., (1969), *Czech. J. Phys.*, **B19**, 1246.

Tauc, J. and Menth, A., (1972), *J. Non-Crystalline Solids*, **8/10**, 569.

Tauc, J., Abrahám, A., Pajasová, L., Grigorovici, R. and Vancu, A., (1965), Proc. Int. Conference Physics of Non-Crystalline Solids, p. 606. North Holland, Amsterdam.

Tauc, J., DiSalvo, F. S., Peterson, G. E. and Wood, D. L., (1972), Amorphous Magnetism (H. O. Cooper and A. M. de Graaf, editors), p. 119, Plenum Press, New York and London.

Tauc, J., Grigorovici, R. and Vancu, A., (1966), *Phys. Stat. Sol.*, 15, 627.
Tauc, J., Štourač, L., Vorlíček, V. and Závětová, M., (1968), Proc. Int. Conf. Physics of Semiconductors p. 1251, Nauka, Moscow.
Tauc, J., Abrahám, A., Zallen, R. and Slade, M., (1970a), *J. Non-Crystalline Solids*, 4, 279.
Tauc, J., Menth, A. and Wood, D. L., (1970b), *Phys. Rev. Lett.*, 25, 749.
Tausend, A., (1969), The Physics of Selenium and Tellurium, (W. C. Cooper, editor), p. 233. Pergamon Press, New York.
Taylor, D. W., (1967), *Phys. Rev.*, 156, 1017.
Taylor, P. C., Bishop, S. G. and Mitchell, D. L., (1970), *Solid State Communications*, 8, 1783; (1971), *Phys. Rev. Lett.*, 27, 414; (1973), *J. Non-Crystalline Solids*, in press.
Theye, M., (1971a), *Optics Communications*, 2, 329; (1971b), *Mat. Res. Bull.*, 6, 103.
Thorpe, M. F. and Weaire, D., (1971a), *Phys. Rev. Lett.*, 27, 158: (1971b)., *Phys. Rev.*, B4, 3518.
Toyozawa, Y., (1958), *Prog. Theoret. Phys. (Kyoto)*, 20, 53: (1959), *Prog. Theoret. Phys. (Kyoto)*, 22, 445; *Supplement*, 12, 111.
Trousil, Z., Pajasová, L. and Závětová, M., (1971), *Czech. J. Phys.*, B 21, 220.
Tsu, R., Howard, W. E. and Esaki, L., (1968), *Phys. Rev.*, 172, 779; (1970), *J. Non-Crystalline Solids*, 4, 322.
Tynes, A. R. (1970), *Appl. Optics*, 9, 2706.
Urbach, F., (1953), *Phys. Rev.*, 92, 1324.
Van Vechten, J. A., (1972), *Solid State Communications*, 11, 7.
Vaško, A., (1965), *Phys. Stat. Sol.*, 8, K41; (1969), The Physics of Selenium and Tellurium, (W. C. Cooper, editor) p. 241, Pergamon Press, New York; (1970), *J. Non-Crystalline Solids*, 3, 225.
Vaško, A., Ležal, D. and Srb, I., (1970), *J. Non-Crystalline Solids*, 4, 311.
Vorlíček, V., and Zvára, M., (1971), *Phys. Stat. Sol.*, (b) 48, 93.
Wales, J., Lovitt, G. J. and Hill, R. A., (1967), *Thin Solid Films*, 1, 137.
Walsh, P. J., Vogel, R. and Evans, E. J., (1969), *Phys. Rev.*, 178, 1274.
Ward, A. T., (1968), *J. Phys. Chem.*, 72, 4133.
Weiser, K. and Brodsky, M. H., (1970), *Phys. Rev.*, Bl, 791.
Weiser, K. and Stuke, J., (1969), *Phys. Stat. Sol.*, 35, 747.
Weiser, K., Fischer, R. and Brodsky, M. H., (1970), Proc. Internat. Confer. Physics of Semiconductors, p. 667. Cambridge, Mass., US Atomic Energy Commission.
Wemple, S. H., (1973), *Phys. Rev.*, B7, 3767.
Wemple, S. H. and DiDomenico Jr., M., (1969), *Phys. Rev. Lett.*, 23, 1156.
Weyl, W. A., (1951), Coloured Glasses, The Society of Glass Technology, Sheffield, England.
Whan, R. W. and Stein, H. J., (1963), *Appl. Phys. Lett.*, 3, 187.
Weaire, D., (1971), *Phys. Rev. Lett.*, 26, 1541.
Weaire, D. and Thorpe, M. F., (1971), *Phys. Rev.*, B4, 2508.
Weaire, D., Thorpe, M. F. and Heine, V., (1972), *J. Non-Crystalline Solids*, 8/10, 128.
Wiech, G. and Zöpf, E., (1972), Proceedings of the Int. Conf. on Band-Structure Spectroscopy of Metals and Alloys, Glasgow.
Wihl, M., Cardona, M. and Tauc, J., (1972), *J. Non-Crystalline Solids*, 8/10, 172.
Wood, C., Shaffer, J. C. and Proctor, W. G., (1972), *Phys. Rev. Lett.*, 29, 485.
Wood, D. L. and Remeika, J. P., (1966), *J. Appl. Phys.*, 37, 1232; (1967), *J. Appl. Phys.*, 38, 1038.
Wood, D. L. and Tauc, J., (1972), *Phys. Rev.*, 5, 3144.
Young, P. A., (1971), *J. Phys. C*, 4, 93.
Yu, Y. P. and Cardona, M., (1971), *Phys. Stat. Sol.*, (b) 47, 251.
Zallen, R., (1968), *Phys. Rev.*, 173, 824.
Zallen, R., Slade, M. and Ward, A. T., (1971a), *Phys. Rev.*, 3, 4257.
Zallen, R., Drews, R. E., Emerald, R. L. and Slade, M. L., (1971b), *Phys. Rev. Lett.*, 26, 1564.
Zarzycki, J. and Naudin, F., (1960), *Verres Refractives*, 14, 113.
Závětová, M. and Vorlíček, V., (1971), *Phys. Stat. Solid.*, (b) 48, 113.
Závětová M., Koc, S. and Zemek, J., (1972), *Czech. J. Phys.*, B22, 429.
Zeppenfeld, K. and Raether, H., (1966), *Z. Phys.*, 193, 471.

Chapter 5

Electronic Properties of Amorphous Semiconductors

H. Fritzsche

Department of Physics and The James Franck Institute, The University of Chicago, Chicago, Illinois 60637

5.1 INTRODUCTION

Representing systems of higher complexity, amorphous semiconductors require for their characterization a larger number of parameters than their crystalline counterparts. Furthermore, the theoretical difficulty of calculating the distribution of electronic states and their transport properties in

---

*Supported by the Air Force Office of Scientific Research (AFSC), United States Air Force, under contract F44620–71–C–0025, and aided through general support of materials science by the Advanced Research Projects Agency.

222   H. Fritzsche

disordered structures and the resulting need for simplifying approximations provide the theoretical expressions with considerable flexibility. At the same time amorphous semiconductors yield experimental data which are typically less structured and hence less rich in information than the equivalent data obtained with crystalline semiconductors. As a consequence, research on amorphous semiconductors is ironically often hampered by 'good agreement with theory'. Many experiments fail to distinguish between mechanisms of quite different physical origin. This makes the problem of extracting a coherent picture from the experiments particularly difficult.

The characterization of amorphous semiconductors is further complicated by the fact that the same chemical composition can exist in many structural states which depend on the thermal history and preparational variables. Moreover, compositional heterogeneity, voids, cracks, as well as the preservation in some materials of molecular structures with relatively extended short-range order makes *real* amorphous materials quite different from the ideal disordered systems presently studied by theorists.

In the next section of this chapter the amorphous semiconductors are classified according to their chemical bonds and composition. Chemical bonding arguments are used to explain the origin of the valence and conduction bands and the effects of alloying additives in producing donor and acceptor states in the gap. The different models which have been proposed for describing the electronic states and their conduction properties in amorphous semiconductors are then presented.

The third section deals with the electronic properties of amorphous semiconductors. These are compared with and described in terms of transport equations which treat the amorphous solid as a rather ideal homogeneous, uniform material.

In the fourth section the effects of heterogeneities, voids, and potential fluctuations often encountered in real amorphous films are briefly discussed. Some of these are bound to have a strong effect on the electronic properties described in the third section and perhaps account for several discrepancies between the observations and the theories which assume uniform materials.

5.2   CHEMICAL BONDS AND BAND MODELS

It was Ioffe (1951) who first pointed out that the basic electronic properties of a solid are determined primarily by the character of the bonds between nearest neighbors rather than by the long-range order. The chemical bond approach enabled Welker (1952) to predict many of the semiconducting properties of the III–V compounds. Mooser and Pearson (1956, 1960) expanded the chemical approach to semiconductivity into a systematic survey of different compounds and crystal classes. Although in the case of crystals this approach has been replaced by band structure calculations, several properties common to large classes of amorphous semiconductors become plausible if one understands their chemical bonding.

TABLE 5.1 *Classification and Examples of Non-Crystalline Semiconductors.*

1. *Covalent Non-Crystalline Solids*
   A. *Tetrahedral Amorphous Films*
      Si, Ge, SiC, InSb, GaAs, GaSb....................
   B. *Tetrahedral Glasses, $A^{II} B^{IV} C_2^{V}$*
      $CdGe_x As_2$, $CdSi_x P_2$, $ZnSi_x P_2$, $CdSn_x As_2$ ....
   C. *Lone Pair Semiconductors*
      (i) *elements and compounds*
         Se, S, Te, $As_2 Se_3$, $As_2 S_3$ ....
      (ii) *cross-linked networks*

Ge-Sb-Se	Si-Ge-As-Te
Ge-As-Se	$As_2 Se_3$-$As_2 Te$
As-Se-Te	$Tl_2 Se$-$As_2 Te_3$
....	....

   D. *Others*
      B, As, $(Cu_{1-x} Au_x) Te_2$ ....

2. *Semiconducting Oxide Glasses*

$V_2 O_5$-$P_2 O_5$	MnO-$Al_2 O_3$-SiO_2
$V_2 O_5$-$P_2 O_5$-BaO	CoO-$Al_2 O_3$-SiO_2
$V_2 O_5$-GeO_2-BaO	FeO-$Al_2 O_3$-SiO_2
$V_2 O_5$-PbO-$Fe_2 O_3$	TiO_2-$B_2 O_3$-BaO
....	....

3. *Dielectric Films*
   SiO_x, $Al_2 O_3$, ZrO_2, $Ta_2 O_3$, $Si_3 N_4$, BN, ...

5.2.1 Classification of Amorphous Materials

The noncrystalline solids (hereafter referred to as NCS) may be grouped into three major categories as shown in Table 5.1. The first contains the covalently bonded NCS. We discuss primarily these semiconductors in this chapter because they are affected most strongly by disorder. This category contains (A) the tetrahedral semiconductors, which can only be prepared by thin film deposition, (B) tetrahedral glasses, and (C) the lone pair semiconductors. As explained below, the name of this class of materials (Kastner (1972)) stems from the fact that their properties are primarily influenced by the two nonbonding p orbitals of the Group VI chalcogen elements in two-fold coordination.

Among the lone pair semiconductors it is important to distinguish (i) the elements and compounds with a considerable range of local order, and (ii) the three-dimensional network structures. The first of these subgroups contain molecular complexes of larger size which are held together with weak van der Waals forces. The three-dimensionally cross-linked network structures approximate more closely the ideal of structural disorder.

The second main group contains the oxide glasses, which have strong ionic bonds and are usually good insulators but which can be made semiconducting by the addition of transition metal ions in different valence states. The conduction process then proceeds via a charge exchange among the mixed valence transition metal ions. The conduction band of the host glass plays no role in the charge transport, the glass merely acts as the host medium and effects the tunneling conduction only through the phonon and polarization interaction. These materials have been reviewed by Mackenzie (1964), Owen (1967, 1970b), and Mott (1968).

The dielectric films listed in Table 5.1 are sensitive to disorder as their electronic conduction relies on deviations from stoichiometry and the presence of defect centers which act as donors and acceptors. A wealth of data is available, particularly on SiO and Al_2O_3 films because of their importance in device technology. Their electronic properties which have been reviewed by Jonscher (1967) and Hill (1967), clearly distinguish them from the covalent NCS.

For other reviews on the properties of NCS the reader is referred to Kolomiets (1964), Mott (1967), Fritzsche (1972), Adler (1971), Mott and Davis (1971) as well as to the Proceedings of recent conferences on amorphous and liquid semiconductors (Doremus ed. (1970), Mott ed. (1970b), Cohen ed. (1972)).

5.2.2 Chemical Bond Description of Covalent NCS

In crystals the coordination environment of each atomic site is fixed and determined by the long-range order. In vitreous materials, on the other hand, the coordination environment can adjust to satisfy the valence requirements of each atom (Mott (1967)). As a consequence, the composition of vitreous semiconductors can be changed within wide limits without necessarily creating unsatisfied bonds. Hence 'intrinsic' semiconduction is not restricted to stoichiometric compositions as it is in crystals. Shallow donors or acceptors are not likely to occur in amorphous semiconductors; As atoms in a-Se, for instance, will arrange to be in three-fold coordination, thereby satisfying their bonds.

In order to understand the nature of donors and acceptors in amorphous semiconductors, it is helpful to contrast the origin of electronic states in the two classes of semiconductors: (a) the tetrahedral semiconductors and (b) semiconductors containing Group VI elements in two-fold coordination. Figure 5.1 sketches the origin of the states in Ge and Se. The states in the solid are considered to be a broadened superposition of the molecular states. Ge in four-fold coordination has hybridized sp^3 orbitals. These are split into bonding (σ) and antibonding (σ^*) states which form in tetrahedral semiconductors the valence and conduction band, respectively. In Se, only two of the three p states are utilized for bonding when the chalcogen is in two-fold coordination. This leaves one nonbonding electron pair. As sketched in Figure 5.1b in solid Se these lone pair (LP) electrons form a band near the original p state energy. The σ and σ^* band are split symmetrically with respect to this reference energy. The higher lying filled LP band is the valence band in these materials (Mooser and Pearson (1960)). This important difference is used by Kastner (1972) to explain that dangling

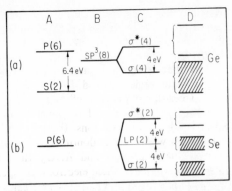

Fig. 5.1. Bonding in (a) Ge and (b) Se. (A) atomic states, (B) hybridized states, (C) molecular states, (D) states are broadened into bands in the solid (after Kastner, 1971).

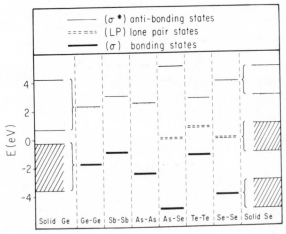

Fig. 5.2. Bonding and antibonding states of various bonds in relation to the energy bands of solid Ge and Se. Note that each group VI element in two-fold coordination is accompanied by lone pair states.

bonds and certain alloying additives form different states in tetrahedral and in LP semiconductors.

When a dangling bond is formed, for example, a filled state is pulled out of the σ band and an unoccupied state from the σ* band. In tetrahedral semiconductors these produce localized states in the gap (Mott (1971)): a donor near the gap center and above it an acceptor which exists only when the donor is occupied. In LP semiconductors the occupied state from σ falls into the LP band and the acceptor somewhat above it. At large dangling bond concentrations also the unoccupied state from σ* may lie within the LP band (Cutler (1971)).

According to Kastner, the effect of alloying, say Group IV, V and VI elements, into an amorphous LP semiconductor such as a-Se can be understood in the following manner. An amorphous Se-As binary alloy, for instance, contains different bonds, each one with a different $\sigma-\sigma^*$ energy separation. It is important to note that LP states are associated only with chalcogen atoms. The molecular state energies estimated from optical absorption spectra (Cardona and Greenaway (1964), Tauc *et al.* (1965)) are sketched for several bonds in Figure 5.2. The center energy of a particular $\sigma-\sigma^*$ bond is shifted with respect to the LP states of Se by a certain amount which one might assume to be approximately the ionization potential difference.

The LP state of other Group VI elements form donors if they fall above the LP band of the host chalcogenide. The energy of LP states depends on the chemical environment. LP electrons next to electropositive atoms will have higher energies than those near electronegative atoms. Hence, the addition of electropositive elements to LP semiconductors raises some LP states as localized donor states into the gap, thus producing or broadening the valence band tail. On the other hand, when a strongly electropositive atom like Cu or Cd is added to say vitreous As_2Se_3, strongly bonded copper or cadmium selenide is formed, and as a result the number of As-As bonds is increased.

Consider next the case of adding Ge to Se. First the Ge atoms act as *bond modifiers;* they strengthen the average bond by cross-linking the Se chain structure, thereby increasing the glass transition temperature and the resistivity (de Neufville (1972)). A completely cross-linked network is reached at the compound composition $GeSe_2$. At higher Ge concentrations the structure gets weakened by the increasing number of weaker Ge-Ge bonds.

It is interesting to note that the general rule, valid for tetrahedral NCS, that each band contains the same number of states, ceases to hold in LP semiconductors. Here the number of states in the LP valence band is equal to twice the number of chalcogen atoms. Consider for example vitreous Se-Ge binary alloys. In Se rich alloys, the LP band is the valence band. In Ge rich alloys the σ band of Ge is the valence band, the few LP states of Se are localized. Because of the different environment of the Se atoms the localized LP states fall into an energy range of finite width. At a certain critical Se concentration the density of LP states is sufficient for delocalization to occur at the center of the LP band. At higher Se concentrations the LP band becomes the valence band. The transition at the critical Se concentration is of the Anderson (1958) type rather than a Mott transition (Mott and Zinamon (1970)) because the LP band is filled.

Most vitreous covalent semiconducting alloys contain a Group VI element as a major constituent and thus are LP semiconductors. The density of states at the top of the LP valence band tails into the gap because of local energy fluctuations of the LP electrons. These occur because of the lack of long-range order but are enhanced with the presence of (i) dangling bonds, (ii) other chalcogens, and (iii) electropositive atoms. The conduction band tail is broadened by the presence in the alloy of weaker bonds.

In amorphous tetrahedral semiconductors, on the other hand, the band edges are broadened only by lack of long-range order. Dangling bonds, however, create localized donor and acceptor states deeper in the gap. LP states of chalcogen impurities may also fall into the gap producing donor states. We note that there seem to be fewer causes for band tailing in tetrahedral than in LP semiconductors.

In describing the LP semiconductors in this manner we assumed that the s-states of the chalcogen atoms lie well below the p-states. Optical absorption and photoemission measurements (Nielsen (1972)) have so far not revealed a separate s-band in Se. Hybridization of the s-states with the p-states may be the reason for this. Such hybridization will affect the bonding and LP valence bands but it will not alter the distinctive difference between these bands and the valence band in tetrahedral semiconductors and the qualitative conclusion drawn from the simplified picture sketched above.

The fact that in an amorphous covalent material the local coordination environment is not fixed by long-range order causes the very different effects of impurities on the electronic properties of NCS as compared to those of crystals. It appears that one cannot shift the Fermi level close to either band or form a p-n junction in an amorphous semiconductor by changing its composition because there is no way to produce shallow donors close to the conduction band or acceptors close to the valence band. This seems to be the most distinctive difference between amorphous and crystalline semiconductors. The Fermi level remains in the center region of the gap in NCS, approximately in the central third of the gap.

5.2.3   Band Models

In view of the large differences in the nature of the various groups of amorphous semiconductors, it is obvious that any single model cannot describe the essential features of all amorphous materials. We first describe a few models which are presently in use and then draw attention to some fundamental assumptions and simplifications which are made in using such models.

Mott-CFO Model

This model sketched in Figure 5.3 is derived from the concept of an ideal covalent random network structure. Simple chemical considerations as well as the observation that amorphous semiconductors are transparent in the infrared and exhibit a thermally activated conductivity suggest that the valence and conduction bands are separated by a gap. Translational and compositional disorder are assumed to cause fluctuations of the potential of sufficient magnitude that they give rise to localized states extending from the conduction and valence bands into the gap. These localized states are not associated with definite imperfections but are the result of the randomness of the potential. Their number and energy spread increases with the degree of the randomness and the strength of the scattering. It is important to note that the valence band tail states are assumed to be neutral when occupied and the conduction band tail states are neutral when empty. This

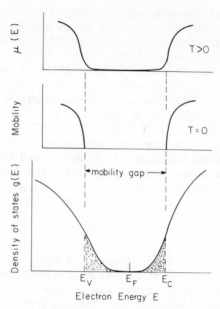

Fig. 5.3. Sketch of Mott-CFO model for covalent semiconductors having three-dimensional cross-linked network structure. The critical energies E_c and E_v define the mobility gap. For $T > 0$ the mobility $\mu(E)$ may be finite in the gap because of thermally assisted tunneling. E_F = Fermi energy. The distribution of localized gap states may be nonmonotonic when defect states of a certain energy are prevalent.

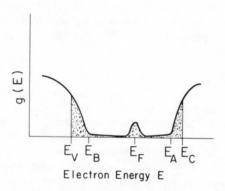

Fig. 5.4. Density of states g(E) suggested by Davis and Mott (1970). $E_c - E_v$ is the mobility gap. The ranges $E_c - E_A$ and $E_B - E_v$ contain localized states originating from lack of long-range order. Thermally assisted hopping may take place in these ranges. A band of compensated levels is proposed to lie near the gap center. Recently, Mott (1971) suggested that the center band of localized levels may be split into a donor band and an acceptor band.

places the Fermi energy somewhere near the gap center. In addition one expects that deviations from the ideal covalent random network, such as vacancies, dangling bonds, and chain ends, contribute localized states in certain energy ranges. These then give rise to a nonmonotonic density of localized states curve. As first noted by Mott (1967, 1969a, 1969b, 1969c, 1970a) the character of the wavefunctions changes at critical energies E_c and E_v which separate the extended and the localized states. Here the electron and hole mobilities drop sharply from a low-mobility band transport with finite mobility at $T = 0$ to a thermally activated tunneling between localized gap states which disappears at $T = 0$. These so-called *mobility edges* (Cohen, Fritzsche, and Ovshinsky (1969)) define a *mobility gap* $E_c - E_v$ which contains only localized states. This model is believed to apply to alloy glasses which contain compositional as well as positional disorder.

Davis-Mott Model

A band model as shown in Figure 5.4 was proposed by Davis and Mott (1970). The mobility edges for electrons and holes lie again at E_c and E_v. A stronger distinction is made between localized states which originate from lack of long-range order and others which are due to defects in the structure. The first kind of localized states extend only to E_A and E_B in the mobility gap. The defect states form longer tails but of insufficient density to pin the Fermi level. Moreover, the authors propose a band of compensated levels near the gap center in order to pin the Fermi level and to account for the behavior of the a.c. conductivity. The center band may be split into two bands (Mott (1971)).

Marshall-Owen Model

In the model of Marshall and Owen (1971), shown in Figure 5.5, the position of the Fermi level is determined by bands of donors and acceptors in the upper and lower halves of the mobility gap, respectively. The concen-

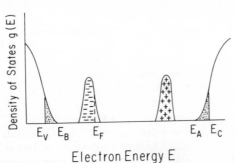

Fig. 5.5. Density of states g(E) suggested by Marshall and Owen (1971) for As_2Se_3. The energies E_v, E_B, E_A, and E_c have the same meaning as in the Davis and Mott model. A band of localized acceptor states lies below and a band of donor states above the gap center. In the cases shown, the acceptors are nearly compensated by the donors. As T is increased E_F moves toward the gap center.

trations of donors and acceptors adjust themselves by self-compensation to be nearly equal so that the Fermi level remains near the gap center. At low temperature it moves to one of the impurity bands because self-compensation is not likely to be complete. This model is based mainly on the observation that the high field drift mobility in $As_2 Se_3$ is of the Poole-Frenkel form presumably because of field-stimulated emission of carriers from charged trapping centers (acceptors).

Sharp Band Edge Model

Molecular solids and tight binding, semi-insulating materials with large band gaps have electronic structures which are relatively insensitive to disorder. In these materials tail states are negligible and a band model not significantly different from that of crystals with sharp band edges is appropriate. Since the molecular units are well defined entities one expects that the energies of localized defect states fall into rather narrow energy ranges deep in the gap. Because the energy to create a defect state is of the order of the band gap, their number will be small and self-compensation as well as association of donor and acceptor type defects are energetically favored. We believe this model is useful for describing oxide films, oxide glasses, a-Se and molecular semi-insulating materials.

5.3   ELECTRICAL PROPERTIES

The purpose of the models described above is to provide the density of extended and localized states g(E) and the mobility $\mu(E)$ as a function energy so that one can proceed to calculate the conductivity from (Kubo 1957, Greenwood 1958)

$$\sigma = e \int g(E) \, \mu(E) \, f(E) \, (1 - f(E)) \, dE. \tag{5.1}$$

The Fermi energy E_F in the distribution function f(E) is determined by the distribution and the charge state of the gap states and the condition that the material be neutral. The task of characterizing a material is then to determine from the experimental data the parameters describing g(E), $\mu(E)$, and among other parameters the capture and scattering cross sections of the localized states.

This approach is based on several assumptions which are not easily justified in amorphous materials. The band models and Eq. (5.1) assume that the energy eigenstates are independent of their occupation. This one particle approximation neglects correlation effects which are particularly important for localized states. But even for extended states correlation effects will influence the electron kinetics because of the short diffusion length in these low mobility states. Correlation effects appear to be of greater importance here than in crystalline semiconductors because the transport phenomena in NCS deal with conduction within the localized states and near the mobility edge, which itself is sensitive to potential fluctuations. A substantial redistribution of electrons over localized states caused for example by strong photon excitation or high electric fields produces large densities of charged centers which then modify the internal potentials and consequently g(E) and the mobility edge.

Moreover, Eq. (5.1) assumes that conduction in each energy shell can be treated separately. This is a doubtful procedure when charge motion involves exchange of energy with the lattice as in the case in phonon assisted hopping and polaron conduction.

One of the least justified assumptions underlying Eq. (5.1) is that the material is homogeneous. Compositional and structural heterogeneity and concommitant potential fluctuations seem to play an important role in real noncrystalline systems whose properties we are going to study. In these, a constant energy surface cuts through regions of greatly different conductivities. Conduction must then be treated as a percolation process favoring paths of high conductance.

With these reservations in mind we proceed with Eq. (5.1) and the one electron band models in describing the electrical properties. Major difficulties in applying these models to the experimental data will hopefully guide us to identify the weaknesses of this simplified approach. A brief discussion of the effects of heterogeneity is deferred to Section 5.4.

5.3.1   d.c. Conductivity

Adopting the notation of Davis and Mott (1970) one distinguishes three principal contributions to the conductivity.

(a) *Band conduction* of electrons excited above E_c or holes below E_v. Written for electrons this yields

$$\sigma = \sigma_0 \exp \left[-(E_c - E_F)/kT \right] \tag{5.2}$$

If g(E) and $\mu(E)$ do not vary too rapidly with kT above E_c one can use their average values μ_c and g(E_c). The separation of E_F from the mobility edge will change with T. Assuming in the temperature range of interest a linear temperature dependence

$$E_c - E_F = \Delta E - \gamma T \tag{5.3}$$

then Eq. (5.2) becomes

$$\sigma = C \exp \left(-\Delta E/kT \right) \tag{5.4}$$

with

$$C = eg(E_c) \, kT \, \mu_c \exp \left(\gamma/k \right) \tag{5.5}$$

If the extended states are not strongly affected by disorder μ_c might describe the motion of nearly free electrons (or holes) with occasional scattering. In most NCS it appears likely, however, that $\mu \leqslant 5 \ cm^2 V^{-1} sec^{-1}$ which corresponds to a mean free path less than the interatomic spacing. In this case, Cohen (1970) suggests that charge transport proceeds (in classical terms) via diffusion or Brownian motion. Adopting this classical picture for estimating the mobility one considers fast jumps between neighboring sites and obtains

$$\mu = \frac{1}{6} \frac{ea^2}{kT} \nu \tag{5.6}$$

where the interatomic spacing is a and the jump frequency $\nu \sim 10^{15}$ sec^{-1} is an electronic hopping frequency associated with the interatomic transfer integral $J = h\nu$. A more detailed analysis of this problem is based on the random phase model which assumes that the extended state wavefunctions can be represented as a linear combination of atomic wavefunctions with coefficients which have no phase relation from one site to the next. The conductivity was calculated by Mott (1967, 1970a), Hindley (1970a), and recently by Friedman (1971).

The mobility in the extended states near E_c or E_v is according to Friedman

$$\mu = \frac{2\pi}{3} \frac{ea^2}{h} z \frac{J}{kT} a^3 \, Jg(E_c) \qquad (5.7)$$

Here z is the coordination number, J is the two site transfer integral ($J \sim 1$ eV). With a = 2.5 Å, z = 4 and $g(E_c) = 10^{21}$ eV^{-1} cm^{-3} Eq. (5.7) yields for room temperature $\mu \sim 5$ cm^2/V sec, a value not significantly different from that obtained by the simple estimate Eq. (5.6).

In the case of ambipolar conduction, the exponential term describing the conduction of holes has to be added to Eq. (5.2). Depending on the relative magnitudes of the prefactors and the activation energies it is then possible that the conductivity curve has two slopes with one type of carrier dominating at low T and the other at high T.

The conductivity does not follow a simple exponential law, Eq. (5.4), when ΔE depends on T as a result of a temperature dependent shift of E_F. Such a shift results from the neutrality condition when g(E) is asymmetric with respect to E_F.

(b) *Thermally assisted tunneling* in the localized gap states near the mobility edges, near E_A and E_B in Figure 5.4 or in the donor or acceptor bands of Figure 5.5. The largest tunnel contribution arises from jumps to unoccupied levels of nearest neighbor centers. Since it is unlikely to find such a level at the same energy the tunneling process is usually inelastic, i.e., it involves the emission or absorption of a phonon. Hence this thermally assisted tunneling process involves a hopping energy ΔW_1 in addition to the activation energy $E - E_F$ needed to raise the electron to the appropriate localized state at E. The conductivity will thus be of the form

$$\sigma = \sigma_1 \exp\left[-(E-E_F + \Delta W_1)/kT\right] \qquad (5.8)$$

For the Davis-Mott Model $E = E_A$ (or E_B). $\mu(E)$ drops sharply below E_c so that $\sigma_1 \ll \sigma_0$. In a more general case the tunnel probability (contained in $\mu(E)$) and g(E) decreases rapidly whereas f(E) increases as one moves in energy toward the gap center. As a consequence the maximum tunneling contribution shifts toward E_F as T is lowered. The conductivity curve is not expected to exhibit a constant activation energy in this range but a decreasing slope as T is lowered.

(c) *Tunneling conduction near E_F* should be of the form

$$\sigma = \sigma_2 \exp\left(-\Delta W_2/kT\right) \qquad (5.9)$$

where ΔW_2 has the same physical meaning as ΔW_1. Since the density of states near E_F and the range of their wavefunctions is probably smaller near E_F than near E_A or E_B one expects $\sigma_2 \ll \sigma_1$. As the temperature is lowered the number and energy of phonons available for absorption decreases so that tunneling is restricted to seek centers which are not nearest neighbors but which instead lie energetically closer and within the range kT. For this so-called variable range hopping process Mott (1969a, 1972) derived the relation,

$$\sigma = \text{const} \, \exp \, [-(T_o/T)^{1/4} \tag{5.10}$$

Ambegaokar, Halperin and Langer (1971) find for T_o

$$T_o = 16\alpha^3/kg\,(E_F) \tag{5.11}$$

where α, the coefficient of exponential decay of the localized state wave-functions, was assumed to be independent of E. The derivation assumes the density of states at the Fermi level $g(E_F)$ to be energy independent up to

$$|E_{max}-E_F| = kT^{3/4} \, T_o^{1/4} \tag{5.12}$$

This restriction and the need for multiphonon processes when the energy separation of two tunneling sites exceeds the largest phonon energy limit the validity of Eq. (5.10) to low temperatures.

The numerical factor in Eq. (5.11) depends on details of summing up the percolation paths. Pollak (1972a) obtains a factor 11, Jones and Schaich (1972) a factor 10. The prefactor in Eq. (5.10), which has been calculated by Mott (1972), Brenig *et al.* (1971), Pollak (1972a), depends on the electron phonon coupling as well as on $g(E_F)$ and α. Pollak (1972a) extended the treatment to include a more general form of g(E). Pollak (1972b), and Knotek and Pollak (1972) began to consider correlation effects in the tunneling process. Last and Thouless (1971) demonstrated that the percolation criterion is irrelevant and that it over-estimates the conductance appreciably. This problem will be discussed further in the last section.

5.3.2 Thermopower
The work of Cutler and Mott (1969) suggests that a general expression for the thermopower S exists (Fritzsche (1971b), Cutler (1972)) which is closely related to that of Eq. (5.1). The thermopower is related to the Peltier coefficient Π by

$$S = \Pi/T \tag{5.13}$$

The Peltier coefficient has a simple physical meaning. It is the energy carried by the electrons per unit charge. The energy is measured relative to the Fermi energy. Each electron contributes to Π in proportion to its relative contribution to the total conductivity. The weighting factor for electrons with dE at E is thus $\sigma(E)dE/\sigma$.

This yields

$$\Pi = -\frac{1}{e} \int (E - E_F) \frac{\sigma(E)}{\sigma} dE \qquad (5.14)$$

and

$$S = -\frac{k}{e} \int \frac{E - E_F}{kT} \frac{\sigma(E)}{\sigma} dE. \qquad (5.15)$$

The sign convention is such that $S < 0$ for electrons at energies $E > E_F$. The magnitude of an individual contribution to S scales roughly with the distance from E_F in units of kT and with its fractional contribution to the total current.

If conduction takes place in one band only, then one obtains the familiar expressions

$$S = -\frac{k}{e} \left[\frac{E_c - E_F}{kT} + A_c \right] \text{ for } E > E_c \qquad (5.16a)$$

$$S = \frac{k}{e} \left[\frac{E_F - E_v}{kT} + A_v \right] \text{ for } E < E_v \qquad (5.16b)$$

where

$$A_c = \int_0^\infty \frac{\epsilon}{kT} \delta(\epsilon) d\epsilon \Big/ \int_0^\infty \delta(\epsilon) d\epsilon \text{ with } \epsilon = E - E_c \qquad (5.17a)$$

$$A_v = \int_{-\infty}^0 \frac{\epsilon}{kT} \delta(\epsilon) d\epsilon \Big/ \int_{-\infty}^0 \delta(\epsilon) d\epsilon \text{ with } \epsilon = E_v - E \qquad (5.17b)$$

are weighted averages over the carriers above the mobility edge E_c and below E_v respectively. $A = 1$ for constant g and μ for example, and $A = 3$ when g and μ both increase linearly with energy above E_c. Substituting Eq. (5.3) into (5.16) yields

$$S = \pm \frac{k}{e} \left[\frac{\Delta E}{kT} - \frac{\gamma}{k} + A \right] \qquad (5.18)$$

Davis and Mott (1970) pointed out that a plot of S against $1/T$ offers a possibility for determining γ if a reasonable estimate for A can be made.

A comparison of Eqs. (5.2) and (5.16) shows that a plot of S and of $\ln \sigma$ against $1/T$ should have the same slope if conduction takes place in only one band. However, one observes quite often that $\ln \sigma$ against $1/T$ has a greater slope than S against $1/T$. This may be caused by the following. Let us assume that the conductivity or the mobility contains an activation term $\exp(-\Delta W/kT)$ as for instance in Eq. (5.8). If ΔW is independent of E in the energy range of dominant conduction, then the activation term cancels in Eq. (5.14) and does not appear in the expression of S. The slope of $\ln \sigma$ is then larger than that of S by $\Delta W/k$.

It is also of interest to consider the case of combined conduction $\sigma = \sigma_c + \sigma_v$ when the position of E_F is fixed by localized states. Let $b = \sigma_c/\sigma_v$ and the center of the mobility gap $E_M = \frac{1}{2} (E_c + E_v)$ then straightforward manipulation yields

$$S = \frac{k}{e} \left[\frac{1-b}{1+b} \frac{E_c - E_v}{2\,kT} + \frac{E_F - E_M}{kT} + \frac{1}{1+b} A_v - \frac{b}{1+b} A_c \right] \qquad (5.19)$$

It should be noted that the slope of S against $1/T$ does not yield $\frac{1}{2} (E_c - E_v)$ as in the case of intrinsic conduction for which

$$(E_F - E_M)/kT = \ln (N_c/N_v) \qquad (5.20)$$

where N_c and N_v are the effective densities of states in the conduction and valence bands, respectively. The slopes of $\ln \sigma$ and of S are again different. This holds true in general whenever the conduction takes place in more than one energy range. Eq. (5.15) appears to be applicable even when the current flows by hopping in the localized states. Although this has not been verified one expects the sign of the thermopower to depend on whether conduction takes place above or below the Fermi level. The cases of polaron hopping and phonon-assisted tunneling must be studied in detail to decide what fraction of the activation energy is transported with the carriers. Thermopower measurements may reveal the answer to these questions.

When only states near E_F contribute to the current it is convenient to write $f(1-f) = -kT\, df/dE$ in Eq. (5.1). Expanding $g(E)\, \mu(E)$ in a Taylor series at $E = E_F$ one finds for the first nonvanishing term in Eq. (5.4) (Mott and Jones, 1958).

$$S = -\frac{\pi^2}{3} \frac{k}{e} kT \left[d \ln (\mu g)/dE \right] E_F \qquad (5.21)$$

This is Mott's formula for metallic conduction. Instead of the bracket in Eq. (5.21) one can write $[d \ln \sigma (E)/dE]_{E_F}$ because of the delta function behavior of df/dE. In this 'metallic' case the sign of the thermopower depends on whether the major contribution to the current lies above or below E_F. Brenig, Döhler, and Heyszenau (1973) calculated the thermopower for variable range hopping conduction which follows Eq. (5.10). In the limit of low T the thermopower is similar to Eq. (5.21) but enhanced by the factor $(T_o/T)^{3/4}$ where T_o is given by Eq. (5.11).

5.3.3   d.c. Conduction in Lone Pair Semiconductors

Over a considerable temperature range the d.c. conductivity of most non-crystalline covalent alloys follows Eq. (5.4), with C lying between 10^2 and 5×10^3 ohm^{-1} cm^{-1}. Since C is proportional to the mobility, the relatively small variation of its value and its similarity in intrinsic crystalline and in amorphous semiconductors has been a puzzle for a long time. (Myuller (1962), (1965); Zertsalova, Fainberg, and Grechanik (1965); Stuke (1970b)). Some examples are shown in Figure 5.6 where the room temperature conductivity is plotted against ΔE. It appears now that the C values cover a considerable range (Davis and Mott (1970)) and that an agreement with some values found in crystals may be fortuitous. In crystals the larger value of the mobility is partly offset by a smaller value of the effective mass entering the

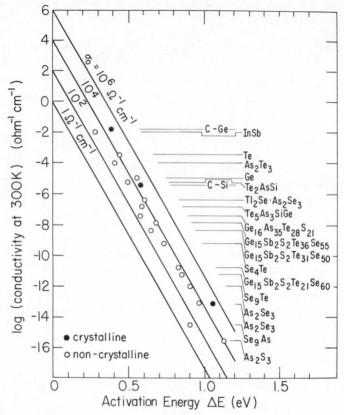

Fig. 5.6. Relation between conductivity at 300 K and activation energy ΔE as defined by Eq. (4) of text for several amorphous and three crystalline semi-conductors. In several cases the conductivity curve was extrapolated to 300 K in order to obtain these values. The values of σ and of ΔE are not characteristic of a given material but may depend on the method of preparation.

prefactor C. Moreover, it appears that one cannot obtain a reliable value of the mobility without knowing the T-dependence of C. Let us nevertheless estimate C from Eq. (5.5) assuming conduction as taking place in one of the bands near the mobility edge E_c with an average mobility $\mu_c = 4$ $cm^2 V^{-1} sec^{-1}$. Taking $\gamma = 3 \times 10^{-4}$ eV/K, similar to temperature shifts of optical gaps, using $kT\, g(E_c) = 5 \times 10^{19}\ cm^{-3}$ similar to the effective density of states at 300 K in crystalline semiconductors, one obtains $C \sim 10^3\ ohm^{-1}$ cm^{-1}, the value observed in most covalent NCS. If we take for γ values between $1-5 \times 10^{-4}$ eV/K, values for C between $10^2-10^4\ ohm^{-1}\ cm^{-1}$ are obtained.

The composition of most covalent NCS alloys can be changed over large ranges. In nearly all cases this produces a gradual change in the physical properties. As an example the change of σ and ΔE, in the system

$Tl_2 SeAs_2 Se_3$ as the selenium is gradually replaced by Te, is shown in Figure 5.7 (Andriesh and Kolomiets 1965). The value of C remains nearly constant. The same authors plot the thermopower S against log σ for different compositions as shown in Figure 5.8. Because of the positive sign of S we assume conduction is predominantly by holes in the valence band. With Eqs. (5.16b) and (5.4) one. obtains for the log σ = 0 intercept the value 0.5 $\times$ 10^{-3} V/degree so that

$$\frac{k}{e}(\ln C + A_v - \gamma/k) = 0.5 \times 10^3 \, V/degree.$$

With C = 2 $\times$ 10^3 ohm^{-1} cm^{-1} and $A_v \sim 2$ this yields $\gamma \sim 4k = 3.5 \times 10^{-4}$ eV/K, a very reasonable value. The present data and the uncertainty in A_v and A_c make it impossible to decide whether carriers of opposite sign contribute significantly, in which case Eq. (5.19) should be applied. Additional work by Uphoff and Healy (1961 and 1962), Vengel and Kolomiets (1957) and by Kolomiets and Nazarova (1960) on the T-dependence of σ and S led one to conclude that in most covalent alloy glasses one deals with conduction by holes near the mobility edge E_v. However, recently Rockstad and Flasck (1973) observed that many chalcogenide glasses have a negative thermopower. Evaporated or sputtered films of the Ge-Te-Se ternary system were found to be n-type for Ge rich compositions. Annealing below the glass transition temperature shifts the thermopower to more negative values and thus increases the composition range of dominant n-type conduction. With E_F near the gap center and both carrier types having similar mobilities one expects ambipolar conduction to be a common occurrence in covalent alloy glasses. A shift to negative values of the thermopower also occurs when the material is illuminated (Rockstad 1973).

In most covalent NCS it is found that ΔE, the thermal activation energy of the conductivity is about half the magnitude of the optical energy gap. This means that E_F is not far from the center of the mobility gap. Does this mean that these materials are 'intrinsic'? In the case of crystalline semiconductors the word intrinsic is used to mean that the conduction properties are not affected by the presence of localized impurity states. The position of E_F is then determined by the equality

$$n = p \tag{5.22}$$

of the mobile electron and hole concentration.

In order to retain intrinsic conduction to temperatures as low as 150 K as observed in Figure 5.7, the gap must be virtually free of localized states. With $\Delta E \sim 0.5$ eV, for example, the free carrier density at 150 K is about n $\sim 5 \times 10^2 cm^{-3}$. There is considerable evidence that the density of gap states far exceeds this number. We therefore conclude that the position of E_F is not determined by Eq. (5.22) but by the distribution and occupation of localized gap states. These localized gap states may very well be inherent to the NCS and thus one of the 'intrinsic' properties. However, it is better to refer to Eq. (5.22) as the condition for intrinsic conduction because a NCS can exist in many structural states and there is no way to operationally define the distribution of localized states which is inherent to the material.

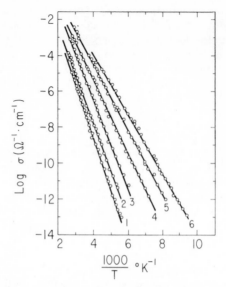

Fig. 5.7. Temperature dependence of the conductivity of several compositions of the system Tl_2SeAs_2 $(Te_xSe_{1-x})_3$ according to Andriesh and Kolomiets (1965). The composition of the samples labelled 1 through 6 correspond to x = 0, 0.2, 0.4, 0.6, 0.8, 1.0, respectively.

The possibility that Eq. (5.22) becomes satisfied at higher temperatures, particularly in the case of NCS having a relatively small gap, cannot be excluded. In most cases, however, ΔE is given by Eq. (5.3). Although E_F may be close to the gap center, the magnitude of the mobility gap cannot be obtained from ΔE. Moreover, ΔE determined experimentally from the conductivity by means of Eq. (5.4) overestimates the value of $E_c - E_F$ (or of $E_F - E_v$) at T = 0 because at low temperatures $E_c - E_F$ is parabolic in T and does not follow Eq. (5.3).

Figure 5.9 shows the change of ΔE in As_2Te_3 with increasing additions of Si, Ge, and Tl, respectively (Štourač et al. (1972)). These changes are accompanied with changes in the optical gap (such that ΔE remains about one half the optical gap) and in most other parameters, which depend on the bonding, such as the glass transition temperature, hardness, sound velocity, internal damping. A comparison of these systematic variations with different compositional variables appears to be most fruitful for gaining insight into the relation between the chemical band structure and the distribution and the nature of electronic states.

A striking example for the role played by the local order in determining the semiconducting properties is the Tl-Te alloy system, studied by Ferrier, Prado, and Anseau (1972). Figures 5.10 and 5.11 show the conductivity at 77 K and the activation energy ΔE, respectively, of flash evaporated or co-evaporated amorphous films as a function of composition. The lowest σ

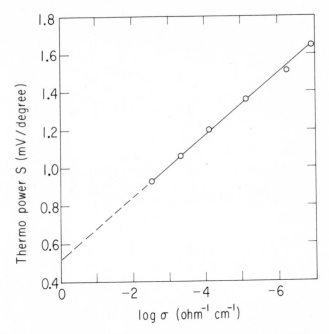

Fig. 5.8. Plot of thermopower S against $-\ln \sigma$ at 300 K for the different compositions of the system shown in previous figure. (After Andriesh and Kolomiets (1965)).

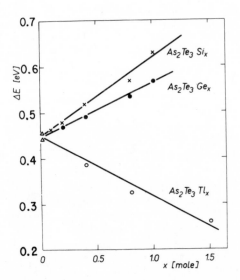

Fig. 5.9. Change of the conductivity activation energy ΔE as Ge, Si, and Tl, respectively, is alloyed to amorphous $As_2 Te_3$ (after Štourač et al. (1972)).

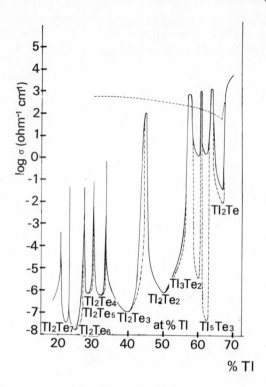

Ferrier, Prado, and Anseau (1972)

Fig. 5.10. Conductivity of co-evaporated films of $Tl_x Te_{100-x}$ as a function of composition x. Solid curve is at 78 K as deposited. Dash-dotted curve: at 78 K after annealing. Dashed curve: conductivity of liquid at 425°C. (After Ferrier *et al* (1972)).

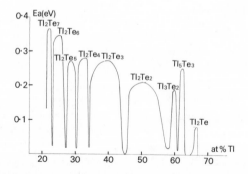

Fig. 5.11. Activation energy ΔE of conductivity obtained from high temperature part of $\ln \sigma$ vs. $1/T$ curves. (After Ferrier *et al*. (1972)).

and highest values of ΔE are found at compound compositions at which one expects a high degree of chemical order and satisfaction of valence bonds. Annealing appears to further improve the local order, particularly at high Tl concentrations. This study shows that compound formation occurs during film deposition and at annealing temperatures below crystallization. The amorphous state can exist with a high degree of local order. In many cases compounds are formed in NCS which do not occur in the crystalline form.

The activation energies ΔE of Figure 5.11 were measured near 300 K where σ follows Eq. (5.2) with holes near E_v dominating the conduction and $C \sim 10^3$ ohm^{-1}cm^{-1}. As the temperature is lowered, ΔE decreases. Between 200 and 78 K, $\ln\sigma$ follows Mott's $T^{-1/4}$ law, Eq. (5.10). In order to interpret this part as tunneling conduction near E_F the authors require for the density of gap states values $6 \times 10^{18} < g(E_F) < 5 \times 10^{19}$ eV^{-1}cm^{-3}.

Lone pair semiconductors are usually in an annealed metastable equilibrium state when they are prepared by cooling from the melt. Thin films prepared by sputtering or flash evaporation show annealing effects. In practically all cases one finds an increase in resistivity, ΔE, and in the optical gap with annealing. These changes come to completion below the glass transition temperature.

Deviations from the simple relation Eq. (5.4) are observed, particularly in small gap materials. It is also necessary to consider conduction by both types of carriers in materials like $Te_{70}Cu_{25}Au_5$ for which Tsuei (1968) found $\Delta E = 0.1$ eV. $Tl_2Te \cdot As_2Te_3$ studied by Nagels et al. (1970) shows exceptional behavior in that $\Delta E = 0.25$ eV from the σ-curve and $\Delta E = 0.13$ eV from the slope of S against $1/T$. The $\frac{1}{T} = 0$ intercept of S is of the same sign as S which is unusual as it yields a positive value for $A - \gamma/k$. It appears that Eq. (5.19) should be applied to these cases.

As the temperature is lowered the thermopower is expected to approach zero as $T \to 0$ according to thermodynamic arguments. According to Eq. (5.15) this can only happen if the dominant conduction process moves toward E_F as T is lowered. Indications of the appearance of a new conduction process at low T have been observed as a break in the conductivity curve at low T (Davis and Mott (1970), Tsuei (1968), Tauc, et al. (1968)). This low T branch depends on the thermal history of the sample. A second phase which may precipitate at the annealing temperatures may, however, strongly affect σ and S as discussed in the Section 5.4 on heterogeneity. In some cases an increase of conductance due to annealing can be traced to a crystallization of a very thin surface film (Fagan and Fritzsche (1970a)).

5.3.4 Tetrahedral Glasses

Tetrahedral glasses of composition $A^{II}B^{IV}C_2^V$ as well as mixtures of these, form the largest group of glasses which contain neither oxygen nor a chalcogen atom. They can be formed by quenching from the melt (Goryunova 1965, 1968, Nikolskaya et al. 1966). These materials are electron analogs of III—V compound and in structure similar to tetrahedral semiconductors. It appears that $A^{II}C_2^V$ is the building unit of these glasses since the concentration of the Group IV element can be varied over wide limits and it can be

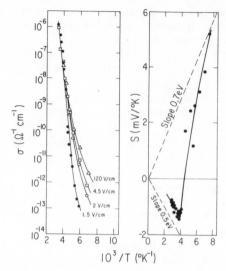

$$10^3 / T \ (^{\circ}K^{-1})$$

Fig. 5.12. Temperature dependence of the conductivity σ at various fields and of the thermopower S of amorphous $CdGeAs_2$ according to Tauc *et al.* (1968). The dotted lines have the slope of 0.7 eV and 0.5 eV, respectively, for A = 0 in Eq. (5.16) of text.

replaced by Sb, Tl, Mg. Al, Ga, and In (Hruby and Štourač (1969); Čerevinka *et al.* (1970)).

The electrical and magnetic properties of the tetrahedral glasses show a larger diversity than those of the lone pair semiconductors. The activation energies span a wide range (Goryunova (1965), (1968)) with $\Delta E = 0.15$ eV for $CdSnAs_2$ to $\Delta E = 1.1$ eV for $ZnGeP_2$ and $CdSiP_2$.

$CdGeAs_2$ is an interesting example of the importance of thermopower measurements. Figure 5.12 shows the results of σ and S measurement by Tauc *et al.* (1968). Except for the lowest T the conductivity can be described by Eq. (5.4) with $\Delta E = 0.6$ eV and $C \sim 5 \times 10^4$ ohm^{-1}cm^{-1}. From the conductivity data alone one would never have suspected the abrupt change from dominant electron conduction to hole conduction shown by the thermopower. The thermopower suggests that one deals with simultaneous conduction of carriers of both signs, electrons dominating above 250 K and holes at low T. The slopes of the two dotted lines drawn into Figure (5.12b) correspond to $E_c - E_F = 0.5$ eV and $E_F - E_v = 0.7$ eV, respectively, in Eq. (5.16). From optical absorption Tauc *et al.* (1968) obtain an optical gap of about 1.2 eV. One concludes from the magnitude of S and of ΔE that E_F remains close to the gap center throughout the temperature range. A small shift in E_F or an increase of the ratio of hole mobility to electron mobility as T is lowered appears to cause the shift to predominant hole conduction at low T. An upward shift of the Fermi level with increasing Ge content in $CdGe_xAs_2$ was observed by Callaerts *et al.* (1970). They find positive thermopower for x = 0.2, a change of sign near

$0.3 < x < 0.4$ and negative values for $x > 0.6$. The discrepancy with the results of Tauc *et al.* (1968) is probably due to differences in preparation conditions.

$CdM_x As_2$ glasses with M = Ge, Si, Mg. Tl, Ga, In, Sb and with $0 < X < 1.2$ were studied by Červinka *et al.* (1970) and Abraham *et al.* (1970). They find that (i) all compositions follow Eq. (5.4) with $C \sim 10^4$ ohm^{-1} cm^{-1}, (ii) ΔE hardly changes as Ge or Si is added to $CdAs_2$, (iii) ΔE decreases appreciably with additions of Sb or Tl, (iv) E_F remains near the gap center, (v) both electron and hole conduction are observed. As many as $4 \times 10^{19} cm^{-3}$ paramagnetic centers appear as Ge, Sb or Tl is added to $CdAs_2$. One expects that localized states near the gap center are occupied by an unpaired electron and thus be paramagnetic (Ball (1971)). In chalcogenide glasses the density of paramagnetic centers must be less than $3 \times 10^{16} cm^{-3}$ according to susceptibility studies (Tauc and Menth (1972)) and less than $10^{15} cm^{-3}$ according to electron spin resonance studies (Agarwal (1973)). It is therefore interesting that in these tetrahedral glasses the density of paramagnetic centers is proportional to the concentration of Ge, Sb, or Tl atoms (Matyas (1972)). This problem has recently been solved by DiSalvo *et al.* (1973) who repeated Matyas' magnetic susceptibility measurements on $CdGeAs_2$ and did not find any paramagnetic centers with a detection sensitivity of $10^{17} cm^{-3}$.

Matyas also reported that the diamagnetism is greatly enhanced by the addition of any of the above elements in contrast to Se, $As_2 Se_3$, and $As_2 S_3$ whose diamagnetism is the same in the crystalline and amorphous state (DiSalvo (1973); Hudgens (1973)).

5.3.5   d.c. Conduction in Tetrahedral NCS Films

Special Structural Problems
The properties of tetrahedrally coordinated amorphous films of Si, Ge, SiC and several III–V compounds have been quite extensively studied because it is attractive to compare them with the well known properties of their crystalline form. Despite the rich literature dealing with amorphous films of Si and Ge no coherent picture appears yet to emerge and the experimental results coming from different laboratories are often seriously at variance. At first sight this is surprising because with the elemental semiconductors Ge and Si one needs not worry about stoichiometry and compositional disorder. The materials, which were thought of as being the simplest amorphous systems and of holding the key for understanding noncrystalline semiconductors in general, turn out to be very exceptional in many respects.

Tetrahedral NCS cannot be prepared by rapid quenching from the melt because the co-ordination number is eight in the liquid. Instead, amorphous films of these materials must be obtained by vacuum deposition, cathode sputtering, electrolytic deposition, or by radio frequency glow discharge decomposition of a gas like SiH_4 or GeH_4. The main difficulty of these films appears to be the fact that they grow with a porous network of microvoids. Their existence was first postulated (Brodsky and Title (1969), Moss and Graczyk (1969)) to explain that amorphous Ge and Si are about

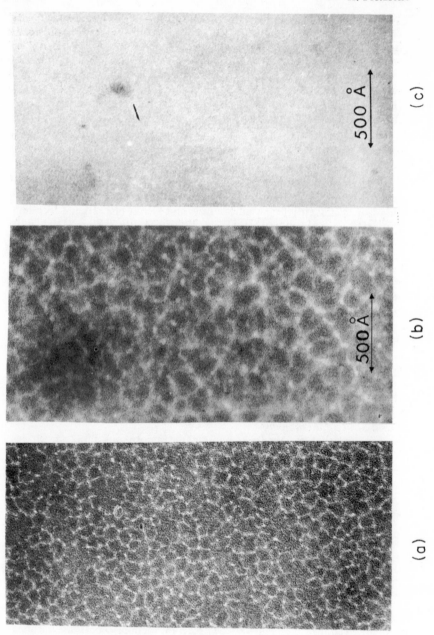

Fig. 5.13. Bright field electron micrograph of an a-Ge film formed at different substrate temperatures T_s. (a) $T_s = 25°C$, (b) $T_s = 150°C$, and (c) $T_s = 250°C$. The void network observed in the lower substrate temperature films is completely absent in the film prepared at $T_s = 250°C$ (after Donovan and Heinemann (1971)).

10–15% less dense than the crystals although the nearest and next nearest neighbor distances are the same. Furthermore, the void network provided a large internal surface area to account for the observed spin resonance signal which was found to be identical to that of surface states in line width and gyromagnetic ratio. Since then the void network was found to be the reason for the increased absorption of water in these films (Kastner and Fritzsche (1970)) and for the anomalous structure in their ultraviolet optical constants (Galeener (1971); Bauer, Galeener, Spicer (1972)).

Films deposited by sputtering or vacuum deposition often start by forming on the substrate a large number of islands which then grow together. A mismatch or some oxide contamination can then lead to interface regions which make up the void network. As an example Figure 5.13 shows electron microscope pictures taken by Donovan and Heinemann (1971) of a-Ge evaporated on substrates held at 25, 150, and 200°C, respectively. Barna, Barna and Pocza (1972) were successful in photographing the microvoids as a Ge film grows at 350°C *in situ* in the electron microscope. Their voids seem however not to be connected.

These microvoids add a further dimension of complexity to the variables which determine the structure of NCS films. The electrical and optical properties must be strongly affected because they are very sensitive to the preparation and annealing conditions. Careful studies were made of the effects on the physical properties of (i) the substrate temperature during deposition (Theye (1971)), Donovan, Ashley, Spicer (1970); Bauer, Galeener, Spicer (1972)), (ii) the deposition rate (Walley (1968), Koc *et al.* (1972a)) (iii) the residual oxygen contamination (Koc *et al.* (1972b)), and (iv) annealing. The morphology of the voids and defects and their ability to coalesce and anneal out may depend on the amount of oxides and hydrides formed and other gases trapped during preparation and on the temperature

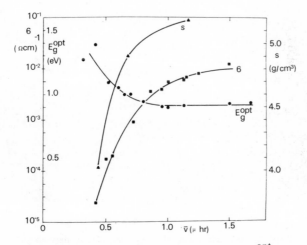

Fig. 5.14. Conductivity σ at 300 K, density s, and optical gap E_g^{opt} of sputtered a-Ge films as a function of sputtering rate. (After Koc *et al.* (1972a)). Density of c-Ge is s = 5.35 g/cm^3.

of the evaporation source which determines the size and kind of molecular complexes in the vapor stream.

The aim of these studies is to find the parameters which yield the tetrahedral films in the form of a voidfree random network structure similar to the model constructed by Polk (1971).

Figure 5.14 shows the effect of the sputtering rate on the conductivity σ, density s, and the optical gap E_g^{opt} of a-Ge (Koc *et al.* (1972a)). Very pure argon was used for sputtering. One observes a great variation in the properties by changing only one of the numerous preparation parameters. Evaporation, electrolytic deposition or glow discharge deposition of GeH_4 or SiH_4 (Chittick *et al.* (1969)) produces further differences in the materials.

We do not know whether these problems are common to all tetrahedral amorphous films. The similarity of their properties and the annealing behavior suggest that this is indeed the case. Most information about the electrical behavior of tetrahedral films has been obtained with material containing internal voids. It appears nevertheless important to review the properties and to analyze to what extent they reveal properties inherent to this group of amorphous semiconductors.

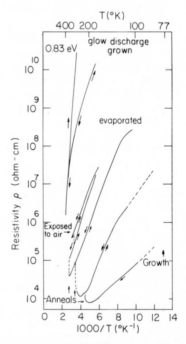

Fig. 5.15. Resistivity curves of evaporated Si films depend strongly on thermal history. (After Brodsky *et al.* (1970)). Arrows indicate direction of temperature sweep. Si film prepared from SiH_4 in a glow discharge is much more resistive. (After Chittick (1970)).

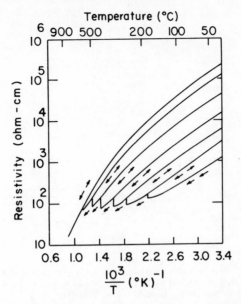

Fig. 5.16. Resistivity curves of sputtered SiC films during temperature cycling. (After Mogab and Kingery (1968)).

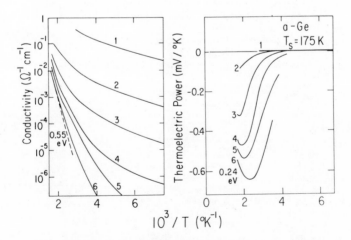

Fig. 5.17. Temperature dependence of conductivity and thermopower S of a-Ge deposited at 175 K. Annealing time is 15 min., annealing temperatures: 1) 300 K, 2) 550 K, 3) 650 K, 4) 700 K, 5) 735 K, 6)) 765 K, (after Stuke, 1971).

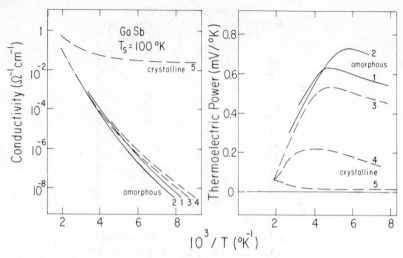

Fig. 5.18. Conductivity and thermopower S of a-GaAs evaporated at 140 K. Annealed
for 15 min. at 1) 300 K, 2) 360 K, 3) 425 K, 4) 460 K, 5) 490 K, 6) 520 K,
7) 540 K, 9) 670 K. Results for a crystalline GaAs are also shown. (After
Stuke (1971)).

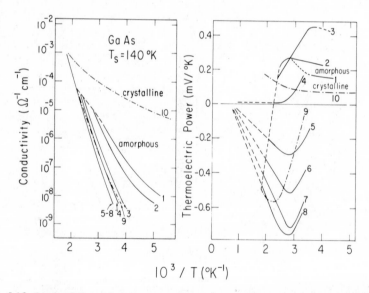

Fig. 5.19. Conductivity and thermopower of a-GaSb evaporated at 100 K, annealed for
15 min. at 1) 300 K, 2) 385 K, 3) 480 K, 4) 510 K, 5) 535 K (after Stuke,
(1971)).

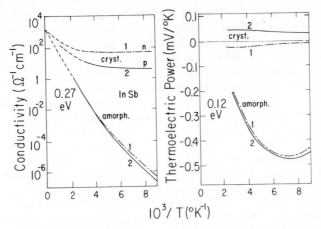

Fig. 5.20. Temperature dependence of (a) conductivity and (b) thermopower of amorphous and polycrystalline InSb. Curves 1: InSb + 1% Te, curves 2: InSb + 1% Zn (after Stuke, (1970a)).

Conductivity and Thermopower
Examples of resistivity curves of evaporated Si, sputtered SiC, and evaporated Ge are shown in Figures 5.15, 5.16 and 5.17, respectively. The resistivity of rf glow discharge grown Si film as obtained by Chittick *et al.* (1969), and Chittick (1970) is included in Figure 5.15 for comparison.

After a film is annealed to completion at a certain temperature it is stable below that temperature but further annealing takes place at higher temperatures. Upon annealing the density deficiency decreases from about 12% to 3% at the amount of low angle scattering, which is a measure of voids, drops considerably. The good agreement between the radial distribution function of Moss and Graczyk (1969, 1970) and model calculations by Polk (1971) suggest that well annealed films of Ge and Si closely resemble ideal random network structures.

The temperature dependence of the conductivity σ and the thermopower S of several tetrahedral NCS measured by Stuke (1971) is shown in Figures 5.17–5.20. T_s denotes the substrate temperature during evaporation. The curves are numbered according to sequentially higher annealing temperatures.

We distinguish a high temperature region where the thermopower is proportional to 1/T, a transition region, and a low temperature region in which ISI decreases with decreasing temperature. The difficulty of detecting a change in conduction process in the gradually decreasing slopes of the resistivity curves emphasizes again the importance of thermopower measurements.

The shapes of the thermopower curves suggest that at higher temperatures band conduction dominates with E_F near the gap center and that the energy separation between E_F and the conducting states diminish at low temperatures. Interpreting the high T region with Eqs. (5.2) and (5.18) valid for conduction in a single band one finds, however, that the activation energies obtained from ln σ are considerably larger than those obtained from S, sometimes by more than a factor of two. This discrepancy may be caused by carriers of both sign contributing to the conduction and by a thermally activated mobility as in Eq. (5.8).

Below 250 K in a-Ge (see Figure 5.17) and below 500 K in a-Si the magnitude of the thermopower becomes very small and nearly T-independent. The conduction process appears to be thermally assisted tunneling through defect states close to E_F. The sign of S may be positive or negative depending on the asymmetric distribution of states above and below E_F.

The magnitude of S of a-GaSb at low T and of a-GaAs at early annealing stages (curves 1 and 2 in Figure 5.18) is too high to be accounted for by Eq. (5.21). Conduction appears to take place through defect states but separated in energy from E_F. If that is so then the slowly decreasing magnitude of S as T is lowered suggests a nearly T-independent concentration of carriers in the conducting states. This in turn leads to a thermally activated mobility if one tries to explain the T-dependence of the σ-curves in that temperature range. The activation energies required depend on temperature and on annealing. Their values, about 0.2 eV in GaSb and 0.3 eV in GaAs seem too high for a hopping energy ΔW, in Eq. (5.8). Approximately the same magnitude is needed if a thermally activated mobility is used to resolve the discrepancy between the slopes of ln σ and S $vs.$ 1/T at high temperature.

In explaining the strong effects of annealing on the electrical properties of these materials one must consider not only a change in the density and distribution of defect centers but also a change of the mobility gap $E_c - E_v$. As a consequence the ratio $b = \sigma_c/\sigma_v$ and the value $E_M - E_F$ can change without necessarily affecting $E_c - E_F$ (or $E_F - E_v$) which governs the

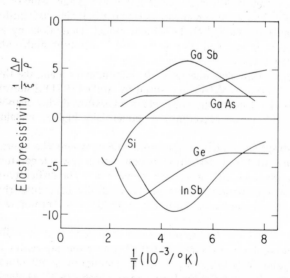

Fig. 5.21. Temperature dependence of the elastoresistivity of amorphous films of tetrahedral semiconductors (after Fuhs and Stuke (1970)).

dominant conduction process. The optical gap of a-Ge and a-Si (Brodsky *et al.* (1970)) was found to increase by as much as 20% with annealing. Although these two quantities are not directly related this increase indicates the kind of change one might expect for $E_c - E_v$. Annealing increases the contribution of electrons to the conduction in a-Ge and GaAs. It is remarkable that the annealing steps 5–8 do not change σ of GaAs but strongly affect S. This again suggests the presence of both types of carriers.

There is still some doubt about the sign of the thermopower in a-Ge. Grigorovici *et al.* (1966) report a large positive peak near 500 K in contrast to the results of Beyer and Stuke (1972) shown in Figure 5.17 and in contrast to the negative peak in a-Si at 600 K (Grigorovici *et al.* (1967)). A positive sign was also observed by Chopra and Bahl (1970) and Buchy, Clavaguera, and Germain (1971). The strong influence of the preparation parameters, of the substrate, and of the degree of oxidation of a-Ge and a-Si are presently being studied (Knotek, Pollak, and Donovan (1973). Beyer and Stuke (1973). Le Comber *et al.* (1973), Elliott and Yoffe (1973), Seager, Knotek and Clark (1973), Bluzer and Bahl (1973)).

Elastoresistance
Figure 5.21 shows the elastoresistivity for uniaxial strain parallel to the direction of current flow for several tetrahedral amorphous semiconductors (Fuhs and Stuke, 1970). The results obtained with a-Ge agree with those found earlier by Grigorovici and Devenyi (1968). Although the state of anneal is not specified it appears that
(i) the peaks of the elastoresistivity and of the thermopower occur at the same temperature,
(ii) At temperatures above the peak the elastoresistivity m is proportional to

$$\frac{\Delta\rho}{\rho\xi} = m = A + B/kT \qquad (5.23)$$

(iii) The sign of m is that of the thermopower.

These measurements show again the transition from the high temperature conduction process to that at low temperatures. The transition takes place over a considerable temperature range. Grigorovici *et al.* (1968) and Devenyi *et al.* (1970) proposed therefore a separate conduction process for the transition region. This does not seem to be necessary. Figure 5.22 shows in its upper half the resistance and the lower half the longitudinal elastoresistance m of a-Ge as measured by Devenyi *et al.* (1970). By extrapolating the low T and the high T conductances into the transition region and assuming a constant m for the low T branch and a straight line continuation of m for the high T branch, one obtains, after combining the two components appropriately, the elastoresistance values for the transition region shown by the crosses in the lower figure. There is qualitative agreement. The remaining discrepancy can probably be removed by choosing a better extrapolation.

Assuming conduction is only one band one obtains by substituting Eq. (5.4) into Eq. (5.23) for a-Ge (Fuhs and Stuke (1970)).

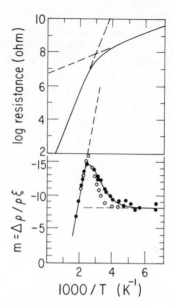

Fig. 5.22. Resistance and longitudinal elastoresistance of a-Ge according to Devenyi *et al.* (1970). The full circles are data points. The crosses were calculated from the extrapolations shown by dashed lines (see text).

$$A = -\frac{\delta C}{C\xi} = 11$$

and   (5.24)

$$B = \frac{\delta (\Delta E)}{\xi} = -0.6 \text{ eV/strain}$$

One would expect the coefficient A to be smaller and B to be larger than the values obtained by an order or magnitude. The reasons for this are the following. The prefactor C of Eq. (5.4) is about the same for most amorphous semiconductors as found in experiments and also as predicted by the random phase model. One therefore expects $\delta C/C$ to be of the same magnitude as the strain ξ, because of the change in volume, and not ten times larger. The factor B, on the other hand, measures the change of electron energies with strain. These deformation potentials are usually considerably larger (Keyes (1960)) because the directed covalent bonds are sensitive to the strain.

The elastoresistivity experiments were carried out in the hope of distinguishing thermally assisted tunneling conduction from band conduction, by the greater sensitivity to strains of tunneling conduction. The tunneling probability is proportional to the overlap term $\exp(-2 \alpha R)$ where R is the separation of tunneling states. This leads to a contribution to the elastoresistivity of the form

$$\frac{\delta\rho}{\rho\xi} = 2 \alpha R \left(\frac{\delta \ln R}{\xi} + \frac{\delta \ln \alpha}{\xi} \right)$$ (5.25)

The first term in the brackets is of order unity. In crystalline Ge, the second term gives rise to the very large elastoresistance effects because the localized state wavefunctions change drastically when the degeneracy of the conduction or valence band is lifted by uniaxial strain (Fritzsche, (1962, 1965)). Such effects are not expected in amorphous materials. It is presently not possible to give an estimate of the second term in the brackets. It may have either sign and the magnitude will probably be of order unity. The observed magnitude of the elastoresistivity at low temperatures as well as the occurrence of either sign is therefore plausible. It may not, however, be used as a criterion for distinguishing tunneling conduction from band conduction.

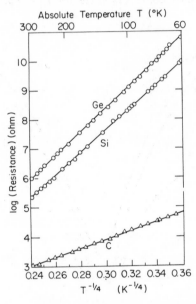

Fig. 5.23. Resistance of amorphous carbon, silicon, and germanium film plotted logarithmically against $T^{-1/4}$ (after Morgan and Walley (1971)).

Evidence for Tunneling Conduction near E_F
The very small value of the thermopower below about 250 K in a-Ge and below 500 K in a-Si suggest that conduction takes place close to E_F. The conductivity is too small and T-dependent for E_F to lie in a band of extended states. The conduction must therefore be by thermally assisted tunneling according to Eq. (5.8) and, as the temperature is lowered and carriers prefer to tunnel to a more distant but energetically more favorable site, by Eq. (5.10). Measurements on Ge by Clark (1967), Walley and Jonscher (1967), Chittick (1970) and of Chopra and Bahl (1970) fit Eq. (5.10) over a large temperature range. Figure 5.23 shows examples of such plots for Ge, Si, and C (Morgan and Walley (1971)). The value of T_o is 7.1×10^7 K for the Ge and Si samples shown in Figure 5.23 and $T_o = 1.4 \times 10^6$ K for C.

254   H. Fritzsche

Surprisingly, it is the good agreement which poses a problem for the validity of the hopping model. Eq. (5.10) was derived under the assumption that both $g(E)$ and α are constant up to a distance $\pm E_{max}$ from the Fermi level. Calculating $|E_{max} - E_F|$ from Eq. (5.12) one obtains a value of 0.57 eV for a-Ge and a-Si and 0.19 eV for C at 300 K. However, tunneling processes involving large energy differences, i.e. multiphonon processes, are rare events. Pollak et al. (1973) showed that expression (5.10) for variable range hopping should also be valid at higher temperatures because only states in the range $\sim 2 kT$ around E_F contribute significantly.

The fact that nearly the same value of T_o was obtained for Si and for Ge as well as for samples in different laboratories is unexpected. If one assumes $\alpha^{-1} = 10 \text{Å}$ then Eq. (5.11) yields for $T_o = 7 \times 10^7$ K a density of

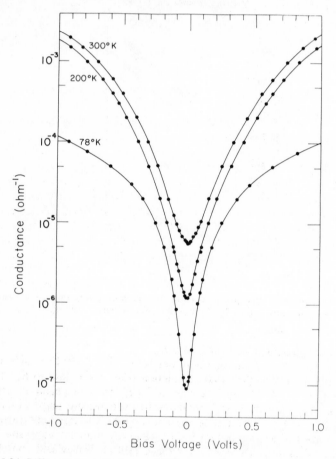

Fig. 5.24. Differential conductance dI/dV as a function of bias voltage of an Al-Al$_2$O$_3$ – amorphous Ge tunneling diode (after Osmun and Fritzsche (1970)). At low temperatures the resistance of the a-Ge electrode becomes comparable with the tunnel resistance. The bias voltage across the oxide barrier is then smaller than the applied bias shown here.

states $g(E_F) = 2.5 \times 10^{18} eV^{-1} cm^{-3}$. The notorious sensitivity to annealing and the preparation conditions should lead to very different values of $g(E_F)$ and hence of T_0.

Brodsky and Gambino (1972) found that annealing increases T_0 in a-Si somewhat. They and others pointed out that the experimental curves fit equally well other power laws T^{-n} with $n \neq 1/4$ in the exponent of Eq. (5.10). Such expressions can be obtained by assuming a more general energy dependence of the density of states $g(E)$ and of α (E) near E_F. Knotek *et al.* (1973) prepared a-Ge films of different thicknesses in ultrahigh vacuum and measured the conductivity *in situ*. Their results agree very well with the analysis of Pollak *et al.* (1973). Using the band model shown in Figure 5.4 they obtain a total density of midgap states of about $5 \times 10^{17} cm^{-3}$, a density of states at the peak $g(E_F) = 1.5 \times 10^{18} eV^{-1} cm^{-3}$, and a decay parameter of the localized state wavefunctions of $\alpha^{-1} = 9.4$ Å. The validity of their analysis is supported by the observation that the conductivity of very thin films (thickness less than 500 Å) is described by $\ln \sigma =$ const./$T^{1/3}$ which is the expression for variable range hopping in 2 dimensions.

Electron Tunneling into Amorphous Tetrahedral Films
In order to obtain some information about the position of E_F in the mobility gap and about the existence of a peak in the distribution of localized states as proposed by Mott (1969c), electron tunneling into a-Ge films was studied using $Al - Al_2O_3 - Ge$ tunnel junctions (Nwachuku and Kuhn, (1968) Osmun and Fritzsche, (1970) Sauvage and Mogab (1972)). In the earlier work a rather sharp conductance well was observed and some structure which was interpreted as tunneling into gap states. The later work showed however, that those features were produced by the measuring circuit. The bias dependence of the tunneling conductance is shown in Figure 5.24. The tunneling conductance has a minimum at exactly zero bias and rises smoothly and symmetrically with bias voltage. The voltage drop in the a-Ge film was found to be negligible except for the highest voltage points at low temperature. As the temperature is lowered, the curves not only shift to lower conductance values but they become steeper.

The essential features of the tunneling conductance curves, that is (i) their symmetry with respect to polarity reversal of the bias and (ii) the increased steepening of the conductance minimum at zero bias with decreasing temperature, have been confirmed for a-Ge and a-Si by Smith and Clark (1972) and for a-Si by Sauvage and Mogab (1972).

Konak and Stuke (1972) found the same features for tunneling into InSb and GaSb films except that the temperature dependence was less pronounced. Neither of these features is expected and it appears that the tunneling measurements pose a new problem rather than yield information about the density of localized gap states.

The most surprising feature is the symmetry. If one describes the tunneling conductance G(V,T) in terms of a T-independent transfer function P(E), which includes the tunnel matrix element and the density of states functions of both sides, as

$$G(V,T) = \int P(E) \left[f_M (E,T) - f_{sc} (E - eV, T) \right] dE \qquad (5.26)$$

then one obtains a symmetrical conductance $G(V,T) = G(-V,T)$ only when $P(E)$ is symmetrical with respect to E_F at $V = 0$. Here f_M and f_{sc} are the Fermi functions in the metal and the semiconductor, respectively. It would indeed be fortuitous if the Fermi level would lie exactly at the minimum of the density distribution of gap states and if the distribution were symmetric in energy with respect to E_F in many samples of amorphous Ge and Si prepared under different conditions. As an alternative explanation it was proposed (Osmun and Fritzsche, (1970)) that tunneling takes place between the metal and an interface layer with interface states which are symmetrically distributed and which remain in equilibrium with the bulk semiconductor as electrons tunnel from and into these states. This does not solve the problem but merely transfers it to the surface, about which we know even less than the bulk. Tunneling experiments on annealed tetrahedral semiconductors would be helpful to resolve the question whether the heterogeneity produced by the network of voids is responsible for the observed tunneling characteristics.

Setting aside the problem of finding the reason for the symmetry of the transfer function $P(E)$, let us choose $P(E)$ of the form

$$P(E) = \text{const. exp} - [a_1 - a_2 \ |E - E_F|] \qquad (5.27)$$

and inquire whether the observed temperature dependence is caused by the Fermi functions f_M and f_{sc}. This case has been solved by Stratton (1962) who obtains for the temperature dependence

$$\frac{G(V,T)}{G(V,O)} = \frac{\pi a_2 \ kT}{\sin(\pi a_2 \ kT)} \qquad (5.28)$$

The coefficient a_2 governs the steepness of the conductance increase with bias; its physical meaning depends on the particular tunneling process considered. These calculations predict that the temperature dependence is independent of bias and becomes less strong at lower T in contrast to the observed sharpening of the conductance curves and the large drop in G from 200 K to 78 K. This discrepancy is not removed by considering the temperature shift of the gap states and of the mobility edges, because these shifts decrease as T is lowered. However, phonon-assisted tunneling is a process which yields a stronger T-dependence at low T than at high T. Sauvage, Mogab, and Adler (1972) suggest that an electron tunneling to a localized gap state must tunnel a certain distance into the semiconductor. This distance is about the average separation of the number of states $g(E)$ kT available at T between E and E + dE. With phonon-assistance, electrons can tunnel to states of different energy. The tunneling electron can thus take advantage of states lying closer to the interface. At low T on the other hand, tunneling must take place on the average to states further from the oxide-semiconductor interface. Thus one expects G to follow an expression similar to Eq. (5.10). Sauvage, Mogab, and Adler analyze their data in this manner and obtain for $g(E_F)$ in a-Si the high value of about $10^{20} \ eV^{-1} \ cm^{-3}$. Improvements in the calculation and accounting for the void structure commonly observed in these materials may yield a more reasonable density of states value. This treatment ignores the problem posed by the observed symmetry of the G(V) characteristic.

Magnetoresistance, Electron Spin Resonances, Porosity
The presence of an anomalous magnetoresistance effect is another unique feature of tetrahedral amorphous semiconductors (Mell and Stuke (1970)). Its anomalous feature is its independence of the relative orientation between magnetic field and current. Its sign is usually negative but for some temperatures and annealing conditions it may become positive. A similar anomalous magnetoresistance is known to occur in crystalline semiconductors at concentration levels spanning the critical concentration for Mott's nonmetal to metal transition. The explanations offered for the anomalous effect in crystalline semiconductors, scattering by localized spins (Toyozawa (1962)) and Kondo condensation (Khosla and Fischer (1971)) require low temperatures and are not directly applicable to the phenomena observed in amorphous semiconductors. A negative magnetoresistance was observed in $Tl_2 Se \cdot As_2 TTe_3$ by Carver and Allgaier (1972) who used a Corbino disc arrangement. Its magnitude at 20 K Oe was $3 \times 10^{-4} - 3 \times 10^{-3}$. It is likely that the heterogeneity resulting from the internal void structure has a strong influence on the magnetoresistance. By considering the increase of α with magnetic field H in directions perpendicular to H, Mikoshiba (1962, 1963) and recently Shklovskii (1971) calculated the magnetoresistance of hopping conduction. They find a magnetoresistance which is positive and dependent on the angle between I and H, in disagreement with observations.

Electron spin resonance has revealed a very large free spin density between 10^{20} and 3×10^{20} cm^{-3} in amorphous Ge, Si, and SiC films deposited at room temperature (Brodsky and Title (1969) Brodsky *et al.* (1970)). The total number of free spins is proportional to the film thickness. It is rather insensitive to the method of preparation; similar results are obtained with a-Si prepared by sputtering, ion bombardment (Crowder *et al.* (1970)), or by evaporation in ultrahigh or moderate vacuum. It seems impossible to place this density of localized unpaired electron states in the mobility gap of the semiconductor, since the majority of the localized states of the valence band tail correspond to paired electrons and those of the conduction band tail are unoccupied. On the other hand, deep states near E_F corresponding to such high concentration would render the amorphous material a semimetal. Brodsky and Title (1969) suggest instead, that the ESR signal arises from dangling bonds on internal surface of the voids in the amorphous films. This interpretation is supported by the fact that the gyromagnetic ratios, line shapes, and line widths of these signals correspond to those found on the surfaces of the respective crystalline materials (Haneman (1968) Haneman *et al.* (1968)). Annealing causes the voids to coalesce or disappear which reduces the density deficiency and, as observed by Brodsky *et al.* (1970), the spin density. Amorphous Si produced by radio frequency decomposition of SiH_4 (Chittick *et al.* (1969) and Chittick (1970)) does not show on ESR signal (Brodsky (1971)). Saturation of the dangling bonds by hydrogen may be the reason for this.

A consequence of the network of voids is the porosity of tetrahedral films. This was studied by Kastner and Fritzsche (1970) and Fritzsche *et al.* (1972) by measuring the absorption of water and organic molecules of different sizes in a Ge film deposited on the surface of an oscillating quartz crystal microbalance. The observation of Barna *et al.* (1972), that a short exposure to oxygen or water changes the crystallization rate of these films and the crystallization mode, whether it is nucleated in the bulk or at the surface, raises the question as to how much the electronic and optical properties as well as the spin resonance are affected by exposure to atmosphere prior to heat treatment. In contrast, chalcogenide alloy films were found to be nonporous and exhibiting spin densities about 4—5 orders of magnitude smaller than those of a-Ge and Si (Agarwal and Fritzsche (1970) Agarwal (1973)).

5.3.6 Hall Effect

The measurement of the Hall coefficient, one of the most useful tools for the study of transport phenomena in crystalline semiconductors, proved to be of little value in amorphous semiconductors in the past. Not only was the experiment exceedingly difficult to perform, because of the small value of the Hall mobility, but the Hall coefficient could not be related to the carrier concentration in the usual manner. The observation that the Hall coefficient was always negative regardless of the sign of the thermopower served as a warning that something was wrong.

Several mechanisms were suggested to explain the negative Hall coefficient observed in materials having a positive thermopower. Pearson (1964) suggested crystalline inclusions and material heterogeneity as the cause. This did not agree with the temperature independence of μ_H and that its value remained the same when the glass was heated to the liquid (Male (1967)). Panus *et al.* (1968) discussed the possibility of ambipolar conduction in which the ratio of drift to Hall mobility for holes far exceed the ratio for electrons. Böer (1970a) obtained a small and negative Hall mobility by requiring proper compensation of the electron and hole currents. The last two requirements seem unlikely to be met by a large number of materials over a wide temperature range. For the same reason we find the suggestion of Queisser (1971) unlikely. He points out that ohmic conduction is often observed even in the presence of space charges which easily occur when the dielectric relaxation time exceeds the diffusion length lifetime of the minority carriers. In this so-called relaxation case (van Roosbroeck and Casey, Jr. (1972)), the Hall effect is a nonlinear problem even at vanishingly small fields. Queisser showed that in this case Hall effect and thermopower can have opposite sign if the electron and hole currents are properly compensated.

Friedman (1971) found a solution to this problem. He calculated the Hall mobility μ_H using the random phase model which served also as the basis for the mobility expression Eq. (5.7). As in the case of hopping conduction of a small polaron (Holstein and Friedman (1968)) Friedman assumed that the applied magnetic field modifies the phase of the transfer integral between sites. A minimum of three sites, which are mutual nearest neighbors and for which there is a finite transfer integral between all pairs of sites, is needed to obtain a Hall effect. When 3-site interactions predominate, the Hall coefficient is negative. The term involving 4 sites yields a positive contribution but it appears to be negligible. Friedman finds for the Hall mobility in the extended states near the mobility edge E_c (or E_v)

$$\mu_H = 4\pi \frac{e\,a^2}{h}\,[a^3 J\,g\,(E_c)]\,(\eta\bar{z}/z) \qquad (5.29)$$

J is the two site transfer integral ($J \sim 1$ eV), and $g(E_c)$ is the density of states at the mobility edge (assumed to be constant over an energy interval kT). The factor η is, in units of a^4, the average square of the projected three site areas on a plane perpendicular to the magnetic field

$$(A_{Kji}^{(z)})^2 = \eta\,a^4. \qquad (5.30)$$

Its value is about $\eta = 1/3$. The coordination number is z and the average number of interacting sites considered in this calculation is $\bar{z} = 3$.

With Eq. (5.7) and (5.29) one obtains for the ratio of Hall to conductivity mobility

$$\mu_H/\mu = \frac{6\,kT}{J}\,\eta\,\frac{\bar{z}}{z^2} \qquad (5.31)$$

With the same values used for estimating the magnitude of C in Eq. (5.5), $a = 3\text{Å}$, $z = 4$, $\bar{z} = 3$, $\eta = 1/3$, $J\,g(E_c) = 8 \times 10^{20}$ cm^{-3} one obtains $\mu_H \sim 10^{-1}$ cm^2/V sec and $\mu_H/\mu \sim 10^{-2}$.

These results agree very well with the experimental data. Figure 5.25 shows the measurements of Male (1967) on $As_2 Se_2 Te$. The Hall mobility is 10^{-1}cm^2/V sec for both the liquid and the solid and independent of temperature between 300–800 K. Roilos (1971), Seager et al. (1972) and Emin et al. (1972) studied several chalcogenide glasses and find thermally

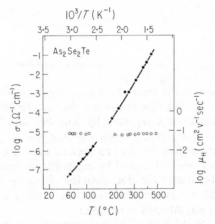

Fig. 5.25. Conductivity and Hall mobility of liquid and vitreous $As_2 Se_2 Te$ as a function of temperature (after Male (1967)).

H. Fritzsche

TABLE 5.2

Material	μ_H (cm^2/V sec)	T (°K)	Source
$Tl_2 Se.As_2(Se, Te)_3$	-0.02		Kolomiets and Nazarova (1960)
$Tl_2 Se-As_2 Te_3$	-0.09	240–340	Ivkin, Kolomiets and Lebedev (1964)
at 37.5% Te	-0.36	300	
As–Te–I	-0.08	300	Peck and Dewald (1964)
	-0.12	365	
As–Te–Br	-0.01	300	
	-0.10	365	
AsTe	-0.065	300	Panus, Ksendov, and Borisova (1968)
$AsGe_{0.1} Te$	-0.053		
$AsGe_{0.2} Te$	-0.050		
$AsGe_{0.3} Te$	-0.049		
$AsGe_{0.4} Te$	-0.043		
$AsGe_{0.3} Te_{0.8}$	-0.045		
$AsGe_{0.3} Te_{1.5}$	-0.056		
$As_2 Se_2 Te$	-0.09	300–800	Male (1967)
$AsSeTe_2$	-0.10		
$As_2 SeTl_2 Se$	-0.10		
$As_2 SeTl_2 Se$	-0.10		
$Tl_2 Te.As_2 Te_3$	$+0.10$	300	Kornfeld and Sochava (1959)
$Tl_2 Te.As_2 Te_3$	$+0.16$	325	Nagals et al. (1970)
	$+0.028$	125	
Se	$+0.37 \pm +0.03$	293	Juska et al. (1969a)
	-7 ± 1.5		Juska et al. (1969b)
Se	-0.32 ± 0.1	295	Dresner 1964
a–Ge	$-(0.02 \pm 0.01)$	300	Clark (1967)

activated negative μ_H with activation energies between 0.03 and 0.05 eV. Their results are shown in Figure 5.26.

Table 5.2 summarizes the values of the μ_H obtained so far except those shown in Figure 5.26. The sign represents that of the Hall coefficient. A positive Hall coefficient was reported by Kornfeld and Sochava (1959). These measurements were recently verified by Nagels et al. (1970) who took special care to ascertain that the Hall coefficient was a genuine property of the amorphous phase and not caused by crystalline inclusions. This exception is remarkable because of the close chemical similarity of all the systems investigated. It will be remembered that this small gap material had a thermopower which was difficult to interpret.

Juska, Matulionis, and Viscakas (1969a) reported a positive Hall mobility $\mu_H = (0.37 \pm 0.03)$ cm^2/V sec at 293 K for Se with evaporated contacts.

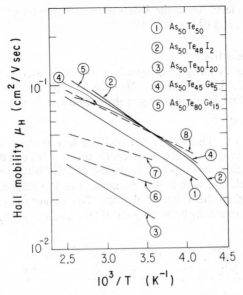

Fig. 5.26. Temperature dependence of Hall mobility of several As-Te-I and As-Te-Ge alloys (after Saeger *et al.* (1972)) and for vitreous $(As_2Se_3)_x$. As_2Te_3 (after Roilos (1971)). Curves 6, 7, 8 correspond to x = 0.5, 1,3, respectively.

They produced a high degree of polymerization in their samples by rapid heating to the melting point and quick cooling to room temperatures. Juska *et al.* (1969b) obtained for the same Se a negative Hall mobility $\mu_H = (-7.0 \pm 1.5)$ cm²/V sec when they fused in Au or In electrodes. Dresner (1964) measured a negative Hall mobility of $\mu_H = -0.32 \pm 0.1$ cm²/V sec on Se at 295 K which was illuminated with a flux of 2×10^{16} photons/cm² sec. These were absorbed in a 0.1 μm thick surface layer of the sample.

The only measurement on a tetrahedral NCS is that on a-Ge by Clark (1967) who also found μ_H to be negative and of order 0.02 cm²/V sec. Since his data were taken at 300 K it is likely that only part of the conduction took place in the extended band states.

The Hall mobilities of all systems studied so far have the same magnitude within a factor of 10. According to Friedman, the origin of this is the same as that of the closely similar values for the pre-exponential factor C of the conductivity (see Figure 5.4). It is the insensitivity of the product a^3 J g(E_c) to structural and compositional variations of the material. The observed temperature dependence of μ_H does not agree with Eq. (5.29). Emin *et al.* (1972) suggests that the mobility activation energy may be associated with polaron effects. If that turns out to be true, then our model for the conduction process must be changed radically, because in contrast to Eq. (5.31), $\mu_H/\mu \geqslant 10$ so that $\mu \sim 10^{-3}$cm²/V sec for small polaron hopping (Friedman and Holstein (1963) Emin (1971a, 1971b, 1973)). With Eq. (5.31) and the measured values of μ_H one obtains on the other hand a conductivity mobility $\mu \sim 1-5$ cm²/V sec. For vitreous $Tl_2SeAs_2Te_3$ a mobility value of $\mu \sim 1$ cm²/

V sec was estimated from the Drude free carrier infrared absorption by
Mitchell, Taylor, and Bishop (1971).

5.3.7 a.c. Conduction

The conductivity of many chalcogenide glasses and oxide films (Argall and
Jonscher (1968) Linsley, Owen, Hayatee (1970) Sayer *et al.* (1971))
exhibits over a large frequency range a dependence on frequency as

$$\sigma_{ac}(\omega) = \text{const } \omega^s \qquad (5.32)$$

with $0.7 < S < 1.0$. This power law and the fact that σ_{ac} increases only
slightly with T is often taken as evidence that σ_{ac} is caused by a kind of
hopping mechanism as proposed by Pollak and Geballe (1961). The loss
mechanism considered here is that component of the polarization which is
lagging the applied field by 90 degrees. The polarization is caused by
phonon assisted tunneling between pairs of localized gap states.

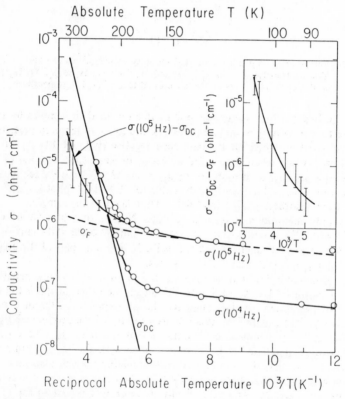

Fig. 5.27 Temperature dependence of the conductivity of evaporated a-As$_2$Te$_3$ for d.c.
and for two frequencies. After Rockstad (1972). The dotted curve marked
σ_F shows the linear T-dependence predicted by Eq. (5.33). Since σ_{DC}
$+ \sigma_F < \sigma_{ac}$, there appears to exist the additional contribution shown in the
insert.

Austin and Mott (1969) adapted the theory of Pollak and Geballe to the case where tunneling conduction takes place near E_F and obtained

$$\sigma_{ac} (\omega) = \frac{\pi}{3} [g(E_F)]^2 \, kT \, e^2 \, \alpha^{-5} \, \omega \, [\ln (\nu_p/\omega)]^4 \qquad (5.33)$$

where ν_p is a phonon frequency ($\sim 10^{13} \text{sec}^{-1}$) and α describes the decay with r of the localized state wavefunction $\exp (-\alpha r)$. The last factor in Eq. (5.33) decreases very slowly with frequency so that this theory yields the frequency dependence required by Eq. (5.32).

The temperature dependence of the total conductivity of $As_2 Te_3$ is shown in Figure 5.27 for two frequencies. These measurements by Rockstad (1972) show that the a.c. conductivity of the small band gap materials may be composed of three contributions. The low temperature part is proportional to T as described by Eq. (5.33). If this portion is extrapolated to higher T and added to the d.c. conductivity, one obtains less than the total a.c. conductivity. The additional component σ_2 is also proportional to ω^s but rises more rapidly with T than Eq. (5.33). Rockstad (1971, 1972) attributes σ_2 to hopping conduction in localized tail states. The insert in Figure 5.27 shows that σ_2 is thermally activated. The slope is always smaller

TABLE 5.3 *Density of Stages $g(E_F)$ from a.c. Conductivity Data and Eq. (37) of Text.*

Material	σ_{DC} (ohm^{-1}cm^{-1}) at 300 K	σ_{AC} (ohm^{-1}cm^{-1}) at 300 K $\omega = 10^6$ Hz	$g(E_F)$ (eV^{-1}cm^{-3}) for $\alpha^{-1} = 8$ Å	Reference
$As_2 Te_3$	5×10^{-4}	2×10^{-6}	5×10^{19}	Rockstad (1971)
$Te_2 AsSi$	5×10^{-5}	2×10^{-6}	5×10^{19}	Rockstad (1971)
$Te_{53} As_{32} Si_{10} Ge_5$	5×10^{-6}	1.6×10^{-6}	4.5×10^{19}	Rockstad (1971)
$As_2 Se_3$	10^{-12}	8×10^{-10}	10^{18}	Owen et al. (1971)
$As_2 Se_3$	6×10^{-13}	8×10^{-11}	3×10^{17}	Crevecoeur et al. (1971)
$As_2 Se_3$	10^{-12}	8×10^{-9}	3×10^{18}	Lakatos et al. (1971)
$As_2 Se_3 + 0.2\%$ Ag	5×10^{-11}	8×10^{-9}	3×10^{18}	Owen et al. (1970)
$As_2 S_3$	3×10^{-16}	2×10^{-10}	5×10^{17}	Owen et al. (1970)
$As_2 S_3$	5×10^{-15}	1.2×10^{-9}	1.2×10^{18}	Lakatos et al. (1971)
$As_2 S_3 + 0.2\%$ Ag	3×10^{-14}	8×10^{-10}	10^{18}	Owen et al. (1970)
Se	2.8×10^{-14}	1.8×10^{-9}	1.5×10^{18}	Lakatos et al. (1971)

than that of the d.c. conductivity curve. An alternative explanation for the strong T dependence σ_2 may be multiple hopping (Pollak (1971)).

The attractiveness of Eq. (5.33) lies in the possibility of determining the density of gap states $g(E_F)$ at E_F. Assuming $\alpha^{-1} = 8$Å Davis and Mott (1970) obtained the quantity $g(E_F)$ from Eq. (5.33) using the values σ_{ac} measured at 1.6×10^5 Hz ($\omega = 10^6 \sec^{-1}$) and 300 K for various materials (Owen (1967) (1970b) Owen and Robertson (1970) Rockstad (1969)). These results are listed in Table 5.3 together with some recent data on Se, As_2Se_3, and As_2S_3 obtained by Lakatos and Abkowitz (1971) and As_2Se_3 obtained by Creveceour and deWit (1971). In the case of small gap materials the $g(E_F)$ values listed in Table 5.3 were obtained by extrapolating the low temperature σ_{ac} to 300 K as illustrated by the dotted curve in Figure 5.27. It should be stressed here that Eq. (5.33) is applicable only when the observed T-dependence of σ_{ac} is linear. A T-independent expression for σ_{av} is obtained by Pollak (1971). His expression yields a considerably smaller $g(E_F)$ for the same conductivity. Several examples listed in Table 5.3 show a smaller than linear T-dependence. Polanco and Roberts (1972) observe a T-independent σ_{ac} in As_2S_3.

The values of $g(E_F)$ obtained from Eq. (5.33) depend strongly on the decay parameter of the localized state wavefunctions. A much larger value than $\alpha^{-1} = 8$Å seems unreasonable for deep gap states; a smaller value would make $g(E_F)$ even larger. It appears therefore that $g(E_F)$ is too large to be accounted for by exponential tails of localized states as sketched in Figure 5.3. Davis and Mott (1970) therefore suggest a peak in $g(E)$ near the gap center as shown in their model Figure 5.4. This interpretation leads, however, to several contradictions which force us to examine other alternatives for explaining the a.c. conductivity. The arguments raised against the general validity of explaining a.c. conduction by the above process are the following.

(i) Such large values of $g(E_F)$ should have been detected by optical absorption but were not.

(ii) These densities would quite effectively pin E_F. However, internal photoemission (Mort and Lakatos (1970)) show that E_F in Se adjusts to the Fermi level of the contacting metal.

(iii) Dangling bonds might give rise to midgap occupied and unoccupied states as shown in Figure 5.4 in some NCS (Mott (1972)). Particularly in chalcogenide glasses this is unlikely, however, because in these lone pair semiconductors (Kastner (1972)) dangling bonds lead to an occupied state in the valence band and an unoccupied acceptor-like state in the gap.

(iv) The order of magnitude, temperature, and frequency dependence of σ_{ac} observed in amorphous semiconductors is very much the same as that of a large number of insulating NCS, e.g. fused quartz, alkali glasses, porcelain, plexiglass, marble, pyres, celluloid, and bakelite (Gevers and DuPre (1946) von Hippel (1954)) and of anodic oxide films (Young (1961)).

As a matter of fact any loss mechanism which gives a wide and flat distribution of relaxation times yields the nearly linear frequency dependence of Eq. (5.32) (Owen (1963) Fritzsche (1971a) Pollak and Pike (1972)). The fact that this behavior is so generally observed in NCS suggests that it may be related to the lack of long-range order rather than to a specific electronic conduction mechanism. Other phenomena which are characteristic of NCS and are observed in a variety of materials are the anomalous thermal conductivity and a specific heat term at low temperatures which is linear in T (Zeller and Pohl (1971)). These anomalous thermal properties are indicative of low energy excitations of constant density. Anderson, Halperin and Varma (1971) associate these with different structural configurations which the solid can attain even at low temperatures by tunneling. If changes in configurations are accompanied by.

Other loss mechanisms must be considered. For examples the background loss in amorphous dielectrics due to anharmonic coupling of the electric field to acoustic phonons may be of importance also in these materials (Amrhein and Mueller (1968) Amrhein (1969) Austin and Garbett (1971)).

The large spread in σ_{ac} values observed in vitreous As_2Se_3, shown in Figure 5.28, may be caused by internal heterogeneities. This material as well as several other wide gap vitreous semiconductors are chemically unstable when illuminated with light beyond the band gap energy (Berkes $et\ al.$ (1971)).

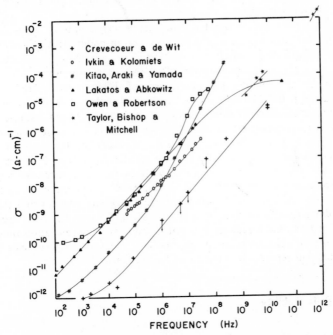

Fig. 5.28. The a.c. conductivity of vitreous As_2Se_3 at 300 K over a wide frequency range. The data are by Lakatos and Abkowitz (1971), Austin and Garbett (1971), Taylor, Bishop, and Mitchell (1970), and by Owen (1970), see Austin and Garbett (1971), after Mitchell and Taylor (1972).

A wide and flat distribution of relaxation times is obtained also when the relaxation times are thermally activated with a statistical distribution of activation energies. Pike (1973) suggests a hopping model for the a.c. conductivity in scandium oxide. The activation energy is that of electrons hopping *over* the barrier between sites associated with oxygen vacancies. He points out that values s $\geqslant$ 1 in Eq. (5.32) are incompatible with Eq. (5.33) but permissible in his theory. Lakatos and Abkovitz (1971) find s > 1 at 77 K in Se.

Other questions which are presently under study are (i) the effect of contacts, (ii) the physical origin of a quadratic ω-dependence of σ, (iii) the lower limit to the distribution of relaxation times, (iv) the low frequency tail of the lattice vibration absorption, and the a.c. loss mechanisms at high fields (Jonscher (1971, 1972)). Some of these questions will be discussed briefly in the following.

Creveceour and deWit (1971) and Lakatos and Abkowitz (1971) made a careful study of the effect of contacts on the loss measurements. Street, Davies, and Yoffe (1971) explained that a failure to consider the relatively small resistance of the evaporated thin film electrodes in analyzing the network equations, yields $\sigma_{ac} \alpha \omega^2$ in a certain frequency interval. Such temperature independent quadratic term was reported for As_2Se_3 (Davis and Shaw (1970) Owen (1967) Kitao *et al.* (1970)). Lakatos and Abkowitz (1971) showed that blocking contacts cause this erroneous ω^2 region to move to lower frequencies.

A temperature independent $\sigma \sim \omega^2$ term, which appears to be characteristic of the materials, is observed, however, in anodized Al_2O_3, evaporated SiO (Argall and Jonscher (1968)), and in polymeric CS_2 (Chan and Jonscher (1969)). In the case of CS_2 measurements were carried out without contacts applied and on samples of different sizes. Mott (1969) explored the possibility that the ω^2 region originates from photon assisted tunneling, that is from direct optical transitions between localized gap states. With a constant $g(E_F)$ over the region of photon energies one obtains $\sigma \alpha \omega^2$. However, the magnitude of this direct absorption term was found by Pollak (1971) to be too small to account for the experimental observations. Although he compared his calculations with the results reported for As_2Se_3, which were found to be in error, his arguments and conclusions nevertheless still hold for Al_2O_3, SiO, and CS_2. Pollak suggested instead that the a.c. conductivity rises faster than linearly (2 > s > 1) because the distribution of relaxation times has a peak near its cutoff value at small relaxation times for which $\omega\tau$<1. He considered correlation effects which increase with decreasing pair separation as the cause for this peak.

5.3.8 Drift Mobility

Method

The long relaxation time $\tau_{rel} = \epsilon_o\kappa/\sigma$ in semi-insulating materials makes it possible to pull a pulse of excess charge through the material of length L and measure the drift velocity

$$V_D = \mu_D F \tag{5.34}$$

by determining the transit time

$$T_r = L/V_D \tag{5.35}$$

before the excess charge becomes neutralized. The excess charge drifting in the applied field F induces a charge at the electrodes. The induced surface charge can be detected either as a change in voltage across the sample or as a conduction current depending on the RC time constant of the detection circuit.

During transit some carriers may get lost into deep-lying traps. This is characterized by a lifetime τ. The number of drifting carriers is then

$$N(t) = N(o) \exp(-t/\tau) \tag{5.36}$$

The redistribution of charge on the electrodes is

$$\frac{dq}{dt} = e N(t) V_D/L \tag{5.37}$$

so that integration yields the induced charge for $0 \leqslant t \leqslant T_r$

$$q(t) = N(o) e V_D \frac{\tau}{L} [1 - \exp(-t/\tau)]. \tag{5.38}$$

This technique was developed by Spear (1957, 1960) who recently reviewed its experimental details (1969). A thin sheet of electron-hole pairs is generated at the sample surface by a pulse of light or a pulsed electron beam. Depending on the polarity of the applied field F a pulse of either electrons or holes can be extracted from the sheet of electron-hole pairs. N(o), the number of carriers extracted from the generation region, may be field and temperature dependent (Tabak and Warter (1968)).

The time-of-flight method described above requires that the injected charge be sufficiently small not to disturb the applied field distribution. Theories including space charge effects for higher injection levels have been developed by Helfrich and Mark (1962) and Many and Rakavy (1962) and recently by Batra *et al.* (1970).

The drift mobility μ_D obtained from Eq. (5.34) is usually not equal to the conductivity mobility μ of Eqs. (5.6) or (5.7) because the drifting carriers may repeatedly be trapped by and thermally released from shallow traps during transit. Shallow traps are those with which the carriers can retain local thermal equilibrium. The demarcation line between deep and shallow traps depends therefore on T. The relative time each carrier spends immobilized in shallow traps τ_t and freely drifting in the band, τ_s, is

$$\frac{\tau_t}{\tau_s} = \frac{N_t}{N_c} \exp(E_c - E_t)/kT \tag{5.39}$$

for a shallow trap density of N_t and trap depth $(E_c - E_t)$. N_c is the effective density of states of the conducting band. The drift mobility is thus related to μ as

$$\mu_D = \mu \frac{\tau_s}{\tau_s + \tau_t} = \mu [1 + \frac{N_t}{N_c} \exp(E_c - E_t)/kT]^{-1} \tag{5.40}$$

which is approximately

$$\mu_D = \mu \frac{N_c}{N_t} \exp\left[-(E_c - E_f)/kT\right] \tag{5.41}$$

The lifetime τ characterizing the loss of carriers to deep traps can be obtained from the slope of $\ln q(t)$ vs. t of Eq. (5.37). In order to observe a break in the q(t) curve at $t = T_r$, the lifetime τ must be of order T_r or longer; otherwise too many carriers get lost. The mean free time τ_s of the carriers with respect to shallow traps, on the other hand, should fulfill either one of the following conditions

$$\tau_s > T_r \text{ (trap-free case)} \tag{5.42}$$

$$\tau_s + \tau_t < T_r \text{ (trap-limited case)} \tag{5.43}$$

when $\tau_s \sim T_r$, one finds a statistical spread of the transit time which makes its determination quite uncertain. Furthermore the delayed release of excess charges from trapping centers near the generation surface may cause the q(t) curve to have a flat top and a long tail extending beyond $t = T_r$ (Silver, Dy, and Huang (1971, 1972)).

Despite these limitations a wealth of useful information has been obtained from the charge drift experiments, particularly for the high resistivity semiconductors Se, As_2Se_3, As_2S_3, S and a-Si obtained by glow discharge decomposition of SiH_4.

Results

(a) Selenium The temperature dependence of μ_D for holes and electrons in Se is shown in Figure 5.29. These results of Dolezalek and Spear (1970), agree very well with those observed earlier by Spear (1957, 1960), Hartke (1962), Grunwald and Blakney (1968), and recently by Juska et al. (1969b) and Schottmiller et al. (1970). This close agreement of different measurements of a transport parameter as sensitive to impurities as the mobility is unparalled even in crystalline semiconductors. Its significance will be discussed below.

The question arises whether these drift mobilities represent microscopic mobilities μ, which are thermally activated, as in the case of polaron conduction and hopping, or whether they are trap limited, in which case the activation energies correspond to effective shallow trap energies. This question has not yet been settled; the pros and cons are the following. The traps are thought to correspond to the localized states between $E_c - E_A$ and between $E_v - E_B$ of Figure 5.4 which originate from the lack of long-range order. The reasons given interpreting the mobility as trap-limited are the following.

(i) Juska et al. (1969a) analyzed the statistical spread of the transit times T_r. When this spread Δt arises from multiple trapping then Δt should be proportional to $F^{-1/2}$ because

$$\frac{\Delta t}{T_r} \sim \left[\frac{\tau_t}{T_r}\right]^{1/2} \tag{5.44}$$

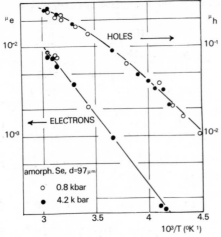

Fig. 5.29. Temperature dependence of the electron and hole drift mobilities μ_D in vitreous Se at two pressures (after Dolezalek and Spear, 1970).

These authors were able to verify this relation for holes at two temperatures.

(ii) The hole drift mobility exhibits a constant activation energy only below 260 K. The whole T-dependence could be well fitted (Grunwald and Blakney (1968) Juska *et al.* (1969a)) with the complete expression Eq. (5.40). Both groups assumed a temperature dependence of the microscopic mobility of the form $\mu \propto T^{-3/2}$ in accordance with lattice scattering in crystals. Although the small magnitude found for $\mu \sim 0.3$ cm^2/V sec at 300 K does not permit such scattering interpretation of the mobility, similar agreement between Eq. (5.40) and μ_D will probably be obtained without making that assumption.

(iii) The drift mobilities are independent of pressure as seen in Figure 5.29. Whether or not this observation is an argument against interpreting μ_D as a microscopic mobility depends on our willingness to accept that a thermally activated μ is necessarily related to thermally activated hopping conduction and secondly that hopping conduction can be identified by its pressure dependence.

(iv) Mort, Schmidlin and Lakatos (1971) find that the number of carriers photoemitted from a metal into amorphous Se is independent of temperature between 220 and 300 K. They show that this is a strong indication for a temperature independent microscopic mobility μ.

What, however, is the origin of the shallow traps which limit the drift mobilities? Their depth and number depend somewhat on the degree of polymerization (Grunwald and Blakney (1968) Juska *et al.* (1969a)). If they indeed originate from the lack of long-range order and constitute the states in the range between $E_c - E_A$ and $E_B - E_v$ in the Davis-Mott model of Figure 5.4, then it is difficult to understand why the drift mobilities in monoclinic Se (Spear (1961)) have about the same activation energies, (0.25

TABLE 5.4 *Drift Mobility and Life Time of Carriers in Se.*

Material	Holes			Electrons		
	μ_D	ΔE	τ	μ_D	ΔE	τ
	$cm^2/V\,s$	eV	μ sec	$cm^2/V\,s$	eV	μ sec
	293 K			293 K		
pure Se						
(Spear (1960))	0.135	0.14		6×10^{-3}	0.25	
(Hartke (1962))	0.13	0.14		7×10^{-3}	0.285	
(Grunwald and Blakney (1968))	0.12	0.14		7×10^{-3}	0.29–0.33	
(Juska et al. (1969))	0.19	0.3				
(Rossiter and Warfield (1971))	0.14	0.095		6×10^{-3}	0.27	16
(Schottmiller et al. (1970))	0.14	0.16	10–50	6×10^{-3}	0.33	50
Se + 2.1% S	0.12		0.5	6×10^{-3}	0.31	60
+ 2.4% S	0.15		0.45	5.7×10^{-3}		80
+ 5% S	0.082		0.8	7×10^{-3}		100
+ 12% S	0.042		0.42	1.8×10^{-3}		150
Se + 1% Te	9×10^{-3}	0.26	20	1.2×10^{-3}	0.33	30
+ 3.2% Te	7×10^{-3}		30	—	—	small
+ 5.4% Te	6×10^{-3}		25	—	—	small
Se + 0.5% As	0.14		0.3	2.6×10^{-3}	0.33	40
+ 2% As			0.3	7.8×10^{-4}	0.33	200
+ 3% As				2×10^{-4}		750
+ 6.6% As				5.4×10^{-5}		2000
+ 9% As				1×10^{-5}		3000
Se + 1% Ge			0.3	3×10^{-3}		
Se + 0.01% Cl	0.14	0.16	10–50	no transit pulse		

eV for electrons and 0.3 eV for holes) as in amorphous Se. Several observations made by Plessner (1951) and Henkels and Maczuk (1953) on crystalline Se are also similar to those of amorphous Se.

The transit time measurements have been used very successfully by Schottmiller et al. (1970) for studying the effect of alloying on the drift mobilities and carrier lifetimes in Se. Table 5.4 lists for electrons and holes in Se the drift mobilities at room temperature, their activation energies, and the carrier lifetime with respect to deep traps. The upper part of the table lists the values obtained in different laboratories on pure Se. The lower part lists the results of Schottmiller et al. (1970) on alloying Se. The same authors furthermore used the intensity of certain Raman scattering peaks to estimate the relative concentration of Se_8 monomer rings and Se_n polymeric chains in vitreous Se. Pure amorphous Se at room temperature contains between 30 and 50 percent of its atoms in ring configurations. Different additives are found to affect rings and chains in a different manner.

(i) Univalent additives like Cl, Br, Tl, and also Na and K are found to reduce the viscosity of liquid Se, probably because these elements act as chain stoppers (Keezer and Bailey (1967)). As a consequence the number of chains is increased without altering the number of rings. In Table 5.4 one notices that Cl has no effect on the hole motility, but completely kills the electron mobility.

(ii) Isoelectronic additives like S and Te are believed to be incorporated in the rings, forming mixed ring species like $Se_{8-x}S_x$. There is evidence that Te reduces the Se_8 ring population to half its value at 12 percent Te concentration. Calculations indicate (Ward and Myers (1969)) that Te increases the chain fraction and S decreases it slightly. The alloying effects on the mobility are not quite as pronounced with S and Te as with univalent additives perhaps because the rings and chains are not too greatly altered.

(iii) Branching additives like As reduce the number of rings. One finds an almost linear reduction to zero at about 20 percent As. At the same time cross links are formed by molecular species of the form $As_1Se_{3/2}$. One observes As to decrease μ_D for electrons. The lifetime for holes is rapidly shortened whereas that for electrons is increased enormously. The possibility of such long lifetimes is remarkable for amorphous solids. It suggests that the gap of a-Se must be rather free of states.

Schottmiller *et al.* conclude that electron transport is associated with the Se_8 rings. A correlation of hole transport with the structure is not yet possible. The crosslinking As appears to act as an effective trapping center.

By using higher injection levels and the theories (Many and Rakavy (1962) Helfrich and Mark (1962)) which take account of the resulting space charges, Rossiter and Warfield (1971) were able to extend the drift velocity measurements to 78 K. As shown in Table 5.4 the observed activation energy for holes is 0.095 eV, considerably smaller than those of other workers. They find, moreover, a break in the T-dependence of μ_D for holes at 150 K. Below this temperature the activation energy is only 0.0093 eV. In addition, they report a small field dependence of the hole drift mobility for fields in excess of 10^4 V/cm. This had not been seen by the earlier groups.

(b) As_2Se_3 and As_2S_3 A well defined transit time, Eq. (5.35), determines the end point of the transit pulse only when the injected carriers travel together with about the same drift mobility. For this to occur it is necessary that the injected carriers (i) start their travel at the same time and (ii) do not get dispersed by experiencing a wide distribution of trapping and release times.

In contrast to the case of Se, the hole transit pulse in As_2Se_3 shows a long tail and there is often no discontinuity in the pulse discernable. Scharfe (1970) observed a break in the transient pulse only when he used Au contacts with evaporated As_2Se_3. Al contacts in contrast yielded a transit pulse decaying without structure. Using Au contacts and taking the breakpoint as a measure of the transit time T_r Scharfe obtained a drift mobility for holes which increases linearly with field. A hole drift mobility of 4 ×

10^{-5} cm^2/V sec independent of sample thickness was found for F = 10^5 V/cm. The drift mobility of electrons was too small to be detected.

Marshall and Owen (1971) studied bulk films. They found rounded and structureless transit pulse shapes for both Al and Au contacts. They concluded that the rounded pulse shape is caused by a large statistical variation in T_r of the individual carriers as a result of rather deep traps which trap the carriers only a few times during transit. The transit pulse shape allows one to define a response time t_o from which Marshall and Owen obtained a nominal drift mobility by setting $t_o = T_r$. These values were found to be independent of film thickness and proportional to exp $(\beta F^{1/2} /kT)$ in contrast to $\mu_D(F) \alpha F$ observed by Scharfe. The zero field extrapolations of μ_D exhibit on activation energy of 0.43 eV. Because the observed field dependence is characteristic of the Poole-Frenkel field-induced lowering of Coulomb potentials (see later section) Marshall and Owen conclude that the hole traps are acceptors, lying about 0.43 eV above the valence band as sketched in Figure 5.5. In order to explain the conductivity activation energy of 0.95 eV they propose that the acceptors are compensated by nearly the same number of donors in the upper half of the gap. Knowing the temperature dependence of μ_D and assuming a conductivity mobility $\mu \sim 10$ cm^2/V sec, Marshall and Owen estimate a donor density of about 10^{19} cm^{-3} from Eq. (5.4). However, the Poole-Frenkel exponent of the form $\beta F^{1/2}$ /kT should be observed at fields which are sufficiently high to achieve field lowering of the Coulomb wells over a distance small compared to the Coulomb center separation. This condition sets an upper bound of 10^{18} cm^{-3} to the donor density in the experiment of Marshall and Owen. Furthermore, one might ask whether thermal equilibrium with such deep traps (0.43 eV) can be maintained by the carriers during transit to justify the use of Eq. (5.41).

By comparing the field dependence of μ_D with that of σ, one can distinguish the Poole-Frenkel effect from the mechanisms which affect only μ_D. Marshall and Owen find σ (F) to rise somewhat less rapidly with field than μ_D. DeWit and Crevecoeur (1971), on the other hand, find $\sigma \alpha$ exp (F/F_o) and a considerably smaller field dependence than that observed by Marshall and Owen.

The problem of extracting the field dependence from drift mobility experiments is far from straightforward. The field changes the demarcation line between shallow traps $(\tau_t < T_r)$ and deep traps $(\tau_t > T_r)$ by shortening T_r. As a consequence there are fewer shallow traps effective at high field and, as Fox and Locklar (1972) point out, this increases the drift mobility. Very instructive are computer calculations which generate pulse shapes for a simple trapping model and a given μ_D. An appropriate break in the calculated curve is then interpreted as signaling T_r in the accustomed fashion. The μ_D values thus obtained can then be compared with those used in the calculations. Silver, Dy, and Huang (1971, 1972) find that even simple models often yield 'observed' μ_D values which have wrong magnitudes and field dependencies. Gill and Kanazawa (1972) analyzed the pulse shape resulting from a power law $\mu_D \alpha F^n$ and an exponential relation $\mu_D \alpha$ exp $(CF^{1/2})$. They find that they could not obtain the functional form of the field dependence of μ_D from the calculated pulses.

As$_2$S$_3$ was studied by Ing, Neyhart, and Schmidlin (1971). Also in this material, holes are more mobile than electrons. However, since the majority of photogenerated holes do not reach the negative electrode within the measurement time even in the thinnest samples, no value for μ_D could be obtained from these measurements. This material as well as the multi-component covalent alloys contain a distribution of traps which make it impossible to analyze the transient response in a straightforward manner.

(c) Amorphous Silicon Of the tetrahedral NCS, drift mobility measurements were so far successful only in a-Si prepared by decomposition of silane in a radio frequency glow discharge, a method developed by Chittick, Alexander, and Sterling (1969). Figure 5.30 shows the electron drift mobility as a function of 1/T as measured by LeComber and Spear (1970). They were unable to detect the transit pulse of holes. No field dependence of μ_D was observed. Spear and LeComber (1972), LeComber, Madan and Spear (1972) attribute region (1) above 250 K to trap limited drift of electrons in extended states above E_c. Although all tail states distributed over a range of about 0.2 eV are believed to act as trapping centers, see insert Figure 5.31 (b), the electron transit pulse showed a rather distinct discontinuity from which T_r could be determined. Spear and LeComber (1972) attribute the good coherence of the electron pulse despite the distribution of trapped electrons to the fact that this distribution

$$N_t(E) = g(E)\, f(E) \tag{5.45}$$

is peaked at the energy E_A in Figure 5.4., at which g(E) drops suddenly by more than an order of ten, in this case at 0.19 eV below E_c.

Figure 5.31 summarizes the results of LeComber, Madan and Spear (1972) on silane decomposed a-Si. The magnitude of C in Eq. (5.4) distinguishes band conduction $(1 < C < 10^4 \text{ ohm}^{-1}\text{cm}^{-1})$ from hopping conduction $(C \sim 10^{-4} \text{ohm}^{-1}\text{cm}^{-1})$. Starting at high substrate temperatures T_s, the Fermi level E_F is above the gap center and conduction is at room temperature by electrons near E_c with activation energy ϵ_1 and at lower T by hopping near

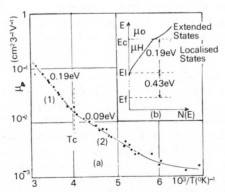

Fig. 5.30. (a) T-dependence of the electron drift mobility μ_D of an a-Si film, 1.3 μm thick, produced by glow discharge decomposition of silane. (b) Model of the electronic state distribution (after Le Comber and Spear (1970)).

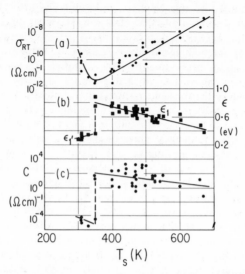

Fig. 5.31. Conductivity parameters of a-Si films deposited by r.f. decomposition of SiH$_4$ as a function of substrate temperature T$_s$. (a) conductivity at 300 K, (b) activation energy ϵ in high temperature region 1, and (c) pre-exponential factor C of Eq. (5.4) of text. After Le Comber, Madan, and Spear (1972).

E$_A$. As T$_s$ is lowered E$_F$ is pulled toward the valence band by an increased density of states in the lower half of the gap. This causes ϵ_1 to increase. Below T$_s$ = 350 K, E$_F$ has moved down so far that conduction shifts abruptly to hopping by holes, presumably near E$_B$ of Figure 5.4, with a hopping energy $\Delta W = \epsilon_1'$.

LeComber *et al.* (1972) suggest that sputtered and evaporated a-Si and a-Ge are essentially similar to low T$_s$ films (200 < T$_s$ < 300 K) of Figure 5.31. These decrease their conductances with annealing, start out as hole conductors and switch to electron conductors only at higher T.

The advantage of drift mobility measurements in conjunction with conductivity is that one can distinguish changes in the equilibrium Fermi level (which affects σ) from changes in g(E) and μ(E) of band and trapping centers which affect μ_D.

5.3.9   Contacts and Surfaces

The early research in the field of crystalline semiconductors was to a large extend inspired by the useful rectifying and injecting properties of contacts. These properties not only led to a variety of semiconducting devices but also yielded information about the electronic states near and at the interface and about the dynamics of charge carriers.

We know close to nothing about the properties of contacts to NCS. It is essential to know these properties in order to properly interpret the increased conduction at high fields and the mechanisms governing electronic switching which is discussed in Chapter 6. Furthermore, Van Roosbroeck and Casey (1972) explained recently that, under certain circumstances, even ohmic conduction at low fields can be dominated entirely by space charge and contact effects. This happens in the so-called relaxation case, when a semiconductor, whose diffusion length lifetime is short compared to the dielectric relaxation time, is contacted by a barrier space charge contact.

Phenomenologically we can distinguish three quite different sets of contact phenomena. (i) Medium and small band gap NCS ($\Delta E < 0.8$ eV) contacted by metals. These show hardly any rectification or polarity asymmetry. The contact resistance is usually a small fraction of the total resistance. (ii) Metal contacts to wide gap and semi-insulating NCS such as Se, As_2Se_3, and As_2S_3. Very high contact resistances and rectification ratios are usually encountered but can be avoided by special material selection. (iii) Heterojunctions between crystalline and amorphous semiconductors. Only heterojunctions to medium band gap NCS have been studied (Grigorovici *et al.* (1964, 1968) Henisch *et al.* (1972)). They show some rectification.

Contacts

One expects that contacts to NCS are accompanied by space charge regions just as in crystalline materials. We first discuss the form of the screening potential predicted by the band models of Figures 5.3 and 5.4. We then try to answer the question why contacts to medium band gap amorphous semiconductors often have no or a very small rectification ratio as well as a small resistance relative to that of the bulk. Finally some remarks will follow about special difficulties associated with real contacts and the study of their properties.

A space charge region arises from the equilibrium condition that the Fermi level be the same in the metal, the interface, and the NCS. In the absence of a significant number of interface states the relative electron affinities of the metal and NCS in contact determine the polarity of the contact dipole layer. When an appreciable density of interface states ($\sim 10^{14} cm^{-2} eV^{-1}$) is present, the NCS on the one hand, and the metal electrode on the other, come into equilibrium with the interface layer essentially independently. The work function of the metal has then no effect on the sign of the space charge layer in the NCS. The properties of the interface layer are governed by the diffusion and alloying which has taken place while forming the contact.

Figure 5.32 compares the screening potential obtained from a band model such as shown in Figure 5.3 with that predicted by the Davis-Mott model shown in Figure 5.4. In both cases an interface potential $U(0) = 1$ eV was assumed and a dielectric constant $\kappa = 16$. Letting tails of localized states in the first model drop exponentially toward E_F as

$$g(E_v)\exp\left[-(E - E_v)/\epsilon\right] \text{ and } g(E_c)\exp\left[-(E_c - E)/\epsilon\right] \qquad (5.46)$$

respectively, one obtains with $\epsilon = 0.14$ eV and $E_c - E_v = 1$ eV a density of states $g(E_F) = 10^{18} eV^{-1}cm^{-3}$ at the Fermi level. The screening length

$$\lambda = (\kappa/4\pi e^2 g(E_F))^{1/2} \qquad (5.47)$$

is then $\lambda = 300$ Å. The calculation of the barrier potential proceeds in a

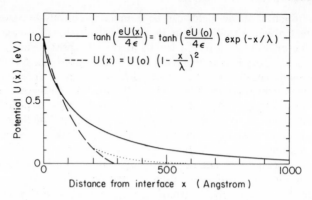

Fig. 5.32. Potential distribution in a semiconductor near a contact according to Eqs. (5.49) and (5.50) of text based on Figures 5.3 and 5.4, respectively. $\lambda = 300$ Å in both cases. The dotted line represents an exponential tail not described by Eq. (5.50). It is found for large X where E_F is within the band of midgap states.

manner similar to that for an intrinsic semiconductor (Sze (1969)), so that one obtains U(x) from

$$\tanh (U(x)/4\epsilon) = \tanh (U(0)/4\epsilon) \exp (-x/\lambda). \qquad (5.48)$$

The model of Davis and Mott yields near the interface a constant space charge and hence a parabolic Schottky barrier

$$U(x) = U(0) (1 - x/\lambda)^2 \qquad (5.49)$$

as long as E_F at x is below (or above) the hump of midgap states. Further away from the interface U(x) decreases more slowly as shown by the dotted line in Figure 5.32 because at large x the Fermi level lies in the hump of midgap states. As a numerical example we chose a hump of height $g(E_F) = 2 \times 10^{17} eV^{-1} cm^{-3}$ and half width 0.05 eV, so that again $\lambda = 300$ Å is obtained. As expected the screening is better the closer the localized states are concentrated near E_F. In contrast to the above examples, it is interesting to note that if the gap (~ 1 eV) were free of states, the only charges available for screening would have been the 5×10^{10} free carriers (at 300K) yielding a screening length of about 15 μm. Such a large screening length would strongly influence the properties of thin film samples when the thickness is of this magnitude or less.

We estimate the magnitude of λ to lie between 300 Å and 600 Å for the tetrahedral NCS and the chalcogenide alloys having band gap 2 $\Delta E < 1.5$ eV. The fact that the space charge region within λ from the surface or contact is either a depletion or an accumulation region for the dominant charge carriers can be used to establish an upper limit to λ by measuring the conductivity as the film thickness is decreased. Furthermore, the varying absorption depth of light of different wavelength enables one to probe the width of the barrier layer behind a semitransparent contact. Wey and Fritzsche (1972a, 1972b) studied the barrier photovoltage, the light

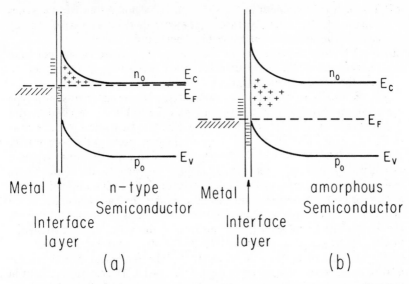

Fig. 5.33. Potential distribution near metal-semiconductor contacts. The sign of the charge in the interface layer is chosen arbitrarily. (a) n-type crystalline semiconductor. The resistance in the bulk is low on account of the large value of n_0. In comparison, the Schottky barrier presents a large contact resistance. (b) amorphous semiconductor with E_F near gap center. Ambipolar current flow allows the barrier region to have a negligible resistance compared to that of bulk.

induced change of the resistance as a function of wavelength, and the barrier capacitance of a chalcogenide alloy glass of composition $Ge_{16}As_{35}Te_{28}S_{21}$ sandwiched between semitransparent electrodes. They found that Au, Al, and nichrome contacts produce a négative space charge and that Sb and Te produce a nearly neutral contact.

The space charge regions in NCS are only a few hundred Angstrom units wide when they contain a large density of gap states near the Fermi level. This width, however, seems still too large to explain the frequent occurrence of low contact resistance and small rectification ratios in medium band gap amorphous semiconductors.

The contact resistance is of course low when the contact potential is zero or when the space charge has the same sign as that of the dominant current carriers (ohmic or injecting contact). This, however, should only happen in special cases. In the following discussion we assume electrons to be the dominant current carriers and the contact to have a positive space charge region as sketched in Figure 5.33a. This is a typical rectifying contact for an n-type crystalline semiconductor. In amorphous materials, a low contact resistance may still result on account of one of the following factors:

(i) The reverse bias saturation current is very much larger and the bulk conductivity much smaller when E_F is close to the gap center than when it is close to one of the band or mobility edges.

(ii) The high field strength in the space charge region delocalizes the states adjacent to the mobility edges and moves these edges toward the gap center. This lowers the effective blocking barrier.

(iii) Discrete localized charges in the barrier may cause strong potential fluctuations in the plane of the contact permitting charge carriers to pass the barrier at certain spots.

Let us discuss these points in more detail. If the material is an n-type *cystalline* semiconductor, as shown in Figure 5.33(a), the bulk conductivity is determined by the electron density n_o which is large because E_F is close to E_c. The contact resistance in that case is high because (a) the hole density p_o is exceedingly small and (b) the electron current in the barrier region is exponentially limited by the barrier height. Consequently, the contact resistance is large compared to the bulk resistance and a high rectification ratio is observed. On the other hand, let us consider the case when the above material is an *amorphous* semiconductor with the same barrier height and gap width. The Fermi level E_F is then close to the gap center as sketched in Figure 5.33(b). As a consequence, n_o is decreased by a factor exp $[-(E_c - E_v)/2 \, kT]$ and p_o is increased by the reciprocal of this factor relative to the respective values in the n-type crystal. The electron current in the bulk has therefore no greater conductance than the hole current in the space charge region except for the difference in electron and hole mobilities. With a barrier thickness small compared to the electrode separation, even an appreciable difference in mobilities can exist before a barrier resistance becomes noticeable. In this case the generation and recombination rates have to be sufficiently large near the onset of the barrier region so that the conduction takes place in one band in the barrier region and in the other band in the bulk. These are the same processes which cause an intrinsic crystalline semiconductor to have a small relative contact resistance and a small rectification ratio.

Although this first point (i) is likely to be the most effective in reducing the contact resistance and the rectification ratio of an amorphous semiconductor contact, it should be pointed out that the mobility edges are not shifted by an electrostatic potential in the same way as band edges are. The high fields associated with the space charge region near a contact delocalizes shallow states and effectively moves the mobility edges closer toward each other.

Another factor which makes a barrier more transparent to carriers of both signs is point (iii) mentioned above. The space charge regions in some of the amorphous semiconductors may be too narrow for the conventional description of contacts to be valid which assumes uniform charge distribution and smooth potentials having equipotential surfaces parallel to the contact interface. Discrete localized charges in the barrier may cause strong potential fluctuations. The average separation of $10^{18} \, cm^{-3}$ centers is approximately 100 Å which is not too different from λ. On account of statistical fluctuations in the density of charged centers and in the properties of the interface layer the barrier may be heterogeneous and easily penetrable for charge carriers at certain spots which have a reduced barrier potential or one of opposite sign. It is then possible that one and the same contact is capable of injecting charge carriers of either sign. In this case the capacitive and conductive barrier properties will be governed by entirely different circumstances. This picture will be explained further in the section on heterogeneity.

In a relaxation-case semiconductor a linear current-voltage characteristic can occur even though most of the applied voltage drops at the contact (van Roosbroeck and Casey (1972)). This situation can be detected by testing the scaling of resistance with electrode separation.

It is not easy to ascertain the presence and the sign of the contact space charge in NCS. The space charge capacitance is difficult to measure when the contact resistance is low. One of the most direct, although still qualitative methods is the measurement of the barrier photovoltage (Wey and Fritzsche (1972a, b)) and of the thermostimulated depolarization. High resistance contacts, which show no polarity effects can often be traced to the presence of a thin insulating oxide layer. Even though such an oxide layer would hardly offer a resistance to tunneling carriers between two mental electrodes, it produces a tunneling resistance of many Megohms if one side is a NCS semiconductor (Osmun and Fritzsche (1970)).

The properties of contacts to dielectric films, particularly of thermally grown or anodized oxide films are hard to analyze because of a gradual transition from the metal to a fully oxidized film. Driven by internal fields some impurities drift preferentially to the interface or surface giving rise to parallel opposing layers of charged defects or ions within the NCS (Kuper (1969) Fawkes and Burgess (1969)). As a consequence it is usually difficult to vary the thickness of a thin film specimen without changing the composition or the structure of the film. A further complication is introduced by the little understood process of 'forming' the contact before taking measurements. This often consists of applying a current pulse which briefly produces a hot spot and probably most likely alloying. The formation of a selenide layer at the interface between Cu, Ag, and Al electrodes and amorphous selenium was studied by Nielsen (1973). He interrupted the Se deposition after every few angstrom and measured the energy distribution of photoemitted electrons and the photocurrent. Whereas the reaction with the Al electrode is self-limiting (about 40 Å thick) similar to that of oxidation, rather thick selenide layers result with Cu and Ag.

The high solubility of many metals in NCS appears to enhance the alloying tendency of contact materials which form low temperature eutectics with one of the major constituents in the NCS. Oki et al. (1969) and Bosnell and Voisey (1970) found a strong effect of the contact materials on the crystallization temperature and the electrical properties of amorphous Ge, Si, and B films. Carbon or graphite contacts are usually found to be satisfactory after the trapped gases are removed.

Problems with contacts occur most often with wide gap insulators and semi-insulating NCS such as Se, As_2Se_3, and As_2S_3. In these high resistive materials it becomes very difficult to make an electrical contact which will not perturb the small carrier concentration initially present so that the space charge region is expected to become very wide.

Figure 5.34 illustrates a few points which are quite commonly observed in wide band gap material (Carruthers (1973), Polanco, Roberts and Myers (1972)). Shown are I–V characteristics plotted logarithmically of three different contact combinations to a 2 μm thick evaporated As_2Se_3 film. The first metal was evaporated first, then As_2Se_3 with the second metal electrode on top. The voltage is that of the top electrode with respect to the bottom electrode. The results are the following:

(i) when the bottom electrode is a nonoxidizing metal, the low voltage ohmic region corresponds to the bulk resistance of the NCS (diodes I and II).

(ii) when the bottom electrode is allowed to oxidize before it is covered with NCS, it gives rise to a large contact resistance (diode III).

(iii) At larger biases $V > 0.5$ Volt this oxidized contact passes electrons from the metal into the NCS but not in reverse. This is often called a hole blocking contact.

(iv) The choice of metal for the top electrode is not so critical because oxidation is less likely but the top electrode will differ in its behavior from the bottom electrode even when the same metal is used (see diode 1). After all, the contacts are prepared quite differently.

(v) When there is no oxide layer (diodes I and II) the $I - V$ is nearly linear up to 10^4 V/cm in the top positive polarity and sublinear in the reverse direction. Such a sublinear regime is expected in relaxation-case semiconductors (Queisser, Casey Jr., van Roosbroeck (1971), Döhler and Heyszenau (1973)).

A powerful method for studying the space charge regions and trap distributions at the interfaces and in the bulk of semi-insulating materials was recently developed by Simmons and Taylor (1972a, 1972b, 1973), Simmons and Nadkarni (1972), Nadkarni and Simmons (1972, 1973).

Surfaces and Field Effect

Although amorphous semiconductors can readily be deposited as thin films, only a few accounts have been published on the effect of adsorbed gases and of surface charges on the thin film conductance. In both cases the surface potential and hence the space charge region near the surface is changed. This in turn produces a change in conductance from which information about the density of state distribution of surface states and localized gap states may be obtained (Kastner and Fritzsche (1970) Barbe (1971)).

In the conventional insulated gate field effect geometry, the amorphous semiconductor film forms one plate of a parallel plate capacitor in which mylar, mica, Al_2O_3, or SiO_2 is used as an insulator. The dielectric strength of a typical insulator permits about $\Delta N_t = 2 \times 10^{12}$ cm^{-2} carriers to be deposited on the surface. When $g(E_F)$, the density of states near the Fermi level is large, most of the excess charge introduced by the applied field is trapped near E_F and $g(E_F)$ can then be assumed constant in calculating the conductance change. Let ΔN_t be the total number of electrons transferred per unit area to the semiconductor of thickness d, and $\Delta N_t - \Delta N$ the number of electrons trapped in surface states. Then

$$e\, \Delta N = \int_o^d q(x)\, dx \qquad (5.50)$$

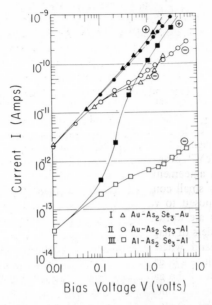

Fig. 5.34. Current voltage characteristics at 300 K of a 2 μm thick evaporated As2Se3 film sandwiched between different evaporated metal electrodes. The code name of the diodes names the materials in sequence of deposition. The polarity is that of the top with respect to the bottom electrode. After Carruthers (1973).

Where the charge density $q(x)$ is related to $\Delta E_F(x)$ the change of energy between E_F and E_v by

$$q(x) = - e\, g(E_F)\, \Delta E_F(x) \qquad (5.51)$$

if $g(E_F)$ is constant within ΔE_F. Using the Poisson equation one obtains

$$\Delta E_F(x) = \frac{\Delta N}{g(E_F)\,\lambda}\, \exp(-x/\lambda) \qquad (5.52)$$

when the screening length λ is given by Eq. (5.47). With $\Delta E_F(x)$ governing the changes in concentrations of mobile carriers one obtains for the relative conductivity changes

$$\frac{\Delta\sigma(x)}{\sigma_o} = \frac{b}{1+b}\,[\exp(\Delta E_F/kT) - 1] + \frac{1}{1+b}\,[\exp(-\Delta E_F/kT) - 1] \qquad (5.53)$$

The parameter b is the ratio of electron to hole current when $\Delta E_F = 0$. The relative conductance change is obtained by integrating Eq. (5.52) over the film thickness d

$$\frac{\Delta G}{G_o} = \frac{\lambda}{d}\left[\frac{b}{1+b}\,F(-v_s) + \frac{1}{1+b}\,F(v_s)\right] \qquad (5.54)$$

where

$$F(v_s) = \int_0^{v_s} (e^{-v}-1) \, dv/v = \sum_{m=1}^{\infty} \frac{(-v_s)^m}{m \cdot m!} \qquad (5.55)$$

and

$$v_s = \frac{\Delta E_F(o)}{kT} \qquad (5.56)$$

is the reduced surface potential. The conductance is shown in Figure 5.35 for various current ratios b. The same figure shows the results of measurements (50 Hz modulation) by Egerton (1971) on a chalcogenide alloy film. The small magnitude of the conductance change yields $g(E_F) \sim 10^{19} \, \mathrm{cm}^{-3} \mathrm{eV}^{-1}$ even if one assumes a surface state density of $2 \times 10^{13} \, \mathrm{cm}^{-2} \mathrm{eV}^{-1}$. In agreement with this explanation Kastner and Fritzsche (1970) found very small conductance changes when evaporated amorphous Ge films were exposed to various adsorbing gases. In crystalline Ge these same experiments are known to produce conductance changes which are larger by many orders of magnitude.

On the other hand, when the density of surface and gap states is small it is possible to produce a large change in surface potential and to measure the energy distribution of the gap states. Spear and LeComber (1972) reported such measurements for amorphous Si prepared in a radio-frequency glow discharge from SiH_4. They found that in this material the conductance of the film could be changed by several orders of magnitude with the field effect. As shown in Figure 5.36 these authors find a nonmonotonic distribution of gap states whose density decreases with increasing substrate temperature during deposition. The density varies between 10^{16} and $5 \times 10^{17} \, \mathrm{cm}^{-3} \mathrm{eV}^{-1}$. The Fermi level lies 0.55 eV below the conduction band near one of the density of states maxima. The large magnitude of the field effect as well as the fact that LeComber and Spear (1970) and LeComber, Madan and Spear (1972) were able to measure the drift mobility in this material shows that the glow discharge of SiH_4 produces an amorphous Si film which has a particularly small density of gap states. These a-Si films are in many respects different from those prepared conventionally by evaporation or cathode sputtering. They do not contain the high density of free spins and have a resistivity which is many orders of magnitude higher than that of conventional a-Si films (see Figure 5.15). They might possibly contain hydrogen which saturates the dangling bonds.

The few surface studies reported suggest the presence of a considerable density of surface and gap states in the amorphous semiconductors studied. Kastner and Fritzsche (1970) found that one monolayer of H_2O adsorbed on a 1000 Å thick film of chalcogenide alloy increases its conductance by less than one percent. Amorphous Ge evaporated at room temperature is porous to H_2O so that a large area of internal surfaces can be covered with water. A density of $5 \times 10^{20} \, H_2O/cm^3$ absorbed in a 1000 Å thick Ge film produced an increase in conductance by only 10 percent. These small changes contrast strongly with the behavior of crystalline semiconductors. They suggest a large density of surface and gap states.

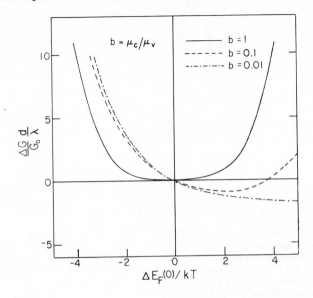

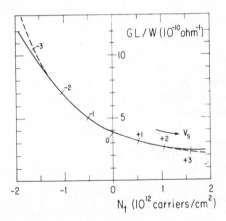

Fig. 5.35. (a) The product of the relative change of conductance and the ratio of sample thickness d to screening length λ as a function of surface potential $\Delta E_F(o)/kT$ for various ratios b of electron to hole mobilities, after Kastner and Fritzsche (1970).

(b) Field induced change of conductance G of a 450 Å thick film of chalcogenide alloy glass measured with 50 Hz modulation. W = film width. GL/W is plotted against ΔN_t, the surface density of electrons deposited. Dotted line shows fit of Eq. (5.54) to experimental curve assuming no surface states. After Egerton (1971).

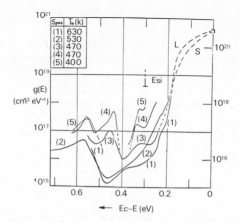

Fig. 5.36. The density of localized states g(E) plotted against $E_c - E$ for a number of a-Si specimens produced by glow discharge decomposition of silane. The samples were prepared on substrates held at different temperatures as noted in the figure. The point E Si indicates the lower limit for g(E) set by these experiments for an evaporated Si film. L and S denote linear and square low interpolations of g(E) to a value at E_c estimated from drift mobility (after Spear and LeComber (1972)).

5.3.10 Conduction at High Fields

The low conductivity of many NCS makes it possible to apply large electric fields and to explore the conduction properties beyond the ohmic regime without the danger of excessive heating. High field effects can be classified into those which produce or result from a nonuniform field distribution and others which can occur in a uniform field. The two cases can be distinguished in principle by testing the dependence of the current on the electrode separation. In practice this test is hindered by the difficulty mentioned above, of changing the specimen thickness without affecting the composition or structure.

Nonuniform Fields

Contacts must supply a reservoir of carriers ready to enter the semiconductor at the rate required by the ohmic currents. At low fields, the balance is maintained by dielectric relaxation which quickly compensates any deviation from equilibrium by a charge supply from the interior and from the other contact. At high fields departures from this balance occur. When the contact region has an abundance of carriers the ohmic extrapolation is exceeded by space charge supplied currents. These are called by the misnomer 'space charge limited currents'. When, on the other hand, the demand for carriers exceeds the supply, the current becomes emission limited. A very comprehensive account of injection in insulators has been given recently by Lampert and Mark (1970), Lampert and Schilling (1970), and by Baron and Meyer (1970). The injection processes in insulators and amorphous semiconductors differ markedly from minority carrier injection in crystalline semiconductors where the majority carrier density adjusts quickly, on account of the short dielectric relaxation time, to assure local charge neutrality.

Let us consider what happens when an excess minority carrier charge is injected into a volume element. Relaxation and recombination processes will proceed to restore neutrality and equilibrium, respectively. The first process, which involves charge flow between the volume element and its environment, proceeds within the dielectric relaxation time $\tau_{rel} = \epsilon_0 \kappa/\sigma$. It renders the volume element neutral but in general with carrier concentrations $n \neq n_0$ and $p \neq p_0$. These can be described with two quasi-Fermi levels E_F^n and E_F^p, respectively. Recombination, on the other hand, causes the quasi-Fermi levels to coalesce to a single Fermi level such that

$$np = n_0 p_0 \qquad (5.57)$$

within an average lifetime τ_0, termed diffusion length lifetime (Zitter, (1958)). Neutrality, however, is not obtained by this. Both processes are therefore required for reestablishing equilibrium. Van Roesbroeck (1960, 1961) and van Roosbroeck and Casey (1970, 1972) pointed out that the kinetics of injection as well as the resulting $I - V$ characteristic are entirely different in the following two limiting cases, which are named after the slower process. In the lifetime case ($\tau_{rel} \ll \tau_0$) neutrality is first restored and then the excess carriers decay slowly via recombination. In the relaxation case ($\tau_0 \ll \tau_{rel}$) the law of mass action quickly restores the np product to its equilibrium (and minimum) value (Eq. 5.57), while a space charge e(n-p) remains, slowly decaying with a time constant τ_{rel}.

Amorphous semiconductors are relaxation case materials which means that carrier transport may be strongly affected by space charge effects. The transport equations become rather complicated and the consequences for amorphous semiconductors have not been fully evaluated. The fact that a large fraction of the space charge resides in deep traps with often very slow release times complicates the analysis. The diffusion length lifetime will therefore depend on the injection level similar to the dependence of the photoconductive decay time on the density of excess carriers. Decay times of hours are commonly observed (after an initial rapid decay) when $\Delta E/kT$ is large. Furthermore, a large density of gap states present in some amorphous alloys will prevent any large excursion of E_F from its equilibrium value.

Unfortunately, space does not permit us to treat the work of van Roosbroeck and Casey (1970, 1972) in accordance with its importance.

In insulators and amorphous semiconductors one distinguishes the following injection models:

(i) one carrier injection, a case in which all defect states are in communication only with the band into which the carriers are injected. The presence of the other band is neglected.

(ii) one carrier injection into a band of a high resistance semiconductor in which the communication with the other band characterized by a carrier lifetime τ_0 is very much faster than the dielectric relaxation time τ_{rel}.

(iii) double injection, where both electrodes are injecting, the cathode electrons and the anode holes. The current is limited by recombination and neutralization. These injection phenomena disturb the charge neutrality and produce nonuniform fields. They require injecting contacts, i.e., contacts having a space charge of the same sign as that of the injected carriers. In the simplest picture usually treated, the contact space charge consists predominantly of excess free carriers. A number of field lines of the applied voltage terminate on these mobile charges causing them to be injected into the bulk. The number of these excess carriers injected into the semiconductor as well as their drift velocity is proportional to the field in this simple picture. Consequently the one carrier space charge limited current (SCLC) of type (i) rises with the square of the voltage as expressed by Eq. (5.58). A transition from ohmic to the square-law region occurs when the density of injected carriers equals that of the thermal carriers or, what amounts to the same, when the transit time of the injected carriers equals the low field dielectric relaxation time. With traps present, many of the injected carriers will be captured by traps and a much higher voltage is needed to yield SCLC. At even higher voltages, the traps become filled and the current rises rapidly to the trap-free SCLC value. A thorough discussion of this topic is given by Mott and Gurney (1940), Rose (1955), Lampert (1964), Tredgold (1966) and by Lampert and Mark (1970). The SCLC density j depends on the voltage V and electrode spacing L as

$$j = \frac{9}{32\pi} \mu\kappa\Theta V^2/L^3 \tag{5.58}$$

This relationship holds with $\Theta = 1$ for a trap-free semiconductor and with

$$\Theta = (N_c/N_t) \exp - (\frac{E_c - E_t}{kT}) \tag{5.59}$$

when a density of N_t traps is located at E_t. N_c is the effective density of conduction band states with kT of E_c. The onset of SCLC, i.e., the voltage at which the injected current equals the current of thermal carriers, shifts to higher voltages as T is lowered or as the trap depth is increased because of Eq. (5.59). From this it can easily be seen that any distribution of traps, as expected in NCS, will lead to a voltage dependence different from Eq. (5.58). Rose obtained for an exponential distribution of localized states such as described by Eq. (5.44) a SCLC which depends on voltage as

$$j = const\ V^{n+1}/L^{2n+1} \text{ with } n = \epsilon/kT \tag{5.60}$$

The current in thin semi-insulating films is often found to have a steep voltage dependence. However, an attempt to deduce the distribution of traps from a study of the current-voltage characteristic and its dependence on film thickness and temperature appears futile because the Eqs. (5.58)–(5.60) are based on a very simplified model.

For the case of a uniform trap density $g(E_F)$ at the Fermi level Lampert and Mark (1970) derived a SCLC of the form

$$j = e\ n_o\mu_o F \exp (F/F_o) \tag{5.61}$$

with

$$F_o = \frac{e \, L \, kT}{\epsilon_o \, \kappa} \, g\,(E_F) \qquad (5.62)$$

Hall (1970) interpreted the high field data on chalcogenide alloys with this relaxation and obtained $g(E_F) = 2 \times 10^{17} \, eV^{-1} cm^{-3}$ for low fields and $6 \times 10^{17} \, eV^{-1} cm^{-3}$ for high fields. Koc, Zavetova and Zemek (1972) obtained $g(E_F) = 1.5 \times 10^{17} \, eV^{-1} cm^{-3}$ for a-Ge in the same manner. An exponential field dependence of the conductance is often observed. But it can also arise from a decrease of the thermal activation energy ΔE by a uniform field in the bulk as discussed in the next section. The crucial test for SCLC, the dependence of j and F_o on the electrode separation L has not been performed.

Croitoru et al. (1970) checked the L dependence and observed SCLC in As_2Te_3 according to Eqs. (5.58) and (5.59) with a trap depth $E_t = 0.28$ eV. Se was studied by Hartke (1962) and Lanyon (1963) and Sb_2S_3 by Budinas et al. (1969).

The one carrier injection of type (ii) was first discussed by van Roosbroeck and Casey (1970, 1972). In this relaxation regime it is assumed that the dielectric relaxation time $\tau_{rel} = \epsilon_o \kappa / \sigma$ is very long compared to the carrier lifetime τ_o. The recombination rate is sufficiently large that the Fermi level cannot split into two quasi-Fermi levels, as is the case in normal minority carrier injection, but instead is shifted by injection from its equilibrium position. As a consequence the product of mobile carrier concentrations remains constant, Eq. (5.57). For example, if injection doubles the minority carrier concentration, the majority carrier density and hence the conductivity is *reduced* by a factor of two. The predicted I–V characteristic is therefore sublinear instead of superlinear as for the other injection mechanisms. This was observed by Queisser, Casey, and van Roosbroeck (1971) in high resistivity crystalline GaAs. This effect may be responsible for the sublinear I–V curve seen in Figure 5.34 for negative polarity. In contrast to minority carrier injection, injection of dominant current carriers in the van Roosbroeck limit yields an *increase* in conductance and is, as shown by Parrott (1971), indistinguishable from SCLC discussed above, which leads to Eq. (5.58).

Double injection (iii) has been postulated in connection with switching in amorphous semiconductors by Henisch et al. (1970) but no quantitative analysis has yet been undertaken.

The injection processes (i) – (iii) discussed so far were fed from a reservoir of excess carriers adjacent to the contact, i.e. they were controlled by the space charge which screens the contact potential. If, on the other hand, the contact is neutral (flat bands) or blocking, or when the electrode separation is smaller than the screening length λ, then the high field carrier injection is not governed by the space charge but by the emission process. In this limit one considers two additional high field effects.

(iv) Schottky emission of carriers over a barrier which is formed by the image potential reduced by the applied field.

(v) Tunneling through the barrier which becomes lower and narrower as the applied field is increased.

These phenomena discussed in detail by Hill (1967) and Simmons (1970) will be mentioned here only very briefly.

Schottky (1914) emission is governed by the Richardson-Schottky equation of thermal emission over the image-force barrier

$$j = A \ T^2 \ \exp\left(-\frac{\Delta E_B}{kT}\right)$$ (5.63)

where A is the Richardson constant

$$A = 4 \ \pi \ em \ k^2/h^2 = 120 \ A/cm^2 \ °K^2$$

and the barrier height ΔE_B is reduced by the field

$$\Delta E_B = \Delta E_B(o) - \beta_s \ F^{1/2}$$ (5.64)

where

$$\beta_s = e^{3/2}/\kappa^{1/2}$$ (5.65)

is the Schottky coefficient and κ the high frequency dielectric constant. $\beta_s = 3.8 \times 10^{-4}/\kappa^{1/2}$, when ΔE in (eV) and F in (V/cm). As the temperature is lowered, Schottky emission becomes rapidly small and tunneling emission through the barrier will dominate. Although details of the expression describing tunneling emission depend on the nature of the barrier, i.e. on whether it results from a blocking contact or the image potential modified by the high field, it has the general form (Fowler and Nordheim (1928) Gomer 1961)).

$$j = const. \ F^2 \ \exp\left(- F_o/F\right).$$ (5.66)

F_o contains the electron mass and the workfunction of the electrode with respect to the semiconductor.

It is difficult to find indisputable experimental evidence for these processes in noncrystalline semiconductors and insulators. Although in principle the dependence of the current on electrode spacing should distinguish between uniform and nonuniform fields and the dependence on temperature and voltage between the different injection mechanisms, very little has been learnt so far from these experiments about the nature of contacts and the distribution and the density of localized states in NCS.

Uniform Fields

When the NCS is sufficiently thick, the high field effects should be determined by the uniform field in the bulk. The independence of the current on the polarity of the applied voltage is usually taken as an experimental proof that conduction is not determined by nonuniform fields near the contact. This requirement is, however, not sufficient since most processes, discussed in the previous chapter, are insensitive to the parameters of the electrode material. The injection mechanisms are essentially controlled by the properties of the semiconductor too and Queisser *et al.* (1971) showed that ohmic conduction can be space charge controlled over a wide voltage range in a relaxation case semiconductor. Moreover, the rectification ratio of NCS contacts are expected to be small even for blocking

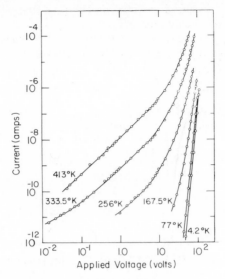

Fig. 5.37. Current voltage characteristic at various temperatures of evaporated silicon oxide films (after Servini and Jonscher (1969)).

contacts having a barrier for majority carriers. It is therefore difficult to decide whether or not one deals with the uniform field case. Only in a few cases quoted below has the thickness dependence been studied to make sure that the fields are indeed uniform.

Figure 5.37 shows as a typical example of high field conduction data measured by Servini and Jonscher (1969) on SiO at different temperatures. The conductivity $\sigma = j/F$ increases for electric fields F larger than 10^4 V/cm as

$$\sigma(F, T) = \sigma(o, T) \exp\left(\frac{\beta F^{1/2}}{kT}\right) \qquad (5.67)$$

At low temperatures a new process sets in: the I–V characteristic becomes temperature independent and the current rises very rapidly with voltage.

Similar behavior has been found in SiO (Hartman et al. (1966) Jonscher (1967) Stuart (1967) Hill et al. (1970)), Ta_2O_3 (Mead (1962)), Si_3N_4 (Sze 1967; Hu et al. (1967) Yeargan and Taylor (1968) Deal et al. (1968)) BN (Rand and Roberts (1968)) SiC (Weinreich and Ribner (1968)), Se (Croitoru and Vescan (1969)), Si (Brodsky et al. (1970)), Ge (Croitoru and Vescan (1969)), and As-Te-Ge-Si (Croitoru et al. (1970)).

The field dependence of Eq. (5.67) is commonly attributed to the Poole-Frenkel effect (Frenkel (1938)) and derived in the following manner. Let us assume the concentration of mobile carriers is governed by the ionization of donors. The Coulomb potential of each donor is modified by the field F in the x-direction as follows

$$\frac{-e^2}{\kappa r} - e F x \qquad (5.68)$$

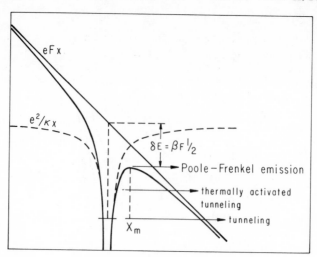

Fig. 5.38. Sketch of an isolated Coulomb potential in the presence of a large uniform field F.

As shown in Figure 5.38 the electron can escape the donor preferentially in the x-direction, where a potential maximum occurs at x_m when

$$\frac{-e^2}{\kappa x_m^2} = eF \tag{5.69}$$

The binding energy of the Coulomb center (donor or acceptor) is lowered by

$$\delta E \beta F^{1/2} \text{ with } \beta = 2 \beta_s \tag{5.70}$$

The Poole-Frenkel coefficient β is larger than β_s of Eq. (5.65) because in contrast to the image charge, the charge of the donor remains of course fixed in space. Lowering of the donor binding energy by δE leads to Eq. (5.67).

This simple model has been extended by Jonscher (1967) and Hartke (1968) to account for the escape of the carrier in all directions. This yields the expression (Hill (1971))

$$\sigma(F,T) = A(T) \phi^{-3} (\phi \cosh \phi - \sinh \phi) \tag{5.71}$$

where ϕ depends on the field and the temperature in one of two ways. If the ionized Coulomb centers are sufficiently far apart so that x_m of Eq. (5.69) is less than this average separation R then one is dealing with the Poole-Frenkel effect and

$$\phi = \beta F^{1/2}/kT \tag{5.72}$$

At larger concentrations of ionized centers, $R < x_m$ and the position of the barrier maximum remains halfway between centers, independent of F. The lowering of the barrier due to the field is then proportional to e RF/2 so that

$$\phi = e \, FR/2 \, kT \tag{5.73}$$

Details of the analysis and the function $A(T)$ in Eq. (5.71) depend further on whether the distance traveled by the released carrier before recapture is field dependent or determined by the next Coulomb center. Furthermore, the question whether the carriers remain in local thermal equilibrium has been raised (Jonscher and Ansari (1971)) but has not been analyzed sufficiently.

Camphausen *et al.* (1972) and Connell *et al.* (1972) argue that after escape, a carrier makes a positive contribution to the current irrespective of the direction of escape. They obtain in place of Eq. (5.71)

$$\sigma(F,T) = \sigma(o,T) 2\phi^{-2} \, (1 + \phi \sinh \phi - \cosh \phi) \tag{5.74}$$

Good agreement between experimental results and expressions such as Eqs. (5.71) or (5.74) with ϕ determined by Eq. (5.72) or (5.73) are usually obtained. Figure 5.39 shows how well the high field data obtained by Servini and Jonscher (1969) obtained with SiO between 140 K and 410 K can be fitted by Eq. (5.71). Figure 5.40 shows the agreement between data for amorphous Ge obtained by Camphausen *et al.* (1972) with Eq. (5.74). Morgan and Walley (1971), on the other hand, obtained good agreement of their high field results on a-Ge with Eq. (5.71).

Bagley (1970) considers the case in which the probability for electrons to jump between neighboring potential wells is changed by the field. He considers a one-dimensional case, subtracts backward from forward jumps and obtains

$$j = n_t \, e \, R \, \nu \exp \left(- E_i/kT \right) \sinh \phi \tag{5.75}$$

where n_t is the density of trapped carriers at the energy E_i from E_c (or E_v) and is given by (5.73). The attempt to escape frequency ν is estimated to be of order 10^{13} Hz. With the choice of $R = 150$ Å he finds a good fit of Eq. (5.75) to the high field data taken by Walsh *et al.* (1969) on chalcogenide alloy glass.

Before commenting on the different paths taken to describe the high field conductance data, let us turn our attention to the low temperature regime where $\sigma (F,T)$ becomes temperature independent. This can be seen in the example shown in Figure 5.37. Hill (1971) suggests that this conduction is field ionization by tunneling into the conduction band. This process is indicated by a horizontal arrow in Figure 5.38. His explanation is plausible if one considers that the potential maximum according to Eq. (5.69) occurs at about $x_m \sim 100$ Å for $F = 10^4$ V/cm and near $x_m \sim 10$ Å for 10^6 V/cm in a material of dielectric constant $\kappa = 10$.

At intermediate temperatures Hill (1971) succeeds in analyzing the data in terms of a thermally activated field emission.

Comments
All expressions (5.71), (5.74), and (5.75), discussed so far for describing high field conduction, have in common that at low fields the conductivity rises with field as

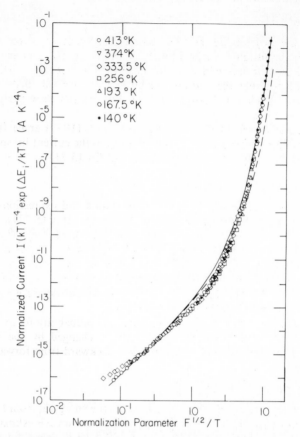

Fig. 5.39. Same data as shown in Figure 5.37. Here, the normalized current with $E_i =$ 0.35 eV is plotted against the parameter $F^{1/2}/T$. The curves represent $2 \sinh \phi$ and $\cosh \phi - \phi^{-1} \sinh \phi$, respectively (after Jonscher (1971b)).

$$\sigma(F,T) = \sigma(o,T) + A F^2 \qquad (5.76)$$

In most cases one observes, however, that the first nonohmic term is proportional to F (Fagen and Fritzsche (1970a) de Wit and Crevecoeur (1972) Marshall and Owen (1972) Owen (1973)). De Wit and Crevecoeur (1972) suggest that the discrepancy between Eq. (5.76) and the observations might be aleviated by considering the statistical distribution of the center separations R.

The underlying assumption leading to the above expressions has been that the electrons and holes remain in thermal equilibrium. In other words, there is no splitting of E_F into quasi-Fermi levels, except perhaps implicitly in the case of tunneling discussed by Hill (1971). The high field problem as treated above furthermore deals with discrete donor or acceptor levels and with Coulomb potentials which arise only from those levels which yield the carriers. Except for Simmons (1967) and Connell *et al.* (1972) no attention

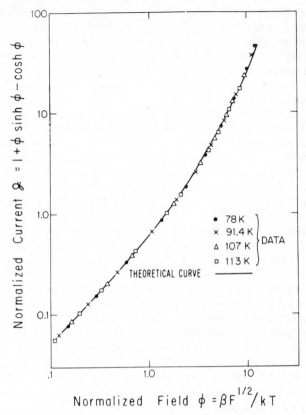

Fig. 5.40. Normalized current j versus normalized field ϕ for a sample of electrolytically prepared a-Ge (after Camphausen, Connell and Paul (1972)).

is paid to the effect of F on the Fermi level. In this spirit the thermal activation energy is often interpreted as that of the discrete levels, and densities of Coulomb centers are obtained from R in Eqs. (5.73) and (5.75).

This approach might have some validity in SiO and other dielectric films with well defined centers and rather sharp band edges. However, in amorphous semiconductors, for which band models of the type shown in Figures 5.3 and 5.4 are appropriate, a different approach must be taken.

At the outset one should distinguish situations in which there is no splitting of E_F into quasi-Fermi levels from cases where the distribution of carriers gets out of equilibrium. Experimental evidence for the latter case was found (Fagen and Fritzsche (1970b)) in the slow decay of an excess conductance in a chalcogenide alloy glass after a brief exposure to high electric fields at 78 K. We defer this case until later and first assume that E_F does not split into quasi-Fermi levels. One then can take E_F as a reference energy and consider the field induced shift of the mobility edges E_v and E_c toward E_F. One expects this shift to be proportional to F in first order. This yields the observed first order change of σ with field. The characteristic length, which determines the shift of the mobility edges with field, is then the screening length λ which determines also the long wavelength limit of the spectrum of potential fluctuations discussed in more detail in the following section. According to Eq. (5.47) the magnitude of λ is determined by the density of states close to E_F. It is very different from R, the separation of Coulomb centers.

In the presence of potential fluctuations, the mobility edges are more properly viewed as percolation thresholds. In addition to a shift of these thresholds one expects the field to affect the so-called channel states near the threshold and therefore the mobility μ. Such a change should affect μ_D and σ in the same manner because μ is the only common factor. Marshall and Owen (1972) report such a case. The observation, however, that the photocurrent remains ohmic to much higher fields than the dark current (Fagen and Fritzsche (1970a) (1970c)) suggests that the carrier density and not the mobility changes with field in chalcogenide alloy glasses.

In this picture an exponential dependence on F is expected at high fields similar to the previous field expressions. However, the origin of the potential fluctuations is not directly related to the states, which yield the electrons. The observation of a well defined thermal activation energy E_i does not signify the presence of discrete states at that energy; E_i is the energy separation of E_F from E_v or E_c, respectively.

Unfortunately, these many-parameter expressions for $\sigma(F,T)$ are rather similar in their ability to fit the data so that one is unable to positively identify in this manner the transport process involved. It is, however, fair to emphasize that estimates of donor and acceptor concentrations and of their activation energies from high field conductivity experiments are often based on unrealistically simplified models.

If, on the other hand, a nonequilibrium situation is reached, whether by avalanche, internal field emission or tunneling, then the recombination and relaxation kinetics, which limits the amount of disequilibration, need be included in the calculation before the excess conductance can be obtained. Hindley (1970b) and Mott (1971) estimated that the probability for avalanche excitation of carriers at high fields is competitive with the other processes. Additional experiments such as the measurement of photoconduction and a.c. conduction at high fields (Jonscher (1972) Jonscher and Loh (1971)) are needed to identify and separate the different processes.

5.4   HETEROGENEITIES

An ideal amorphous semiconductor is thought of as a uniform three-dimensional random network structure in which the valence requirements of each atom are satisfied. Any real material contains, however, imperfections and deviations from the ideal disordered network in the form of dangling bonds, voids, or density fluctuations. Furthermore, some vitreous semiconductors contain structural units which are bonded together by weak van der Waals type forces. This section deals in particular with structural and compositional inhomogeneities whose own size or whose effect (by forming a depletion or accumulation layer around it) extends over dimensions of the order of the dielectric screening length. Such inhomogeneities are associated with nonuniform charge distributions which lead to internal potential fluctuations. These are bound to have a strong effect on the electronic properties. It is possible that some of the unexplained features of amorphous semiconductors are caused by heterogeneities.

Since most vitreous alloy glasses contain compositions and mixtures which far exceed the limits of thermodynamic equilibrium, the metastable vitreous state is prone to separate into two or more finely divided phases which may be all vitreous or partially crystalline. In some cases the process of phase separation may not have reached completion so that the phase boundaries are soft and ill defined. As one deals with nonequilibrium structural states these become frozen in below a certain temperature because of the low diffusivity of the constituent atoms.

Single phases too can show inhomogeneities and deviations from ideal disorder. This is expected to occur in materials with strong intramolecular binding and weak forces between molecular complexes. They can easily be recognized by their low melting or softening temperature, large optical band gap, and high compressibility. The molecular complexes can be in the form of chains or rings as in Se or in smaller units as $AsSe_{3/2}$ and $AsS_{3/2}$ in As_2Se_3 and As_2S_3. Sharp Gaussian peaks in the far infrared absorption spectra observed in these materials by Taylor, Bishop, and Mitchell (1971) are indicative of vibrations with well defined optical phonon branches and hence of the preservation of these structures in the vitreous and liquid state. Figs. 5.10 and 5.11 demonstrate the strong tendency in covalent glasses for forming structural units of chemical compound compositions. The size of these structural complexes, their mixture and statistical deviations from them, depend on the thermal history, light exposure (Berkes et al. (1971), and the presence of structure modifying impurities (Schottmiller et al. (1970)). The sharp infrared absorption peaks are entirely absent in three-dimensional covalent network glasses. Further evidence for the possibility of a high degree of chemical ordering is the large excess enthalpy observed by de Neufville (1972) in several Te-Ge glasses which suggests an order-disorder transformation occurring around the melting point of the corresponding crystalline material.

The submicroscopic network of voids observed in Ge and Si is another kind of heterogeneity. Depending on the preparation conditions it may form a connected network or isolated swiss cheese-like holes (Ehrenreich and Turnbull (1971)) as observed by Barna et al. (1972). Amorphous carbon is known to be porous. The presence of voids might be a common property of amorphous tetrahedral films deposited on low temperature substrates. Voids and defects also occur in chalcogenide glasses as shown by infrared Schlieren microscopy by Vasko (1968).

The surface potential of the internal void structure must be screened within the characteristic length λ and thus leads to internal potential fluctuation. It will be recalled that the tetrahedral amorphous semiconductors have properties quite dissimilar from those of chalcogenide glasses. They exhibit, for instance, a high density of free spins, on anomalous magnetoresistance, and a conductvity-temperature relation described by Eq. (5.10). All tetrahedral materials show very strong annealing effects at temperatures far below crystallization. At these low temperatures diffusion must be very small. The annealing effects are thus indicative of a very loose structure which is in all likelihood heterogeneous. If this is the case, any attempt to explain the properties of these materials with a model appropriate for a homogeneous material will be futile.

Even without microscopic voids, charge and potential fluctuations are expected to result from variations in electronegativity which in turn are a consequence of spatial fluctuations in density and composition. Furthermore, the difference in electronegativity between the elements in the respective columns of the periodic table may cause a local charge transfer when isoelectronic elements are substituted even without a change in coordination. In crystalline material this electronegativity difference gives rise to isoelectronic donor centers (Thomas *et al.* (1965)).

Another cause for charged centers was recently discussed by Pollak (1970). Consider a localized valence bond state V (occupied and hence neutral) and a localized conduction bond state C (unoccupied and therefore neutral), being moved together. One has to imagine the structural environment which gives rise to these localized states to be moved along. At large distances the energy differences $E_i - E_a$ between the ionization energy of V and the electron affinity of C is required to form the ionic pair V^+C^-. As soon as the Coulomb energy of V^+C^- exceeds $E_i - E_a$ at small distances, the ionic pair is energetically favored. This rough model of the formation of an ionic diatomic molecule has to be amended naturally to include screening and the interaction of the structural environments which gives rise to these localized states. It illustrates, however, the importance of Coulomb effects in the self-consistent treatment of the localized states.

In the following we first discuss the consequences of spatial heterogeneities in terms of the Mott-CFO model and then several phenomenological aspects of heterogeneity.

5.4.1 Model with Potential Fluctuations

One can distinguish short-range variation of the potential over distances of interatomic separations from long-range variations associated with screened Coulomb potentials and heterogeneities (Fritzsche (1971a)). The local potential varies by many electron volts. The long scale fluctuations with which we will be concerned here have probably a mean amplitude of only a few 0.1 eV. We limit ourselves to fluctuations over a few hundred angstrom, so that the potential contour presents the classical turning points of the charge carriers. Not the same potential is felt by electrons in different bands. Local density fluctuations, strains, and compositional variations will cause a spatial variation in the local band gap, whereas Coulomb potentials affect both bands symmetrically. If one imagines the long-range potential fluctuations to be decomposed into an antisymmetric and a symmetric part with respect to the two bands, then the latter is similar to the one sketched in Figure 5.41. Here the lines drawn do *not* represent band edges or mobility edges. It is the symmetric part of the local potential averaged over a region

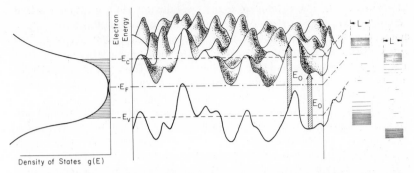

Density of States g(E)

Fig. 5.41. Sketch of the symmetric part of long wavelength potential fluctuations. The short-range fluctuations as well as variations which shift valence and conduction band states in opposite directions, are omitted for clarity. E_0 corresponds to an average optical gap. E_c and E_v are the percolation thresholds or mobility edges. The left hand side shows the density of states. The distribution of localized gap states between E_c and E_v is not necessarily smooth and monotonic. On the right hand side the states in two volume elements of size L^3 are sketched. The length of the horizontal bars illustrate schematically whether the states are localized or extended within L^3. Note the coexistence of localized states and channel states near E_c and E_v.

of order the coherence length of an electron taking part in an optical transition. The two vertical arrows represent optical transitions of energy E_0 taking place in two different volume elements. In order to fix our thoughts, let us assume that each volume element contains about 2×10^3 atoms and measures 200 Å in diameter. The energy levels of these two volume elements which should be thought of as still being situated in the material, are sketched on the far right hand side of Figure 5.41. The purpose of this sketch is to show that some band-like states are extended over the volume element and other states originate from short-range potential fluctuations and are localized. The mean potential of the volume elements are shifted with respect to one another by the symmetric part of the long-range potential. The long-range potential produces additional localized states but their range of localization is larger and determined by the classical turning points of the potential.

Let us now inquire which states provide conduction from one side of the amorphous semiconductor to the other. As shown by the work of Cohen (1970), Economou *et al.* (1970), and Eggarter *et al.* (1970), from a classical point of view, this is a problem to be solved by percolation theory. This theory, which in the past dealt almost entirely with percolation on lattices (Frisch *et al.* (1962)), was recently extended to continua and applied to amorphous semiconductors by Zallen and Scher (1971), This subject is reviewed by Shante and Kirkpatrick (1971). Zallen and Scher describe the percolation behavior in a two-dimensional continuum such as the one shown in Figure 5.41 in the following manner. Referring to allowed regions as water regions and forbidden regions as land areas, these authors explain that at low E the allowed areas comprise isolated lakes in an infinitely extended continent. While at high E there is the opposite situation in which isolated islands are imbedded in an infinitely extended ocean of allowed territory. Somewhere in between at a critical E_c there is critical percolation threshold as the extended ocean makes its debut on the scene. For energies $E < E_c$ the electrons are thus confined to various regions in the volume. At a certain critical energy E_c the probability for finding a continuous path extending throughout the material becomes finite. The probability for continuous channels increases rapidly for $E > E_c$. The same argument holds for holes for which one defines a critical percolation energy E_v. These percolation thresholds E_c and E_v represent the mobility edges of the Mott-CFO model at T = 0 (Fritzsche (1971a)).

For the interpretation of transport processes as well as for the trapping and recombination kinetics and for calculation optical transition matrix elements it is important to note that in this classical picture:

(i) the first extended states close to E_c and E_v are channeling states and occupy only a fraction of space;

(ii) electrons are localized predominantly where there are few holes and most holes are localized at places different from the electrons;

(iii) the position of the percolation thresholds E_c and E_v depend self-consistently on the occupation of the localized states. They will therefore be changed with the electron distribution at high injection or photo excitation levels or by high electric fields;

(iv) the optical gap E_o bears no direct relationship with the mobility gap $E_c - E_v$;

(v) the matrix element for optical transitions from localized to channeling states and between channeling states should be smaller than that between extended band states because of the spatial relationship of these different states;

(vi) localized gap states associated with a specific defect structure and thus having a certain energy will fall into an energy interval which is broadened by the long-range potential fluctuations.

A note of warning seems in order. The application of percolation theory to amorphous semiconductors containing internal potential fluctuations is far from straightforward. The definition of E_c and E_v as percolation thresholds as done by Fritzsche (1971a), Zallen and Scher (1971) and Cohen (1970) is possible only if one neglects the generation and recombination kinetics. Ambipolar current flow in conjunction with finite generation and recombination rates makes d.c. conduction involving electrons below E_c and holes above E_v possible without hopping. However, in this classical picture, the space allowed for electrons at E_c does not overlap the space allowed for holes at E_v. Thus recombination involves excitation to higher energies or lateral charge transport.

Other interesting consequences of this classical picture is that the electrical gap $E_c - E_v$ is always smaller than the optical width E_o (Shklovskii (1971)), and that the matrix elements for optical transitions between localized and extended states are less than those between extended states (Fritzsche (1971a) Dow and Redfield (1972) Dow and Hopfield (1972) Sak (1972)).

The amplitude and the average wavelength of the potential fluctuation vary considerably between materials and with heat treatment. This classical picture loses its validity when the average wavelength becomes small and appreciable tunneling is possible.

This model with potential fluctuations does not contradict the findings of a very low current noise in chalcogenide glasses by Main and Owen (1970). The current flow goes around the regions of high potential and does not cross potential barriers or other internal discontinuities. It would be interesting to see whether conduction across or along the void network in tetrahedral semiconductors gives rise to excess noise.

5.4.2 Percolation Theory and Mobility

The task of calculating the macroscopic conductivity, the quantity observed experimentally from the three-dimensional network of conductances which relate to the local processes of current transport, is a classical percolation problem. Ambegaokar *et al.* (1971) and Pollak (1972a, 1972b) used it to derive the hopping conductivity, Eq. (5.10), from the jump probabilities between pairs of localized centers. Eggarter and Cohen (1970) extended it to analyze the mobility as a function of energy near the mobility edge. In the previous section the energy E_c was defined as the percolation threshold for electrons, which means that at this energy the probability of finding an extended conductance path through the material becomes finite. The probability $P(x)$ for continuous channels increases rapidly for $E > E_c$ because the fractional volume, x, accessible for electrons, increases with energy as seen from Figure 5.41.

Percolation theory yields $P(x)$. By treating the motion of an electron in a randomly fluctuating potential one obtains the fraction of accessible volume $x(E)$ as a function of energy. The task remains to find the average conductance G. Last and Thouless (1971) demonstrated that $G(x)$ cannot be equated with $P(x)$. They did this by measuring the conductance of a sheet of conducting paper while punching an increasing number of holes into it. While $P(x)$ increases very rapidly from zero as x exceeds the percolation threshold x_c, the conductance rises much slower on account of the fact that the first connected paths are very long and convoluted and that they occupy only a small fraction of the volume. Kirkpatrick (1971) carried out an equivalent experiment by Monte Carlo sampling a disordered three-dimensional resistor network. His result, which agrees very well with that of Last and Thouless is shown in Figure 5.42. It is assumed here that the local mobility is μ_o in the allowed regions and $\mu_o = 0$ elsewhere. The total mobility μ, averaged over the total volume, is plotted in units of μ_o in Figure 5.42. As comparison the percolation probability $P(x)$ is also shown. $P(x)$ is the fraction of volume made up of infinitely extended channels as a function of x the fraction of allowed volumes, P and x are identical for $x > 0.5$ but differ significantly for $x < 0.5$ because more allowed volume is found in isolated islands. Figure 5.42 shows the rather slow rise of μ/μ_o above the threshold x_c. Kirkpatrick (1971) finds $\mu/\mu_o \propto (x - x_c)^{1.6}$.

From these results it is not difficult to estimate (Kirkpatrick, 1972) the functional form of μ (E) near the mobility edge E_c. The extent of the

density of states tail increases with the size of the potential fluctuations. If
one assumes a Gaussian distribution of potentials with a width $\Delta\epsilon$ as shown
in Figure 5.43(a) then the tail states extend approximately $2\Delta\epsilon$ below the
sharp $\Delta\epsilon = 0$ band edge. At a certain energy E_c the fraction of allowed
volume is $x_c = 0.3$ which is the percolation threshold, and above a certain
energy all volume is allowed, i.e. $x = 1$. These energies are marked in Figure
5.43(b). The relation $x(E)$ for $0.3 < x < 1$ can be determined for a given
potential. If one assumes a linear relationship, one can use the data of
Figure 5.42 directly and obtains μ/μ_o as a function of E as sketched in
Figure 5.43(c). The ratio μ/μ_o increases with E because of the increased
connectedness of the medium as E is raised. Of course, the local mobility μ_o
might itself be a function of E. However, this will not change the essential
result of this classical treatment, which predicts a rather soft mobility edge
near E_c and E_v in contrast to the sharp increase assumed in the models of
Figures 5.3, 5.4, and 5.5. The conclusions drawn from these models do not
actually rest on the sharpness of the mobility edge. The only consequences
of a soft edge would be that C in Eq. (5.5) acquires an additional temperature factor T or $T^{3/2}$. Moreover, it increases the terms A_c and A_v in the
expression for the thermopower Eq. (5.16). One would expect $A = 3-4$
when $\mu \alpha E^{3/2}$. The present observations are not in conflict with these
predictions.

5.4.3 Phenomenological Aspects of Heterogeneity
The presence of internal fields and heterogeneities in defect semiconductors
has been proposed by many authors to explain one or the other anomalous
effect. For example, the unusually large photovoltages observed in thin
films of polycrystalline PbS, ZnS and obliquely evaporated films of Ge, Si,
GaAs, and CdS have been associated with inhomogeneities (see Chopra
(1969) for detailed references). Stuke (1965) proposed an internal barrier
model to explain the apparent change in the mobility activation energy in
hexagonal selenium with illumination and deformation. A.D. Pearson
(1964) suggested that n-type crystalline inclusions in a p-type glass are the
cause for the opposite sign of the thermopower S and the Hall effect R in
vitreous semiconductors. The Hall effect in inhomogeneous materials was
further analyzed by Heleskivi and Salo (1972) and by deWit (1972). They
find that opposite signs of S and R are expected only when both phases of
the material have similar conductivities over an extended temperature range.

Heterogeneities in amorphous semiconductors have been discussed more
recently by Mott (1967), Mott and Allgaier (1967), Owen (1967), Böer
(1969, 1970b, 1972), Tauc (1970), Heywang and Haberland (1970),
Fritzsche (1971a), Inglis and Williams (1970, 1971) and Williams (1972). It
appears that the exponential absorption tail (Urbach tail) in many noncrystalline and crystalline materials can be attributed according to Dow and
Redfield (1972) to the dissociation of band edge excitons via tunneling into
adjacent potential wells produced by internal potential fluctuations. Except
for this Urbach effect near the onset of interband absorption, the potential
fluctuations do not greatly affect the optical transitions because both the
initial and the final states are essentially at the same potential. DC transport

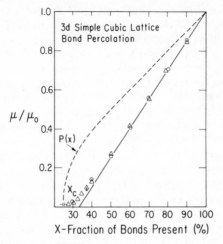

Fig. 5.42. The data points represent the average mobility relative to that in the classically allowed region as a function of the fraction x of volume allowed. The full line represents the result of the effective-medium theory of conduction in mixtures. The dotted curve is the percolation probability $P(x)$, which is the fraction of material that is both allowed and connected to contacts. After Kirkpatrick (1972).

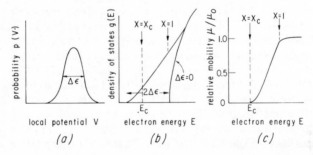

Fig. 5.43. (a) Assumed Gaussian distribution of potentials in disordered material. (b) Electron with energy E in a fluctuating potential finds fraction $x(E)$ allowed. (c) Relative mobility according to Figure 5.42 assuming $x(E)$ is linear between percolation value $x_c = 0.3$ and $x = 1$. The mobility increases slowly above the percolation threshold E_c. After Kirkpatrick (1972).

processes, on the other hand, depend on carriers traversing larger distances and thus are more sensitively affected by internal potential fluctuations.

Most authors who deal with effects of inhomogeneities on the transport properties consider inhomogeneous aggregates consisting of two or more distinct phases separated by sharp boundaries. The conductance of various geometries are discussed by Bruggeman (1935), Volger (1950), Landauer (1952), Juretschke et al. (1956) and Tick and Fehlner (1972). The effects of random and continuous inhomogeneities were treated by Herring (1960).

Although most of the galvanomagnetic effects discussed by Herring should be very small in low mobility semiconductors, his mathematical approach is relevant to amorphous semiconductors.

There is a wealth of information on so-called granular metals which is most relevant for electrical, optical, and magnetic properties of hetero-geneous materials. These materials are made by simultaneous sputtering or evaporation of a metal and a dielectric such as SiO_2. The good control over the size and density of the metal grains enables one to simulate many heterogeneous systems (Cohen and Abeles (1968) Cohen and Gittleman (1968) Gittleman et al. (1972) Goldstein and Gittleman (1971) Sheng and Abeles (1972) Sheng, Abeles and Arie (1973) Hauser (1973)).

Dielectric properties, both the dielectric constant and the loss as a function of frequency are strongly affected by inhomogeneities because the applied a.c. field interacts with internal space charges and in mixtures of components having different resistivities space charge polarizations occur. Several simple geometries are treated by Volger (1960) and van Beek (1967). Pollak (1962) showed that variations in conductivity give rise to a conductance which increases with frequency even when conduction is pure band conduction (no hopping).

The thermoelectric power seems to react rather sensitively to inhomo-geneities. It has often been noticed that annealing changes the thermopower strongly while the resistivity is hardly affected (Stuke (1971)). Thermo-electric effects in inhomogeneous films have been treated by Volger (1950) and Lipskis et al. (1971). Circulating thermoelectric currents between dif-ferent phases produce among other effects an excess thermal conductivity (Airapetiants (1957) and Airapetiants et al. (1958)).

Heterogeneities are expected to affect contact properties. Any region at the interface with a low barrier or with injecting properties essentially short-circuits regions having a high barrier. If the interface potential spatially fluctuates from positive to negative values with respect to the flat band energy, then both electrons and holes can be injected.

A void structure of the type shown in Figure 5.13(a) may contain at the interior surfaces a high density of surface states. Phonon-assisted tunneling to these internal surface states may be the mechanism responsible for the tunneling characteristic such as that shown in Figure 5.24. Surface states often show a symmetric density distribution (Chiarotti et al. (1971)) with E_F pinned at its center.

5.5   CONCLUSIONS

A critical look at the experimental data which form the basis for our under-standing of and the distribution of electronic states and their transport properties in amorphous semiconductors reveals some major problems which need be solved before further progress is made.

The porosity and void structure observed in some of the most thoroughly studied amorphous tetrahedral NCS cast serious doubts on the usefulness of models which assume homogeneous materials for interpreting their transport properties. The elastoresistance, magnetoresistance as well as the annealing behavior of the thermopower and of the conductivity may in tetrahedral NCS be closely associated with the heterogeneous void structure. If this turns out to be true then these measurements reveal little about the bulk properties.

Do we have experimental proof of or strong evidence for any of the essential features which are contained in the three models described in one of the first sections? The answer is no. The mobility edge is not discernable optically unless the matrix element for optical transitions changes substantially as one of the states (initial or final) is taken to lie just outside or just within the mobility gap. Although the matrix element will decrease with increasing depth of the localized states (Fritzsche (1971a) Dow and Redfield (1972) Dow and Hopfield (1972) Sak (1972)) no *abrupt* change is expected at the mobility edge (Mott (1969a)). The concept of a mobility edge is meaningful and useful only if one tries to explain the 'intrinsic' temperature dependence of the conductivity, Eq. (5.4) and Figure 5.7, with a model which assumes that a tail of a considerable concentration of localized states is butting against the density of extended states. There is no clear experimental evidence for or against the presence of such tail states. We wish to point out, however, that the present experimental results can also be described with a more conventional model, in which the density of extended states falls essentially to zero at band edges. The model requires then deep gap states near the gap center to assure, by pinning E_F, that 'intrinsic' conductivity extends to low T. Such a model allows for traps close to the band edges, i.e. donors close to the valence band and acceptors close to the conduction band.

Can we estimate the density of localized gap states? Their presence is noticed in measurements for instance of drift mobility, photoconduction, (Arnoldussen, Bube, Fagen and Holmberg (1972a, b) Weiser (1972)), thermostimulated currents (Kolomiets and Mazets (1970)), recombination radiation (Kolomiets, Mamontova, and Babaev (1970)), and high field conduction. A density of gap states cannot be obtained from these experiments though, without having values for the appropriate trapping and recombination cross sections.

Although earlier reports on magnetic susceptibility studies (Tauc, Menth, and Wood (1970) Tauc and Menth (1972)) suggested between 10^{17} and 10^{18} cm^{-3} paramagnetic centers in $As_2 S_3$ and $As_2 Se_3$, these could be traced to magnetic impurities and the upper limit of pure vitreous $As_2 S_3$ was set at 3×10^{16} cm^{-3} (Bagley *et al.* (1972)). According to electron spin resonance measurements (Agarwal (1973)) the unpaired spin density in many chalcogenide glasses is less than 10^{15} cm^{-3}. Optical absorption studies at photon energies quite far below the optical gap suggest that gap state densities must be smaller than 10^{17} cm^{-3}eV^{-1} unless the optical absorption cross sections are unusually small. Photoemission studies set 3×10^{19} cm^{-3}eV^{-1} as an upper limit for the density of tail states. Estimates of the screening length λ, Eq. (5.47) from field effect and contact capacitance measurements yield $g(E_F) \sim 5 \times 10^{18}$ cm^{-3}eV^{-1} in chalcogenide glasses. The best measurements of the density of localized states function are those of Spear and LeComber (1972) and LeComber *et al.* (1972) on silane decomposed a-Si shown in

Figure 5.31. After having measured g(E) with the field effect one now can measure the a.c. conductivity, the $T^{-1/4}$ regime of ln σ, and optical absorption, and check the corresponding theoretical expressions. A measurement of paragmagnetic susceptibility would elucidate the question of the spin state of the localized centers. It is worth noting that $g(E_F)$ cannot be obtained from a linear specific heat term at low temperatures, because the electron gas will fail to equilibrate over the localized states within a reasonable time span at low temperatures. Furthermore, the anomalous linear specific heat term of Zeller and Pohl (1971) might be much larger than an electronic contribution.

There is very little certainty about the magnitude and the temperature and field dependence of the carrier mobility. If the Hall mobility is interpreted on the basis of the random phase approximation Eq. (5.29), then the conductivity mobility is with Eq. (5.31) about $\mu \sim 5-10$ cm^2/V sec for many of the chalcogenide glasses. The observation that μ_H is thermally activated with about 0.05 eV does not agree with Eq. (5.29) however. Such an activation energy is expected when polaron effects become important. For that case, however, Friedman and Holstein (1963) and Emin (1971) find that the conductivity mobility is more than an order of magnitude smaller than μ_H and that it is thermally activated by an energy several times larger than that of μ_H. Only recently the question has been raised as to what extent polaron effects and their influence on the transport properties and the optical transition matrix elements, have to be considered in amorphous semiconductors (Klinger (1970); Emin (1971, 1973) Emin *et al.* (1972)).

The problems associated with nonlinear conduction processes are even more involved. The rapid current increase at high fields can be fitted with various expressions. The models on which these are based, however, fail to distinguish between quite different conduction and trapping mechanisms which occur once the carriers are released. As a result no quantitative data about the density and energy location of gap states have been obtained from these measurements. Moreover, the criteria used for distinguishing uniform and non-uniform field distributions at high fields and for disregarding hot electron and contact effects appear often unjustified, particularly at low temperatures.

Large excess carrier concentrations produced either by high fields or photon excitation are associated with an even larger non-equilibrium density of trapped electrons and holes. These in turn will modify the spatial potential fluctuations which determine the energy distribution of localized states and the position of the percolation thresholds. A self-consistent treatment of the electron states for different occupation functions has not yet been carried out.

There are obviously plenty of challenging problems.

ACKNOWLEDGEMENTS

I am grateful to my students S. C. Agarwal, J. Freeouf, L. Gordy, S. Hudgens, M. Kastner, J. W. Osmun, M. Paesler and H.-Y. Wey for many lively discussions.

REFERENCES

Abrahám, A. et al. (1970), Proc. Tenth Intl. Conf. Physics Semicond., p. 784. Cambridge, Mass.: U.S. At. Energy Commission.
Adler, D., (1971), Critical Reviews in Solid State Sciences. The Chemical Rubber Co.
Agarwal, S., (1973), Phys. Rev., B7, 685.
Agarwal, S. C. and Fritzsche, H., (1970), Bull. APS II, 15, 244.
Airapetiants, C. V., (1957), Soviet Phys., Tech. Phys., 2, 429.
Airapetiants, C. V. and Bresler, M. S., (1958), Soviet Phys., Tech. Phys., 3, 1778.
Ambegoakar, V., Halperin, B. I. and Langer, J. S., (1971), Phys. Rev., B4, 2612.
Amrhein, E. M., (1969), Phys. Letters, 29A, 329.
Amrhein, E. M. and Mueller, F. M. (1968), Trans. Faraday Soc., 64, 666.
Anderson, P. W., (1958), Phys. Rev., 109, 1492.
Anderson, P. W., Halperin, B. I. and Varma, C. M., (1972), Phil. Mag., 25, 1.
Andriesh, A. M. and Kolomiets, B. T., (1965), Soviet Physics-Solid State, 6, 2652.
Argall, F. and Jonscher, A. K., (1968), Thin Solid Films, 2, 185.
Arnoldussen, T. C., Bube, R. H., Fagen, E. A. and Holmberg, S., (1972a), Proc. IV Intl. Conf. on Amorphous and Liquid Semiconductors, Ann Arbor, Mitch, J. Non-Cryst. Solids, 8–10, 933.
Arnoldussen, T. C., Bube, R. H., Fagen, E. A. and Holmberg, S., (1972b), J. Appl. Phys., 43, 1798.
Austin, I. G. and Garbett, E. S., (1971), Phil. Mag., 23, 17.
Austin, I. G. and Mott, N. F., (1969), Adv. in Phys., 18, 41.
Bagley, B. G., (1970), Solid State Comm., 8, 345.
Bagley, B. G., Di Salvo, F. J. and Waszczak, J. V., (1972), Bull. Am. Phys. Soc., 17, 322.
Ball, M. A., (1971), J. Phys., C4, L 107.
Barbe, (1971), J. Vacuum Science and Techn., 8, 102.
Barna, A., Barna, P. B. and Pocza, J. F., (1972), J. Non-Cryst. Solids, 8/10, 36.
Baron, R. and Mayer, J. W., (1970), 'Double Injection in Semiconductors' in Semiconductors and Semimetals, (R. K. Willardson and A. C. Beer, editors) Vol. 6, p. 202. Academic Press, New York.
Batra, I. P., Schechtman, B. H. and Seki, H., (1970), Phys. Rev., B2, 1592.
Bauer, R. S., Galeener, F. L. and Spicer, W. E., (1972), J. Non-Cryst. Solids, 8–10, 196.
Berkes, J. S., Ing, S. W. Jr. and Hillegas, W. J., (1971), J. Appl. Phys., 42, 4908.
Beyer, W. and Stuke, J., (1972), J. Non-Cryst. Solids, 8–10, 321.
Beyer, W. and Stuke, J., (1973), Proc. Fifth Intl. Conf. Amorphous and Liquid Semicond., Garmisch, Germany
Blakney, R. M. and Grunwald, H. P., (1967), Phys. Rev., 159, 664.
Bluzer, N. and Bahl, S. K., (1973), Proc. Fifth Intl. Conf. Amorphous and Liquid Semicond., Garmisch, Germany.
Böer, K. W., (1969), Phys. stat. sol., 34, 721, 733.
Böer, K. W., (1970a), J. Non-Cryst. Solids, 2, 444.
Böer, K. W., (1970b), Phys. stat. sol., (a) 3, 1007.
Böer, K. W., (1972), J. Non-Cryst. Solids, 8–10, 586.
Bosnell, J. R. and Voisey, U. C., (1970), Thin Solid Films, 6, 161.
Brenig, W., Döhler, G. H. and Wölfle, P., (1971), Z. Physik, 246, 1.
Brenig, W., Wölfle, P. and Döhler, G., (1971), Phys. Rev. Lettters., 35A, 77.
Brenig, W., Döhler, G. H. and Heyszenau, H., (1973), Phil. Mag., (to be published).
Brodsky, M. H., (1971), J. Vacuum Science and Techn., 8, 125.
Brodsky, M. H. and Gambino, R. J., (1972), J. Non-Cryst. Solids, 8–10, 739.
Brodsky, M. H. and Title, R. S., (1969), Phys. Rev. Letters, 23, 581.
Brodsky, M. H., Title, R. S., Weiser, K. and Pettit, G. D., (1970), Phys. Rev., B1, 2632.
Bruggeman, D. A. G., (1935), Ann. Physik, 24, 636.
Buchy, F., Clavaguers, M. T. and Germain, P., (1971), Proc. Intl. Conf. on Conduction in Low Mobility Materials, Eilat, Israel, Adv. Phys.
Budinas, T., Mackus, P., Smilga, A. and Viscakas, J., (1969), Phys. stat. sol., 31, 375.

Callaerts, R., Denayer, M., Hashmi, F. H. and Nagels, P., (1970), *Discussions Faraday Soc.,* **50**, 27.
Camphausen, D. L., Connell, G. A. N. and Paul, W., (1972), *J. Non-Cryst. Solids,* **8—10**, 223.
Cardona, M. and Greenaway, D. L., (1964), *Phys. Rev.,* **133**, A 1685.
Carruthers, T., (1973), to be published.
Carver, G. P. and Allgaier, R. S., (1972), *J. Non-Cryst. Solids,* **8—10**, 347.
Červinka, L. *et al.* (1970), *J. Non-Cryst. Solids,* **4**, 258.
Chan, W. S. and Jonscher, A. K., (1969), *phys. stat. sol.,* **32**, 749.
Chiarotti, G., Nannarone, S., Pastore, R. and Chiaradia, P., (1971), *Phys. Rev.,* **B4**, 3398.
Chittick, R. C., (1970), *J. Non-Cryst. Solids,* **3**, 255.
Chittick, R. C., Alexander, J. H. and Sterling, H. F., (1969), *J. Electrochem. Soc.,* **116**, 77.
Chopra, K. L., (1969), Thin Film Phenomena, McGraw Hill Book Co., New York.
Chopra, K. L. and Bahl, S. K. (1970), *Phys. Rev.,* **B1**, 2545.
Clark, A. H., (1967), *Phys. Rev.,* **154**, 750.
Cohen, M. H., (1970), *J. Non-Cryst. Solids,* **4**, 391.
Cohen, M. H., ed., (1972), Proc. IV Intl. Conf. on Liquid and Amorphous Semiconductors, Ann Arbor, Mich., *J. Non-Cryst. Solids,* **8—10**.
Cohen, M. H., Fritzsche, H. and Ovshinsky, S. R., (1969), *Phys. Rev. Letters,* **22**, 1065.
Cohen, R. W. and Abeles, B., (1968), *Phys. Rev.,* **168**, 444.
Cohen, R. W. and Gittleman, J. I., (1968), *Phys. Rev.,* **176**, 500.
Connell, G. A. N., Camphausen, D. L. and Paul, W., (1972), *Phil. Mag.,* **26**, 541.
Crevecoeur, C. and de Wit, H. J., (1971), *Solid State Comm.,* **9**, 445.
Croitoru, N. and Vescan, L., (1969), *Thin Solid Films,* **3**, 269.
Croitoru, N., Vescan, L., Popescu, C. and Lazarescu, M., (1970), *J. Non-Cryst. Solids,* **4**, 493.
Crowder, B. L., Title, R. S., Brodsky, M. H. and Pettit, G. D., (1970), *Appl. Phys. Letters,* **16**, 205.
Cutler, M., (1971), *Phil. Mag.,* **24**, 381.
Cutler, M., (1972), *Phil. Mag.,* **25**, 173.
Cutler, M. and Mott, N. F., (1969), *Phys. Rev.,* **181**, 1336.
Davis, E. A. and Mott, N. F., (1970), *Phil. Mag.,* **22**, 903.
Davis, E. A. and Shaw, R. F., (1970), *J. Non-Cryst. Solids,* **2**, 406.
Deal, B. E., Fleming, P. J. and Castro, P. L., (1968), *J. Electrochem. Soc.,* **115**, 300.
deNeufville, J., (1972), *J. Non-Cryst. Solids,* **8—10**, 85.
Devenyi, A., Beln, A. and Korony, G., (1970), *J. Non-Cryst. Solids,* **4**, 380.
de Wit, H. J., (1972), *J. Appl. Phys.,* **43**, 908.
de Wit, H. J. and Crevecoeur, C., (1972), *J. Non-Cryst. Solids,* **8—10**, 787.
DiSalvo, F. J., Bagley, B. G., Tauc, J. and Waszczak, J. V., (1973), Proc. Fifth Intl. Conf. Amorphous and Liquid Semicond., Garmisch, Germany.
DiSalvo, F. J., Menth, A., Waszczak, J. V. and Tauc, J., (1973), *Phys. Rev.,* **B6**, 4574.
Döhler, G. H. and Heyszenau, H., (1973), *Phys. Rev. Letters,* **30**, 1200.
Dolezalek, F. K. and Spear, W. E., (1970), *J. Non-Cryst. Solids,* **4**, 97.
Donovan, T. and Heinemann, K., (1971), *Phys. Rev. Letters,* **27**, 1794.
Donovan, T. M., Spicer, W. E., Bennett, J. M. and Ashley, E. J., (1970), *Phys. Rev.,* **B2**, 397.
Donovan, T. M., Ashley, E. J. and Spicer, W. E., (1970), *Phys. Lett.,* **32A**, 85.
Doremus, W., ed., (1970) 'Semiconductor Effects in Amorphous Solids', *J. Non-Cryst. Solids,* **2**, 1—575.
Dow, J. D. and Hopfield, J. J., (1972), *J. Non-Cryst. Solids,* **8—10**, 664.
Dow, J. D. and Redfield, D., (1972), *Phys. Rev.,* **B5**, 594.
Dresner, J., (1964), *J. Phys. Chem. Solids,* **25**, 505.
Economou, E. N., Kirkpatrick, S., Cohen, M. H. and Eggarter, T. P., (1970), *Phys. Rev. Letters,* **25**, 520.
Egerton, R. F., (1971), *Appl. Phys. Letters,* **19**, 203.
Eggarter, T. P. and Cohen, M. H., (1970), *Phys. Rev. Letters,* **25**, 807.

Ehrenreich, H. and Turnbull, D., (1970), *Comments on Solid State Physics*, **3**, 75.
Elliott, P. J. and Yoffe, A. D., (1973), Proc. Fifth Intl. Conf. Amorphous and Liquid Semicond., Garmisch, Germany.
Emin, D., (1971a), *Ann. Phys.*, **64**, 336.
Emin, D., (1971b), *Phys. Rev.*, **B4**, 3639.
Emin, D., Seager, C. H. and Quinn, R. K., (1972), *Phys. Rev. Letters*, **28**, 813.
Emin, D., (1973), Scottish Universities Summer School in Physics, Aberdeen, Scotland 1972, (P. Le Comber, editor), Academic Press, London and New York.
Fagen, E. A. and Fritzsche, H., (1970a), *J. Non-Cryst. Solids*, **2**, 170.
Fagen, E. A. and Fritzsche, H. (1970b), *J. Non-Cryst. Solids*, **2**, 180.
Fagen, E. A. and Fritzsche, H., (1970c), *J. Non-Cryst. Solids*, **4**, 480.
Ferrier, R. P., Prado, J. M. and Anseau, M. R., (1972), *J. Non-Cryst. Solids*, **8–10**, 798.
Fawkes, F. M. and Burgess, T. E., (1969), *Surface Science*, **13**, 184.
Fowler, R. H. and Nordheim, L. W., (1928), *Proc. Roy. Soc.*, *(London)*, **A119**, 173.
Fox, S. J. and Locklar, H. C., Jr., (1972), *J. Non-Cryst. Solids*, **8–10**, 552.
Frenkel, J., (1938), *J. Exp. Theor. Phys. (USSR)*, **8**, 1242.
Friedman, L., (1971), *J. Non-Cryst. Solids*, **6**, 329.
Friedman, L. and Holstein, T., (1963), *Ann. Phys.*, **21**, 494.
Frisch, H. L., Hammersley, J. M. and Welsh, D. J. A., (1962), *Phys. Rev.*, **126**, 949.
Fritzsche, H., (1962), *Phys. Rev.*, **125**, 1560.
Fritzsche, H., (1965), in: Physics of Solids at High Pressures (C. T. Tomizuka and R. M. Emrick, editors), p. 184. Academic Press, New York.
Fritzsche, H. (1971a), *J. Non-Cryst. Solids*, **6**, 49.
Fritzsche, H. (1971b), *Solid State Comm.*, **9**, 1813.
Fritzsche, H., (1972), 'Non-Crystalline Materials', *Bussei (Japan)*, **13**, 59.
Fritzsche, H., Kastner, M. and Leung, C., (1972), *Bull. Am. Phys. Soc.*, **17**, 321.
Fuhs, W. and Stuke, J., (1970), *Mat. Res. Bull.*, **5**, 611.
Galeener, F. L., (1971), *Phys. Rev. Letters*, **27**, 1716.
Gevers, M. and Dupre, F. K., (1946), *Trans. Faraday Soc.*, **42A**, 47.
Gill, W. D. and Kanazawa, K. K., (1972), *J. Appl. Phys.*, **43**, 529.
Gittleman, J. I., Goldstein, Y. and Bozowski, S., (1972), *Phys. Rev.*, **B5**, 3609.
Goldstein, Y. and Gittleman, J. I., (1971), *Solid State Comm.*, **9**, 1197.
Gomer, R., (1961), Field Emission and Field Ionization, Harvard University Press, Cambridge, Mass.
Goryunova, N. A., (1965), *J. Anorg. Mat., (USSR)*, **1**, 885; (1968), Ninth Intl. Conf. Phys. Semcond., p. 1198. Moscow.
Greenwood, D. A., (1958), *Proc. Phys. Soc. London*, **71**, 585.
Grigorovici, R., (1968), *Mat. Res. Bull.*, **3**, 13.
Grigorovici, R., Croitoru, N. and Devenyi, A., (1966), *Rev. Roum. Phys.*, **11**, 869.
Grigorovici, R., Croitoru, N. and Devenyi, A., (1967), *phys. stat. sol.*, **23**, 621.
Grigorovici, R., Croitoru, N., Devenyi, A. and Teleman, E., (1964), Proc. of the VII Intl. Conf. on the Phys. of Semiconductors, p. 423. Dunod, Paris.
Grigorovici, R., Croitoru, N., Marina, M. and Nastase, L., (1968), *Rev. Roum. Phys.*, **13**, 317.
Grigorovici, R. and Devenyi, A., (1968), Proc. IX Intl. Conf. on the Phys. of Semiconductors, Moscow, p. 1267. Nauka, Leningrad.
Grunwald, H. P. and Blakney, R. M., (1968), *Phys. Rev.*, **165**, 1006.
Hall, J. E., (1970), *J. Non-Cryst. Solids*, **2**, 125.
Haneman, D., (1968), *Phys. Rev.*, **170**, 705.
Haneman, D., Chung, M. F. and Taloni, A., (1968), *Phys. Rev.*, **170**, 719.
Hartke, J. L., (1962), *Phys. Rev.*, **125**, 1177.
Hartke, J. L., (1968), *J. Appl. Phys.*, **39**, 4871.
Hartman, T. E., Blair, J. C. and Bauer, R., (1966), *J. Appl. Phys.*, **37**, 2468.
Hauser, J. J., (1973), *Phys. Rev.*, **B7**, 4099.
Heleskivi, J. and Salo, T., (1972), *J. Appl. Phys.*, **43**, 740.
Helfrich, W. and Mark, P., (1962), *Z. Physik.*, **166**, 370.
Henisch, H. K., Fagen, E. A. and Ovshinsky, S. R., (1970), *J. Non-Cryst. Solids*, **4**, 538.
Henisch, H. K., Pryor, R. W. and Vendura Jr., G. J., (1972), *J. Non-Cryst. Solids*, **8–10**, 415.

Henkels, H. W. and Maczuk, H., (1953), *Phys. Rev.*, **91**, 1562.
Herring, C., (1960), *J. Appl. Phys.*, **31**, 1939.
Heywang, W. and Haberland, D. R., (1970), *Solid State Electronics*, **13**, 1077.
Hill, A. W., Phahle, A. M. and Calderwood, J. H., (1970), *Thin Solid Films*, **5**, 287.
Hill, R. M., (1967), *Thin Solid Films*, **1**, 39.
Hill, R. M., (1971), *Phil. Mag.*, **23**, 59.
Hindley, N. K., (1970a), *J. Non-Cryst. Solids*, **5**, 17.
Hindley, N. K., (1970b), *J. Non-Cryst. Solids*, **5**, 31.
Holstein, T. and Friedman, L., (1968), *Phys. Rev.*, **165**, 1019.
Hrubý, A. and Štourač, L., (1969), *Mat. Res. Bull.*, **4**, 769.
Hu, S. M., Kerr, D. R. and Gregor, C. V., (1967), *Appl. Phys. Letters.* **10**, 97.
Hudgens, S. J., (1973), *Phys. Rev.*, **B7**, 2481.
Ing., S. W., Jr. Neyhart, J. H. and Schmidlin, F., (1971), *J. Appl. Phys*, **42**, 696.
Inglis, G. B. and Williams, F., (1970), *Phys. Rev. Letters*, **25**, 1275.
Inglis, G. B. and Williams, F., (1971), *J. Non-Cryst. Solids*, **5**, 313.
Ioffe, A. F. (1951), *Bull. Acad. Sci. (USSR)*, **15**, 477.
Ioffe, A. F. and Regel, A. R., (1960), *Progr. Semiconductors*, **4**, 239.
Ivkin, E. B., Kolomiets, B. T. and Lebedev, E. A., (1964), *Bull. Acad. Sci., (USSR) Phys. Ser.*, **28**, 1190.
Jones, R. and Schaich, W., (1972), *J. Phys.*, **C5**, 43.
Jonscher, A. K., (1967), *Thin Solid Films*, **1**, 213.
Jonscher, A. K. and Ansari, A. A., (1971), *Phil. Mag.*, **23**, 205.
Jonscher, A. K., (1971a), *J. Phys.*, **C4**, 1331.
Jonscher, A. K., (1971b), *J. Vac. Science and Tech.*, **8**, 135.
Jonscher, A. K., (1972), *J. Non-Cryst. Solids*, **8–10**, 293.
Jonscher, A. K. and Ansari, A. A., (1971), *Phil. Mag.*, (in press).
Jonscher, A. K. and Hill, R. M., (1971) (to be published).
Jonscher, A. K. and Loh, C. K., (1971), *J. Phys.*, **C4**, 1341.
Juretschke, H., Landauer, R. and Swanson, J., (1956), *J. Appl. Phys.*, **27**, 838.
Juska, G., Matulionis, A. and Viscakas, J., (1969a), *phys. stat. sol.*, **33**, 533.
Juska, G., Matulionis, A., Sakalas, A. and Viscakas, J., (1969b), *phys. stat. sol.*, **36**, K 121.
Kastner, Marc, (1972), *Phys. Rev. Letters*, **28**, 355.
Kastner, Marc. and Fritzsche, H., (1970), *Mat. Res. Bull.*, **5**, 631.
Keezer, R. C. and Bailey, M. W., (1967), *Mat. Res. Bull.*, **2**, 185.
Keyes, R. W., (1960), *Solid State Physics*, **11**, 149.
Khosla, R. P. and Fischer, J. R., (1971), *Phys. Rev.*, **B2**, 4084.
Kiess, H. and Rose, A., (1973), *Phys. Rev. Letters*, **31**, 153.
Kirkpatrick, S., (1971), *Phys. Rev. Letters*, **27**, 1722.
Kitao, M., Araki, F. and Yamada, S., (1970), phys. stat. sol. 37 [2], K119.
Klinger, M. I., (1970), *J. Non-Cryst. Solids*, **5**, 78.
Knotek, M. L. and Pollak, M., (1972), *J. Non-Cryst. Solids*, **8–10**, 505.
Knotek, M. L., Pollak, M., Donovan, T. M. and Kurtzmann, H., (1973), *Phys. Rev. Letters*, **30**, 853.
Knotek, M. L., Pollak, M. and Donovan, T. M., (1973), Proc. Fifth Intl. Conf. Amorphous and Liquid Semicond., Garmisch, Germany.
Koc, S., Renner, O., Zavetova, M. and Zemek, J., (1972a), *Czech. J. Phys.*, **B22**.
Koc, S., Zavetova, M. and Zemek, J., (1972), *Thin Solid Films*, **10**, 165.
Kolomiets, B. T., (1964), *phys. stat. sol.*, **7**, 359, 713.
Kolomiets, B. T., (1968), Proc. IX Intl. Conf. on the Physics of Semiconductors, Moscow, p. 1259. Nauka, Leningrad.
Kolomiets, B. T. and Lebedev, E. A., (1967), *Sov. Phys. Semiconductors*, **1**, 244.
Kolomiets, B. T., Mamontova, T. N. and Babaev, A. A., (1970), *J. Non-Cryst. Solids*, **4**, 289.
Kolomiets, B. T. and Mazets, T. F., (1970), *J. Non-Cryst. Solids*, **3**, 46.
Kolomiets, B. T. and Nazarova, T. F., (1960), *Sov. Phys. Sol. State*, **2**, 369.
Konak, C. and Stuke, J., (1972), *phys. stat. sol.*, **(a) 9**, 333.
Kornfeld, M. I. and Sochava, L. S., (1959), *Fiz. Tverd. Tela.*, **1**, 1370.

Kubo, R., (1957), *J. Phys. Soc. Jap.,* 12, 570.
Kuper, A. B., (1969), *Surface Science,* 13, 172.
Lakatos, A. I. and Abkowitz, M., (1971), *Phys. Rev.,* 3, 1791.
Lampert, M. A., (1964), *Rept. Progr. Phys.,* 27, 329.
Lampert, M. A. and Mark, P., (1970), Current Injection in Solids, Academic Press, New York.
Lampert and Schilling, (1970), in Semiconductors and Semimetals, Vol. 6, (R. K. Willardson and A. C. Beer, editors), p. 1. Academic Press, New York.
Landauer, R., (1952), *J. Appl. Phys.,* 23, 779.
Lanyon, H. P. D., (1963), *Phys. Rev.,* 130, 134.
Last, B. J. and Thouless, D. J., (1971), *Phys. Rev. Letters,* 27, 1719.
Le Comber, P. G. and Spear, W. E., (1970), *Phys. Rev. Letters,* 25, 509.
Le Comber, P. G., Madan, A. and Spear, W. E., (1972), *J. Non-Cryst. Solids,* 11, 219.
Le Comber, P. G., Loveland, R. J., Spear, W. E. and Vaughan, R. A., (1973), Proc. Fifth Intl. Conf. Amorphous and Liquid Semicond., Garmisch, Germany.
Linsley, G. S., Owen, A. E., Hayatee, F. N., (1970), *J. Non-Cryst. Solids,* 4, 208.
Lipkis, K., Sakalas, A. and Viscakas, J., (1971), *phys. stat. sol.,* (a) 3, K217.
Mackenzie, J. D., (1964), in: Modern Aspects of the Vitreous State 3, 126, (J. D. Mackenzie, editor), Butterworths, London.
Main, C. and Owen, A. E., (1970), *phys. stat. sol.,* (a) 1, 297.
Male, J. C., (1967), *Br. J. Appl. Phys.,* 18, 1543.
Many, A., Rakavy, G., (1962), *Phys. Rev.,* 126, 1980.
Marshall, J. M. and Owen, A. E., (1971), *Phil. Mag.,* 24, 1281.
Marshall, J. M. and Owen, A. E., (1972), *phys. stat. sol,* 12, 181.
Matyás, M., (1972), *J. Non-Cryst. Solids,* 8–10, 592
Mead, C. A., (1962), *Phys. Rev.,* 128, 2088.
Mell, H. and Stuke, J., (1970), *J. Non-Cryst. Solids,* 4, 304.
Mikkor, M. and Vassell, W. C., (1970), *Phys. Rev.,* B2, 1875.
Mikoshiba, N., (1962), *Phys. Rev.,* 127, 1962; (1963), *J. Phys. Chem. Solids,* 24, 341.
Miller, A. and Abrahams, E., (1960), *Phys. Rev.,* 120, 745.
Mitchell, D. L., Bishop, S. G. and Taylor, P. C., (1970), *J. Non-Cryst. Solids,* 8–10, 231.
Mitchell, D. L., Taylor, P. C. and Bishop, S. C., (1971), *Solid State Comm.,* 9, 1833.
Mogab, C. J. and Kingery, W. D., (1968), *J. Appl. Phys.,* 39, 3640.
Mooser, E. and Pearson, W. B., (1956), *Phys. Rev.,* 101, 1608.
Mooser, E. and Pearson, W. B., (1960), in Progress in Semiconductors, Vol. 5, p. 104, Heywood and Co., Ltd., London.
Morgan, M. and Jonscher, A. K., (1971), *Thin Solid Films,* 9, 67.
Morgan, M. and Walley, P. A., (1971), *Phil. Mag.,* 23, 661.
Mort, J. and Lakatos, A. I., (1970), *J. Non-Cryst. Solids,* 4, 117.
Mort, J., Schmidlin, F. W. and Lakatos, A. I., (1971), *J. Appl. Phys.,* 42, 5761.
Moss, S. C. and Graczyk, J. F., (1969), *Phys. Rev. Letters,* 23, 1167.
Moss, S. C. and Graczyk, J. F., (1970), Proc. X Intl. Conf. on the Phys. of Semiconductors, p. 658. Cambridge, Mass.
Mott, N. F., (1967), *Adv. in Phys.,* 16, 49.
Mott, N. F., (1968), *J. Non-Cryst. Solids,* 1, 1.
Mott, N. F., (1969a), *Phil. Mag.,* 19, 835.
Mott, N. F., (1969b), *Contemp. Phys.,* 10, 125.
Mott, N. F., (1969c), *Festkorperprobleme,* 9, 22.
Mott, N. F., (1970a), *Phil. Mag.,* 22, 7.
Mott, N. F., (1970b), *J. Non-Cryst. Solids,* 4,
Mott, N. F., (1971), *Phil. Mag.,* 24, 911.
Mott, N. F., (1972), *J. Non-Cryst. Solids,* 8–10, 1.
Mott, N. F. and Allgaier, R. S., (1967), *phys. stat. sol,* 21, 343.
Mott, N. F. and Davis, E. A., (1971), Electronic Processes in Non-Crystalline Materials, Clarendon Press, Oxford.
Mott, N. F. and Gurney, R. W., (1940), Electronic Processes in Ionic Crystals, Oxford Univ. Press, London.

310   H. Fritzsche

Mott, N. F. and Jones, H., (1958),, Theory of Properties of Metals and Alloys, Ch. VII (Dover).
Mott, N. F. and Zinamon, Z., (1970), *Rep. Prog. Phys.*, 33, 881.
Müller, L. and Müller, M., (1970), *J. Non-Cryst. Solids*, 4, 504.
Myuller, R. L., (1962), *J. Appl. Chemistry of the USSR*, 35, 519.
Myuller, R. L., (1965), Electrical Properties and Structure of Glass, (O. V. Mazurin, editor), p. 64. Consultants Bureau, New York.
Nadkarni, G. S. and Simmons, J. G., (1972), *J. Appl. Phys.*, 43, 3650.
Nadkarni, G. S. and Simmons, J. G., (1973), *Phys. Rev.*, B7, 3719.
Nagels, P., Callaerts, R., Denayer, M. and De Coninck, R., (1970), *J. Non-Cryst. Solids*, 4, 295.
Nielsen, P., (1971), *Solid State Comm.*, 9, 1745.
Nielsen, P., (1972), *Phys. Rev.*, B6, 3739.
Nielsen, P., (1973), *Thin Solid Films*, 15, 309.
Nikolskaya, G. F., Berger, L. I., Shchikina, I. K. and Kovaluva, J. K., (1966), *J. Anorg. Mat. USSR*, 2, 1876.
Nwachuku, A. and Kuhn, M., (1968), *Appl. Phys. Letters*, 12, 163.
Oki, Ogawa, and Fujiki, (1969), *Jap. J. Appl. Phys.*, 8, 1056.
Osmun, J. W. and Fritzsche, H., (1970), *Appl. Phys. Letters*, 16, 87.
Owen, A. E., (1963), *Progress in Ceramic Science*, 3, 77.
Owen, A. E., (1967), *Glass Industry*, 48, 632, 695.
Owen, A. E., (1970a), *J. Non-Cryst. Solids.*, 4, 78.
Owen, A. E., (1970b), *Contemp. Phys.*, 11, 227, 157.
Owen, A. E. and Robertson, J. M., (1970), *J. Non-Cryst. Solids*, 2, 40.
Owen, A. E. and Robertson, J. M., (1973), IEEE Trans. on Electron Devices, *ED--20*, 105.
Panus, V. R., Ksendzov, Ya. M. and Borisova, Z. U., (1968), *Inorg. Mater. (USSR)*, 4, 778.
Parrott, J. E., (1971), *Solid State Comm.*, 14, 885.
Pearson, A. D., (1964), *J. Electrochem. Soc.*, 111, 753.
Peck, W. F. and Dewald, J. F., (1964), J. Electrochem. Šoc., 111, 561.
Pike, G. E., (1973), *Phys. Rev.*, B6, 1572.
Plessner, K. W., (1951), *Proc. Phys. Soc.*, 864, 671.
Polanco, J. I., Roberts, G. G. and Myers, M. B., (1972), *Phil. Mag.*, 25, 117.
Polk, D. E., (1971), *J. Non-Cryst. Solids*, 5, 365.
Pollak, M., (1962), Proc. Intl. Conf. Phys, of Semiconductors, p. 86. Exeter,
Pollak, M., (1965), *Phys. Rev.*, 138, A 1822.
Pollak, M., (1970), *Discuss. Faraday. Soc.*, 50, 13.
Pollak, M., (1971), *Phil. Mag.*, 23, 519.
Pollak, M., (1972a), *J. Non-Cryst. Solids*, 8–10, 486.
Pollak, M., (1972b), *J. Non-Cryst. Solids*, 11, 1 (1972).
Pollak, M. and Geballe, T. H., (1961), *Phys. Rev.*, 122, 1742.
Pollak, M. and Pike, G. E., (1972), *Phys. Rev. Letters*, 28, 1449.
Pollak, M., Knotek, M. L., Kurtzmann, H. and Glick, H. (1973), *Phys. Rev. Letters*, 30, 856.
Queisser, H. J., (1971), *J. Appl. Phys.*, 42, 5567.
Queisser, H. J., Casey, Jr., H. C. and van Roosbroeck, W., (1971), *Phys. Rev. Letters*, 26, 551.
Queisser, H. J., (1972), Proc. of European Solid State Device Research Conf. Sept. 15, 1972, (P. N. Robson, editor), Lancaster, England.
Rand, M. J. and Roberts, J. F., (1968), *J. Electrochem. Soc.*, 115, 423.
Rockstad, H. K., (1969), *Solid State Comm.*, 7, 1507.
Rockstad, H. K., (1971), *Solid State Comm.*, 9, 2233.
Rockstad, H. K., (1972), *J. Non-Cryst. Solids.*, 8–10, 621.
Rockstad, H. K. and Flasck, R., (1973), (to be published).
Rockstad, H. K., (1973), Proc. Fifth Intl. Conf. on Amorphous and Liquid Semicond., Garmisch, Germany.

Roilos, M., (1971), *J. Non-Cryst. Solids*, 6, 5.
Rose, A., (1955), *Phys. Rev.*, 97, 1538.
Rossiter, E. L. and Warfield, G., (1971), *J. Appl. Phys.*, 42, 2527.
Sak, J., (1972), *J. Phys.-C Solid State*, 5, 1335.
Sauvage, J. A. and Mogab, C. J., (1972), *J. Non-Cryst. Solids*, 8–10, 607.
Sauvage, J. A., Mogab, C. J. and Adler, D., (1972), (to be published).
Sayer, M., Mansingh, A., Reyes, J. M. and Rosenblatt, G., (1971), *J. Appl. Phys.*, 42, 2857.
Scharfe, M. E., (1970), *Phys. Rev.*, B2, 5025.
Scharfe, M. E. and Tabak, M., (1969), *Appl. Phys.*, 40, 3230.
Scher, H. and Zallen, R., (1970), *J. Chem. Phys.*, 53, 3759.
Schottky, W., (1914), *Physik Zf.*, 15, 872.
Schottmiller, J., Tobak, M., Lucovsky, G. and Ward, A., (1970), *J. Non-Cryst. Solids*, 4, 80.
Seager, C. H., Emin, D. and Quinn, R. K., (1972), *J. Non-Cryst. Solids*, 8–10, 341.
Seager, C. H., Knotek, M. L. and Clark, A. H., (1973), Proc. Fifth Intl. Conf. Amorphous and Liquid Semicond., Garmisch, Germany.
Servini, A. and Jonscher, A. K., (1969), *Thin Solid Films*, 3, 341.
Shante, V. K. S. and Kirkpatrick, Scott, (1971), *Adv. in Phys.*, 20, 325.
Sheng, P. and Abeles, B., (1972), *Phys. Rev. Letters*, 28, 34.
Sheng, P., Abeles, B. and Arie, Y., (1973), *Phys. Rev. Letters*, 31, 44.
Shklovski, B. I., (1971), *JETF*, 61, 2033.
Silver, M., Dy, K. S. and Huang, I. L., (1971), *Phys. Rev. Letters*, 27, 21; (1972), *J. Non-Cryst. Solids*, 8–10, 773.
Simmons, J. G., (1963), *J. Appl. Phys.*, 34, 1793.
Simmons, J. G., (1967), *Phys. Rev.*, 155, 657.
Simmons, J. G., (1970), in Handbook of Thin Film Technology, Ch. 14, (L. I. Maissel and R. Glang, editors), McGraw Hill Book Co., New York.
Simmons, J. G. and Nadkarni, G. S., (1972), *Phys. Rev.*, B6, 4815.
Simmons, J. G. and Taylor, G. W., (1972a), *Phys. Rev.*, B6, 4793.
Simmons, J. G. and Taylor, G. W., (1972b), *Phys. Rev.*, B6, 4804.
Simmons, J. G. and Taylor, G. W., (1973), (to be published).
Smith, C. W. and Clark, A. H., (1972), *Thin Solid Films*, 9, 207.
Spear, W. E., (1957), *Proc. Phys. Soc., (London)*, 870, 1139.
Spear, W. E., (1960), *Proc. Phys. Soc., (London)*, 76, 826.
Spear, W. E., (1961), *J. Phys. Chem. Solids*, 21, 110.
Spear, W. E., (1969), *J. Non-Cryst. Solids*, 1, 197.
Spear, W. E. and LeComber, P. G., (1972), *J. Non-Cryst. Solids*, 8–10, 727.
Stourac, L. A., Abraham, A., Hruby, A. and Závetová, M., (1972), *J. Non-Cryst. Solids*, 8–10, 353.
Stratton, R., (1962), *J. Phys. Chem. Solids*, 23, 1177.
Street, R. A., Davies, G. and Yoffe, A. D., (1971), *J. Non-Cryst. Solids*, 5, 276.
Street, R. A. and Yoffe, A. D., (1972), *J. Non-Cryst. Solids*, 8–10, 745.
Stuart, M., (1967), *phys. stat. sol.*, 23, 595.
Stuke, J., (1965),, Recent Advances in Selenium Physics, (European Selenium Tellurium Committee), Pergamon Press, New York.
Stuke, J., (1970a), Proc. 10th Intl. Conf. on the Physics of Semiconductors, p. 14. Cambridge, Mass.
Stuke, J., (1970b), *J. Non-Cryst. Solids*, 4, 1.
Stuke, J., (1971), Proc. of Intl. Conf. on Low Mobility Materials, p. 193. Eilat, Israel.
Sze, S. M., (1967), *J. Appl. Phys.*, 38, 2951.
Sze, S. M., (1969), Physics of Semiconductor Devices, J. Wiley, New York.
Tabak, M. D., (1970), *Phys. Rev.*, B2, 2104.
Tabak, M. D., Pai, D. M. and Scharfe, M. E., (1971), (to be published).
Tabak, M. D. and Warter, P. J., Jr., (1968), *Phys. Rev.*, 173, 899.
Tauc, J., (1970), *Mater. Res. Bull.*, 5, 721.
Tauc, J., *et al.* (1965), in Proc. Conf. Physics of Non-Crystalline Solids, Delft, 1964, (J. A. Prins, editor), p. 606. North-Holland, Amsterdam.

Tauc, J., Štourač, L., Vorlícek, V. and Závětová, M., (1968), Proc. of the IX Intl. Conf. on the Physics of Semiconductors, p. 1251. Moscow.

Tauc, J., Menth, A. and Wood, D. L., (1970), *Phys. Rev. Letters*, 25, 749.

Tauc, J. and Menth, A., (1972), *J. Non-Cryst. Solids*, 8–10, 569.

Taylor, P. C., Bishop, S. G. and Mitchell, D. L., (1971), *Phys. Rev. Letters*, 27, 414.

Theye, M.-L., (1971), *Mat. Res. Bull.*, 6, 103.

Thomas, D. G., Hopfield, J. J. and Frosch, C. J., (1965), *Phys. Rev. Letters*, 15, 857.

Tick, P. A. and Fehlner, F. P., (1972), *J. Appl. Phys.*, 43, 362.

Toyozawa, Y., (1962), *J. Phys. Soc. Japan*, 17, 986.

Tredgold, R. H., (1966), Space Charge Conduction in Solids, Elsevier Publishing Co., New York.

Tsuei, C. C., (1968), *Phys. Rev.*, 170, 775.

Uphoff, H. L. and Healy, J. H., (1961), *J. Appl. Phys.*, 32, 950.

Uphoff, H. L. and Healy, J. H., (1962), *J. Appl. Phys.*, 33, 2770.

vanBeek, L. K. H., (1967), *Progress in Dielectrics*, 7, 69.

van Roosbroeck, W., (1960), *Phys. Rev.*, 119, 636; (1961), *Phys. Rev.*, 123, 474.

van Roosbroeck, W. and Casey, Jr., H. C., (1970), Proc. 10th Intl. Conf. on Physics of Semiconductors, p. 832, U.S. AEC Division of Technical Information, Springfield, Va.

van Roosbroeck, W. and Casey, Jr., H. C., (1972), *Phys. Rev.*, B5, 2154.

van Roosbroeck, W., (1973), *J. Non-Cryst. Solids*, (to be published).

Vaško, A., (1968), *Mat. Res. Bull.*, 3, 209.

Vengel, T. N. and Kolomiets, B. T., (1957), *Sov. Phys. Tech. Phys.*, 2, 2314.

Viscakas, J. K., Montrimas, E. A. and Payera, A. A., (1969), *Appl. Opt. Suppl.*, 3, 79.

Volger, J., (1950), *Phys. Rev.*, 79, 1023.

Volger, J., (1960), *Progress in Semiconductors*, 4, 205.

von Hippel, A. R., ed. (1954), Dielectric Materials and Applications, J. Wiley, New York.

Walley, P. A., (1968), *Thin Solid Films*, 2, 327.

Walley, P. A. and Jonscher, A. K., (1967/68), *Thin Solid Films*, 1, 367.

Walsh, P. J., Vogel, R. and Evans, E. J., (1969), *Phys. Rev.*, 178, 1274.

Ward, A. T. and Myers, M. B., (1969), *J. Phys. Chem.*, 73, 1374.

Weinreich, O. A. and Ribner, A., (1968), *J. Electrochem. Soc.*, 115, 1090.

Weiser, K., (1972), *J. Non-Cryst. Solids*, 8–10, 922.

Welker, H., (1952), *Z. Naturf.*, 7a, 744.

Wey, H. Y. and Fritzsche, H., (1972a), *J. Non-Cryst. Solids*, 8–10, 336; (1972b), (to be published).

Williams, F., (1972), *J. Non-Cryst. Solids*, 8–10, 516.

Yeargan, J. R. and Taylor, H. L., (1968), *J. Electrochem. Soc.*, 115, 273.

Young, L., (1961),, Anodic Oxide Films, Academic Press, New York.

Zallen, R., Drews, R. E., Emerald, R. L. and Slade, M. L., (1971), *Phys. Rev. Letters*, 26, 1564.

Zallen, R. and Scher, H., (1971), *Phys. Rev.*, B4, 4471.

Zeiler, R. C. and Pohl, R. O., (1971), *Phys. Rev.*, B4, 2029.

Zertsalova, I. N., Fainberg, E. A. and Grechanik, L. A., (1965), (O. V. Mazurin, editor), p. 74. Consultants Bureau, New York.

Zitter, R. N., (1958), *Phys. Rev.*, 112, 852.

Chapter 6

Switching and Memory in Amorphous Semiconductors*

H. Fritzsche

Department of Physics and James Franck Institute, The University of Chicago, Chicago, Illinois 60637

6.1 INTRODUCTION

It usually happens that fundamental research in the sciences of materials follows rather than precedes technological application. Thus the commercial success of xerography (Dessauer and Clark, 1965) retroactively stimulated extensive basic investigation of the optical and transport properties of

*Supported by the Air Force Office of Scientific Research (AFSC), United States Air Force, under contract F44620–71–C–0025, and aided through general support of materials science by the Advanced Research Projects Agency.

selenium and selenide alloys. Similarly, the discovery of nondestructive switching in thin films of certain chalcogenide alloys (Ovshinsky (1959, 1963, 1967, 1968)) has recently provoked an explosive increase of interest in the physics of noncrystalline semiconductors generally. In particular, the utilization of reversible structural transformations in these alloys for binary information storage (Ovshinsky (1966, 1969) Evans, Helbers and Ovshinsky (1970)) has motivated intensive study of the chemistry, morphology, and kinetics of phase separation and vitrification. Moreover, the use of similar alloys for photographic (Ovshinsky (1969) Ovshinsky and Klose (1972) Berkes, Ing, Jr. and Hillegas (1971)) and holographic (Pearson and Bagley (1971) Keneman (1971)) imaging has to a large degree revived photo-crystallization, photonucleation, and photopolymerization as fields of active research.

The continuous dynode electron multiplier (channeltron) is another device (Adams and Manley (1965)) which takes advantage of unique properties of amorphous semiconductors and considerable applied research is devoted to the utilization of vitreous materials for optical mass memories (Ovshinsky (1969) Feinleib *et al.* (1971)) high energy particle detectors (Srinivasan *et al.* (1971)) ultrasonic delay lines (Krause *et al.* (1970)) and magnetic bubble domain memories (Chaudhari *et al.* (1973)).

In this chapter we shall concentrate primarily on electrical switching and memory effects, which, because of their relatively early discovery, have been most influential in stimulating contemporary research and have attained a high degree of reproducibility in practice. Nevertheless, the physics underlying the operation of these devices is still the subject of vigorous research. Our principal aims will be to review critically the various mechanisms of device operation proposed in the literature, and to describe the decisive experiments which elucidate them. Section 6.2 reviews briefly a variety of switching and memory effects which are observed in many different materials. The remainder of the chapter discusses exclusively switching and memory devices made of cholcogenide alloy glasses because they are in the most advanced state of development.

6.2 SWITCHING AND MEMORY EFFECTS, GENERAL REMARKS

Most materials cannot withstand electric fields in excess of about 10^6 V/cm. In insulators, such high fields lead to a destructive breakdown. In a number of materials, crystalline and amorphous, switching and negative resistance effects have drawn particular attention because these phenomena can be non-destructive. A variety of phenomena are commonly referred to in different materials under the loosely defined heading of 'switching' and 'memory'. For the purpose of this review it is important to distinguish the five cases sketched in Figure 6.1.

Case (a). The *negative resistance* device has a I–V characteristic which is retraceable except for some hysteresis which is observed when the current is changed too rapidly for maintaining thermal equilibrium. An example of this is the well known thermistor. By using a small load resistor R_L this negative resistance device can be made to switch from a point (A) where the

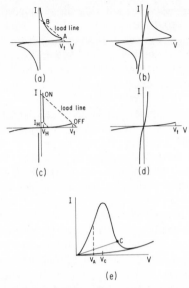

Fig. 6.1. Classification of switching and memory characteristics. (a) Negative resistance device. (b) Negative resistance device with memory. (c) Switching device. (d) Switching device with memory. (e) Voltage controlled negative resistance device with memory.

load line is tangential to the I–V curve to the point of intersection (B) of the load line and the characteristic.

Case (b). The *negative resistance* device *with memory* has two stable states. The first state resembles that of case (a). The second is conductive. It is established at higher currents and remains without decay. The first state can be reset by increasing the current above a certain value and switching it off rapidly. These effects were observed in oxides and chalcogenides by Dewald *et al.* (1962), Pearson *et al.* (1962, 1964), and Eaton (1964). In some cases the polarity of the reset current must be opposite to that of the set-on current. Such polar effects appear to be strong whenever electro-migration dominates as will be discussed later.

Case (c). The *switching device* has no stable operating point between the high resistance OFF state and the conductive ON state to which the device switches when the voltage exceeds the threshold voltage V_t. The device switches to its original OFF state when the current is decreased below the holding current I_H. The characteristic is essentially symmetric. The threshold switch described by Ovshinsky (1967, 1968) is an example of this device.

Case (d). The *switching device with memory* also has two stable states. The high resistance state and the mode of switching resemble those of case (c). The conducting ON state, which is established after switching by means of a setting current, remains even if the voltage is removed entirely. The OFF state can be re-established by a short current pulse of either polarity.

Examples of memory switches have been described by Ovshinsky (1966, 1969) and Evans *et al.* (1970).

Case (e). The *voltage controlled negative resistance* device described by Simmons and Verderber (1967) is obtained after a forming process of wide band gap material like SiO. Starting in a low resistance state, the device passes through a negative resistance region above a voltage V_c to a state of high resistance. The I–V characteristic is retraced when the voltage is changed slowly. However, the high resistance state is retained at low voltages when the voltage is reduced rapidly. The initial low resistance state can be re-established by a voltage pulse exceeding V_A. A unique feature of these devices is their capability of existing in a large number of resistance memory states depending on the choice of point C from which the voltage is rapidly reduced. These memory states are associated with electronic trapping.

One or the other of these characteristics has been observed in a large number of different materials. Examples are amorphous germanium Feldman and Moorjani (1970)), silicon (Feldman and Moorjani (1970) Fulenwider and Herskowitz (1970)) and liquid selenium (Gobrecht and Mahdjuri (1971, 1972a, 1972b) Busch *et al.* (1970) Regel (1972)). Oxide glasses exhibit a rich variety of phenomena and both current controlled and voltage controlled negative resistances. These have been reviewed by Dearnaley *et al.* (1970) and Simmons (1970). Switching in transition metal oxide films is observed after a forming process. This work has been summarized by Chopra (1970). Other oxide switching devices based on the three-component system $CuO: P_2O_5: V_2O_5$ show promise for practical applications (Drake, Scanlan and Engel (1969) Drake and Scanlan (1970) Regan and Drake (1971)). Some of these devices can be switched 10^8 times before failure at high rates (1 MHz), others have current-carrying capabilities of 1A.

Switching or memory effects have also been observed in several crystalline and polycrystalline materials. One of the earliest studies is that of Sb_2S_3 by Gildart (1965, 1970). In some of these crystalline materials, for instance in Nb_2O_5 (Hiatt and Hickmott (1965) Herrell and Park (1972) Basavaiah and Park (1973)), an appropriate forming process and device polarity are important for the setting and reset operations. A number of the Nb_2O_5 devices were found to be stable for more than 10^6 switching cycles and a limited number for more than 10^9 cycles. Very interesting work is carried out also on thin film switches which utilize the large discontinuous resistivity change occurring in VO_2 at the first-order phase transition at $68°C$ (Morin (1959) Bongers and Enz (1966) Adams and Duchene (1972)). Switching and memory effects have recently been observed also in magnetic semiconductors, such as crystalline $CdCr_2Se_4$ (Lugschneider and Zinn (1972)) and in Si-doped Yttrium-Iron Garnet (Bullock and Epstein (1970)).

This review confines itself to the discussion of switching and memory devices made of chalcogenide alloy glasses because these devices are much further developed than the oxide devices and have been studied more intensively in recent years.

Before describing details of the switching and memory phenomena observed with chalcogenide glasses, it is necessary to ask how it comes that such a large variety of materials exhibit, sometimes after an appropriate forming process, very similar I–V characteristics or memory behavior. This fact is surprising in view of the very different electronic properties of these materials and the different physical mechanisms leading to the memory effects. The reason for this is that there are only a few options open when breakdown occurs in an insulating material. During breakdown a transverse spatial instability develops which constricts the current flow to a filament of high current density (Ridley (1963)). The material in the filament region can remain intact only when the current can pass without excessive heat generation. This is possible when some kind of positive feedback mechanism increases the conductance of the filament region essentially instantaneously. If that fails to happen, the filament material burns out leaving an open circuit or a short circuit. In the open circuit case the weak spot is removed and the rest of the material is left with a higher breakdown voltage. In the short circuit case a conducting bridge connects the electrodes in the filament region. This in turn can be burned out like a fuse by a current pulse. Repetition of this process at the same or a different place in the material gives then a kind of memory action. If, on the other hand, a high conductance state is established after switching the voltage drops to its smaller sustaining value. These general features of the switching and memory characteristics do not yield any information about the physical processes involved in breakdown, in establishing a high conductance state, or in the material changes associated with memory. As a consequence one has to investigate finer details of the electronic behavior and the material structure in the respective memory states to elucidate the operative mechanism in each individual case.

In addition to the ambiguity caused by the fact that the major features of switching may fit a number of different theories one has to be aware of complications which are purely experimental. Serious problems are introduced by the forming process. This process changes the composition and structure of the semiconducting film or the nature of the contacts. If this happens it appears impossible to separate electrode effects from bulk effects or to associate details of the switching mechanism with bulk properties of the undisturbed semiconductor film. In the chalcogenide devices such a forming process is not necessary. When there are indications of a material change during normal device operation, special care must be taken in interpreting the results. In the following we shall distinguish first two ideal cases of amorphous materials: type (A), a stable glass which does not change its structure or composition significantly during device operation, and type (B) a material whose structure can be changed in a controlled and reproducible manner. Material (A) is used for the so-called threshold switch and material (B) for the memory switch. How far the materials presently in use realize these ideal cases will be discussed in a later section.

6.3 THRESHOLD SWITCH AND MEMORY SWITCH

Figures 6.1(c) and (d) show the current-voltage characteristics of the threshold switch (Ovshinsky (1967a, 1967b, 1968) Kolomiets et al. (1969a, 1969b) Shanks (1970) Haberland (1970a, b)) and of the memory switch (Ovshinsky (1967b) Evans, Helbers and Ovshinsky (1970)). In both cases a sputtered amorphous semiconductor film of approximately 1 μm thickness is placed between electrodes of about $5 \times 10^{-6} cm^2$ cross-sectional area made of carbon or of metals which do not alloy with the semiconducting material. Type (A) materials chosen for the threshold switch may for example be stable cross-linked alloy glasses such as $Te_{50}As_{30}Si_{10}Ge_{10}$, which have a high glass transition temperature T_g or which crystallize only very slowly at temperatures above T_g. Examples of a type (B) memory material are compositions like $Te_{81}Ge_{15}X_4$ where X represents one or two Group V or VI elements of the periodic table. These compositions are close to the $Te_{83}Ge_{17}$ eutectic of the Ge-Te binary. In a later section the materials will be discussed further.

Common to both types of devices is the switching process which occurs when the voltage drop exceeds a threshold voltage V_t. Once switching has commenced it proceeds very rapidly within a switching time $t_s \leqslant 10^{-10}$ sec to the conducting branch of the characteristic. Even this short time might not constitute the inherent switching speed because the shunt capacitance of a few picofarads and a lead inductance of about 10 nH make it likely that they limit the measurements of t_s. In the high resistance OFF state the resistance is typically in the range $10^5 - 10^7$ ohm at 300 K. In the conducting ON state the voltage drop is about $V_{ON} \sim 1$ volt and nearly independent of current. The principal difference between the two devices is the existence in the threshold switch of a holding current value I_h (See Figure 6.1c), typically between 0.1 and 0.5 mA, below which the ON state cannot be maintained, i.e., when the current falls below this value the threshold switch returns to its original high resistance OFF state. In contrast, the memory switch retains its conductive ON state even when the current level approaches zero or is reversed.

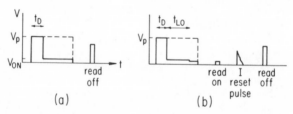

Fig. 6.2. Sketch of the time response to a voltage pulse $V_P > V_t$ and to interrogating voltage pulses $V < V_t$ for (a) a threshold switch and (b) a memory switch. Switching from OFF to ON occurs after a delay time t_D. The memory is SET after a lock-on time interval t_{LO} after switching. The threshold switch (a) returns to OFF after the end of the V_P pulse. The memory switch (b) requires a RESET current pulse to return to the high resistance OFF state.

For the purpose of defining a few more terms which help describe the device behavior the time response of the two devices to an applied voltage pulse is shown in Figures 6.2a and b, respectively. When a voltage pulse V_P exceeding the threshold voltage V_t is applied, switching to the ON state occurs after a delay time t_D. The constriction of the current flow to a current filament, common to all current controlled characteristics, Figure 1a-d, seems to occur during the brief switching interval t_s. The memory switch retains its conductive state only after it has been kept in this state for a sufficient period of time, the lock-on time t_{LO}, to 'set' the memory state. If the set pulse V_P is shorter than $t_D + t_{LO}$ then the memory switch reverts back to its high resistance state as if it were a threshold switch. However, after 'lock-on' the memory switch is in a permanent conductive state as can be verified by an interrogating read pulse of either polarity as shown in Figure 6.2(b). Evidence will be presented below which shows that in the memory devices using chalcogenide semiconductor films, the lock-on period represents the time needed to allow the material in the current channel to devitrify. A brief reset pulse causes the memory switch to return to its high resistive ('off') state. This is verified by the high voltage level of the interrogating 'read off' pulse as shown in Figure 6.2b. The reset pulse appears to melt and homogenize the devitrified material in the current channel. Subsequent rapid cooling of the melt restores the original non-crystalline state.

In the following we discuss the important device parameters to provide the experimental basis for the understanding of the underlying physical mechanisms. In order to obtain meaningful data which are reproducible in different laboratories, one must be aware of a number of factors which influence the measurements. Any forming process which leads to a material change in the high current density channel must be avoided in the threshold switch. Although acceptable for switches in practical applications, the forming process and the resulting material changes complicate the physical interpretation of the device properties. A forming process is likely to occur when (i) one of the electrodes is oxydized, (ii) the electrodes alloy with the semiconductor material, or (iii) the material in the current filament becomes overheated and tends to phase separate. By a proper choice of materials and electrodes and by limiting the current in the ON state these problems can be avoided. It is also important to keep the capacitance of the device and of the parallel circuit small to minimize the heat spike resulting from the discharge of the capacitively stored energy on switching. Whether or not switching has altered the material properties in the region of the current filament can be tested by monitoring the threshold voltage and the OFF resistance of the device after the first and the subsequent switching processes (Stocker (1969) Coward (1971)). The I–V characteristic of the OFF state at low temperatures responds sensitively to any material changes in the current filament (Buckley and Holmberg (1972)). The consequences of a material change in threshold switches will be discussed in a later section. The memory action of the memory switch depends, of course, on a structure change. This is reversed, however, by the reset current pulse.

6.3.1   Threshold Voltage V_t

By utilizing different materials and electrode separations, V_t can be adjusted to lie between 3 and 1000 Volts. V_t is proportional to electrode separation, at least for separations of a few microns and less (Ovshinsky (1968) Kolomiets, Lebedev, and Taksami (1969) Stocker (1970)). Threshold fields near 5×10^5 V/cm are usually found. A unique functional dependence cannot be given for large separations since it is influenced by the thermal and electric boundary conditions determined by the shape and material of the contacts. The threshold field decreases for larger film thicknesses as shown in Figure 6.3 (Kolomiets *et al.* (1969a)). In some cases the V_t versus thickness curve extrapolates to a finite voltage intercept at zero thickness as shown in Figure 6.4 (Buckley (1972)) for a memory material $Te_{81}Ge_{15}Sb_2S_2$. The finite intercept is observed only when V_t is measured with voltage waveforms increasing slowly with time. This allows heating of the center region to take place so that the applied field is not uniform at threshold. Faster voltage ramps lead to larger V_t values and a zero intercept of the linear relation between V_t and the electrode separation. It is therefore important to specify in each case the conditions under which V_t was measured.

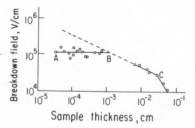

Fig. 6.3.   Average switching field as a function of electrode separation for a Ge-Si-As-Te glass switch. After Kolomiets *et al.* (1969a).

The magnitude of V_t increases with decreasing temperature as shown in Figure 6.5 (Buckley and Holmberg (1972)) for an electrode separation of 1.5 μm. Below room temperature V_t becomes nearly temperature independent. The onset of this low temperature region shifts to higher T as the electrode separation is decreased, so that devices can be made which show only a small temperature dependence near 300 K. V_t is measured by increasing the applied voltage at a certain rate. In the low temperature region V_t is essentially independent of this rate, at higher temperatures V_t is first constant and then increases with the rate.

Walsh *et. al.,* (1970) pointed out that V_t becomes zero close to the glass transition temperature T_g. As shown in Figure 6.5 this is not observed in general. By placing point contacts 0.5 mm apart on the surface of a chalcogenide glass Iizima *et al.* (1970) find that V_t at 300 K increases exponentially with T_g of the chalcogenide glass used. Although some relationship with T_g may result from the fact that many material properties

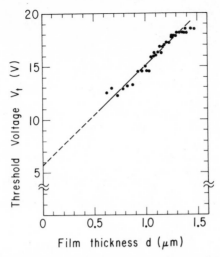

Fig. 6.4. Threshold voltage as a function of electrode separation for memory material $Te_{81}Ge_{15}Sb_2S_2$ using molybdenum electrodes. After Buckley (1972).

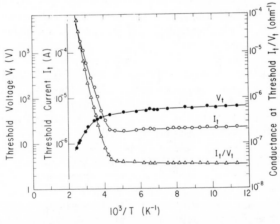

Fig. 6.5. Voltage, current, and conductance at threshold as a function of reciprocal temperature for $Te_{81}Ge_{15}Sb_2S_2$ with 1.5 μm electrode separation After Buckley and Holmberg (1972).

of vitreous semiconductors scale to some extent with T_g, it should be noted that switching has been observed in semiconducting alloys of Te with Se and S some hundred degrees above the melting point (Klose 1967; Busch et al. (1970) Regel (1972)). Furthermore, the inefficient cooling in the device structure used by Iizima et al. causes these to behave very differently from devices whose electrode separation is less than 1 μm.

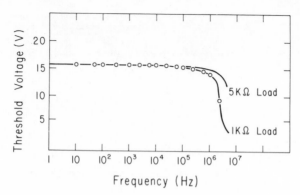

Fig. 6.6. Frequency dependence of a threshold switch with carbon electrodes separated by 1 μm. After Sieja (1966).

The frequency dependence of V_t for sinusoidal excitation was studied by Haberland, Karmann and Repp (1970). An example is shown in Figure 6.6 (Sieja (1966)). At high frequencies V_t drops because of the finite recovery time t_R discussed in a later section. The low frequency and d. c. region is at the present time accessible only for switches which have carbon electrodes and whose amorphous semiconductor films are free of loose ions which would be able to electromigrate. The failure to assure this made earlier switches d. c. unstable.

When the magnitude of the applied voltage is close to V_t one observes a statistical variation of several percent of V_t. This variation essentially disappears by applying a 20 percent overvoltage as shown in Figure 6.7.

The I–V characteristics of threshold and memory switches are essentially symmetric with respect to the polarity of the applied voltage even when electrodes of different materials are used. Strong asymmetries in the switching characteristics have recently been observed (Vendura and Henisch (1970) Henisch and Vendura (1971) Henisch, Pryor and Vendura (1972)) when n-type or p-type crystalline Ge was chosen as electrode material. These asymmetries cannot be explained by merely assuming that a rectifier is placed in series with the switch. The observations indicate that the injection properties of electrodes are essentially the same for different metals but different for n-type and p-type semiconductors.

Furthermore, Henisch et al. (1972) showed that it is possible to initiate switching by a light pulse directed at the crystalline semiconductor electrode when the threshold switch is suitably biased.

6.3.2 Delay Time t_D

As shown in Figure 6.2. the delay time t_D measures the interval between the application of a voltage pulse V_P and the switching process. The delay time t_D decreases rapidly as V_P is increased beyond V_t, and for $V_P > 1.2 V_t$ it follows the relation (Ovshinsky (1968) Henisch (1969) Shanks (1970) Lee

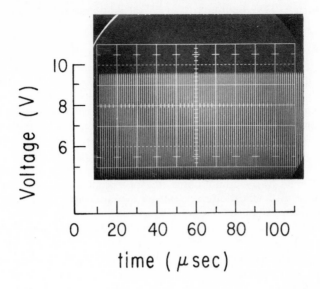

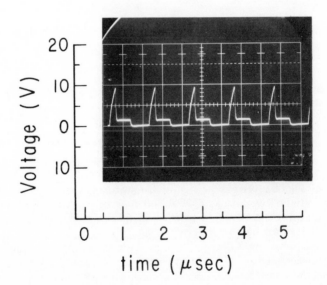

Fig. 6.7. Oscilloscope pictures of voltage drop across threshold switch operated at 1 M Hz and 20 percent overvoltage. Rectifier in series with device eliminates negative voltage signal (courtesy of Energy Conversion Devices, Inc.).

and Henisch (1973) Bunton *et al.* (1971); Bunton and Quilliam (1973))

$$t_D = t_{DO} \exp [- \alpha V_P] \tag{6.1}$$

The prefactor t_{DO} decreases with increasing ambient temperature (Böer, Döhler and Ovshinsky (1970)) and decreasing electrode separation. For T = 300 K and d = 1 μm, t_D decreases typically from 10^{-6} sec to 10^{-9} sec as $(V_p - V_t)/V_t$ is raised from 0.2 to 1.0. Lee, Henisch and Burgess (1972) studied the statistical spread of t_D. They found relatively large statistical fluctuations of t_D for $(V_p - V_t)/V_t < 0.2$ and at larger overvoltages a rather abrupt transition to a regime in which the fluctuations are less than 10^{-3} t_D. A typical example is shown in Figure 6.8 where the length of the bars indicate the statistical spread. For a constant V_p close to V_t fluctuations of t_D between 3–30 μ sec were observed. These fluctuations in t_D are of no consequence in practical circuits because digital devices are operated with sufficient overvoltages to increase their speed, they are, however of interest for elucidating the switching mechanism as discussed below. For small overvoltages the current remains constant. For larger voltages it increases noticeably during t_D (Lee and Henisch (1972) Shaw *et al.* (1972)).

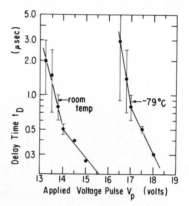

Fig. 6.8.   Delay time as a function of applied voltage for two temperatures showing the statistical spread of t_D increase with decreasing overvoltage. After Lee, Henisch, and Burgess (1972).

The effect of reversing the sign of the switching pulse V_p on the delay time was studied by Shanks (1970) and Balberg (1970) in the belief that this is a sensitive test to explore whether charge carrier injection plays a major role in the switching process and in determining t_D. They found t_D to remain the same even though V_p was reversed during the delay period. Polarity effects due to space charges can only be observed, however, when the dielectric relaxation time is sufficiently long compared to the time scale of the experiment. This was not the case in the experiments of Shanks and Balberg. These experiments were therefore repeated by Henisch and Pryor (1971) at low temperatures. Care was taken to reduce the internal heating as

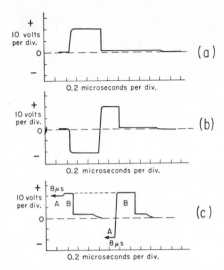

Fig. 6.9. Effect of pre-switching pulse reversal and of a non-switching A pulse on the delay time of a switching pulse: (a) no pre-switching reversal: $t_D \sim 0.6\ \mu$ sec; (b) lengthening of delay time due to pulse reversal: $t_D \sim 0.9\ \mu$ sec; and (c) asymmetric shortening of a switching B pulse by a non-switching A pulse for two relative polarities of A and B. (After Henisch and Pryor (1971)).

much as possible by using low pulse repetition rates and small overvoltages which were sufficiently large, however, to avoid the statistical fluctuation range of t_D. Figure 6.9 shows their results obtained with a threshold switch at $-70°C$. In Figure 6.9(a) one observes $t_D \sim 0.6\ \mu$ sec. This value is independent of polarity. In Figure 6.9(b) the pulse is reversed before it has time to cause switching. The reversal is found to lengthen the total delay to about 0.9 μ sec. The longer the reversal is postponed, the larger is the lengthening of t_D.

Although the total delay is lengthened by the reversal to about 0.9 μ sec, the delay of the positive pulse is somewhat shortened (to about 0.4 μ sec) by the presence of the negative pulse. It appears, therefore, that the conditions established by one polarity are not totally opposed to those established by the other polarity. If that were the case the negative pulse condition would have to be undone before the positive pulse condition could cause switching; this would have resulted in a lengthening of the positive pulse delay in Figure 6.9(b) beyond the original 0.6 μ sec observed in the absence of the negative pulse. In Figure 6.9(c) one observes that a nonswitching pulse A shortens the delay of a switching pulse B more when A and B have the same polarity than when they have opposite polarity. A shortening, however, occurs in both cases which indicates again that only part of the effect of the A pulse on the delay time is polarity dependent.

Lee and Henisch (1972) point out that isothermal conditions do not prevail at larger overvoltages $V_p - V_t$ and higher pulse repetition rates. One would expect the internal device temperature to increase under these conditions. This should result in shorter delay times. The opposite is observed. Haberland (1970), Csillag (1973), Lee and Henisch (1973) report instead an increase of t_D with increasing pulse repetition frequency or if the pulse in question is preceded by a higher switching pulse. These observations are important for distinguishing between thermal and electronic processes leading to V_t.

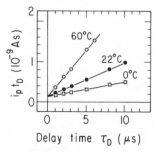

Fig. 6.10. Total charge flow during delay period t_D for various temperatures. Threshold switch made with 1 μm thick Te-As-Ge glass and Ta electrodes. After Haberland and Stiegler (1972).

Haberland (1970) explored the possibility that the parameter governing the onset of switching is the charge flowing to the device during the delay time. Figure 6.10 shows the accumulated charge as a function of t_D for a threshold switch. The delay time was changed by varying the pulse height V_p. The charge accumulated is not constant but increases with t_D and ambient temperature. This is to be expected because there must be a discharge process present which causes the system to return to equilibrium as the charges accumulate. The $t_D = 0$ intercept represents then the minimum charge needed for switching. More information on the effective lifetime of nonequilibrium distributions of charge carriers could provide an independent test of Haberland's interpretation. Lee and Henisch (1972) believe that this lifetime can be obtained from studies of the transient behavior of the ON state discussed below.

6.3.3   ON-State of Threshold Switch

As shown in Figure 6.1(c) the conducting branch of the threshold switch is nearly vertical. The precise I–V relationship of the ON-state is obtained by subtracting the spreading resistance in the electrodes adjacent to the current filament and, in thin film devices, by subtracting the additional strip resistance of the thin metal films. Lee (1972) reports that he obtained $I \propto V^2$ for the ON-state characteristic.

The value of the holding voltage V_h for 1 μm thick chalcogenide films is typically $V_h \sim 1$ volt with electrodes of Mo, W, Ta, and nichrome and $V_h \sim 1.5$ volt for carbon electrodes. A small magnitude of V_h is desirable because it governs for a certain current value the Joule heating generated.

The small voltage drop in the conducting state and the large resistance ratio between the OFF and ON state make these materials particularly suitable for switching and memory applications.

The reason for the near identity of the ON voltage in threshold and memory switches is not yet understood as there is no coherent quantitative description of the conductive state at present. One observes that V_h is practically independent of (i) temperature, (ii) electrode separation for $0.4 < d < 5$ μm, and (iii) electrode area. We conclude from the second point that the potential drop V_h occurs predominantly near one or both electrodes. The third point tells us that the current flow is not uniform over the contact area but instead is restricted to a region whose diameter is independent of the area as predicted by Ridley (1963) for characteristics of the type shown in Figure 6.1 (a-d). Electron microscopic evidence presented below shows that the conducting region in memory switches has a diameter of about 2 μm.

It was pointed out earlier that these devices are of practical interest not because they exhibit breakdown but because of the self-sustaining conducting state which protects them from destructive burnout. The sustaining mechanism of the ON state of the threshold switch is therefore of particular interest. Henisch and his group have started a series of experiments aimed at elucidating the nature of the ON state. The first group of experiments deals with the transient response of the ON state, the second with the effect of electrodes having different injecting properties.

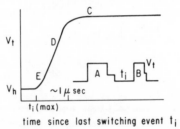

time since last switching event t_i

Fig. 6.11. Typical recovery curve of threshold voltage after switching process. The ON state can be re-established after an interruption lasting less than t_i (max) without requiring a new switching process. After Henisch and Pryor (1971).

If the ON state is completely interrupted for a short time t_i it can be re-established without a new switching process. If the interruption is slightly longer than the maximum value t_i (max), a new switching process takes place upon reimposition of the voltage but with a lower than normal threshold voltage. For sufficiently long interruptions, V_t approaches its normal value as illustrated in Figure 6.11. Henisch and Pryor (1971) find that the recovery curve Figure 6.11 remains essentially unchanged, except for the magnitude of t_i (max), as the ON current of the first pulse is increased by a factor of 10. This indicates that processes other than cooling influence the equilibration rate. The magnitude of t_i (max) is of the order of

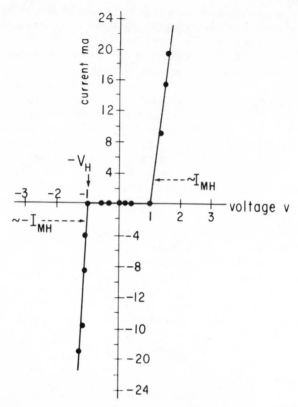

Fig. 6.12. Current-voltage characteristic during interruption period $t_i < t_i$ (max) of the ON state is called transient ON state. After Pryor and Henisch (1972a).

0.3 μ sec and decreases to zero as the current in the ON state approaches the minimum holding current I_h. By keeping $t_i < t_i$ (max) so as to avoid renewed switching and applying a voltage V_i of adjustable magnitude and polarity during the interval t_i one can trace out a transient ON characteristic as shown in Figure 6.12 (Pryor and Henisch (1972a, 1972b)). This transient characteristic coincides with the stable one above I_h. For $-V_h < V < V_h$ the transient current is exceedingly small. In this so-called blocked ON state the current does not equal the current in the OFF state. The transient characteristic in the third quadrant appears to be somewhat steeper than that in the first quadrant. The sharpness of the kinks in the transient ON characteristic and its asymmetry depend more noticeably on the contact material than the gross features of the stable I–V characteristic (Pryor and Henisch (1972a, 1972b)).

Henisch *et al.* observe that the transient ON characteristic is independent of temperature between 200–300 K whereas V_t and I_h are temperature sensitive. It has been found impossible to freeze in the ON state at temperatures as low as 4.2 K (Adler (1972)). These results are important elements in the later discussion of the ON state in terms of an electronic disequilibrium in the bulk of the film and space charge regions adjacent to the contacts.

6.3.4   Set and Reset of Memory

The electrical response of a typical memory switch is shown in Figures 6.1(d) and 6.2(b). The switching process and filamentary conduction of a memory switch is essentially the same as that of the threshold switch. In fact, when the voltage is removed immediately after switching, the memory switch reverts to the OFF state. On the other hand, the memory switch retains its conductive state after it has been kept in this state (5–10 mA current level) for about 10^{-3} sec, the lock-on time t_{LO}, to SET the memory state. Once in the ON-state, the device is turned off by the application of a short, intense RESET current pulse, typically of 5 μ sec duration and 120 mA current. The impedance of the device can be ascertained by a low level (0.5 V, 5 μ sec) read pulse as sketched in Figure 2(b). A typical SET – RESET sequence with interrogating READ pulses is shown in Figure 6.13 (Bunton and Quilliam (1972)). 18 traces are superimposed in this

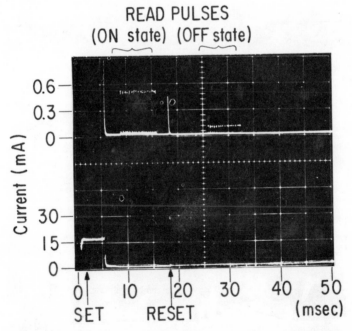

Fig. 6.13. Memory device characteristics during 18 successive SET-RESET cycles (lower trace). The values of the high and low impedance states, 10^3 ohms and 20 ohms, respectively, are shown in the upper trace. The READ pulses are triggered at different times to show the stability of the impedance values under repetitive cycling. Memory material is $Ge_{10}As_{40}Te_{50}$. After Bunton and Quilliam (1973).

figure, but the position of the READ pulse has been varied to demonstrate the stability of the two impedance levels under repetitive cycling.

The memory action in the chalcogenide glass memory devices is based on the reversible crystallization and revitrification of the material in the current filament of the switched-on state. The size and duration of the SET pulse governs the diameter of the SET filament and thus the impedance of the ON state. In the example of Figure 6.13 the OFF and ON impedance states were 10^3 ohms and 20 ohms, respectively.

More device characteristics will be presented in the following section which summarizes the present device development. A later section describes details of the physical changes in the current filament which are responsible for the memory action.

6.3.5 Device Structures

Of great practical interest is the possibility of miniaturizing these devices and of integrating their thin film deposition and photolithographic processing technique with that of conventional silicon diodes. As an example, Figure 6.14 shows a part of a 256-bit memory matrix on a 3 X 3 mm Si chip (Neale, Nelson and Moore (1970) Neale and Aseltine (1973)).

Fig. 6.14. The RM-256 diode isolated memory array. The chip size is 0.3 X 0.3 cm. After Neale, Nelson, and Moore (1970).

In this 16 X 16 array each memory switch is in series with and thus isolated from the next by an integrated silicon p-n junction diode. The memory switches measuring about 30 μ m in diameter consist of a film of amorphous memory material between two molybdenum electrodes. The x-address lines are seen to run over the round memory switches. The y-address lines are buried in the Si wafer in the form of an n^+ silicon layer.

These memory arrays are designated 'Electrically Alterable Read Mostly Memory' (RMM) because the relatively long time of 1 ms presently required for setting the memory into its low resistance state suggests applications which demand only intermittent reset of the information content. The read speed is very fast, in contrast, because each memory cell behaves like a nonvolatile bistable resistor with an OFF – ON resistance ratio in the range $10^3 - 10^6$. The access time is about 50 n sec per bit. This memory is available commercially from Energy Conversion Devices, Inc., in a 256 bit array (RM-256) and as a 2048 bit hybrid package (HRM-2048) which consists of eight 256-bit arrays on a chip with internal 'x'-line decoding.

Examples for applications of such reprogrammable, nonvolatile memories are (i) telephone dial codes, (ii) machine instructions, (iii) microprogramming, (iv) automatic polling, (v) encoded security pass keys.

The major advantages of the RMM are the following.

(a) No power is required to maintain the information.
(b) The information does not degrade during storage.
(c) The stored information can be interrogated without perturbing the memory state.
(d) The RMM is directly compatible with the DTL and TTL logic levels.
(e) Individual bits are electrically alterable.
(f) The RMM is operable throughout the 0–70°C temperature range.
(g) The RMM is resistant to high energy particle radiation.

Two threshold switch structures are shown in Figure 6.15(a) and (b). The substrate is a polished silicon wafer with a thin insulating SiO_2 layer. The 'vertical' configuration is used to obtain $V_t < 50$ volts since the electrode separation can be made small. The gap structure yields devices with high threshold voltages because the resolution of the etching process limits the electrode separation to values greater than 5 microns.

Figure 6.16 shows the layout of a four terminal threshold switch (Van Landingham (1973)). A 0.15 micron thick carbon film is placed between the silicon substrate and an ordinary threshold gap device. The carbon film is isolated from the switch and the substrate by a one micron thick film of SiO_2. The isolation resistance is greater than 10^9 ohms. The input-output coupling capacitance is of the order of 0.01 pF. The carbon film is used as a heater to control V_t of the threshold switch. When the threshold switch is biased with an a.c. voltage below threshold, the electrical pulse to the heater triggers switching. The present design permits one to switch the device with 10 μs pulses of digital circuit voltage level applied to the control resistor. The device then remains in the switching mode until interrupted for a period of order 5 μ sec because the threshold voltage under load is less than that under low duty cycle operation.

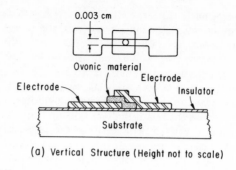

(a) Vertical Structure (Height not to scale)

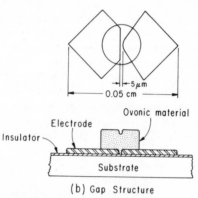

(b) Gap Structure

Fig. 6.15. Electrode configurations of threshold switch. (a) vertical structure. (b) gap structure.

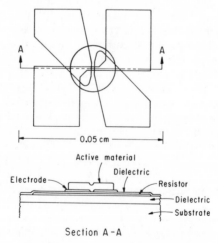

Section A-A

Fig. 6.16. Four terminal threshold switch.

Circuit applications have been described by Perschy (1967), Fleming (1970), Shanks *et al.* (1970a, 1970b), Kobylarz (1970), Nelson (1970), and Van Landingham (1973).

6.3.6 Device Performance

The stability, life, and reproducibility of typical threshold and memory switches have recently been reviewed by Ovshinsky and Fritzsche (1973). The statistical fluctuation of V_t is about a few percent of V_t for threshold switches and ± 5% for memory switches. Individual memory switches were subjected to more than 10^6 set-reset pulse trains as those shown in Figure 6.4(b) without failure. A whole 16 × 16 memory array of Figure 6.13 has presently an average failure free life of 2×10^7 set-reset cycles. Memory devices were found to remain in the set or in the reset condition without change while being subjected to 10^{12} read cycles over the period of one year (Neale and Aseltine (1973)).

Threshold switches have been kept continuously switching for longer than four years using 60 Hz a.c. voltage, i.e. for more than 10^{10} operations (Ovshinksy (1972) Bunton and Quilliam (1973)). For successful operation of threshold switches under d.c. conditions it is important to use carbon electrodes, to select a nearly ideally crosslinked network glass, and to eliminate gases trapped in the amorphous layer during the sputtering process. Poor reliability and life result when alloying contacts are used such as Au, Ag, Cu, or other low melting eutectic forming elements are used (Bunton *et al.* (1971)). The electrode area which determines the self-capacitance of the device has to be kept low (less than a few picofarad) because the energy stored in this capacitance is released at switching and shortens the operating life (Neale (1970)).

6.3.7 Radiation Resistance of Amorphous Semiconductor Devices

Amorphous semiconductor devices offer distinct potential for radiation hardened circuits and applications (Bobrova and Lobenov (1963) Edmond *et al.* (1968) Shanks *et al.* (1970a)). Threshold and memory devices, combined with passive devices, can perform the digital logic functions and other electrical circuit functions required by computer circuitry (Shanks *et al.* (1970b)).

Memory devices were tested in the set and reset state (Smith *et al.* (1972)). As a result of neutron exposure of 10^{16} n/cm² (1 MeV equivalent) typical changes of the low 'set' resistance were less than 50 percent and of the high 'reset' resistance about 25 percent. These changes are insignificant in view of the fact that the ratio of reset to set resistance is larger than 10^3. Ionizing radiation in excess of 10^6 rads at rates of about 10^{11} rads/sec did not cause any significant effects.

Several circuits containing standard threshold switches were chosen to test the feasibility of operating amorphous semiconductor devices in a radiation environment (Ovshinsky *et al.* (1968) Shanks *et al.* (1970a)). Among these circuits were a relaxation oscillator, an astable multivibrator, an AND/OR gate with complementary outputs and a J-K flip-flop. All circuits exhibited a transient ionizing radiation tolerance of at least 10^{11} rads/sec (referred to silicon) and a neutron radiation tolerance of at least 10^{16} n/cm^2 (1 MeV equivalent) without circuit malfunction.

6.4 SWITCHING MECHANISMS

The switching process observed in amorphous semiconductors is characterized not only by the breakdown of the high resistance state of the material but very importantly by the presence of a positive feedback mechanism which provides the high conductance ON state so that the breakdown is nondestructive, and repetitive switching is possible. The strong influence of the choice of contacts and of the amorphous semiconductor material on the success of obtaining stable and reproducible devices, which have a very long life, indicates a number of failure mechanisms which must be avoided before the switching characteristic of the device can be discussed in terms of the physical properties of the semiconductor film and of the contacts. Such failure mechanisms involve in particular alloying of the semiconductor with the electrode material, large scale electromigration, and phase separation or partial crystallization.

In many cases failure mechanisms lead to a change of the device characteristics rather than an outright destruction of the device. It is very important to ascertain that the data to be explained is obtained on devices whose material has not undergone a change as the result of switching.

It was hoped that much information could be gained from measurements on the current carrying filament in the ON state. For that purpose attempts were made to obtain the same switching process by placing two electrodes on the surface of the semiconductor material or for larger electrode separations in bulk samples. A number of interesting experiments on surface filaments will be discussed at the end of the chapter. In general, there is no assurance that the same switching process is taking place when the dimensions of the device are greatly scaled up. Because of the low heat conductivity of amorphous semiconductors overheating and material changes are likely to occur when the electrodes are more than a few microns apart and not well connected to a heat sink.

These remarks may suffice as warning against accepting experimental data on switching without explicit cross-checks of the physical condition of the amorphous material after switching.

The various models proposed to explain the switching process may be categorized into (A) homogeneous and (B) heterogeneous models. In the former the semiconducting film is assumed to remain essentially homogeneous and amorphous during switching. In the latter, a heterogeneous structure change in the region of the current filament plays a dominant role. Two avenues of approach are taken in the homogeneous model: switching is either explained as a thermal process with electronic corrections or as an electronic process with thermal corrections applied to the relevant parameters. These theories will be discussed in the following subsections. It will become apparent that different theories emphasize aspects which may be realized in different regimes of temperature, electrode separations, and material properties.

6.4.1   Heterogeneous Model

As mentioned already, the first switching event experienced by a threshold switch may alter the structure of the amorphous material in the region of the high current density filament of the ON state in a manner such that it does not return to its original amorphous state when the device is switched off. Alloying with the contact material and electromigration may in addition lead to compositional changes which will get enhanced during subsequent switching events. After such a forming process the device usually exhibits a lower threshold voltage and a lower OFF resistance than the virginal device. Coward (1971), Bosnell and Thomas (1972a), Thomas, Fray and Bosnell (1972) observed reductions in V_t by a factor between 5 and 15 and substantial decreases of the OFF resistance. At the same time large changes in the I–V characteristic of the OFF state are observed. Whereas the resistance of the virginal device decreases exponentially with increasing voltage (Walsh, Vogel and Evans (1969)), this decrease almost does not occur in formed devices. The temperature dependence of the device resistance is also greatly reduced by the forming process. This leads to resistance ratios of virginal to formed devices of many orders of magnitude at low temperatures.

All evidence suggests that the forming process creates in the filament region a new material whose properties have little relation to those of the original material. In fact, the new material usually does not behave like an homogeneous amorphous semiconductor.

Bosnell and Thomas (1972b) were able to isolate the filament region of formed devices which were made of Te rich chalcogenide alloy glasses. They observed by electron diffraction and electron microscopic studies crystalline tellurium embedded in a glassy matrix. The Te crystals in form of needles were found to be oriented with their axis along the current direction. In the virgin amorphous film no phase separation was observed. The Te needles will have the high conductivity of a degenerate semiconductor. This fact and the field concentration at the ends of the needles explain the high conductance and the low breakdown voltage of the formed region.

Bosnell and Thomas (1972b) explain switching and the ON-state of threshold and memory devices in the following manner. During the very first switching operations, oriented crystals of Te phase separate from a glassy matrix as a result of the relatively high temperature and electric field in the region of the current filament. These crystals remain *isolated* in the formed OFF state. The ON state is stabilized by the formation of a *connected* crystalline matrix of high conductivity. Excessively high temperatures are thus no longer required to obtain the high conductance state. In memory switches this connected crystalline matrix remains even when the current is removed after a lock-on period. It can be reset into the formed state by a current pulse which revitrifies a sufficient amount of material to render the remaining crystals isolated. In threshold devices the connected crystalline ON state is less stable and returns to the formed OFF state when the current falls below the minimum holding current. The similarity of the ON state in threshold and memory devices finds in this model an easy explanation.

The authors suggest general validity of their model of switching and point out that phase separation is essential for nondestructive and repetitive switching since it provides the conductive crystalline matrix and thus the positive feedback which sustains the conductive state.

Large changes of the material properties in the filament region are associated with phase separation and partial crystallization. The switching properties of the device are then determined by the morphology of this conducting phase. However, by a proper choice of contacts and the semi-conducting glass, phase separation in the filament region and the concom-mitant forming process can be avoided. The threshold voltage is then observed to remain the same for the first and the later switching events. The most stable and long lived devices are found to be those whose threshold voltage changes by less than 10 percent from the initial value (Verderber (1972) Henisch (1972)). Weirauch (1970) reported that he could not detect any crystallites before or after switching in $As_{20}Ge_{30}Se_{50}$.

A small material change in the filament region does not manifest itself in a noticeable change of the OFF resistance at room temperature if the contact area is very much larger than the filament area. A sensitive test for the occurrence of a material change is the comparison of the OFF resistance before and after switching at low temperatures where the resistance changes are larger by several orders of magnitude.

6.4.2 Thermal and Electrothermal Theories

A thermistor-type negative resistance characteristic caused by Joule self-heating of the semiconducting material is generally expected to be possible in all materials whose resistivity decreases rapidly with increasing tempera-ture. The physical process is essentially the following. Joule heating raises the temperature inside the semiconducting material. The resulting increase in conductivity allows more current to flow through the heated regions with the consequence of enhanced Joule heating and an increased concentration of the current flow in the heated region. A new stationary state is established when the heat conducted away from the current filament equals the Joule heat generated in that region. This is possible if the Joule heat generated reaches a plateau or begins to decrease at higher temperatures. This is the case in materials having a large negative temperature coefficient of resistivity. The establishment of a stationary state depends on the external circuit limiting the current and the heat flow to the environment. These necessary external constraints determine to a large extent the stationary state solution and the criteria for its stability. Ignoring at this time the important influence of the external circuit, then the time dependent I–V characteristic is obtained as a solution, with the appropriate boundary conditions, of the heat transport equation.

$$\nabla \cdot (\kappa \nabla T) + j^2/\sigma = C_v \, dT/dt, \tag{6.2}$$

the charge conservation equation

$$\nabla \cdot j = -d\rho/dt, \tag{6.3}$$

and Maxwell's equations together with the temperature dependence of the heat conductivity $\kappa(T)$ and electrical conductivity $\sigma(T)$ of the materials involved. Here ρ is the charge density and C_v the heat capacity.

A general solution of Eqs. (6.2) and (6.3) cannot be found with realistic boundary conditions. Different kinds of simplifications must be introduced which explains the rich literature on thermal breakdown dating back to the turn of the century. Comprehensive studies of this problem are discussed by Fock (1927, 1928), Lueder and Spenke (1935a, 1935b), Franz (1956) and more recently by Boer et al. (1961), Nebauer and Jahnke (1965), Popescu (1970a, 1970b), Klein (1971a, 1971b), Kroll and Cohen (1972). Adler (1971) summarizes the early work and classified the various assumptions made by the different authors. The recent work of Kroll (1973) presents the most thorough theoretical analysis of the thermal breakdown process.

The principal results of these studies are the following:

No negative resistance region and hence no switching occurs when the electrodes are perfect heat sinks and when effects which enhance the conductance at high fields are excluded, i.e. when $\sigma = \sigma(T)$ (Lueder and Spenke (1935a, 1935b), Boer 1970)). Keyes (1971) gave an elegant proof of this statement for one-dimensional heat and current flow toward the electrodes. Kaplan and Adler (1971) demonstrated its validity in the three-dimensional case. The reason for this is simply that the semi-conducting material adjacent to the electrodes remains cold and its high resistance prevents the appearance of a negative resistance region. In a number of studies (Stocker, Barlow Jr., and Weirauch (1970) Warren (1970)) the heat was assumed to flow perpendicular to the current instead of toward the electrodes. A negative resistance region is then obtained because the entire current path between the electrodes heats up in this case. This heat flow geometry however appears unrealistic for a thin film device because the thermal conductivity of chalcogenide glasses is of the order of 0.002 W/cm K which is about 1000 times smaller than the heat conductivity of metal or carbon electrodes. On the other hand, the electrodes never act as perfect heat sinks despite their large heat conductivity so that a negative resistance can in principle be obtained.

As pointed out above the lack of thickness dependence of V_{ON} indicates that in the ON state most of the voltage drop is at the electrodes. In the thermal model this voltage drop arises from the cold layers adjacent to the electrodes.

The breakdown voltage and the ON impedance are greatly reduced when the conductivity of the semiconductor is assumed to rise with increasing field strength or when space charge injection or tunneling processes are introduced to reduce the impedance of the cold layers near the electrodes. (Boer, Döhler, and Ovshinsky (1970) Böer and Ovshinsky (1970) Warren and Male (1970) Kaplan and Adler (1971)). Theories which include such electronic correction terms are called electrothermal theories in contrast to

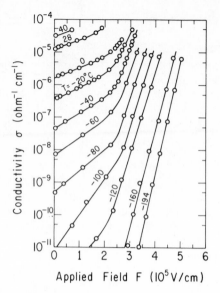

Applied Field F (10^5V/cm)

Fig. 6.17. Field dependence of conductivity of $Te_{81}Ge_{15}Sb_2S_2$ at various temperatures. Applied field was assumed to be uniform. After Buckley and Holmberg (1972).

the thermal thermistor theories. Because of the field-enhanced conductivity, the cold layer effect is diminished in electrothermal theories and a negative resistance region can be obtained even with perfectly heat sinking electrodes and at lower voltages (Rogowski (1924) Kaplan and Adler (1971)).

The conductivity of amorphous semiconductors increases exponentially with field strength (see Chapter 5). Near breakdown these field effects play a dominant role in the conduction process, particularly at low temperatures, and therefore must be included in calculating the breakdown and current sustaining process. An example of the field dependent conductance of the OFF state is shown in Figure 6.17. The high field region of σ can often be expressed (Walsh, Vogel and Evans (1969) Fagen and Fritzsche (1970)) by

$$\sigma = C \exp\left(-\Delta E/kT\right) \exp\left(F/F_o\right) \tag{6.4}$$

or by

$$\sigma = C \exp\left(-\Delta E + eFS/2\right)/kT \tag{6.5}$$

where F_o is about 2.5×10^4 V/cm or 5 percent of the field at threshold. One finds approximately S = 200 Å.

The electro-thermal model is not solved by merely finding a stationary state solution to Eqs. (6.2) to (6.5). If one solves the time independent equations with numerical methods there exists the danger that one follows one branch of solutions without being aware of the existence of other branches. Furthermore, the solution in question may not be stable. It appears that Spenke (1936a, 1936b) was the first to point out the existence of more than one branch of solutions to Eqs. (6.2) and (6.3). Kroll and Cohen (1971) and Kroll (1973) investigated the stability and uniqueness of the solutions to Eqs. (6.2) and (6.3) using Eq. (6.4) and floating thermal boundary conditions at the contacts. For a quantitative treatment of the ON state, Kroll (1973) finds it important to consider the electronic contribution to the heat conductivity, which enhances K at high temperatures, and the faster increase and subsequent saturation of σ as the temperature is raised above the glass transition temperature. If the time dependent heat flow equation, linearized around a given stationary state solution, has a non-zero solution, then the stationary state solution is considered unstable by these authors.

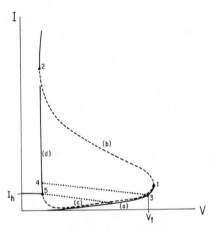

Fig. 6.18. Sketch of solutions obtained by Kroll (1973) with the electrothermal model. Full: stable solution; dashed: unstable, solution; dotted: possible load line.

Figure 6.18 illustrates the results of Kroll and Cohen (1972) and Kroll (1973). They find the stationary state solutions representing the negative resistance region between points (1) and (2) to be unstable and thus physically unattainable. Point (1) can lie on the positive resistance branch below the turnover voltage. Points (1) and (2) are points of bifurcation. At point (1) a second branch (c) of unstable stationary state solutions takes off. It lies close to and somewhat above the stable branch (a). The current is nearly the same for the two branches but in branch (a) the current is essentially uniform whereas in (c) it is concentrated in a narrow filament. Branch (c) turns upward at small voltages and joins branch (b) at bifurcation point (2). The current channel broadens as the current is increased along the nearly vertical part of the I–V curve. The temperature distribution is nearly flat inside the channel and drops exponentially outside that region. The channel temperature was found to be near values at which the curvature $d^2 \sigma/dT^2$ turns negative. This is about $400–500°C$ for $Ge_{15} Te_{81} X_4$. The solutions are stable above point (5) and unstable below it. This holding current below which the conducting branch cannot be sustained has not yet been calculated. Since the calculation of Kroll and Cohen was not designed to find all unstable solutions and bifurcation points, there may be more than those indicated. Particularly point (5) is suspected to be another point of bifurcation.

With increasing thickness of the semiconductor points (1) and (2) in Figure 6.18 move toward each other until all points of the characteristic are stable. The characteristic is then unique and the two branches coalesce.

Kroll (1973) summarizes the electrothermal interpretation of switching as follows: The proximity of the unstable branch (c) to (a) (See Figure 6.18) makes it possible that the system moves from (a) to (c) on account of a statistical fluctuation in one of the parameters T, κ, σ, or in the current. The delay time is explained as the time it takes a fluctuation to reach the size of about the electrode separation. As point (3) moves closer to (1) the required size of the fluctuation decreases until for larger overvoltages an infinitesimal perturbation suffices and the switching ceases to be statistical. This agrees with the dependence of τ_D on overvoltage (Henisch and Pryor (1971)). Once the device is on the unstable branch (c), it switches rapidly along the load line, say from point (3) to point (4) on the stable portion (d) of the branch corresponding to a hot current channel. The holding current may be larger or smaller than the current at threshold. Switching off depends on nucleation and growth of a cold region. The transient ON state and the recovery of V_t shown in Figures 6.12 and 6.11, respectively, may be interpreted in the following manner. In the ON state the field is limited to the cold regions adjacent to the electrodes. For very short interruption intervals t_i the temperature distribution essentially does not change. The I–V characteristic is then determined by Eq. (6.4), the field dependence of the conductance in the cold regions. This yields:

$$I = \text{const. } V \exp{(V/d_c F_o)} \tag{6.6}$$

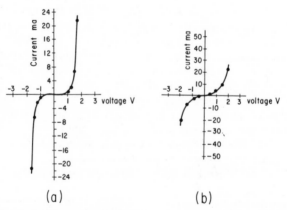

(a)   (b)

Fig. 6.19. Transient ON state as obtained from Eq. (6.6) of text. (a) $d_c = 0.1 L$ and (b) $d_c = 0.25 L$. After Kroll (1973).

where d_c is an effective thickness of the cold layers. Using $F_o = 2.2 \times 10^4$ V/cm Kroll obtains the I–V curves shown in Figures 6.19(a) and (b) for d_c = 10 percent and 25 percent of the electrode separation, respectively. Large values of d_c correspond to electrodes having smaller thermal conductivity. These figures are in good agreement with the observations of Pryor and Henisch (1972) shown in Figure 6.12.

The gradual increase in V_t in region D and E of Figure 6.11 is interpreted to be determined by the extent of diffusion of the hot core. Region C is due to cooling of the central region after the hot core has diffused away.

Not included in this discussion are the effects of the external circuit parameters and of the package capacitance and inductance on the device behavior. Upon switching and collapse of the voltage, the stored electrostatic energy is dissipated in the current channel which is being formed (Klein, Gafui and David (1965) Fritzsche and Ovshinsky (1970b) Berglund and Klein (1971)). This causes a momentary heat spike which must decay before the new stationary filamentary state is reached. Because of the heat spike the current channel actually cools immediately after switching in contrast to the behavior in the absence of the thermal transient. In the case of thin dielectric films which have high breakdown fields, this heat spike is the major cause for material decomposition in the breakdown channel. This point emphasizes the importance of reducing the self-capacitance of the device as well as the parallel circuit capacitance.

Upon switching off, thermal transients contribute to the recovery time which has to elapse before the stationary high resistance state is re-established.

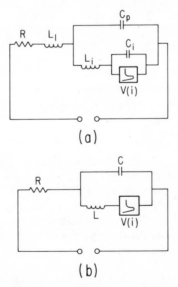

(a)

(b)

Fig. 6.20. (a) Equivalent circuit of switching device with C_i and L_i representing internal capacitance, L_l = lead inductance, and R = load.
(b) simplified circuit.

The rate processes during filament formation as well as the question of stability are not only determined by device parameters but in an important manner by the external circuit. Figure 6.20(a) shows the device represented by a given I–V characteristic V(i), an internal capacitance C_i and an internal inductance L_i in an external circuit having a load resistor R and a lead inductance L_1. A device package capacitance C_P must also be considered which includes the parallel circuit capacitance and that part of the device capacitance whose field lines do not pass through the semiconducting film. The nonlinear differential equation governing the temporal response of the conduction current will be of fourth order. Shaw and Gastman (1971, 1972) pointed out, however, that the equation can be reduced to second order because C_i dV/dt, the amplitude of the intrinsic displacement current is much less than the conduction current through the device. Moreover, the voltage drop across R is much greater than L_1 di/dt, the induced voltage drop across the lead inductance. The problem is then reduced to the simplified circuit shown in Figure 6.20(b). Here L includes also terms which represent the thermal inertia of the device because the current changes associated with temperature changes lag the changes in applied voltage (Landauer and Woo (1972)). The stability conditions of Figure 6.20(b) have been discussed in some detail by Minorsky (1947) and extensively by Shaw, Grubin, and Gastman (1973) who analyzed the time and space evolution of the current density filament in the region of negative differential conductivity. These authors note the importance of distinguishing two cases, depending on whether the I–V characteristic in the negative resistance region is assumed to be independent or dependent on the rate of change of the current. The first case is obviously simpler. One finds as condition for stability

$$R|dV/di| < L/C. \tag{6.7}$$

This means that large load resistances R do not stabilize the circuit, even though the load line intersects the negative resistance characteristic at only one point. When condition (6.7) is not satisfied one usually observed relaxation oscillations.

6.4.3   Electronic Theories

Many of the features of the I–V characteristics discussed by Kroll and Cohen (1972) and illustrated in Figure 6.18 remain unchanged when electronic breakdown precedes thermal or electrothermal breakdown, or when the conducting ON state of the I–V characteristic is dominated by electronic processes which are not contained in the prebreakdown expression for σ (T,F). As pointed out above, the gross features of the I–V characteristic are not sufficiently unique to serve as a criterion for one or the other mechanism.

Because of the large field enhancement of the conductivity before breakdown one suspects the electric processes such as carrier avalanche, double injection, or high field tunneling, can initiate a breakdown process before the thermal or electrothermal breakdown proceeds. This should happen in particular when the latter process is delayed by using thin amorphous semiconductor films and electrode configurations which facilitate cooling. In the case of pulses it is the question whether electronic breakdown processes occur with a shorter delay time than the electrothermal mechanism discussed above. In the pulse mode large overvoltages can lead to fields between 10^6 and 5×10^6 V/cm and thus to an increased probability of electronic breakdown.

Even when switching is initiated by an electronic process, a current channel will form upon switching and a considerable temperature rise will occur in this channel because the Joule heat is dissipated in a rather small volume. Hence any electronic theory must include these thermal effects and associated thermal time constants under certain circumstances. Such thermal correction terms might become as important as the electronic correction terms turned out to be in the electrothermal model discussed earlier. Still, they are considered incidental rather than crucial in the following. We exclude from this discussion thick amorphous layer devices in which thermal effects dominate.

In the ON-state of the threshold switch most of the voltage drop occurs at one or both electrodes, the rest of the semiconductor is well conducting and almost field free. This field distortion indicates the presence of space charges adjacent to the electrodes and a region free of space charge in the center. The electronic theories differ in the process which lead to the conducting and neutral center region and in the mechanisms which sustain the space charge regions.

Hindley (1970a, b) and Mott (1971) estimate the critical field beyond which the mobile carriers gain energy from the field at a rate faster than they loose it via phonon emission. Beyond this critical value avalanche occurs which leads to a conducting plasma which can be sustained by the field near the electrodes. This critical field is approximately given by (Mott (1971)) as

$$e\mu F^2 = \hbar\omega_{ph}^2 \tag{6.8}$$

where $\hbar\omega_{ph}$ is the energy of the phonon involved. Using for the phonon energy 0.030 eV and the mobility $\mu \sim 3$ cm^2/V$_{sec}$ one obtains for the critical field $F \sim 3 \times 10^5$ V/cm. This is of the same magnitude as the switching fields of thin threshold devices.

This estimate leads to a threshold field which is nearly temperature independent. This is indeed observed at lower temperatures as shown in Figure 6.5. The decrease of V_t at higher temperatures seems to be associated with some heating of the center region which causes the field near the electrodes to increase above the average value before switching occurs (Böer, Döhler and Ovshinsky (1970)). This heating effect is larger at elevated temperatures where the conductivity is higher. Mott (1971) pointed out that avalanche production of secondary carriers in amorphous semiconductors differs from that in crystals in an important way. The mobility of the carriers always increases with energy above the mobility edge so that an avalanche should be produced if once the rate at which the electron gains energy from the field exceeds the rate of energy loss to the phonons.

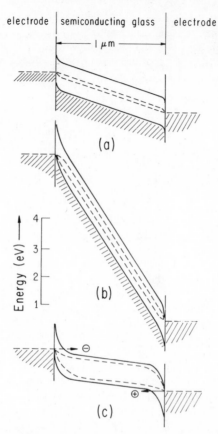

Fig. 6.21. Sketch of potential profile. (a) and (b) before switching representing applied
fields of roughly 10^4 and 6×10^4 V/cm, respectively. The switching field is
about 5×10^5 V/cm. (c) potential profile of the ON state after switching.
The dashed lines represent the quasi Fermi levels for electrons and holes
respectively. The arrows indicate injection.

   Field enhanced hopping of carriers from occupied states below to empty
states above the Fermi level E_F (Fritzsche and Ovshinsky (1970a) Mott
(1971)) causes a nonequilibrium distribution of carriers which can be described by a splitting of E_F into two quasi Fermi levels as shown in Figure
6.21(a). This increases the conductance exponentially with field strength. In
the model of Fritzsche and Ovshinsky (1970a) this excess current cannot be
replenished completely by the electrodes. This exclusion of carriers (Gibson
(1954) Low (1955)) will produce a positive space charge near the cathode and
a negative space charge at the anode. The fields build up at the electrodes,
Figure 6.21b, so that tunnel injection occurs which furnishes the excess
carriers needed to sustain the non-equilibrium, high conductance bulk state,
Figure 6.21c. Not all traps need be filled but the bulk conductance has to be
kept sufficiently high to concentrate the field near the electrodes so that the

tunnel injection is sustained. The essential idea of this switching model is not restricted to field enhanced hopping of carriers. Any process which produces a high excess concentration of carriers will lead to space charges at the electrodes which then increase until carriers can tunnel and replenish the charge carriers lost by recombination.

Mott (1969, 1971) discusses in greater detail how the space charges can be maintained in the ON state. He points out that the carriers during the tunneling process are moving much faster than the carriers in the bands. As a consequence the sign of the space charge adjacent to the cathode is positive and that near the anode negative. It is not clear whether the space charge is associated predominantly with free carriers or with traps. This is an important point in connection with the transient ON state of Pryor and Henisch (1972) shown in Figure 6.12. Here it is possible to reverse the currents and voltages from the first to the third quadrant within nanoseconds quite independently of temperature (Henisch, Pryor, and Vendura (1972)). Equilibration via traps should be slower and temperature dependent. Furthermore, the ON state could not be frozen in at helium temperatures (Adler (1972)). Lee (1972) concludes therefore that mobile carriers or at most very shallow traps produce the space charges.

Boer and Ovshinsky (1970) also suggest that the conductive ON state is sustained by electrons and holes tunneling through the high field regions near the contacts. In their model switching may be initiated by an electro-thermal mechanism, but they argue that the center filament region cannot be conducting merely because of its elevated temperature. From evidence presented below one estimates a filament cross section of about $5 \times 10^{-8} cm^2$. If only one-tenth of V_{ON} falls across the center region, its conductivity must be at least 200 ohm^{-1}cm^{-1} to pass a current of about 10 mA. To achieve this by heating alone the temperature has to be in excess of 1000 K which appears to be too high to explain reproducible switching over 10^{10} times without material destruction. The presence of a conducting and neutral plasma similar to that sketched in Figure 6.21(c) would resolve this difficulty.

Since the equilibrium Fermi level is close to the center of the mobility gap, any nonequilibrium distribution of carriers produces an increased conductivity. Fritzsche (1969) and Fagen and Fritzsche (1970) showed that such disequilibration is indeed produced at voltages shortly before break-down.

Henisch, Fagen, and Ovshinsky (1970) suggest double injection as the primary cause for switching. In their model a negative space charge builds up near the cathode and a positive space charge near the anode because of trapping of the injected carriers. The space charge regions act as virtual electrodes and increase the field in the center. Switching occurs as soon as the two space charge regions overlap leading to a region in the center in which all positive electron traps and negative hole traps are filled. If, as suggested by the CFO-model (Cohen, Fritzsche and Ovshinsky (1969)), the density of positive and negative traps are equal, this leads to a neutral and thus highly conducting region. This is an unstable situation. The high field near the center collapses and the overlap region spreads until the high fields are concentrated near the electrodes. The nonequilibrium state in the bulk is then sustained by tunneling injection as illustrated in Figure 6.21(c) and explained by Mott (1969, 1971).

In order to sustain the ON state by double injection or by tunneling it is necessary, as Lucas (1971) points out, that the carrier lifetime is longer than the transit time. Her model of switching is based on the idea that beyond a critical injection current both electron and hole traps are neutralized and as a consequence recombination is sharply decreased and the diffusion length becomes of the order of the film thickness. This sharply increases the bulk conductance and the sustaining field is again concentrated at the electrodes as in Figure 6.21(c). Lucas obtains for the critical current density at which the so-called recombination instability occurs

$$i_c = 2 \, L \, \mu e N / \tau_o \tag{6.9}$$

Here the diffusion length $L\mu$ is set equal to the electrode separation when switching occurs. N is the density of charged recombination centers and τ_o the excess carrier recombination time. Lucas estimated $\tau_o = 10^{-6}$ sec, assumed $N = 10^{19} \, cm^{-3}$, and obtained $i_c = 10 \, A/cm^2$, which is a reasonable value for the current density at the onset of switching. The estimate of τ_o is least certain. In crystalline semiconductors τ_o is usually much smaller. Heywang and Haberland (1970) point out, however, that potential fluctuations in amorphous semiconductors cause electrons and holes to be trapped at different locations in the material. This greatly decreases the probability for recombination.

To explain the switching process, Heywang and Haberland (1970) use the potential fluctuation model shown in Figure 5.41 of Chapter 5. This figure was taken from their paper with small modifications. They stress that the injected electrons fill predominantly positive traps along the electron percolation paths and injected holes fill their respective traps along the hole percolation paths. A consequence of this is a leveling and decrease of the potential fluctuations and thus an increase in mobility and a widening of the percolation paths. This in turn will facilitate double injection and thereby provide the feedback mechanism which is needed for the high conductance ON state. A minimum charge of the order of 10^{-10} C is required to initiate their feedback mechanism. In this manner they account for the observation shown in Figure 6.10. It appears important to consider the essential elements of the heterogeneous potential fluctuation model in any double injection and electronic breakdown theory.

The electronic switching models discussed so far predict that the voltage in the ON state is always somewhat larger than the band gap. This is usually observed except for a few cases reported by Altunyan and Stafeev (1970). They observed an ON state voltage of about 3 V when they used copper or brass electrodes and 20 V with 0.04 ohm cm crystalline silicon electrodes. In the former case alloying with the chalcogenide glass is probable. Crystalline semiconductor contacts usually yield a large ON-voltage which can be decreased by illuminating the semiconductor contact with electronhole pair producing light (Henisch and Vendura (1971)). The observed characteristics cannot be explained by assuming a rectifier in series with a threshold switch. This configuration offers a good opportunity to study the role played by the electrode material and by injection. It should be interesting to find out whether crystalline semiconductor contacts give rise to a proportionate increase also of the threshold voltage.

The electronic switching models discussed so far assume (i) that an excess electron and hole concentration is established by field dependent processes in the bulk or by injection from the electrodes and (ii) that these excess concentrations in the bulk can be sustained after switching by high field regions near the contacts. These excess concentrations, which cause the bulk to be highly conducting (but neutral), can be described by quasi-Fermi levels for electrons and holes, split apart from the equilibrium Fermi level. The equilibration time τ_0 for the two quasi-Fermi levels to return to the equilibrium level after removal of the voltage is assumed to be long.

The opposite point of view has been taken by van Roosbroeck (1972a). He assumes that amorphous semiconductors are characterized by recombination time τ_0 which is very much shorter than the dielectric relaxation time

$$\tau_{rel} = \epsilon \epsilon_o / \sigma \tag{6.10}$$

where ϵ = dielectric constant and $\epsilon_o = 8.85 \times 10^{-14}$ F/cm is the permittivity of free space. In that case minority carrier injection does not lead to splitting into quasi-Fermi levels but to a shift of the Fermi level and thus to a considerable space charge which extends from the injecting contact to a depth $\mu_D F_m \tau_{rel}$ where μ_D is the trap limited drift mobility of the majority carriers and F_m is the maximum field which occurs at the end of this space charge region, the so-called recombination front. This recombination front, whose width is small when $(\tau_t + \tau_s) \tau_0/\tau_s \ll \tau_{rel}$, plays an essential role in the switching process. Here $\tau_s/(\tau_t + \tau_s)$ is the fraction of time a minority carrier is free. In the ON-state the Fermi level has moved close to the conduction band of the injected carriers rendering the space charge layer highly conductive. The resistance and voltage drop is limited then to the recombination front. For switching to occur it is therefore necessary that the electrode separation is not much larger than the length of the space charge region, van Roosbroeck and Casey, Jr. (1970, 1972). van Roosbroeck (1972, 1973), Queisser et al. (1971), and Queisser (1972) discuss in detail charge transport and switching in these relaxation-type semiconductors. Döhler and Heyszenan (1973) extended the treatment of van Roosbroeck et al. by including the carrier diffusion term but neglecting the trapping of carriers by recombination centers. They found that under these conditions a depletion region of predominant drift and a space charge of majority carriers at the end of the depletion region do not occur. Such a compensating majority carrier space charge region is not stable as also pointed out by Kiess and Rose (1973).

It appears that the nature of contacts and the effects of trapping and of the space charge residing in localized states on relaxation space charge injection have to be further explored before one can decide how far the switching process in amorphous semiconductors is associated with these phenomena.

The question whether amorphous semiconductors fulfil the condition for relaxation-case semiconductors was raised (Fagen (1972)) because the photoconductive decay time is found to be much longer than the lifetime τ_o assumed by van Roosbroeck and coworkers (about 10^{-13} sec). However, Ryvkin (1964) and also van Roosbroeck pointed out that in the relaxation regime the photoconductive decay is not governed by the lifetime but by the much longer relaxation time.

6.5 MEMORY MECHANISM

In chalcogenide glass memory devices the memory action is the result of a reversible structure change between an amorphous state, which is of high resistance, and a small grain crystalline state, which is of low resistance. This structure change takes place in the region of the ON state filament. The transformation to the microcrystalline state occurs during the lock-on period after the device has switched to the ON state and the current filament has formed. The crystallization is believed to be caused by the Joule heating and the high field and excess carrier concentration in the current path (Ovshinsky and Fritzsche (1971) Fritzsche (1969) Cohen, Neale and Paskin (1972) Adler and Moss (1972)). The short reset pulse revitrifies the filament region probably by heating it briefly so that the crystallites redissolve into a homogeneous, disordered material. Rapid cooling, which follows the brief reset pulse, restores the high resistance amorphous state of the material.

Rapid reversal and reproducibility of the structure changes are found in alloy composition for which the tendency to crystallize and the ability to form a glass are properly balanced. Moreover, it is important to understand and control the selective electromagnation of the different species because of the high current densities which occur during device operation.

In general one expects the details of the structure change to be different depending on whether it is accompanied (i) by phase separation, (ii) by a significant change in bonding upon crystallization, or (iii) by crystallization without a change in local order.

Most information about the memory mechanism and the details of the structure changes occurring has been gained from a study of the chemistry (de Neufville (1972)), the crystallization behavior (Moss and de Neufville (1972)), the calorimetric response of the memory glasses (Fritzsche and Ovshinsky (1970c)), and the morphology and composition of the crystallized filaments. These will be discussed in the following subsections.

Studies have also been performed on surface filaments which form between more widely separated electrodes placed on the surface of the amorphous material. This permits easy access to the filament region for direct physical measurements. However, it is likely that entirely different phenomena dominate in such a scaled up model. For example, in the geometries used for the study of surface filaments, diffusion and thermal effects are enhanced by inefficient cooling and long thermal time constants. Nevertheless, these studies discussed below have contributed valuable insight into the dynamics of structure changes in these materials.

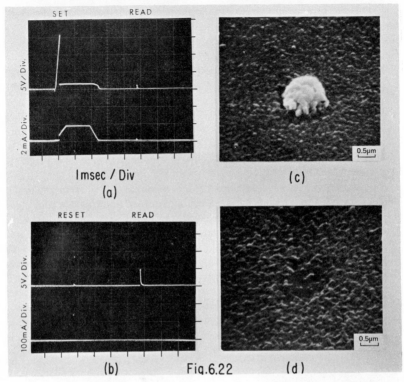

Fig. 6.22. (a) and (b) show the SET and RESET pulses applied to a memory device. (c) and (d) show photomicrographs of the etched amorphous memory material of a device left in the SET and RESET state, respectively. After Sie *et al.* (1972).

It is important to distinguish the reversible structure transformation discussed here from other memory phenomena which can be observed intermittently in a variety of contacts which carry an oxide film. In these cases the ON state is often a metal bridge in the breakdown path. The metal originates from the electrodes or by reduction from an oxide film (Whitehead (1951) Sliva *et al.* (1970)). In this case a reset pulse can restore the high resistance state by burning out the metal bridge. Reproducible memory switch action is unlikely to persist in such a case because the material deteriorates progressively.

6.5.1 Bulk Filaments

The filament region of memory devices can be made visible by chemically removing the upper electrode and selectively etching the amorphous layer. Figure 6.22 shows scanning electron microscope (SEM) pictures of the etched amorphous layers of two memory devices left in the SET and RESET states, respectively (Sie *et al.* (1972)). In the same figure, oscilloscope traces of the voltage and current waveforms during the SET (4 mA, 10 m sec) and RESET (120 mA, 5 μsec) operations are shown on the left hand side. The film thickness of the memory device was 1.5 μm.

These pictures demonstrate that controlled crystalline filament growth during the 'lock-on' process yields the high conductance memory state while revitrification of this polycrystalline region occurs during the reset process (Fritzsche *et al.* (1970c) (1970b) Ovshinsky *et al.* (1971) Cohen *et al.* (1972)). The morphology of the polycrystalline region is visible in Figure 6.22(c) because the etchant attacks the amorphous material more rapidly than the crystalline phase. The etchant was allowed to reduce the thickness of the amorphous layer by about 30% giving rise to the mottled appearance of the layer. The memory-set areas of devices after 30 set-reset operations did not vary appreciably from that of Figures 6.22(c) and (d). The polycrystalline cylinder was found to consist predominantly of dendritic Te crystallites of 300–500 Å diameter imbedded in amorphous Ge – Te and pointing circumferentially outward to the cooler zone of surrounding amorphous material (Sie *et al.* (1972)). The diameter of the filament is about 2 micron and somewhat less near the electrodes. The reset pulse effects the elimination of the crystalline conducting paths and homogenizes the Te rich and Ge rich regions. The heating produced by the reset pulse must be sufficient to ensure complete mixing. Incomplete homogenization reduces the threshold voltage for subsequent 'set' operations slightly. Moreover it stabilizes the filament location which is desirable for reproducible device operations (Cohen *et al.* (1972)). Stabilization of the filament location may, however, be achieved more effectively by proper electrode shaping.

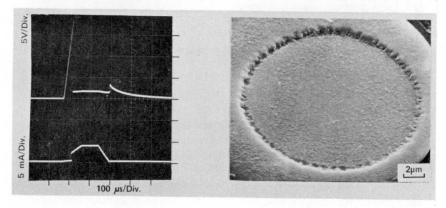

Fig. 6.23. Photomicrograph of the amorphous material in the 20 μm diameter pore of a memory device after 20 switching operations with pulses which were too short to produce 'lock-on' to the memory ON state. After Sie *et al.* (1972).

Figure 6.23 shows in contrast that a memory film remains unperturbed after being switched 20 times into the conducting state but without allowing sufficient time for lock-on to occur. Figure 6.22(a) shows the trapezoidal voltage and current waveforms applied to the memory switch. During each operation a current of 4 mA flowed in the 'on' state for 120 μsec which is insufficient for the permanent memory to form. Figure 6.22(b) shows a scanning electron micrograph (SEM) of the amorphous film after the 20 switching operations. No damage, alloying, or structural changes resulting from the switching operations are observed.

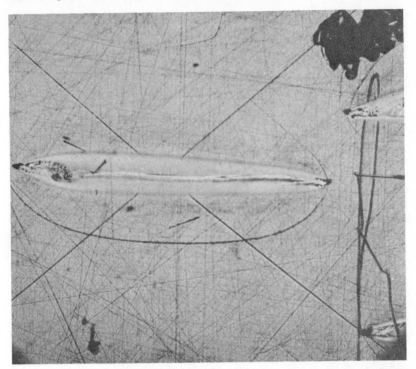

Fig. 6.24. Surface filament in glass of composition $As_{55}Te_{35}Ge_{10}$. After Uttecht *et al.*
(1970).

6.5.2 Surface Filaments

When a voltage is applied to point contacts placed on the surface of a piece
of amorphous semiconductor one observes surface melting accompanied by
switching or a negative resistance. The high resistance state is restored when
the current is interrupted right away. A crystalline filament forms and
bridges the electrodes after a certain lock-on period. An example is shown
in Figure 6.24. This filament can be molten down and quenched into a glass
by a RESET pulse.

Such surface filaments have been studied extensively because the
dynamics of filament growth can be followed cinematographically and
microscopically, temperatures can be measured radiometrically, and compositions determined with a microprobe (Uttecht *et al.* (1970) Sie *et al.* (1970)
Kikuchi *et al.* (1969, 1970) Stocker (1969, 1970) Weirauch (1970) Pearson
et al. (1969) Tanaka *et al.* (1970a, 1970b, 1970c) Sugi *et al.* (1970) Okada
et al. (1970) Iizima *et al.* (1970) Fulenwider *et al.* (1970) Armitage *et al.*
(1971) Eusner *et al.* (1972)), Matsushita *et al.* (1972).

The switching process is purely thermal because of poor heat conduction
and occurs at rather low fields of about 10^4 V/cm. The spreading resistance
of the point contacts and their poor heat sinking capability cause the
temperature to be highest near the contacts. No further insight into the
details of the thermal breakdown process has been gained from these experiments because of the ill defined thermal boundary conditions at the

contacts. Delay times of the order of seconds and lock-on times of the order of minutes are observed. Depending on the cooling conditions filament temperatures after switching between 200°C (Iizima *et al.* (1972)) and 500°C (Weirauch (1970)) above ambient have been measured. Armitage *et al.* (1971) reported a heat spike reaching 800–900°C at the time of switching. Drastic material changes, oxidation and alloying unit occurs at such high temperatures. Recently, Kolomiets *et al.* (1972) found that the intensity of the emitted radiation increases linearly with the current in the ON state. This speaks against a thermal origin of the radiation unless the current increases the diameter of the current filament without changing its temperature. Moreover, they observed in addition to the main radiation power, which is concentrated in the wavelength range 1.5–3 μ, a smaller radiation component below 1.5 μ with a different rise and fall time kinetics. More evidence is needed before one can associate one or the other component with radiative recombination.

In different materials, quite different mechanisms are operative in the formation of a crystalline filament. For instance, in $As_{55}Te_{35}Ge_{10}$ and glasses of similar composition one observes directional growth of a nearly single phase polycrystalline As_2Te_3 filament from the anode (Uttecht *et al.* (1970) Kikuchi *et al.* (1970) Okada *et al.*, (1970)). The growth velocity is about 10^{-3} cm/sec and the width of the filament increases with current such that the current density remains at about $3 \times 10^4 A/cm^2$. In the molten channel and under the action of the field, the more electronegative Te ions drift toward the anode. The higher melting point composition As_2Te_3 will crystallize at the anode. The high thermal and electrical conduction of the As_2Te_3 crystalline filament prevents growth at any place except the tip. When the voltage is reversed, the ions at the tip are forced into solution and the filament retracts. No filament starts at the cathode since the excess of As accumulating there lowers the melting temperature. The growth velocity is expected to depend on the diffusion constant and the relative electromigration coefficient of the Te ions migrating toward the anode or of the other constituents moving away from it. Diffusion should be decreased by the presence of cross-linking group IV elements.

Phase separation rather than electromigration determines the filament morphology in the Ge-Te eutectic memory materials, such as $Te_{81}Ge_{15}Sb_4$. In this case Te crystallites of about 400 Å diameter form where the filament region of the ON state reaches temperatures above the glass transition temperature T_g. The remaining Ge enriched matrix has a higher value of T_g. It also crystallizes at higher current values. A highly conducting and locked-on filament is established when the crystallites achieve a sufficient degree of connectivity. Memory filament of this type have been studied by Sie *et al.* (1972), Haberland (1970), Cohen *et al.* (1972), and Csillag (1971, 1973).

6.6 MATERIAL CONSIDERATIONS

The two categories of materials useful for threshold and memory devices, respectively, are (A) those whose structure does not change during the device operation and (B) those whose structure can be changed in a controlled and reproducible manner.

Real materials only approximate these idealized situations, and other factors besides composition affect the stability and reproducilibity of non-crystalline solids in devices, such as contact and the substrate materials and preparation and surface conditions. Stable vitreous semiconductors of type (A) are found among the three-dimensionally cross-linked chalcogenide alloy glasses. There appear to be specific compositions in each glass-forming system which are particularly stable (de Neufville (1972)) by having an optimum number of stable bonds and crosslinks. They can be heated to the molten state and slowly cooled without devitrification or phase separation. Increasing the number of components of similar bond strength in a glass stabilizes its structure. At the same time, however, the number of possible compositions increases, which may phase separate.

An extensive study of the stability and the extent of glass forming regions in chalcogenide alloys has been performed by Hilton, Jones, and Brau (1964, 1966a, 1966b), Hilton (1970), Haisty and Krebs (1969a, b), and de Neufville (1972).

Examples of semiconductors of type (B) whose structure can be changed reversibly between two structural states are compositions at the border of glass forming regions and compositions of the type $Te_{81}Ge_{15}X_4$, where X represents one or two elements of Group V or VI of the periodic table. This composition is close to $Te_{83}Ge_{17}$, the eutectic point of the Ge-Te binary. This point is particularly favorable for glass formation because of the relatively low eutectic temperature ($375°C$) and the high viscosity of the melt. The X additives to the pure eutectic composition stabilize the amorphous state but nevertheless permit crystallization to occur as a result of excitation by a light pulse or a current pulse. Long term reversibility of a memory material depends on a number of parameters. Among these are the relative magnitudes of the electromigration coefficients of the constituent atoms and of the melting points of the various crystalline phases.

The characteristic difference between a structure stable material (A) and a reversible material (B) is demonstrated most simply (Fritzsche and Ovshinsky, 1970c), by differential thermal analysis (DTA). By means of two thermocouples the DTA signal measures the temperature difference between the material and a standard of constant heat capacity as both are heated at a constant rate. This signal measures therefore any endothermic or exothermic reactions or changes in heat capacity of the sample.

Figure 6.25(a) shows as an example a DTA trace together with the resistivity curve of a stable material (A) of composition $Te_{50}As_{30}Si_{10}Ge_{10}$. The step in the DTA curve results from an increase in specific heat at the glass transition temperature T_g. The absence of any sharp peak in the DTA curve is a good indicator for the absence of structural changes.

The resistivity and DTA curves of $Te_{81}Ge_{15}Sb_4$, a type (B) material, is shown in Figure 6.25(b). The exothermic peak near T_1 in the DTA curve is caused by divitrification. Above this temperature the material consists of small (100–300 Å) Te and GeTe crystallites. These are degenerate semi-

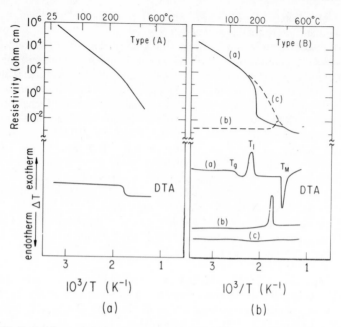

Fig. 6.25. Resistivity and differential thermal analysis trace of (a) a stable material of composition $Te_{50}As_{30}Si_{10}Ge_{10}$ and (b) a memory material $Te_{81}Ge_{15}Sb_4$.

conductors on account of defects and impurities and cause the resistivity to drop sharply at this temperature. At the same time many other physical properties undergo changes, for example, reflectivity, optical transmission (Feinleib and Ovshinsky (1970)), density wettability, dielectric relaxation, and secondary electron emission efficiency.

The eutectic melts near $T_M = 375°C$ and the resistivity increases slightly. In the liquid the structure is similar to that of the vitreous state except for the larger thermal motion. Hence the original vitreous and high resistivity state can be restored by rapid cooling (see DTA curve C). Slow cooling on the other hand leads to the crystalline eutectic (after some supercooling). The crystallization manifests itself by the exothermic solidification peak (see DTA curve b) and the low resistivity at room temperature. The structure transformations, controlled in this example by heating and different rates of cooling, are believed to be initiated or accelerated by light-generated or field-induced excess carriers (Feinleib et al. (1971)). These processes are presently being investigated (Ovshinsky (1969) Evans et al. (1970) Brandes et al. (1970) Berkes et al. (1971)).

ACKNOWLEDGEMENTS

I wish to express my deep gratitude to S. R. Ovshinsky for many clarifying discussions and for introducing me to the field of amorphous semiconductors. This work could not have been carried out without the help, criticism, and important contributions from E. A. Fagen, H. K. Henisch, J. deNeufville, R. G. Neale, C. H. Sie and M. P. Shaw, I wish to thank W. D. Buckley, G. V. Bunton, D. M. Kroll, R. Landauer, W. Lugschneider, A. Csillag, D. R. Haberland, and H. K. Henisch for sending me manuscripts prior to publication. Clarifying discussions with W. van Roosbroeck, H. J. Queisser and G. Döhler about relaxation space charge injection are gratefully acknowledged.

REFERENCES

Adam, G. B. and Duchene, J. C., (1972), *IEEE Trans. Electr. Dev.*, ED **19**, 820.
Adams, J. and Manley, B. W., (1965), *Electron. Eng.*, **37**, 180.
Adler, D., (1971), Critical Reviews in Solid State Sciences *2*, 317, Chem. Rubber Pub. Co., Cleveland, Ohio.
Adler, D., (1972), private communication.
Adler, D. and Moss, S. C., (1972), *J. Vac. Science and Techn.*, **9**, 1182.
Altunyan, S. A. and Stafeev, V. I., (1970), *Sov. Phys. Semicond.*, **4**, 431.
Armitage, D., Brodie, D. E. and Eastman, P. C., (1971), *Can. J. of Phys.*, **49**, 1662.
Balberg, I., (1970), *Appl. Phys. Lett.*, **16**, 491.
Basavaiah, S. and Park, K. C., (1973), *IEEE Trans. on Electron Devices*, **ED-20**, 149. published).
Berglund, C. N. and Klein, N., (1971), *Proc. IEEE*, **59**, 1099.
Berkes, J. S., Ing. Jr., S. W., Hillegas, W. J., (1971), *J. Appl. Phys.*, **42**, 4908.
Bobrova, A. N. and Lobenov, E. M., (1963), Radiation Effects in Solids, Moscow, Akad. Nauk, U.S.S.R.
Böer, K. W., (1962), *Festkörperprobleme*, **1**, 38.
Böer, K. W., (1970), *phys. stat. sol*, (a) **2**, 817.
Böer, K. W., Döhler, G. and Ovshinsky, S. R., (1970), *J. Non-Cryst. Sol.*, **4**, 573.
Böer, K. W., Jahnke, E. and Nebauer, E., (1961), *phys. stat. sol.*, **1**, 231.
Böer, K. W. and Ovshinsky, S. R., (1970), *J. Appl. Phys.*, **41**, 2675.
Bongers, P. F. and Enz, U., (1966), *Philips Res. Rep.*, **21**, 387.
Bosnell, J. R. and Thomas, C. B., (1972), *Solid State Electronics*, **15**, 1261.
Bosnell, J. R. and Thomas, C. B., (1973), *Phil. Mag.*, **27**, 665.
Brandes, R. G., Laming, F. P. and Pearson, A. D., (1970), *Appl. Optics*, **9**, 1712.
Buckley, W. D., (1972), private communication.
Buckley, W. D. and Holmberg, S., (1972), private communication.
Bullock, C. C. and Epstein, D. J., (1970), *Appl. Phys. Lett.*, **17**, 199.
Bunton, G. V., Day, S. C. M., Quilliam, R. M. and Wisbey, P. H., (1971), *J. Non-Cryst. Solids*, **6**, 251.
Bunton, G. V. and Quilliam, R. M., (1973), *IEEE Trans. on Electron Devices*, **ED-20**, 140.
Busch, G., Güntherodt, J. H. Kunzi, H. U. and Schwiger, A., (1970), *Phys. Lett.*, **33A**, 64.
Chaudhari, P., Cuomo, J. J. and Gambino, R. J., (1973), *Appl. Phys. Letters*, **22**, 337.
Chopra, K. L., (1970), Proc. Intl. Congress Thin Films, p. 351. Cannes, France.
Cohen, M. H., Fritzsche, H. and Ovshinsky, S. R., (1969), *Phys. Rev. Lett.*, **22**, 1065.
Cohen, M. H., Neale, R. G. and Paskin, A., (1972), *J. Non-Cryst. Solids*, **8–10**, 885.
Coward, L. A., (1971), *J. Non-Cryst. Solids*, **6**, 107.
Csillag, A., (1971), Proc. of the Second Intl. Conf. on Conduction in Low Mobility Materials, Eilat, Israel, (N. Klein, D. S. Tannhauser, M. Pollak, editors), p. 319. Taylor and Frances, Ltd.
Csillag, A., (1973), *IEEE Trans. on Electron Devices*, **ED-20**, 157.

356 H. Fritzsche

Dearnaley, G., Stoneham, A. M. and Morgan, D. V., (1970), *Rep. Progr. Phys.*, **33**, 1129.
DeNeufville, J. P., (1972), *J. Non-Cryst. Solidss*, **8–10**, 85.
Dessauer, J. H. and Clark, H. E., editors, (1965),, Xerography and Related Processes, The Focal Press, New York.
Dewald, J. F., Pearson, A. D., Northover, W. R. and Peck, W. F., (1962), *J. Electro-chem. Soc.*, **109**, 243c.
Döhler, G. H. and Heyszenau, H., (1973), *Phys. Rev. Letters*, **30** 1200.
Drake, C. F. and Scanlon, I. F., (1970), *J. Non-Cryst. Solids*, **4**, 234.
Drake, C. F., Scanlan, I. F. and Engel, A., (1969), *phys. stat. sol.*, **32**, 193.
Eaton, D. L., (1964), *J. Am. Ceram. Soc.*, **47**, 554.
Edmond, J. T., Male, J. C. and Chester, P. F., (1968), *J. Sci. Instrum.*, **1**, 373.
Eusner, P. R., Durden, L. R. and Slack, L. H., (1972), *J. Am. Ceram. Soc.*, **55**, 43.
Evans, E. J., Helbers, J. H. and Ovshinsky, S. R., (1970), *J. Non-Cryst. Solids*, **2**, 334.
Evans, E. J., (1972), *J. Non-Cryst. Solids*, **8–10**, 702.
Fagen, E. A. and Fritzsche, H., (1970), *J. Non-Cryst. Solids*, **2**, 170.
Fagen, E. A., (1972), private communication.
Feinleib, J. and Ovshinsky, S. R., (1970), *J. Non-Cryst. Solids*, **4**, 564.
Feinleib, J., de Neufville, J., Moss, S. C. and Ovshinsky, S. R., (1971), *Appl. Phys. Lett.*, **18**, 254.
Feldman, D. and Moorjani, K., (1970), *J. Non-Cryst. Solids*, **2**, 82.
Fleming, G. R., (1970), *J. Non-Cryst. Solids*, **2**, 540.
Fock, V., (1927), *Arch. Electrotech.*, **19**, 71.
Fock, V., (1928), *Arch. Electrotech.*, **30**, 411.
Franz, W., (1956), *Encyclopedia of Physics*, **17**, 155.
Fritzsche, H., (1969), *IBM J. Res. and Development.*, **13**, 515.
Fritzsche, H. and Ovshinsky, S. R., (1970a), *J. Non-Cryst. Solids*, **2**, 393.
Fritzsche, H. and Ovshinsky, S. R., (1970b), *J. Non-Cryst. Solids*, **4**, 464.
Fritzsche, H. and Ovshinsky, S. R., (1970c), *J. Non-Cryst. Solids*, **2**, 148.
Fulenwider, J. E. and Herskowitz, G. J., (1970), *Phys. Rev. Lett.*, **25**, 292.
Gibson, A. F., (1954), *Physica*, **20**, 1058.
Gildart, L., (1965), *J. Appl. Phys.*, **36**, 335.
Gildart, L., (1970), *J. Non-Cryst. Solids*, **2**, 240.
Gobrecht, H. and Mahdjuri, F., (1971), *phys. stat. sol.*, (a) **7**, K83; ibid, (1972a), *phys. stat. sol.*, (a) **9**, K111; ibid, (1972b), *J. Phys. C.: Solid State Phys.*, **5**, 366.
Haberland, D. R., Karman, R. and Repp, F., (1970), *Frequenz*, **7**, 212.
Haberland, D. R., (1970a), *Solid State Electronics*, **13**, 207.
Haberland, D. R., (1970b), *Solid State Electronics*, **13**, 451.
Haberland, D. R. and Stiegler, H., (1972), *J. Non-Cryst. Solids*, **8–10**, 408.
Haisty, R. W. and Krebs, H., (1969a), *J. Non-Cryst. Solids*, **1**, 399; ibid, (1969b), J. Non-Cryst. Solids, **1**, 427.
Henisch, H. K., (1969), *Sci. American*, **221**, 30.
Henisch, H. K., (1972),, private communication.
Henisch, H. K., Fagen, E. A. and Ovshinsky, S. R., (1970), *J. Non-Cryst. Solids*, **4**, 538.
Henisch, H. K. and Pryor, R. W., (1971), *Solid State Electronics*, **14**, 765.
Henisch, H. K., Pryor, R. W. and Vendura Jr., G. J., (1972), *J. Non-Cryst. Solids*, **8–10**, 415.
Henisch, H. K. and Vendura, Jr., G. J., (1971), *Appl. Phys. Lett.* **19**, 363.
Herrell, D. J. and Park, K. C., (1972), *J. Non-Cryst. Solids*, **8–10**, 449.
Heywang, W. and Haberland, D. R., (1970), *Solid State Electr.* **13**, 1077.
Hiatt, W. R. and Hickmott, T. W., (1965), *Appl. Phys. Lett.*, **6**, 106.
Hilton, A. R., Jones, C. E. and Brau, M. J., (1964), *Infrared Phys.*, **4**, 213; ibid, (1966a), *Infrared Phys.*, **6**, 183, ibid, (1966b), *Phys. Chem. Glasses*, **7**, 105.
Hilton, A. R., (1970), *J. Non-Cryst. Solids*, **2**, 28.
Hindley, N. K., (1970a), *J. Non-Cryst. Solids*, **5**, 17; ibid, (1970b), *J. Non-Cryst. Solids*, **5**, 31.
Iizima, S., Sugi, M., Kikuchi, M. and Tanaka, R., (1970), *Solid State Commun.*, **8**, 153.

Kaplan, T. and Adler, D., (1971), *Appl. Phys. Lett.,* **19**, 418.

Keneman, S., (1971), *Appl. Phys. Lett.,* **19**, 205.

Keyes, R. W., (1971), *Comments on Solid State Phys.,* **4**, 1.

Kiess, H. and Rose, A., (1973), *Phys. Rev. Letters,* **31**, 153.

Kikuch, M. and Iizima, S., (1969), *Appl. Phys. Lett.,* **15**, 323.

Kikuchi, M., Iizima, S., Sugi, M., Tanaka, K., (1970), *Jap. J. Appl. Phys. Suppl.,* **39**, 203.

Klein, N., (1969), *Adv. Electr. Phys.,* **26**, 309.

Klein, N., (1971a), *Thin Solid Films,* 7, 149; ibid, (1971b), Proc. 2nd Intl. Conf. on Conduction in Low Mobility Materials, p. 299. Eliat, Israel.

Klein, N., Gafui, H. and David, H. J., (1965), *Phys. Failure Electron. (Proc. Symp.)* **3**, 315.

Klose, P., (1967), private communication.

Kobylarz, T. S., (1970), *J. Non-Cryst. Solids,* **2**, 515.

Kolomiets, B. T., Lebedev, E. A. and Taksami, I. A., (1969a), *Soviet Physics Semicond.,* **3**, 267; (1969b), ibid, **3**, 621.

Kolomiets, B. T., Lebedev, E. A., Rogachev, N. A. and Shpunt, V. Kh., (1972), *Soviet Phys. Semiconductors,* **6**, 167.

Krause, J. T., Kurkjian, C. R., Pinnow, D. A. and Sigety, E. A., (1970), *Appl. Phys. Lett.,* **17**, 367.

Kroll, D. M. and Cohen, M. H., (1972), *J. Non-Cryst. Solids,* **8–10**, 544.

Kroll, D. M., (1973), to be published.

Landauer, R. and Woo, J. W. F., (1973), to be published.

Lee, S. H., (1972), *Appl. Phys. Lett.,* **21**, 544.

Lee, S. H. and Henisch, H. K., (1973), *Solid State Electr.,* **16**, 155.

Lee, S. H., Henisch, H. K. and Burgess, W. D., (1972), *J. Non-Cryst. Solids,* **8–10**, 422.

Low, G. G. E., (1955), *Proc. Phys. Soc. (London),* **B68**, 31.

Lucas, I., (1971), *J. Non-Cryst. Solids,* **6**, 136.

Lueder, H. and Spenke, E., (1935a), *Physik. Z.* **36**, 766; ibid, (1935b), *Z. techn. Physik,* **11**, 373.

Lugschneider, W. and Zinn, W., (1973), *IEEE Transactions on Magnetics,* to be published.

Matsushita, T., Yamagami, T. and Okuda, M., (1972), *Jap. J. Appl. Phys.,* **11** 923.

Minorsky, N. (1947), Introduction to Nonlinear Mechanics, J. W. Edwards, Ann Arbor, Mich., p. 55.

Morin, F. J., (1959), *Phys. Rev. Lett.,* **3**, 34.

Moss, S. C. and de Neufville, J., (1972), *J. Non-Cryst. Solids,* **8–10**, 45.

Mott, N. F., (1969), *Contemp. Phys.,* **10**, 125.

Mott, N. F., (1971), *Phil. Mag.,* **24**, 911.

Neale, R. G., (1970), *J. Non-Cryst. Solids,* **2**, 558.

Neale, R. G. and Aseltine, J. A., (1973), *IEEE Transactions on Electron Devices,* **ED–20**, 195.

Neale, R. G., Nelson, D. L. and Moore, G. E., (1970), *Electronics,* **43**, 56.

Nebauer, E. and Jahnke, E., (1965), *phys. stat. sol.,* **8**, 881.

Nelson, D. L., (1970), *J. Non-Cryst. Solids,* **2**, 528.

Okada, Y., Iizima, S., Sugi, M., Kikuchi, M. and Tanaka, K., (1970), *J. Appl. Phys.,* **41**, 5341.

Ovshinsky, S. R., (1959), *Electronics,* **32**, 76.

Ovshinsky, S. R., (1963), *Automation,* **10**, 45.

Ovshinsky, S. R., (1966), *U. S. Patent 3,* **271**, 591.

Ovshinsky, S. R., (1967a), *Bull. Sci. USSR,* Edited by B. T. Kolomiets, **9**, 91.

Ovshinsky, S. R., (1967b), Intl. Colloq. on Amorphous and Liquid Semicond., Acad. Soc. Repub. Rumania, Bucharest.

Ovshinsky, S. R., (1968), *Phys. Rev. Lett.,* **21**, 1450.

Ovshinsky, S. R., (1969), Proc. Fifth Annual Natl. Conf. on Industrial Res., p. 68. Sept. 18.

Ovshinsky, S. R., Evans, E. J., Nelson, D. L. and Fritzsche, H., (1968), *IEEE Trans.*

Nucl. Science, **NS-15**, 311.

Ovshinsky, S. R. and Fritzsche, H., (1971), *Metallurgical Trans.*, **2**, 641.

Ovshinsky, S. R. and Klose, P., (1972), *J. Non-Cryst. Solids*, **8–10**, 892.

Ovshinsky, S. R. and Fritzsche, H., (1973), *IEEE Transactions on Electron Devices*, **ED-20**, 91.

Pearson, A. D., (1964), Modern Aspects of the Vitreous State, Vol. 3, (J. D. Mackenzie, editor).

Pearson, A. D. and Bagley, B. G., (1971), *Mat. Res. Bull.*, **6**, 1041.

Pearson, A. D. and Miller, C. E., (1969), *Appl. Phys. Lett.*, **14**, 280.

Pearson, A. D., Northover, W. R., Dewald, J. F. and Peck, W. F., (1962), *Adv. in Glass Tech.* **357**.

Perschy, J. A., (1967), *Electronics*, **40**, 74.

Popescu, C., (1970a), *Solid State Electronics*, **13**, 441. ibid, (1970b), **13**, 887.

Popescu, C. and Croitoru, N., (1972), *J. Non-Cryst. Solids*, **8–10**, 531.

Pryor, R. W. and Henisch, H. K., (1972a), *J. Non-Cryst. Solids*, **7**, 181.

Queisser, H. J., Casey Jr., H. C. and van Roosbroeck, W., (1971), *Phys. Rev. Letters*, **26**, 551.

Queisser, H. J., (1972), Proc. of European Solid State Device Research Conf., Sept. 15, 1972. (P. N. Robson, editor), Lancaster, England.

Regan, M. and Drake, C. F., (1971), *Mat. Res. Bull.*, **6**, 487.

Regel, A. R., Andreev, A. A. and Mamadalier, M., (1972), *J. Non-Cryst. Solids*, **8–10**, 455.

Ridley, B. K., (1963), *Proc. Phys. Soc.* (London), **82**, 954

Rogowski, W., (1924), *Arch. Electrotechn.*, **13**, 153.

Ryvkin, S. M., (1964), Photoelectric Effects in Semiconductors, Consultants Bureau, New York.

Shanks, R. R., (1970), *J. Non-Cryst. Solids*, **2**, 504.

Shanks, R. R., Helbers, J. H., Fowler, R. L., Chambers, H. C. and Niehaus, D. J., (1970a), Radiation Hardening Circuitry Using New Devices, Techn. Report, AFAL–TR–70–15.

Shanks, R. R., Helbers, J. H. and Nelson, D. L., (1970b), Ovonic Computer Circuits Development, Technical Report AFAL–TR–69–309.

Shaw, M. P. and Gastman, I. J., (1971), *Appl. Phys. Lett.*, **19**, 243; ibid, (1972), *J. Non-Cryst. Solids*, **8–10**, 999.

Shaw, M. P., Grubin, H. L. and Gastman, I. J., (1972), *IEEE Trans. Electronic Devices*, **ED-20**, 169.

Shaw, M. P., Moss, S. C., Kostylev, S. A. and Slack, L. H., (1973), *Appl. Phys. Lett.*, **22**, 114.

Sie, C. H., Dugan, M. P. and Moss, S. C., (1972), *J. Non-Cryst. Solids*, **8–10**, 877.

Sieja, N. F., (1966), private communication.

Simmons, J. G., (1970), *Contemp. Phys.*, **11**, 21.

Simmons, J. G. and Verderber, R. R., (1967), Proc. Roy. Soc., A301, 77.

Sliva, P. O., Dir, G. and Griffiths, C., (1970), *J. Non-Cryst. Solids*, **2**, 316.

Smith, R. A., Sanford, R. and Warnock, F. E., (1972), *J. Non-Cryst. Solids*, **8–10**, 862.

Spenke, E., (1936a), *Wiss. Veröff. Siemens-Werk.*, **15**, 92; ibid, (1936b), *Arch. Electrotech.*, **30**, 15.

Srinivasan, G. R., Younger, S., Macedo, P. B. and Litovitz, T. A., (1971), *Phys. Rev. Lett.*, **27**, 1084.

Stocker, H. J., (1969), *Appl. Phys. Lett.*, **15**, 55.

Stocker, H. J., (1970), *J. Non-Cryst. Solids*, **2**, 371.

Stocker, H. J., Barlow Jr., C. A. and Weirauch, D. F., (1970), *J. Non-Cryst. Solids*, **4**, 523

Sugi, M., Okada, Y., Iizima, S., Kikuchi, M. and Tanaka, K., (1970), *Solid State Comm*, **8**, 829.

Tanaka, K., Iizima, S., Sugi, M. and Kikuchi, M., (1970a), *Solid State Comm.*, **8**, 75; ibid, (1970b), *Solid State Comm*, **8**, 387; ibid, (1970c), *Solid State Comm.*, **8**, 1333.

Thomas, C. B., Fray, A. F. and Bosnell, J. R., (1972), *Phil. Mag.*, **26**, 617.

Uttecht, R., Stevenson, H., Sie, C. H., Griener, J. D. and Raghavan, K. S., (1970), *J. Non-Cryst. Solids*, **2**, 358.

Van Landingham, K. E., (1973), *IEEE Trans. Electronic Devices*, **ED–20**, 178.

van Roosbroeck, W. and Casey, Jr., H. C., (1970), Proc. Tenth Intl. Conf. on Physics of Semicond., (S. P. Keller, J. C. Hensel and F. Stern, editors), U.S. AEC Div. of Techn. Inf., Springfield, VA., 1970, p. 832.

van Roosbroeck, W., (1972), *Phys. Rev. Lett.*, **28**, 1120.

van Roosbroeck, W. and Casey, Jr., H. C., (1972), *Phys. Rev.*, **B5** 2154

van Roosbroeck, W., (1973), *J. Non-Cryst. Solids*, to be published.

Vendura Jr., G. J. and Henisch, H. K., (1970), *J. Non-Cryst. Solids*, **3**, 29.

Verderber, R. R., (1972), private communication.

Verderber, R. R., Simmons, J. G. and Eales, B., (1967), *Phil. Mag.*, **16**, 1049.

Walsh, P. J., Vogel, R. and Evans, E. J., (1969), *Phys. Rev.*, **178**, 1274.

Walsh, P. J., Hall, J. E., Nicolaides, R., Defeo, S., Calella, P., Kuchmas, J. and Doremus, W., (1970), *J. Non-Cryst. Solids*, **2**, 107.

Warren, A. C., (1970), *J. Non-Cryst. Solids*, **4**, 613.

Warren, A. C. and Male, J. C., (1970), *Electronics Lett.*, **6**, 567.

Weirauch, D. F., (1970), *Appl. Phys. Lett.*, **16**, 72.

Whitehead, S., (1951), Dielectric Breakdown in Solids, Oxford University Press, London.

Chapter 7

Structure and Electronic Properties of Liquid Semiconductors

J. E. Enderby

Department of Physics, University of Leicester

7.1   THE STRUCTURE OF PURE LIQUIDS

7.1.1   The Interference Function

It is now quite clear that a good knowledge of $a(q)$, the interference function, is necessary for a fundamental understanding of the electronic properties of liquid conductors. The purpose of this section is to describe the use of neutron and X-ray diffraction techniques in determining this quantity. In most scattering experiments the measured intensities are proportional to differential scattering cross sections. Let b_c and b_i represent respectively the bound atom scattering lengths for coherent and incoherent scattering. Then the differential scattering cross section for coherent scattering is $Nb_c^2 a(q)$ whilst that for incoherent scattering is Nb_i^2, provided the scattering is perfectly elastic (see, for example, Bacon (1962)), $a(q)$ here is defined in the usual way as the expectation value of $N S(q) S^*(q)$ where

$$S(q) = \sum_i e^{-i q \cdot r_i}$$

N is the number of scatterers, q is a wave number and r_i refers to the positions of the nuclei.

Contrary to what is frequently supposed, $a(q)$ is not directly accessible through X-ray or neutron diffraction experiments either in principle or in practice. Indeed, until comparatively recently marked discrepancies existed between $a(q)$ data derived by different techniques and these in part can be traced to a failure to recognize the inherent limitations of neutron or X-ray diffraction methods.

In a conventional diffraction experiment an integrated intensity $\mathscr{J}$ is measured at a given angle of scattering θ. The integration over the energy is performed by the detector so that an effective differential scattering cross section $(\frac{d\sigma}{d\Omega})_{eff}$ rather than a true static cross section is measured. In terms of the generalized quantity $S(q, \omega)$ which measures the probability that the radiation absorbs energy $\hbar\omega$ from the liquid and imparts momentum $\hbar q$ to it, Placzek (1952) showed that

$$\left(\frac{d\sigma}{d\Omega}\right)_{eff} \propto \int_{-E_o/\hbar}^{\infty} \frac{\kappa}{\kappa_o} S(q, \omega) f(\omega) \, d\omega \qquad (7.1)$$

where E_o is the incident energy of the neutron κ_o and κ are the incident and final wave numbers, and $f(\omega)$ measures the energy dependence of the detection system. If E_o greatly exceeds the energy transfers, the integral on the right-hand side of Eq. (7.1) may be approximated by a(q). This, the so-called static approximation (Lomer and Lowe, 1965) applies for X-rays but not, in general, for neutrons.

There may be contributions from multiple scattering and incoherent effects and these must be allowed for since a(q) involves only single coherent scattering processes. We therefore write

$$\mathscr{J} \propto \alpha\,(\theta) \left\{ \left(\frac{d\sigma_c}{d\Omega} \right)_{eff} + \left(\frac{d\sigma_i}{d\Omega} \right)_{eff} + \Delta_c + \Delta_i \right\}$$

where $\alpha(\theta)$ is the usual absorption correction and Δ represents the multiple scattering. If we introduce a constant of proportionality which depends on the strength of the incident beam, the number of scatterers, etc., and divide by $\alpha(\theta)$ to obtain I, the observed intensity corrected for self-absorption by the sample, we have

$$I = \frac{\mathscr{J}}{\alpha\,(\theta)} = \beta(\theta) \left\{ \left(\frac{d\sigma_c}{d\Omega} \right)_{eff} + \left(\frac{d\sigma_i}{d\Omega} \right)_{eff} + (\Delta_c + \Delta_i) \right\}$$

$$= \beta(\theta) \left[N \left\{ b_c^{\,2} \, a(q)_{eff} + (b_i^{\,2})_{eff} \right\} + (\Delta_c + \Delta_i) \right]. \qquad (7.2)$$

$$= \gamma(\theta) \left[a(q)_{eff} + \delta \right]$$

To extract a(q) from the diffraction data α, γ and δ must be determined and the relationship between the effective cross section and the true cross section established. In addition a correction must be applied to the observed scattered intensity to allow for scattering from the sample holder (or 'cell') when data are taken in the transmission mode.

7.1.2   The Neutron Method

The total scattered intensity of thermal neutron from a liquid sample and its cell, $\mathscr{J}_{tot}$, can be measured by means of conventional techniques which do not require detailed description here (see, for example, Bacon (1962)).

For slab geometry $\alpha(\theta)$ is known analytically and is given by

$$\alpha(\theta) = t \sec \theta \exp\,(-\sigma_T t) \qquad (7.3)$$

where t is the thickness of the specimen and σ_T is the total cross section per unit volume for scattering and absorption. $\mathscr{J}_{cell}$, can, for metallic sample holders, be as little as 4% of $\mathscr{J}_{tot}$ at angles away from the Bragg peaks of the cell material and is readily determined by measuring the scattered intensity from the empty cell and applying a small correction to allow for the attenuation of the neutron beam by the liquid sample.

For cylindrical geometry $\alpha(\theta)$ is no longer expressible in a simple analytic form (Bond (1962)). However, for thin metallic cells $\mathscr{J}_{cell}$ is small enough to be reliably determined by the method described above and $\alpha(\theta)$ turns out to be almost isotropic and can be evaluated from the tables given by Bond. In the case of thick silica or alumina cells $\mathscr{J}_{cell}$ is a significant fraction of $\mathscr{J}_{tot}$ ($\sim 35\%$) and these methods are not sufficiently accurate. The basic problem has been fully discussed by Paalman and Pings (1962) who present numerical methods which allow the absorption in the sample to be properly allowed for in calculating $\mathscr{J}_{cell}$.

There are two principal ways in which $\gamma(\theta)$ can be determined. A commonly used method is to assume that $a(q) = a(q)_{eff}$ and then exploit the known behaviour of $a(q)$ at high and low q (see, for example, Gingrich and Heaton, 1961). A better method is to calibrate the apparatus at each value of q with a piece of vanadium (for which $b_c^2 = 0$) of similar size and shape to that of the liquid sample. The advantages of this latter method have been discussed in detail by North *et al.* (1968a). The accuracy is limited at the present time by our knowledge of b_c^2. It should be noted, however, that both b_c and b_i are q independent.

The multiple scattering terms have been discussed by Vineyard (1954), Cocking and Heard (1965) and Blech and Averbach (1965) for a variety of geometries. Let J_n* represent the scattered neutron current per incident transmitted through the sample per unit solid angle; the suffix n (1, 2, 3 . . .) denotes the once, twice, thrice, etc., scattered components. It turns out that provided $\sigma_T t < 1$ both Δ_c and Δ_i ($\equiv \Sigma J_n / \alpha(\theta)$) are *independent of q*. The fact that the multiple scattering is isotropic for thin samples (and this has been checked experimentally by North *et al.* (1968a)) greatly simplifies the problem of data reduction.

To establish the connection between $a(q)$ and $a(q)_{eff}$ we note that the neutron counter records *all* neutrons scattered at a given angle irrespective of the energy changes which may have occurred in the scattering process. Small changes in energy on scattering cause changes in the momentum transfer between the neutron and the specimen, and in the case of the energy transfer being small compared with the incident energy the corrections may be calculated from formulae due to Placzek (1952). For a $1/\kappa$ detector these may be written as

$$\left(\frac{d\sigma}{d\Omega}\right)_{eff} = Nb^2 \left[P_o - \frac{q^2 P_1'}{2\epsilon} + \frac{1}{16\epsilon^2} \left\{ (u - 2q^2) P_2' + 2q^4 P_2'' \right\} + \dots \right] \quad (7.4)$$

where

$$P_n = \left(\frac{\hbar}{k_B T}\right)^n a_n(q)$$

$$P_n' = \frac{\partial P_n}{\partial (q^2)} \text{ etc.}$$

$$\epsilon = \frac{E_o}{k_B T}$$

$$u = 4\kappa_o^2$$

---

*The distinction between coherent and incoherent scattering will be dropped in this section.

and

$$a_n(q) = \int_{-\infty}^{\infty} \omega^n \, S(q,\omega) \, d\omega.$$

The square brackets are to be evaluated at the value of q for elastic scattering, i.e., $4\pi \, (\sin\theta)/\lambda_o$ where λ_o is the incident wavelength. Since the first few energy moments of $S(q, \omega)$ are known (Placzek (1952)) Eq. (7.4) can be evaluated to order m/M where m and M are the masses of the neutron and nucleus respectively.

The result is that effective cross sections in Eq. (7.2) are related to the true cross sections by

$$a(q)_{eff} = a(q) + f_p(q)$$

$$(b_i{}^2)_{eff} = b_i{}^2 \left\{ 1 + f_p(q) \right\}$$

where

$$f_P(q) = \frac{\overline{K}m}{3\epsilon M} - \frac{\alpha}{2\epsilon}$$

$$\alpha = \frac{\hbar^2 q^2}{2Mk_B T}$$

and $\overline{K}$ is the average kinetic energy of an atom in units of $k_B T$, which at high temperatures tends to $\frac{3}{2}$. Tabulated values of the Placzek corrections have been given by North et al. (1968a). They are particularly significant for the lighter elements like Li, Be and Al.

7.1.3 The X-ray Method
X-ray measurements are usually made in a reflection mode although recent experimental work (see, for instance, North and Wagner (1969)) has shown that transmission methods are also feasible. The advantage of the latter relates primarily to the measurement of a (q) at low values of q.

The corrections which must be applied to the experimental data are similar to those previously described although there are significant differences. The multiple scattering, for example, can be usually neglected except, possibly, for data taken in the transmission mode. The calibration constant, $\gamma(\theta)$, contains a polarization correction and has not so far been determined absolutely. The usual approach is to use the high and low q method together with the sum rule requirement discussed in Section 7.1.4 (Eq. (7.7)).

A further difference concerns the bound atom cross sections. The coherent cross section is not isotropic since the scattering processes involve the extra-nuclear electrons. Conventionally $\frac{d\sigma_c}{d\Omega}$ is expressed as Nf^2 a(q) and f, the so-called X-ray form factor, has to be calculated with the aid of the Hartree (or Hartree-Fock) theory and is a function of q. The incoherent scattering arises from the Compton effect and is also q dependent. Both of these facts limit the accuracy with which a(q) may be extracted from the experimental data.

A useful summary of the X-ray method and of the problems involved in data reduction has been given by Pings (1968).

7.1.4   The Radial Distribution Function

It is now convenient to introduce the quantity $n_2(R_k, R_j) \, d R_k \, d R_j$ which is the probability that the volume elements $d R_k \, d R_j$ are both occupied. In a liquid $n_2(R_k, R_j)$ depends only on $r = |R_k - R_j|$. A *radial distribution function* may be defined through $g(r) = n_2(r)/n^2$ where n is the mean number density of atoms; since the occupancy of two distant volume elements is uncorrelated, $g(r)$ tends to unity at large r. The connection between $a(q)$ and $g(r)$ may, with certain reservations which are outside the scope of this chapter, be expressed in the form of a Fourier transform either as:

$$a(q) = 1 + \frac{4\pi n}{q} \int_0^\infty dr \left\{ g(r) - 1 \right\} r \sin q \, r \qquad (7.5)$$

or

$$g(r) = 1 + \frac{1}{2\pi^2 nr} \int_0^\infty dq \left\{ a(q) - 1 \right\} q \sin q r \qquad (7.6)$$

In principle, measuring $a(q)$ enables $g(r)$ to be evaluated. Difficulties in practice arise from the inherent uncertainties in $a(q)$ and the existence of truncation errors coming from the finite upper limit (typically $\sim 12 A^{-1}$) which restricts the evaluation of Eq. (7.6).

For small r, $g(r)$ tends to zero because of the finite size of atoms. It follows from Eq. (7.6) that $a(q)$ must satisfy the sum rule

$$-2\pi^2 n = \int dq \, q^2 \, (a(q) - 1) \qquad (7.7)$$

This equation often forms the basis for normalizing experimental data and is a useful supplement to the high and low q method referred to above.

7.1.5   Experimental Results

It will be useful to focus particular attention on three liquid types classified according to their electrical conductivity, σ, as discussed by Mott (1971).

(a) *Metallic Liquids* ($\geqslant 3,000 \ \Omega^{-1} \ cm^{-1}$)

Reliable data are now available for a wide variety of metallic liquids. Figures 7.1 and 7.2 illustrate the interference function for liquid Pb (a typical quadrivalent liquid metal) at four temperatures. The corresponding radial distribution functions are shown in Figures 7.3 and 7.4 and it should be noted that $a(q)$ depends on temperature and at high q oscillates about the asymptotic value of unity.

$a(o)$ is different from zero and can be related to the isothermal compressibility χ_T through

$$a(o) = nk_B T \chi_T$$

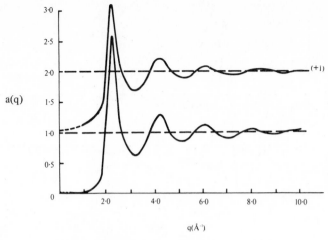

Fig. 7.1. The interference function for liquid Pb
(a) 340°C   (b) 600°C.

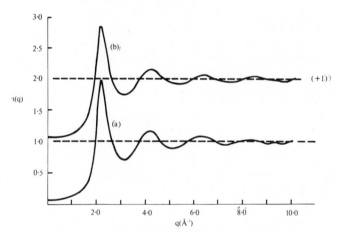

Fig. 7.2. The interference function for liquid Pb
(a) 780°C   (b) 1100°C.

For metallic liquids close to the melting point $a(o)$ is in the range 0.01 to 0.03 (Enderby and March (1965)). Between this thermodynamic limit and q values of $\sim 1.0A^{-1}$ our knowledge of $a(q)$ is, in general, limited.

(b) *Semi-metallic liquids* $(3,000 \leqslant \sigma \leqslant 300 \ \Omega^{-1} \ cm^{-1})$
The only pure liquid which falls into this category is liquid Te. Neutron investigations of $a(q)$ have recently been carried out by Tourand, Cabane and Breuil (1972) and by Hawker, Howe and Enderby (1973a). Although a rigorous comparison of the two sets of data are not possible because the temperatures at which they were taken were different, the results of Hawker

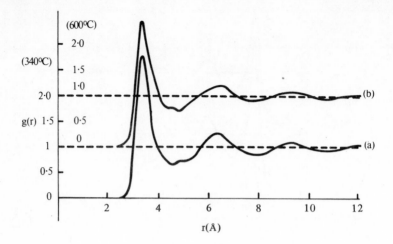

Fig. 7.3. The radial distribution function for liquid Pb
(a) 340°C   (b) 600°C.

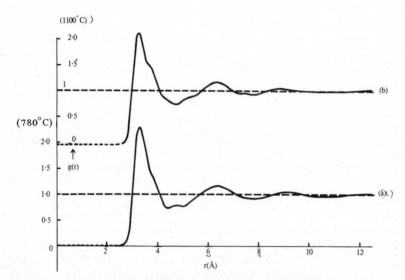

Fig. 7.4. The radial distribution function for liquid Pb
(a) 780°C   (b) 1100°C.

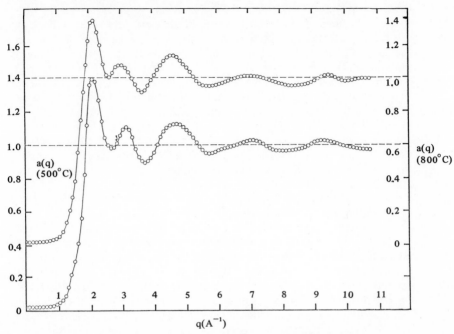

Fig. 7.5. The interference functions for liquid Te at 500°C and 800°C.

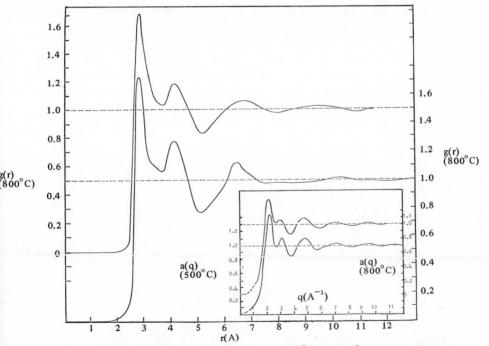

Fig. 7.6. The radial distribution for liquid Te at 500°C and 800°C. The interference functions calculated by back transferring these g(r) are shown in the inset.

et al. for a(q) at 800°C shown in Figure 7.5, are in broad agreement with those of Tourand *et al.* at 900°C, although there are differences in detail. A difficulty arises when the data are transformed to yield a radial distribution function (Figure 7.6). Tourand *et al.* argue that there are three resolved peaks at 3.01, 3.82 and 4.52Å in g(r) and that they all represent real structure. Hawker *et al.*, on the other hand, consider it dangerous to assume that truncation errors are unimportant in the region of $4A^{-1}$ for liquid Te in view of the tendency of a(q) to damp down rather slowly at high q. A numerical inversion of g(r) (the insert in Figure 7.6) shows that an a(q), consistent with the experimental data, can be regenerated from a g(r) in which the only peaks resolved are at 3.01 and 4.52. Hawker *et al.* conclude that neither sets of a(q) data reported support the existence of a clearly defined peak at 3.82Å.

(c) *Insulating and Semi-conducting Liquids* ($\sigma \leqslant 300 \ \Omega^{-1} \ cm^{-1}$)
It is necessary to subdivide this type according to whether or not the interatomic forces may be conveniently represented by a central pairwise interaction. Data for liquid argon in which the forces are essentially pairwise in character have been obtained by Mikolaj and Pings (1967). In general, a(q) for this type of liquid is not unlike that for metallic liquids. There are detailed differences, however, and these will be discussed in Section 7.3.

An example of a pure liquid in which there are highly non-central forces and for which a pairwise interaction potential would be completely inappropriate is afforded by liquid Se. Comparatively little work of a structural nature has been done on this liquid. The neutron investigation by Scott (1968, unpublished) is interesting because it predicts an interference

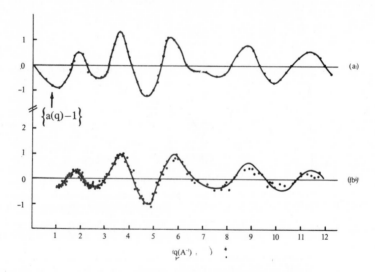

Fig. 7.7.   Structure of (a) amorphous Se at 298°K
                              (b) liquid Se at 553°K.

function which almost exactly coincides with that obtained by Henniger *et al.* (1967) for amorphous Se. It is instructive to compare these data (Figure 7.7) with those for the simpler liquids Ar and Pb, and we return to this comparison in Section 7.3.

7.2 THE STRUCTURE OF MULTICOMPONENT LIQUIDS

We introduce the partial radial distribution function $g_{\alpha\beta}$ which measures the average distribution of type β atom observed from an α atom at the origin. Here α and β are dummy suffices which may take the values $1, 2, \ldots \ldots j$ for a liquid containing j components. The differential scattering cross-section for the coherent scattering of monochromatic radiation may be written as

$$\frac{d\sigma}{d\Omega} = \left\langle \sum \sum f_p f_q \, \text{expiq} \, (\mathbf{r}_p' - \mathbf{r}_q) \right\rangle$$

$$\equiv N \left\{ \sum_{\alpha=1}^{j} c_\alpha f_\alpha^{\,2} + \sum_{\alpha_1} \sum_{\beta=1} c_\alpha c_\beta f_\alpha f_\beta (a_{\alpha\beta} - 1) \right\} \tag{7.8}$$

where c_α is the atomic concentration of the α atom type and $f_p, f_q, \mathbf{r}_p$ and $\mathbf{r}_q$ represent respectively the coherent scattering amplitudes and the position co-ordinate of the pth and q^{th} atoms. $a_{\alpha\beta}(q)$ are known as the *partial interference functions* and are defined through

$$a_{\alpha\beta} \, (q) = 1 + \frac{4\pi n}{q} \int dr \, [g_{\alpha\beta} - 1] \, r \sin q \, r \tag{7.9}$$

The basic problem facing the experimentalist is to derive the fundamental quantities $a_{\alpha\beta}$ from diffraction data. As we have seen, there are quite severe experimental difficulties in relating the observed intensity patterns to $\frac{d\sigma}{d\Omega}$ even for pure liquids; for multivalent systems, however, further difficulties confront us.

Let us focus attention on a liquid which contains two species denoted by 1 and 2. For this case Eq. (7.8) reads

$$\frac{d\sigma}{d\Omega} = N \left\{ c_1 f_1^{\,2} + c_2 f_2^{\,2} + F(q) \right\}$$

where

$$F(q) = c_1^{\,2} f_1^{\,2} (a_{11} - 1) + c_2 f_2^{\,2} (a_{22} - 1) + 2 c_1 c_2 f_1 f_2 (a_{12} - 1). \tag{7.10}$$

Thus three partial interference functions are involved and in order to determine them by experiment *three* separate diffraction experiments on the same liquid must be carried out.

The first experiment designed to determine a_{11}, a_{22} and a_{12} was carried

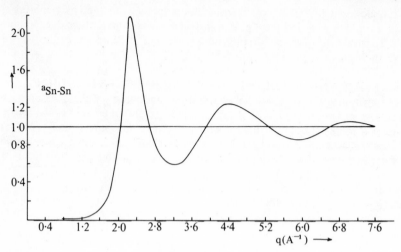

Fig. 7.8 (a)

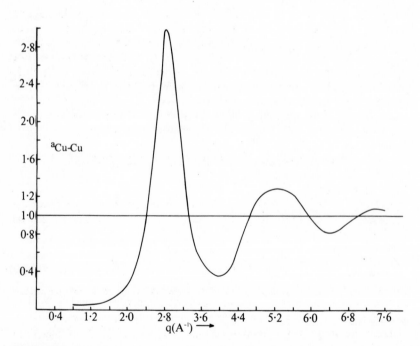

Fig. 7.8 (b)

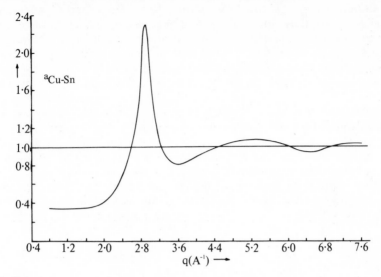

Fig. 7.8 (c)

Fig. 7.8. The partial interference functions for liquid Cu-Sn.
(a) Sn-Sn   (b) Cu-Cu   (c) Cu-Sn.

out by Enderby *et al.* (1966) for liquid alloy, Cu_6Sn_5. The techniques of neutron diffraction were used and the neutron scattering amplitude for copper was changed by alloying pure Sn with Cu enriched to 99% [63]Cu with Cu enriched to 99% [65]Cu and with natural Cu. In addition use was made of the X-ray diffraction data obtained by Williams and co-workers (1966, unpublished). The results are shown in Figure 7.8 and it is important to notice that the main peak in a [a]Cu-Sn does not fall midway between that for [a]Cu-Cu and [a]Sn-Sn. These data, together with comparable results for liquid Cu_2Te and Ag_2Te, will be discussed further in Section 7.3.

7.3   STRUCTURE AND INTERATOMIC FORCES

7.3.1   Introduction
It is useful to focus attention on *simple* liquids. By simple we mean the intermolecular potential energy function $\Phi(r_1, r_2 \ldots r_N)$ has the special form given by

$$\Phi(r_1, r_2 \ldots r_N) = \underset{i<j}{\Sigma'} \phi(r_i, r_j)$$

where ϕ, the so-called pair potential, is spherically symmetric. For a mixture composed of α and β atoms, the interactions are characterised by three pair potentials, each spherically symmetric, given by $\phi_{\alpha\alpha}(r)$, $\phi_{\beta\beta}(r)$, $\phi_{\alpha\beta}(r)$.

It is quite clear that a great many liquids which are of direct interest to us may be excluded by such a severe constraint on the potential energy function. It might be asked whether any real liquids ever conform to the above definition of 'simple' and the answer is, strictly speaking, no. However, there is a considerable body of evidence which suggests that ϕ for the liquid form of the rare gases (Ar, Kr, Ne, and Xe) approximates very closely to a sum of spherically symmetric pair potentials. An example of the type ϕ used for liquid rare gases is given in Figure 7.9. Ionic liquids like NaCl will also satisfy our criteria and we illustrate the type of pair interactions which are expected for this class of liquid in Figure 7.10. Perhaps, more surprisingly, a model based on pair potentials can also be justified for liquid non-transition metals like Na and Al and certain liquid alloys. This follows from the notion of the pseudo-potential, an idea which we now consider in some detail.

The electrons in a conductor may be divided into two types: the *core* electrons which are tightly bound within the atom and the *conduction* electrons which are relatively loosely bound and which are responsible for

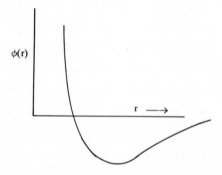

Fig. 7.9.   Pair potential for liquid argon (schematic).

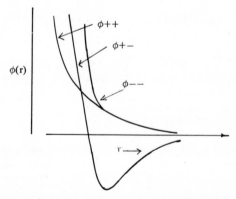

Fig. 7.10. Pair potentials for an ionic liquid A^+B^- (schematic).

most of the characteristic properties of the conducting state. The aim of the pseudopotential method is to transform the equation

$$[T + V(r)] \, \psi_k = E_k \, \psi_k$$

into

$$H\phi_k = E_k \, \phi_k$$

where ϕ_k is a *smooth* wave function and has none of the violent oscillations close to the nucleus which characterise ψ_k, the set of conduction electron wave functions. It turns out that this transformation is easy to accomplish provided the new operator is written $H = T + W(r, k)$ where $W(r, k)$ the so-called pseudopotential is given by

$$W(r, k) = V(r) + \sum_\alpha (E_k - E_\alpha) \, |\alpha> <\alpha| \qquad (7.11)$$

and $|\alpha>$, E_α, E_k are the core states, core eigenvalues and conduction electron eigenvalues respectively, and $V(r)$ is the self-consistent potential.

The advantages of considering $W(r, k)$ rather than $V(r)$ have been discussed many times in the literature. The most significant is that W can be chosen to be much weaker than $V(r)$ and may be decomposed into spherically symmetric (momentum dependent) single ion-core potentials through

$$W(r) = \sum_i' w(r - r_i)$$

Moreover, each Fourier component of W can be screened independently. This forms the basis for the 'dielectric constant' method of dealing with the problem of electron screening.

The matrix elements between two plane wave states $|k>$ and $|k + q>$ may be expressed as

$$<k + q \,|\, W \,|\, k> \equiv W_q(k) = NS(q) <k + q \,|\, w \,|\, k> \equiv NS(q) \, w_q(k)$$

where $w_q(k)$ is known as the form factor. The factorization of the matrix element into a part which depends only on the structure of the system and a part which depends only on the pseudopotential of a single ion is of crucial significance to the theory of electron transport.

Pseudopotential theory enables the total energy per ion to be split up into a part which depends only on the total volume of the conductor and a part which may be written as

$$\frac{1}{2N} \sum \phi(r_i - r_j)$$

where ϕ, the *effective interionic potential*, takes the form

$$\phi(r) = \frac{Z^*e^2}{r} \left[1 \left\{ - \frac{2}{\pi} \int F(q) \, \sin qr \, dq \right] \right. \qquad (7.13)$$

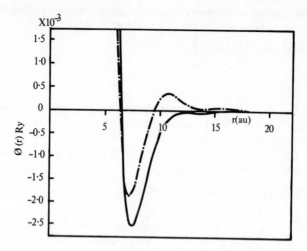

Fig. 7.11. Pair potentials for metallic sodium (Shyu and Gaspari (1967))
——————————— Calculated using a model pseudopotential
----·-·-·-·-·-·-·-·-·-·-·---- Obtained from the phonon spectra.

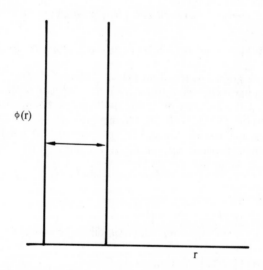

Fig. 7.12. Pair potential for a hard sphere liquid (one component).

Here Z^* represents an effective valence (Harrison (1966)) and $F(q)$ is determined completely by $w_q(k)$.

Tabulated values of $w_q(k)$ obtained by different approximate schemes are available for most elements (see, for example, Ashcroft (1966) Shaw (1969)). A typical effective interionic potential, calculated by Shyu and Gaspari (1967), and based on the $w_q(k)$ given by Ashcroft is shown at Figure 7.11.

7.3.2   The Hard-Sphere Liquid
One liquid system of considerable heuristic value is a hypothetical one composed of a dense assembly of hard spheres. We shall find that such a system represents a useful reference liquid with which to compare real liquids. A hard sphere liquid is clearly a simple liquid within the framework outlined in Section 7.2.1.; the form of ϕ is given in Figure 7.12 and the corresponding potentials for a binary mixture are illustrated in Figure 7.13.

7.3.3   Approximate Theories of the Liquid State
There are three approximate theories of the liquid state in frequent use (see, for example, Enderby and March (1965)). Their common feature is that they attempt to relate the radial distribution function $g(r)$ to the interatomic pair potential $\phi(r)$. For convenience we list the theories and the relevant equations below as applied to a pure liquid. The generalisation to include multi-component liquids is straightforward.

Percus-Yevick (PY):

$$\phi_{PY}(r) = k_B T \ln \left\{ 1 - [c(r)/g(r)] \right\} \tag{7.14}$$

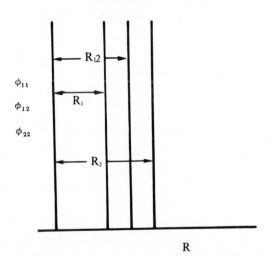

Fig. 7.13. Pair potentials for a hard sphere binary liquid.

Hypernetted chain (HN):

$$\phi_{HN}(r) = k_B T [g(r) - c(r) - 1 - \ln g(r)] \tag{7.15}$$

Born-Green (BG):

$$\phi_{BG}(r) = k_B T \left(-n \int E(|\mathbf{r} - \mathbf{r}'|)h(\mathbf{r}')d\mathbf{r}' - \ln g(r) \right) \tag{7.16}$$

Here $c(r)$ represents the direct correlation function introduced by Ornstein and Zernike, whilst $E(t)$ is given by

$$E(t) = \int_\infty^t drg(r) \frac{\phi'_{BG}(r)}{k_B T} \, ,$$

where $\phi_{BG}'(r)$ is the derivative of $\phi_{BG}(r)$ with respect to r. $c(r)$ can be calculated from the observed interference function $a(q)$ through

$$c(r) = (2\pi^2 nr)^{-1} \int_0^\infty dq \left(\frac{a(q)-1}{a(q)} \right) q \sin qr \tag{7.17}$$

The Eqs. (7.14–7.16) must in general be solved by numerical methods. It is unfortunately not yet possible to say which of these theories form the most reliable means of relating $\phi(r)$ to $a(q)$. The difficulties involved have been discussed at length by Gehlen and Enderby (1969) and need not be repeated here. All the theories seem to be capable of distinguishing between metals and insulators; there is also strong evidence that the HN theory is to be preferred over the BG theory for metallic liquids (see, for example, Howells and Enderby (1972)).

7.3.4 The PY Theory for the Hard-Sphere Liquid
This is the only case for which an analytical solution of the basic Eq. (7.14) has so far been obtained. (Wertheim (1963) Thiele (1963) Lebowitz (1964)).

(a) *Pure Liquids*
As Ashcroft and Lekner (1966) pointed out, the interference function can be obtained immediately from the analytic solution provided by Wertheim and Thiele; its form depends (for a given density) on the hard-sphere diameter, R.

The direct correlation function is given in this model by

$$-c(r) = a + br + dr^3 \quad r \leqslant R$$
$$= 0 \qquad\qquad\quad r \geqslant R \tag{7.18}$$

where

$$a = (1 + 2\xi)^2 / (1 - \xi)^4$$
$$b = -6\pi(1 + \xi/2)/(1 - \xi)^4 /R$$
$$d = (\tfrac{1}{2}) \pi (1 + 2\xi)^2 /(1 - \xi)^4 /R^3$$

The quantity ξ, the packing density, is defined by

$$\xi = (\pi/6)nR^3$$

and since n is fixed by the density, the behaviour of $c(r)$ is completely determined by the choice of R.

$c(q)$ can now be obtained directly from the inversion of Eq. (7.18)

i.e., $c(q) = \dfrac{4\pi n}{q} \displaystyle\int dr\ c(r)r \sin qr$

and hence, through Eq. (7.17), $a(q)$ can be completely determined. The extent to which the structure of pure liquids may be thought of in terms of the random packing of hard spheres is of considerable importance.

(b) *Binary Liquids*
The generalised direct correlation functions $c_{\alpha\beta}(r)$ are given (Pearson and Rushbrooke (1957)) by equations of the form

$$h_{\alpha\beta}(q) = c_{\alpha\beta}(q) + \sum_{j=1,2} \rho_j c_{\alpha j}(q) h_{j\beta}(q) \tag{7.19}$$

where, for instance,

$$c_{\alpha\beta}(q) = \frac{4\pi}{q} \int c_{\alpha\beta}(r)\, r \sin qr\, dr \tag{7.20}$$

For a binary alloy Eq. (7.19) represents four equations, but since $c_{\alpha\beta} = c_{\beta\alpha}$ only three have to be considered. The solutions for $h_{\alpha\beta}(q)$ are

$$h_{11}(q) = [c_{11}(q)(1 - \rho_2 c_{22}(q) + \rho_2 c_{12}^2(q))]\, P^{-1}(q)$$

$$h_{22}(q) = [c_{22}(q)(1 - \rho_1 c_{11}(q) + \rho_1 c_{12}^2(q))]\, P^{-1}(q) \tag{7.21}$$

$$h_{12}(q) = h_{21}(q) = c_{12}(q)\, P(q)^{-1}$$

with

$$P(q) = 1 - \rho_1 c_{11}(q) - \rho_2 c_{22}(q) + \rho_1 \rho_2 c_{11}(q)\, c_{22}(q) - \rho_1 \rho_2 c_{12}^2(q)$$

The partial interference functions are related to $h_{\alpha\beta}(q)$ through:

$$a_{\alpha\beta} = 1 + n h_{\alpha\beta}(q) \tag{7.22}$$

The PY equation for a binary hard sphere liquid has been considered by Lebowitz (1964). Lebowitz obtains the Laplace transforms of $rg_{ij}(r)$ exactly, and these have been inverted and evaluated numerically by Throop and Bearman (1965). However, contact between theory and experiment is most readily established in q space and so, following Enderby and North (1968), we return to an intermediate step in Lebowitz's argument in which the direct correlation functions are rigorously derived. These are as follows:

$$-c_{\alpha\alpha}(r) = a_\alpha + b_\alpha r + dr^3 \quad r \leqslant R_i. \tag{7.23}$$

$$= 0 \qquad\qquad r \geqslant R_i$$

and

$$-c_{12}(r) = -c_{21}(r) = a_1 \quad r \leqslant \lambda$$

$$= a_1 + b(r - \lambda)^2 + 4d(r - \lambda)^3 + d(r - \lambda)^4/r \tag{7.24}$$

$$\lambda \leqslant r \leqslant R_{12}$$

$$= 0 \qquad\qquad r \geqslant R_{12}$$

Here

$$R_{12} = (R_1 + R_2)/2 \text{ and } \lambda = (R_2 - R_1)/2.$$

The coefficients a, b, etc. depend only on the hard sphere diameters R_i, the number density, ρ_α and the generalised packing density, ξ, given by

$$\xi = \pi \rho_1 R_1^3/6 + \pi \rho_2 R_2^3/6$$
$$= \eta_1 R_1^3 + \eta_2 R_2^3 \tag{7.25}$$

Lebowitz gives explicit expressions for a, b, etc., which we need not repeat here.

It is necessary to evaluate the three-dimensional Fourier transforms of Eqs. (7.23) and (7.24) to yield $c_{ij}(q)$. This can be done immediately with the result

$$c_{\alpha\alpha}(q) = \frac{4\pi}{q^2} [A_\alpha \sin qR_\alpha + B_\alpha \cos q R_\alpha + C_\alpha] \tag{7.26}$$

$$c_{12}(q) = \frac{4\pi}{q^2} [A \sin qR_{12} + B \cos qR_{12} + C \sin q\lambda + D \cos q\lambda]$$

Enderby and North give expression for $A_\alpha, \beta_\alpha \ldots \ldots D$ in terms of the original coefficients which appear in Eqs. (7.23) and (7.24). Once $c_{\alpha\beta}(q)$ has been computed, it is a straightforward matter to determine $a_{\alpha\beta}(q)$ with the aid of Eqs. (7.21) and (7.22).

7.3.5 Comparison of the Hard Sphere Theory and Experiment

(a) *Pure Liquids*
We follow closely the recent work of Page and co-workers (1969) in order to investigate the detailed differences between a random assembly of hard-spheres and 'real' liquids.

In the case of hard spheres the g(r) has a sharp edge at the hard sphere diameter. In contrast, g(r) for a real liquid has a finite slope at the point where atoms overlap and on taking a Fourier transform to obtain a(q), one obtains a function which exhibits oscillations decreasing in amplitude with increasing q. The oscillations should die out more rapidly for real liquids than for the hard spheres. Moreover, it may be expected that these oscillations will move out of phase with respect to the hard sphere oscillations as the damping increases. Conversely the greater these effects in a(q) are found to be, the greater the level of the atomic overlap implied by them. It is usually accepted that the pair potential for metals is a good deal 'softer' than for unperturbed rare gas atoms in (say) liquid argon. It is, therefore, of interest to demonstrate the change in magnitude of the overlap between argon and rubidium.

Page *et al.* (1969) measured the interference function from q = 0.7 to 13.5 Å^{-1} for liquid argon at a pressure of 1 atmosphere and a temperature of 85°K. These data are compared with the hard sphere PY calculations by Ashcroft in Figure 7.14. R and η were chosen at 1.76 Å and 0.485 in order to give the best fit to the height and position of the main peak of the interference function. It can be seen that the oscillations in the argon curve decrease more rapidly than those of the hard sphere curve and also go out of phase as the value of q increases.

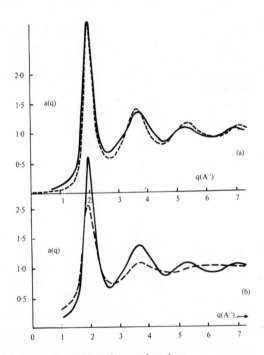

Fig. 7.14. Comparison of interference functions
(a) hard spheres (————) and liquid Ar at 850°K (- - - - -)
(b) hard spheres (————) and liquid Rb at 470°K (- - - - -)
(taken from Page *et al,* 1969).

In addition, Page *et al.* measured the interference function of liquid rubidium at a pressure of 1 atmosphere and at several temperatures from 325°K to 675°K. To compare these data with those for liquid argon, some form of normalisation was required. The normalisation of the q was achieved by scaling so that the first peak in both sets of data coincide. To normalise a(q) itself Page *et al.* chose to regard a(o) as a fixed parameter. Comparison was then made between rubidium and argon under conditions where the two values of a(o) are equal. It can be seen from studying Figure 7.14 that the oscillations in the rubidium data are damped out more rapidly than those for argon and there is a more pronounced phase shift. Nevertheless, it is clear that the random packing of hard spheres serves a useful first approximation to simple liquids at least so far as structure is concerned.

In Figure 7.15 we saw that the experimental data for liquid Te and Se are qualitatively quite different, particularly around the first peak, from those for simple liquids like Rb or Ar. A further point of significance emerges when we observe that there is even less damping in a(q) for Se than for hard spheres. Since the damping associated with hard spheres represents a minimum for simple liquids we are forced to conclude that neither Te nor Se can be discussed in terms of central pair-wise interactions. The extent to which this conclusion is confirmed by the electrical properties of liquid Te and Se will be considered in Section 7.5.2.

It is therefore instructive to examine carefully the damping of a(q) about the asymptotic value of unity. For simple liquids, a measure of the interpenetration of the atomic cores may be obtained. On the other hand, the absence of appreciable damping indicates that non-central forces may be playing an important role.

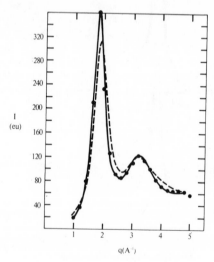

Fig. 7.15. X-ray scattering from liquid NaK.
●●●●● Calculated from hard-sphere theory
─·─·─·─ Experimental.

(b) *Binary Liquids*

Since there are virtually no data available for the partial interference functions for systems other than for certain copper and silver based alloys, the remarks made in this section will necessarily be speculative in character. We shall attempt to squeeze as much information as possible out of the relatively meagre experimental data we have at our disposal.

To test the validity of the hard sphere approach we first consider the equiatomic alloy, NaK, which is a relatively simple liquid alloy and for which X-ray and neutron diffraction data have been obtained (Henniger *et al.*, 1966). In addition data are available for the density and χ_T over the whole range of composition (Abowitz and Gordon (1962)). As for the one component case, the final answer will depend on the choice of ξ, which in turn relates to the choice of R_1 and R_2. Ashcroft and Lekner (1966) found that for pure Na and pure K, a close fit with experiment was obtained with $\xi = 0.45$ so that $R_1 = 3.28A$ (Na) and $R_2 = 4.06A(K)$. Presumably these would alter somewhat in the alloy due to the change in electron density. However, in the absence of any theory of this effect, we shall retain these hard sphere diameters so that if the density is taken as 0.87 g cm^{-3}, which corresponds to $373°K$, ξ (alloy) $= 0.451$. With these assumptions for R_1 and R_2, Enderby and North (1968) calculated $a_{ij}(q)$ from the hard sphere PY theory. In Figure 7.15 we compare the experimental X-ray diffraction data with the theoretical curve derived from the calculated $a_{ij}(q)$ and the published values of the x-ray form factors (Hanson *et al.* (1964)). The measurement of agreement, particularly around the first peak, is encouraging and shows that for this alloy at least the PY hard-sphere description is a very useful first approximation.

Since the position of the first maximum in a_{ij} (q) is virtually independent of c_i, we expect the main maximum in the total X-ray curve to move monotonically from the pure Na peak to the pure K peak as c_i varies from 1 to 0 (Eq. (7.10)). This was indeed observed for Na-K alloys by Orton and Williams (1960), and corresponding behaviour for many other alloy systems has also been reported (e.g., Hg-Tl: Halder *et al.* (1966) Hg-In; Kim *et al.* (1961)). It is also worth emphasising that this model predicts that the main peak in a_{12} falls roughly midway between a_{11} and a_{22}.

The following general comments can be made about liquid alloys of this type:

(i) They are thermodynamically rather simple and are not characterized by marked heats of mixing, etc.

(ii) The X-ray pattern shows very little structure.

(iii) The partial interference functions may, with reasonable accuracy, be obtained from experiments involving one type of radiation and several alloy compositions (see especially Halder and Wagner (1967b).

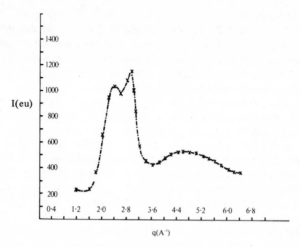

Fig. 7.16. X-ray scattering from liquid $Cu_6 Sn_5$ (taken from Enderby *et al.* 1966).

We next consider liquid alloys which, though metallic in character, have thermodynamic properties which suggest a fair degree of local order. The total X-ray pattern for such liquids appears to have rather more structure than for hard sphere alloys and is frequently characterized by a split main peak. One example is liquid $Cu_6 Sn_5$ and the X-ray data for this liquid is given at Figure 7.16.

If we examine the interference functions shown in Figure 7.8 we can immediately see that the X-ray behaviour follows from the fact that the position for the main peaks in a_{Cu-Sn} and a_{Cu-Cu} is almost identical. In *real* space, a contraction of the Cu-Sn distance has occurred which represents a clear departure from hard sphere behaviour (i.e., the hard sphere radii are not additive). This conclusion is consistent with the known thermodynamic properties of this alloy (e.g., exothermic heat of mixing, negative volume of mixing). In other cases where a double-headed peak has been observed (e.g., Au/Sn Cd/Sb) similar thermodynamic behaviour to that of Cu/Sn is also found. There is evidence (Hendus (1947)) that the definition of the double-headed peak reduces with increasing temperature, which probably means that the position of the peak in a_{12} (q) moves to somewhat lower values of q as the temperature is raised.

As part of a long term programme designed to lead to a fuller understanding of the structure of liquid alloys, Hawker, Howe and Enderby (1973a and 1973b) have recently completed measurements on liquid $Ag_2 Te$, CuTe and $Cu_2 Te$. Three samples of CuTe and $Cu_2 Te$ were prepared using natural Te and (i) natural copper, (ii) copper enriched to 99% with [63]Cu, (iii) copper enriched to 99% with [65]Cu. The coherent scattering cross-sections for [nat]Cu, [63]Cu and Te are in the proportion 1:0.73:1.96:0.51. Similar samples for $Ag_2 Te$ were prepared by using natural Ag, [107]Ag and [105]Ag.

Several conclusions have emerged from this study, the most significant being that the existence of clusters in liquid semiconducting alloys of the Cu-Te type has not been confirmed. Such clusters were first postulated by Hodgkinson (1970) and by Cohen and Sak (1972) as a basis for a general discussion of the electrical properties of liquid semiconductors. As applied to the CuTe system, for example, islands of semiconducting Cu_2Te are dispersed in semi-metallic Te or *vice versa*. Recently, this approach has been applied with apparent success to liquid semiconductors by combining the structural features of the model with ideas derived from classical percolation theory.

The partial interference functions for liquid Cu_2Te are displayed in Figure 7.17, the vertical lines indicating the residual uncertainty due to the inevitable experimental errors in F(q). Hawker *et al.* have drawn attention to three features which they consider to be worthy of special note:

(i) the remarkable similarity between a_{CuCu} and the interference function for pure liquid copper;

(ii) the significant difference between a_{TeTe} and the interference function for pure liquid Te;

(iii) the absence of a definite feature in any of the partial interference functions corresponding to a 'pre-peak' which was found in F(q) at a q value of $1A^{-1}$.

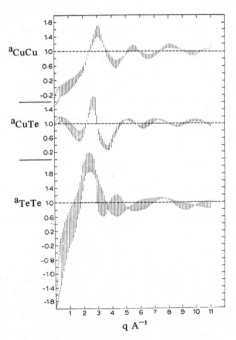

Fig. 7.17. The partial structure factor for liquid Cu_2Te at $1000°C$.

The first feature suggests that the packing of the Cu ions in liquid Cu_2Te is a highly disordered one dominated by a hard-core interaction equal to that for Cu ions in pure liquid Cu, whilst the second one indicates that the covalent character of pure liquid Te disappears as Cu is added. These conclusions are consistent with the view expressed by Enderby and Hawker (1972) that models involving Cu substitutions into covalent Te chains are unlikely to be useful as a starting point for explaining the properties of liquid Cu-Te alloys. The third feature illustrates very clearly the danger in attempting to relate maxima in F(q), particularly at low q, to a particular distance in real space. The pre-peak in F(q) simply reflects the effect of combining three partial interference functions each with a characteristic q-dependence. Hawker *et al.* found no evidence of long range order in any of the partial radial distribution functions.

The extent to which the measured partial interference functions are consistent with a more or less random mixture of Cu and Te ions in which substantial electron transfer has taken place is currently under theoretical investigation.

7.4 ELECTRON TRANSPORT IN LIQUID SEMICONDUCTORS — EXPERIMENTAL

7.4.1 Introduction

We now outline the techniques which have been developed to measure the principal electron transport parameters of liquid semiconductors. The three parameters which are most commonly investigated are the conductivity (σ), the Hall coefficient (R) and the thermoelectric power (S).

This section is primarily concerned with matters arising from the fluidity of the specimens. To measure R (see especially Cusack *et al.* (1965)), the sample has to be constrained to a thin uniform rectangular film equipped with current leads and potential probes. If the container is not completely full the film will move when subject to electrodynamic forces. The motion of tiny voids near the Hall probes causes unpredictable and irreversible increments of voltage. To prevent the formation of small gas bubbles in the specimen it is very desirable to prepare and handle the specimen under vacuum conditions.

Even if the container is well filled, the system of specimen current I and magnetic field H required for the measurement may cause circulating motions governed by magneto-hydrodynamic equations. The nature of these motions is difficult to calculate and will in any case be complicated by ordinary convection if the temperature is not uniform. The change of temperature distribution due to magnetohydrodynamic flow can give rise to a thermal e.m.f. at the Hall probes which depends linearly on both H and I — like the Hall voltage itself. There may also be magnetically induced voltages proportional to H^2. A suitable test for these specimen effects is to vary all the parameters, especially the geometrical ones.

Sample holders or 'cells' have been used with thicknesses varying from 20 to 400 μm. For d.c. methods the body of experience is that, for thickness

< 130 μm, reliable results can be obtained; in general, however, the thinner the specimen the better. Since the thickness of a Hall specimen should be uniform and of known value it is desirable to construct cells of transparent material suitable for measuring the thickness by optical interference methods. An alternative method is to exploit the fact that the Hall coefficient of liquid Hg is now rather well known; the mean of the best results is -75.3×10^{-5} e.m.u. If the thickness is not very uniform or is for some reason difficult to measure, the Hg value can be used to calibrate cells because, other things being equal, the Hall voltage is proportional to the Hall coefficient.

Alumino-silicate glasses have been found useful for work up to 750°C by Enderby and co-workers. These glasses combine a high softening point (800°C) with a medium coefficient of expansion (42.5×10^{-7}deg.C^{-1}) and it is therefore possible to use a lower melting point 'solder' glass with the same expansivity as an adhesive.

The construction of a quartz cell has been described by Busch and Tieche (1962). An upper plane polished plate is sealed to a tube which initially contains the specimen. A lower plate has an ultrasonically drilled rectangular cavity about 22 mm $\times$ 6 mm $\times$ (0.1–0.5) mm and wells for making electrical contact between the specimen and the five graphite plugs. The cell is then held together by tungsten wires and springs. An alternative scheme which involves creating the rectangular cavity by an etching technique has been described by Kendall (1964).

The above discussion has been concerned with cells suitable for measuring R. However, σ may also be measured using the same set up and on the same specimen this has in fact been the common practice in the past. An alternative approach is to use an electrodeless method and such experiments have produced a valuable and impressive collection of conductivity data, particularly for liquid alloys (Roll and Motz (1957) and many other papers in this series).

To measure S, the somewhat severe geometrical constraints which obtain for R and σ are relaxed. Enderby and co-workers have used the cell shown in Figure 7.17, but many variations both in choice of cell material and electrode configuration are possible (see, for example, Davies (1969)). The idea of separating the counter electrode material from the liquid (which may be chemically highly reactive) by a thin sheet of refractory material was first suggested by Cusack *et al.* (1964).

7.4.2 The Electrical Measurements

(a) *The Hall Coefficient and Hall Mobility*
A variety of methods has been used to measure the Hall voltage which is frequently not more than 1 or 2 μV. The d.c. technique gives, with suitable precautions, reliable results although for high resistance materials ($\sigma \leqslant 10\Omega^{-1}$cm^{-1}) one of the a.c. methods is to be preferred. This is because the 'off-set' voltages which arise because of (inevitable) probe misalignments, are difficult to eliminate in the d.c. technique.

In the two-frequency system (Russel and Wahlig (1950)) the Hall voltage has two components of equal amplitude with frequencies equal to the sum and differences of the two primary frequencies.

$$V_H \propto I_o H_o \cos 2\pi \ (F \pm f)t: \qquad\qquad (7.27)$$

where I_o and H_o are the peak values of the current and magnetic field. Spurious voltages due to thermoelectricity or alloy polarization can be ignored so long as the appropriate relaxation time of the sample is long compared with F^{-1} and f^{-1}. The main advantage of the method, however, is that the spurious voltages due to misalignment of the Hall probes and to the effects of the magnetic field do not have the same frequency as V_H and therefore do not have to be compensated for or measured.

The frequencies found useful have been
Busch and Tieche (1962)   H: 50 c/s I: 85 c/s V_H: 135 c/s
Greenfield (1964)   H: 60 c/s I: 77 c/s V_H: 17 and 137 c/s
In these experiments the frequency of H has been determined by the mains frequency and that of V_H is kept low to minimize capacitative coupling.

The technique used by Male (1967) represents an interesting variation of the double a.c. approach. Only the magnetic field is alternating (21 c/s in Male's work), the Hall signal appearing with the same phase and frequency as the field. An apparatus using this principle has been described by Hamer (1962) and a more sensitive version is reported by Rzewuski and Werner (1965). The use of a direct primary current means that the spurious misalignment voltage is readily suppressed without the use of rejection filters. This avoids extra loading in the detector circuit, a factor which can be important with high-impedance samples. All unwanted thermomagnetic and magnetoresistance effects are discriminated against, with the exception of the Nernst effect, where the primary thermal gradient arises from the generation and absorption of Peltier heat at the current electrodes. It is generally assumed that the contribution from this effect may be neglected.

(b) *The Thermoelectric Power and Conductivity*
Potentiometric methods or the use of high impedance d.c. microvoltmeters have proved satisfactory methods of measuring the voltages involved. Some workers (e.g., Enderby and Walsh (1966)) prefer to keep the temperature of the bottom junction shown in Figure 7.18a fixed and vary the temperature of the other junction. The slope of the e.m.f.-temperature curve gives the thermoelectric power P of the thermocouple formed by the specimen and the counter electrode. Given the absolute thermoelectric power of the counter electrode (see specially Cusack and Kendall (1958)) the absolute thermoelectric power of the specimen may be obtained directly from
    S (specimen) = S (counter − P (specimen − counter)
The sign convention is shown in Figure 7.18 where the other method often used (fix Δ T, vary T and T_o) is illustrated. Suitable counter electrodes are Cu, Pt and Mo.

If d.c. methods are used to measure σ for alloys, the prolonged passage of

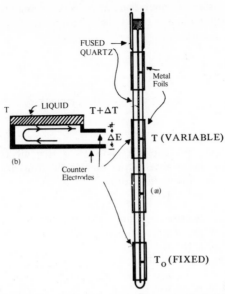

Fig. 7.18. (a) Sample holder for thermopower measurements devised by Enderby and
        Walsh (1966)
        (b) Sign convention for the 'ΔT' method.

the current may create a concentration gradient. Frequent reversals of the
current are therefore to be recommended.

7.5   ELECTRON TRANSPORT IN LIQUID ELEMENTAL SEMICONDUCTORS

7.5.1   Introduction

At the beginning of Section 7.1.5., reference was made to the classification
scheme introduced by Mott (see especially Mott (1971)). The significant
parameter in Mott's approach is the d.c. conductivity because this quantity
enables the magnitude of Λ, the electronic mean free path, to be estimated.
If d represents the average interatomic spacing then the three cases of
interest are:

(a) $\underline{\Lambda > d}$

This is the so-called weak scattering limit and applies for $\Lambda \geqslant 3\text{Å}$ which
corresponds to conductivities in excess of about 3,000 $\Omega^{-1}\text{cm}^{-1}$. Many
liquid metals and alloys fall into this category; the conductivity appears to
be expressible in the simple form

$$\sigma = \frac{k_F^2 e^2 \Lambda}{3\pi^2 h} \tag{7.28}$$

with k_F, the Fermi wave number, calculated according to free electron theory. A prescription which enables Λ to be evaluated in terms of $w_q(k_F)$ has been given by Ziman (1961). The thermoelectric power can be related to σ through

$$S = \frac{\pi^2}{3} \frac{k_B^2 T}{e} \left[\frac{d(\ln \sigma)}{dE} \right]_{E = E_F} \tag{7.29}$$

whilst the Hall coefficient is given by the elementary expression:

$$R = \frac{1}{ne} \tag{7.30}$$

In Eq. (7.30) n represents the density of free electrons and is usually calculated by assuming that all the valence electrons are free to take part in the process of electronic conduction.

(b) $\underline{\Lambda \sim d}$

The theory of electronic conduction in this regime is not yet fully developed and most of the arguments used in this section are due to Mott (1966, 1967, 1970, 1971).

Following the theoretical work of Anderson (1958) Mott concludes that a lower limit for σ *must* exist below which conduction by processes involving extended electronic states is not possible. By making certain plausible assumptions Mott determines σ_{min} to be $\cong 200 \ \Omega^{-1} \ cm^{-1}$. For liquid conductivities above this value, but below $\cong 3,000 \ \Omega^{-1} \ cm^{-1}$, Mott argues that the density of states at the Fermi energy, $n(E_F)$, rather than Λ becomes the controlling factor. Setting $\Lambda \cong d =$ constant, Mott finds that

$$\sigma = \frac{g^2 e^2}{3\hbar d} \tag{7.31}$$

where

$$g \left\{ = \frac{n(E_F)}{n_o(E_F)} \right\} \quad \text{is in the range 1 to} \sim 0.3$$

and $n_o(E_F)$ is the free electron density of states. The thermoelectric power involves the transport of energy (strictly speaking, the transport of entropy) and should therefore be more amenable to interpretation than the Hall effect (see below). Mott argues that Eq. (7.29) can again be used so that

$$S = \frac{\pi^2}{3} \frac{k_B^2 T}{e} \ 2 \left[\frac{d \ln n(E)}{dE} \right]_{E = E_F} \tag{7.32}$$

It is not yet possible to go beyond the simple free electron theory for the Hall coefficient with any degree of confidence. The central difficulty here is that the Hall coefficient involves the transport of *momentum* and this is not a well-defined quantity in situations in which $\Lambda \sim d$ (Edwards (1962)). Banyai and Aldea (1966) and Aldea (1967) conclude that the sign of R depends on the degree of filling of the band whereas Matsurbara and Kaneyoshi (1968) consider that R is always negative.

Recent work by Straub *et al.* (1968), Fukuyama *et al.* (1969) and Friedman (1971), however, support a suggestion made by Ziman (1967) that provided the electric current is carried by electrons close to E_F

$$\frac{R}{R_o} = C(g)^{-\beta} \tag{7.33}$$

where the subscript o means 'calculated according to the free electron theory' and C is a constant of order unity. The value of β is at present in dispute, although most workers are of the opinion that $\beta = 1$.

(c) $\underline{\Lambda \sim d}, g < 0.3$

For g values less than about 0.3 Mott suggests that Anderson localisation occurs. The conductivity arises from the thermal excitation of carriers across the mobility gap or by hopping processes. The semi-empirical relationship for the conductivity in this regime (Stuke (1969)) is

$$\sigma = De^{-E/kT} \tag{7.34}$$

where D is constant and E is a characteristic energy. According to Cutler and Mott (1969) Eq. (7.29) can still be used to calculate the thermoelectric power.

Again, the Hall effect remains a difficulty. Holstein (1961) and Holstein and Friedman (1968) and Friedman (1971) are among several workers who have considered how the Hall coefficient should be calculated for systems in which the electronic wave functions are localised. Friedman has used his theory to obtain a numerical estimate for R in short mean free path situations. He finds that $R\sigma$ is independent of temperature and approximately equal to $k\frac{T}{I}\mu_o$ where μ_o is the conductivity mobility and I is an overlap integral.

7.5.2   Liquid Ge and Si

Liquid Si and Ge are metallic in character. The optical properties, Hall coefficient, thermoelectric power and conductivity are, (where measured), similar to those for other quadrivalent liquid metals like Sn and Pb (Hodgson (1961) Busche and Tieche (1962)). A selection of experimental data for liquid Si Ge, Sn and Pb is presented in Table 7.1.

There have been several useful reviews of the theory for the electrical properties of pure liquid metals and we content ourselves with a brief summary of the present situation. Most of the relevant theoretical ideas to this section are due to Ziman and co-workers (see especially Ziman (1967)).

If we combine the notion of the pseudopotential with the Born scattering theory we obtain explicit expressions for the conductivity and the thermoelectric power. Thus Eqs. (7.28) and (7.29) become

$$1/\sigma = (3\pi^2/e^2 \, \hbar v_F^2) \, \Omega < F(q) > \tag{7.35}$$

$$S = (\pi^2 k_B^2 T\xi)/3eE_F \tag{7.36}$$

where

$$F(q, k) = |w_q(k)|^2 a(q) \tag{7.37}$$

$$<F(q)> = \int_o^{2k_F} F(q, k_F)\, q^3\, dq \tag{7.38}$$

$$\xi = 3 - 2\,F(2k_F)/<F> + \frac{k_F}{2} <\frac{\partial F}{\partial k}>/<F> \tag{7.39}$$

v_F is the Fermi velocity and Ω is the atomic volume.

It' is important to re-emphasise that *no effective mass* correction should be included within the first order theory. A further point of significance is that if the pseudopotential is assumed to be local (i.e., independent of k), the last term in the expression for ξ vanishes.

No attempt is made to compare theory and experiment for liquid Si and Ge, except to consider the validity of Eq. (7.30). The main obstacle is that both $<F(q)>$ and $<\frac{\partial F}{\partial k}>$ are extremely sensitive to the choice made for pseudopotential and the interference function, neither of which are well known for either Si or Ge at the present time. For liquid Pb and Sn, however, the measure of agreement secured in recent calculations is very encouraging (Table 7.1) and there is no reason for believing that liquid Ge or Si are in any way anomalous.

7.5.3 Liquid Te and Se

The electrical properties of liquid Te and Se have been investigated by several workers and a summary of the experimental position is given at Table 7.2. Although no single theory of electron transport in these materials commands general acceptance the model proposed for liquid Te by Cabane and Friedel (1971) is worthy of special comment.

TABLE 7.1 *Electrical Properties of Liquid Si, Ge, Sn and Pb*

Liquid Metal	Temperature °K	$\rho(\mu\Omega\text{-cm})$ Calculated	$\rho(\mu\Omega\text{-cm})$ Observed	$S(\mu V\ deg^{-1})$ Calculated	$S(\mu V\ deg^{-1})$ Observed	R/R₀
Si	1713	25[a]	60[c]			
Ge	1232	39[a]	85[c]		~ 0[c]	1.00[e]
Sn	505	32[a]	48.0[c]		− 0.5[d]	1.00[e]
Pb	613	77[b]	96.5[b]	− 4.2[b]	− 3.5[b]	0.88[e]

References for Table 7.1

(a)   Ashcroft and Lekner (1966).
(b)   North *et al* (1968b).
(c)   Cusack, N. E., (1963), *Repts.Prog.Phys.,* **26**, 361.
(d)   Marwaha, A. S., (1967), *Adv. Phys.,* **16**, 617.
(e)   Busch and Guntherodt (1967).

TABLE 7.2 *Properties of Liquid Te and Se*

Liquid	Melting Point °C	σ Ω^{-1} cm^{-1}	Ref.	R cm^3/C	Ref.	S μV deg^{-1}	Ref.
Te	451	1760	BT	-13.0×10^{-4} -13.0×10^{-4}	EW BT	$+20$	CM
Se	220	10^{-6} 10^{-8} 10^{-7}	AAM GCG HM	~ -100 $\sim -3 \times 10^5$	AAM GCG	$+2000$	HM

References for Table 7.2

CM: Cutler and Mallon (1962).
BT: Busch and Tieche (1962).
EW: Enderby and Walsh (1966).
AAM: Aliev, Abdinov and Mekhtieva (1966).
GCG: Glazov, V. M., Chizhevskaya, S. N. and Glagoleva, N. N., (1969), Liquid Semiconductors Plenum Press, New York.
HM: Henkels, H. W. and Maczuk, J., (1953), *J. App. Phys.*, **24**, 1056.

Cabane and Friedel analyse in detail the structural data described in Section 7.1.4 and conclude that although the interatomic forces are not of a simple pair-wise type, the notion that independent and well-defined chains exist in liquid Te is incorrect. They argue that a better description is to regard the structure of liquid tellurium in terms of a covalently bonded network in which the coordination number of the first shell is between 2 and 3. They then consider in detail the band structure which is to be expected from this model and propose two possibilities (Figure 7.19a, b). If we ignore the theoretical difficulties with respect to the Hall effect for

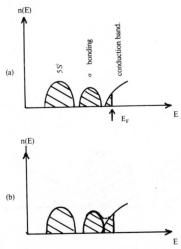

Fig. 7.19. The density of states in liquid Te according to Cabane and Friedel (1970).

systems in which $\Lambda \sim d$, and equate R with R_0, then the model implied by Figure 7.19b can explain both the sign and the magnitude of the Hall coefficient. Moreover, the existence of a substantial Knight shift, K, (Cabane and Froidevaux (1969)) in liquid Te is consistent with a high density of conduction electrons which the model of Cabane and Friedel requires.

Mott (1971), whilst accepting the structural model proposed by Cabane and Friedel, has suggested that the density of states will be influenced by a broadened antibonding band so that E_F will be in a region where $dn(E)/dE$ is negative (Figure 7.20). The conductivity of liquid Te is only 1300 Ω^{-1} cm^{-1} at the melting point which means that equations (7.31) and (7.32) should be used to calculate σ and s. The positive and comparatively large thermo-electric power of liquid Te is readily explained in this manner. If it can be further assumed that $R = R_0 \, g^{-\beta}$, a reasonably complete description of the electrical properties of liquid Te is possible. The careful NMR work of Warren (1972b) shows that K, σ and R are all related by laws of the form

$$K^2 = A\sigma$$

$$R = B\sigma^{-1}$$

where A and B are constants independent of temperature (Figure 7.22). Warren argues that since the temperature dependence of K results mainly from $n(E)$ it is reasonable to set $K \propto g^2$. This implies a β value of 2, in disagreement with the conclusions of Friedman (1971) and Straub *et al.* (1968). The temperature dependence of K σ and R reflect, in this model, an increase in density of state at E_F as the temperature is raised — a physically plausible picture.

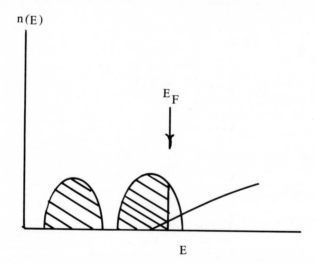

Fig. 7.20. The density of states in liquid Te according to the model proposed by Mott (1971).

A further piece of supporting evidence is to be found by considering the 'entrancement factor' η which is the ratio of the actual nuclear magnetic relaxation rate to that calculated with the aid of the Korringa relationship. The model proposed by Mott predicts that $\eta \propto \sigma^{-1}$ and this is again found to be the case (Figure 7.22).

Liquid Se is not well understood at the present time. The electrical resistivity is extremely high whilst the Hall mobility, μ_H, evaluated at the melting point is an order of magnitude lower than any other known liquid (electronic) conductor. Furthermore, the electrical properties of Se are strongly dependent on impurity and oxygen content as pointed out by Aliev, Abdinov and Mekhtieva (1966). Experimental results for R, σ and μ_H are given in Figure 7.23.

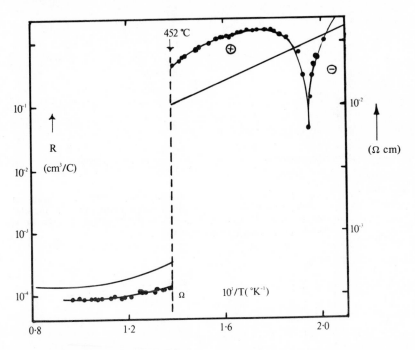

Fig. 7.21. Resistivity and Hall coefficient of solid and liquid Te as a function of temperature (Busch and Tieche (1962)).
——————— : resistivity
●●●●— : Hall coefficient.

7.6   ELECTRON TRANSPORT IN BINARY LIQUIDS I

7.6.1   Introduction

The electrical properties of liquid alloys vary with composition in two distinctive ways. In some cases the Hall coefficient, thermoelectric power

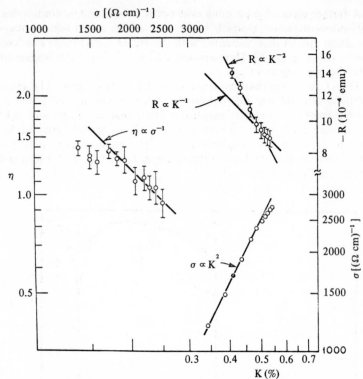

Fig. 7.22. Correlations of NMR and transport parameters with temperature being the
implicit variable: centre left, relaxation enhancement (η) vs electrical
conductivity (σ); upper right, Hall coefficient (R) vs Knight shift (K); lower
right, electrical conductivity vs Knight shift.

(Taken from Warren (1972b)

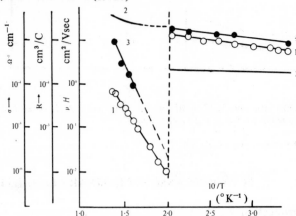

Fig. 7.23. The electrical properties of liquid and solid Se
Curve 1: The Hall coefficient   Curve 2: The conductivity   Curve 3: The Hall mobility.

and conductivity are essentially smooth functions of composition and retain values across the entire phase diagram which are typical of liquid metals. In other cases, however, R, S and σ vary strongly with composition and often achieve values which are substantially different from those typical of the metallic state. The most marked deviations from 'normal' behaviour occur at fixed compositions which are governed by chemical valence considerations. In terms of Mott's classification (discussed in Section 7.5.1.) it appears that alloys of the first sort (type I) correspond to (a) and (b) whilst alloys of the second sort (type II) correspond to those denoted by Mott as (c).

The nearly free electron theory developed by Faber and Ziman (1964) is an obvious starting point for discussing liquid alloys of type I. For those cases in which information is available about the three partial interference functions which characterize the structure of binary alloys, close quantitative agreement between theory and experiment has been obtained. We emphasize that a positive $d\sigma/dT$ is entirely consistent with metallic behaviour in liquid alloys on account of the temperature dependence of the partial interference functions. For this reason many liquid alloys which have in the past been thought of in terms of a semiconducting framework should more properly be regarded as metallic. (It may, in certain cases, be necessary to introduce the Mott g^2 factor but there is little evidence either way on this important point at the present time). Alloys of the second type will form the subject for section 7.7.

7.6.2   The Nearly Free Electron (NFE) Theory for Liquid Alloys

Following Faber and Ziman (1964), we shall assume that the scattering of electrons in a binary alloy can be treated by the Born approximation and evaluated in terms of a pseudopotential. It follows that the resistivity and thermoelectric power S are still given by Eqs. (7.35) and (7.36) but now

$$F(q,k) = |w_1|^2 \left\{ c(1-c) + c^2 a_{11} \right\} + |w_2|^2 \left\{ c(1-c) +) (1-c)^2 a_{22} \right\}$$
$$+ 2 w_1 w_2 c(1-c) (a_{12} - 1). \quad (7.40)$$

In (7.40) the pseudopotentials of the two components are written as w_1 and w_2 whilst c is the atomic fraction of component 1. The Hall coefficient will remain free electron like but may contain the density of states correction factor discussed in Section 7.5.2.

A considerable simplification occurs for alloys in which n and k_F are independent of c and in which the substitutional model (which implies $a_{11} = a_{22} = a_{12} = a$) is valid. Let ξ_1 and ξ_2 denote the thermoelectric parameters for the pure liquids 1 and 2 and ξ the thermoelectric parameter for the alloy; σ_1, σ_2 and σ are the corresponding conductivities. It then follows (Howe and Enderby, 1967) that

$$\xi = 3 - \sigma (\alpha c^2 + 3c + \gamma) \quad (7.41)$$

provided the pseudopotentials are assumed to be local. The constants α, β and γ are given by

$$\alpha = \left\{ \ (3-\xi_1)/\sigma_1 + (3-\xi_2)/\sigma_2 - 2\,(3-\xi_1)\,(3-\xi_2)/\sigma_1\sigma_2)^{\frac{1}{2}}\,(\bar{a}-1)/\bar{a} \right\}$$

$$\beta = \left\{ \ (3-\xi_1)/\sigma_1 + (3-\xi_2)/\sigma_2 - 2\,(3-\xi_1)\,(3-\xi_2)/\sigma_1\sigma_2)^{\frac{1}{2}} \right\} /\bar{a} \quad (7.42)$$

$$+ 2 \left\{ ((3-\xi_1)\,(3-\xi_2)/\sigma\,\sigma_2)^{\frac{1}{2}} - (3-\xi_2)/\sigma_2 \right\}$$

$$\gamma = (3-\xi_2)/\sigma_2$$

where $\bar{a}$ means the interference function evaluated at $q = 2k_F$.

7.6.3   Comparison of Theory and Experiment

(a) *Simple Liquid Alloys*

There are five equi-valence alloys whose behaviour approximates rather closely to the simple alloy system postulated above. Since Eq. (7.41) enables the alloy thermoelectric power to be calculated in terms of $\sigma, \sigma_1, \sigma_2, S_1$ and S_2, a quantitative comparison of theory and experiment can be made.

The first of these alloys to be studied was the monovalent system, liquid Ag-Au and the measure of agreement secured between theory and experiment was very satisfactory (Howe and Enderby, (1967)). Subsequently, data for the four other systems (Mg-Cd, In-Tl, Pb-Sn and Bi-Sb) were obtained by Enderby *et al.* (1968) and all the experimental results are shown in Figure 7.24. The full lines represent the composition dependence of S predicted by the basic theory.

For Ag-Au, Pb-Sn and Bi-Sb, there are no significant disagreements between theory and experiment. By contrast, the composition dependence of S for Mg-Cd (and to a lesser extent In-Tl) is very different from that expected on theoretical grounds. Small departures from the substitutional model and the condition of constant k_F can be allowed for but detailed calculations show that such corrections do not alter the theoretical curve by more than a few per cent. In particular, the discrepancies for liquid Mg-Cd are well outside the experimental errors and the uncertainties introduced by departures from ideal behaviour.

The most plausible explanation of the discrepancies is the failure to take into account the non-locality of the pseudopotentials in setting up the theory for S. To make analytic progress, Enderby *et al.* (1968) assumed that the thermoelectric power is dominated by the last term in equation. It then follows that

$$\xi = 3 - \left\{ \ c^2(3-\xi_1)/\sigma_1 + (1-c)^2(3-\xi_2)/\sigma_2 + c(1-c)I \right\} \quad (7.43)$$

where I is independent of c and is given by

$$I = (3\pi/\hbar e^2 v_F^2)\,\Omega\,(4k_F^4)^{-1}\,[\alpha + \beta]$$

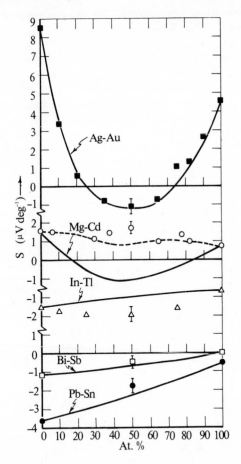

Fig. 7.24. The thermoelectric power of equivalence alloys as a function of composition: ■ Ag-Au; ○ Mg-Cd; △ In-Tl; □ Bi-Sb; ● Pb-Sn. (The first named element is on the left of the diagram.) The predicted composition dependence of S:
————— with energy independent pseudopotentials
— — — — — using Eq. (7.43) with I chosen as 300 $\mu\Omega$-cm.

with

$$\alpha = \int_{0}^{2k_F} [w_1 - w_2] \left\{ \frac{\partial \bar{w}_1}{\partial k} - \frac{\partial \bar{w}_2}{\partial k} \right\} q^3 dq$$

$$\beta = \int_{0}^{2k_F} \left[w_1 \frac{\partial \bar{w}_2}{\partial k} + w_2 \frac{\partial \bar{w}_1}{\partial k} \right] q^3 dq$$

In these equations $\partial \bar{w}/\partial k$ represents the momentum gradient of the pseudo-potential evaluated at the Fermi energy.

The experimental data can be fitted fairly well for Mg-Cd if I is chosen as $300 \, \mu\Omega - cm$ (see Figure 7.24). It will be of interest to see if the k-dependent pseudopotentials proposed by Shaw (1969) yield a value of I which is consistent with the experimental one. Such calculations are still in progress at the present time and no firm conclusion can yet be reported.

No systematic investigation of the Hall effect amongst alloys of type I has been carried out although there is evidence (Busch and Guntherodt, (1967)) that equivalent alloys are among those for which $R = R_o$ at all compositions.

(b) *Alloys in which* $\dfrac{d\sigma}{dT}$ *is Positive*

We now turn to alloys which, though still of type I, are characterised by a positive temperature coefficient of conductivity. A general discussion is hampered by a lack of knowledge of the partial interference functions and for this reason we focus particular attention on the alloy system Cu-Sn. Liquid Cu-Sn is of particular interest because:

(i)   the three partial interference functions, a_{Cu-Cu}, a_{Sn-Sn} and a_{Cu-Sn} are available from experiment at one composition, $Cu_6 Sn_5$ and are substantially different from those expected from a hard-sphere model;

(ii) $2k_F$ changes from $2.60 \, \text{Å}^1$ for pure Cu to $3.13 \, \text{Å}^{-1}$ for pure Sn and therefore spans the main maxima for both a_{Cu-Cu} and a_{Cu-Sn};

(iii) compositions in the range 10—30 at % Sn have a positive temperature coefficient of conductivity (Roll and Motz, (1957)).

(iv) systematic investigations of σ, S and R as functions of composition have been carried out by Enderby and Howe (1968) and by Enderby, Hasan and Simmons (1967).

A selection of the experimental data is shown in Figures 7.25 to 7.28.

It is clear from the Eqs. (7.35) and (7.39) that, provided we ignore effects due to non-locality of w the composite quantity $(3-\xi)/\sigma$ may be evaluated from a knowledge of $F(2k_F)$; this in turn depends on the values of the pseudopotentials and the partial interference function evaluated at $2k_F$.

If the interference functions are taken as approximately independent of c (Enderby, North and Egelstaff, (1967), Halder and Wagner, (1967)) the data provided by Enderby *et al.* (1966) enable $\bar{a}_{Cu-Cu}$, $\bar{a}_{Cu-Sn}$, and $\bar{a}_{Sn-Sn}$ to be evaluated as $2k_F$ increases from $2.50 \, \text{Å}^{-1}$ to $3.13 \, \text{Å}^{-1}$. Inspection of these data shows that $\bar{a}_{Sn-Sn}$ falls monotonically from 1.18 to 0.60. By contrast, $\bar{a}_{Cu-Cu}$ and $\bar{a}_{Cu-Sn}$ rise steeply to maxima of 2.9 and 2.3 respectively as $2k_F$ increases from $2.60 \, \text{Å}^{-1}$ to $2.88 \, \text{Å}^{-1}$; subsequently for $2k_F$ in the range 2.88 to $3.13 \, \text{Å}^{-1}$, both $\bar{a}_{Cu-Cu}$ and $\bar{a}_{Cu-Sn}$ fall to values 1.80 and 1.00 respectively.

w_{Cu} and w_{Sn} were obtained from the measured thermoelectric powers and resistivities of pure liquid copper and pure liquid tin with the aid of Eqs. (7.35) and (7.39) (c = 1 and 0 respectively). $F(2k_F)$ for the liquid alloy can now be evaluated if it is assumed that the k_F dependence of w is small

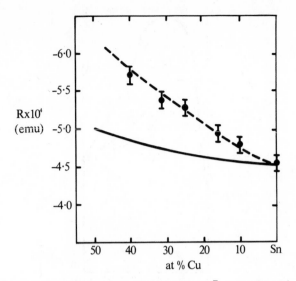

Fig. 7.25. The Hall coefficient of liquid Cu-Sn: ⚹ experimental observation; ———— free electron Hall coefficient; – – – – – – modified free electron behaviour with calculated according to 7.44.

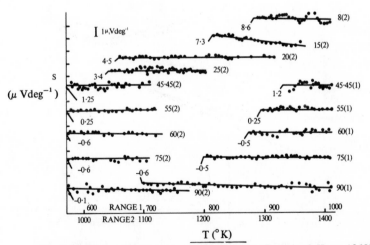

Fig. 7.26. Thermoelectric power of liquid Cu-Sn (Enderby & Howe 1968).

A ⟋————————— B(C)

A is the initial value of S
B is the atomic percentage of Sn
C is the temperature range.

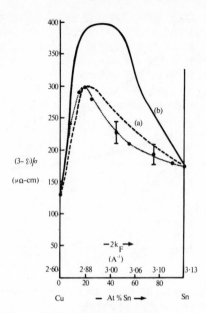

Fig. 7.27. The quantity $(3-\xi)/\sigma$ as a function of composition for liquid
Cu-Sn: ...⬤.....⬤... experiment
(a) -- -- -- -- theory with observed interference functions
(b) ————— theory with hard sphere interference functions

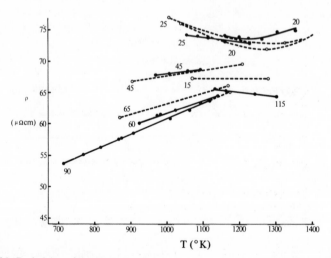

Fig. 7.28. Resistivity of liquid Cu-Sn alloys
————— Enderby and Howe (1968)
-- -- -- -- -- Roll and Motz (1957).
The number in front of each set of data points represents the atomic percentage of Sn.

compared with that for $\bar{a}$. Curve (a) of Figure 7.27 shows the theoretical form of $(3-x)/\sigma$ calculated directly from $F(2k_F)$; the most striking feature is the sharp rise in this quantity as the concentration of Sn increases up to 20 at % $(2k_F - 2.88 \text{ Å}^{-1})$. This behaviour is seen to be very clearly reflected in the experimental data. *It should be emphasised that $F(2k_F)$ is completely determined in this calculation by the end points and the measured interference functions and that there are no adjustable parameters.*

The theoretical form of $(3-x)/\sigma$ was also calculated using interference functions derived from a hard-sphere alloy in the Percus-Yevick approximation (Ashcroft and Langreth, (1967), Enderby and North, (1968)). The results are given in curve (b) of Figure 7.27 and it is clear that the model fails to predict the sharp peak in the experimental data discussed above. The reason for this is that the hard sphere model predicts that a_{Cu-Sn} should fall roughly midway between a_{Sn-Sn} and a_{Cu-Cu}; thus the complementary effect of the concentration dependence of $\bar{a}_{Cu-Cu}$ and $\bar{a}_{Cu-Sn}$ which occurs in evaluation $F(2k_F)$ will be absent for hard spheres.

No detailed calculations for $d\sigma/dT$ can be presented because the temperature dependence of the partial interference functions has not been investigated. It seems probable, however, that if the peak position for a_{Cu-Sn} tends to smaller values of q as the temperature is increased (Section 7.3.5(b)) $<F(q)>$ will be correspondingly reduced.

At temperature just above the liquidus line, R/R_o is significantly greater than unity. The experimental results obtained by Enderby, Hasan and Simmons (1967) and shown at Figure 7.25 can be adequately explained in terms of a model involving bound states such that small amounts of Sn in Cu contribute only 2 electrons to the conduction band. Enderby *et al.* supposed that R remains proportional to n^{-1}; however, composition fluctuations in the liquid alloy will produce a distribution of possible environments so that for any given alloy, the *effective* valences of the Sn atoms may be either two or four. To approximate to this situation, Enderby *et al.* assumed that the effective valence of Cu is always one while that for Sn varies linearly from two at the Cu-rich end to four at the Sn-rich end. It follows from these assumptions that n is given by

$$n = \frac{(1 + c + 2c^2)\, N\rho}{c(M_1 - M_o) + M_o} \qquad (7.44)$$

where M_1, M_o are the atomic weights of Sn and Cu respectively, c is the atomic concentration of Sn, N is Avogadro's number and ρ is the density of the alloy.

The free electron Hall coefficient calculated with the aid of Eq. (7.44) is shown in Figure 7.25 and the close measure of agreement between the calculated and observed values is somewhat surprising. We recognise that the deviation of R/R_o from unity might indicate some structure in the density of states as described in Section 7.6.3.; but in the absence of any complete theory of the Hall effect no conclusion either way is as yet possible. It is, however, interesting to note that for liquid Cu-Sn at high temperatures, R/R_o is unity (Busch and Guntherodt (1967)).

We have dwelt at length on liquid Cu-Sn simply because so much information is available. For other liquids of type I alloys in which $\dfrac{d\sigma}{dT}$ is positive the data are less complete and we do not know what the relevant partial interference functions look like. The indirect evidence cited in Section 7.3.5(b) suggests that we might be able to treat alloy systems like Bi-Te, Sb-Te, Au-Te and Cd-Sb on the same footing. A representative selection of the experimental data for several liquid alloys is contained in Table 7.3 and in the Figures 7.29 to 7.34. Apart from the positive $d\sigma/dT$, the common

TABLE 7.3 *Properties of Type I Liquid Alloys*

Liquid Alloy	Temp °C	σ (Ω^{-1} cm^{-1})	Ref.	R ($\times 10^4$ e.m.u.)	Ref.	S (μV deg^{-1})	Ref.
Bi_2Te_3	585	3360	EW	$-$ 8.7	EW	$+$ 4	EW
Sb_2Te_3	625	1840	EW	$-$ 14.0	EW	$+$ 13.5	EW
SnTe	820	1800	EW	$-$ 12.0	EW	$+$ 10	D
$AuTe_2$	500	3460	ES	$-$ 8.0	ES	$-$	$-$

References for Table 7.3

D: Dancy (1965).
EW: Enderby and Walsh (1966).
ES: Enderby and Simmons (1969).

characteristics of all of these liquids are (i) conductivities which fall in the range 1800 Ω^{-1} cm^{-1} to 5000 Ω^{-1} cm^{-1}; (ii) Hall coefficients and thermoelectric powers which are typically of metals rather than semiconductors; (iii) R/R_0 values which are consistently greater than unity. It should also be noted that the optical reflectivity of liquid CdSb and Bi_2Te_3 is typical of metals rather than semiconductors.

Although no single piece of experimental evidence is decisive, taken together the results presented above argue against the original description by Joffe and Regel (1960) of these materials as conventional semiconductors. Thus the fact that R is negative and in magnitude rather similar to that found for liquid metals suggests that the bulk of the valence electrons are in the conduction band, and that a 'metallic' rather than a 'semiconducting' description of the electron states is appropriate. Both thermoelectric power and the optical work are consistent with such a description. Small Hall coefficients can, of course, be obtained in semiconductors provided a delicate balance is maintained between the conduction processes for electrons and holes. This situation, though not impossible, seems unlikely to occur in all materials of this type. There would, for example, be a wide variation in energy gaps if the liquids had the same energy gaps as those characteristic of the solid state.

If we continue to take R as a measure of the electron density then the fact that R is systematically greater than R_0 suggests that some electrons are in bound or localized states. It has been argued by, for example, Gubanov (1963) and Borland (1963) that in any disordered system some degree of

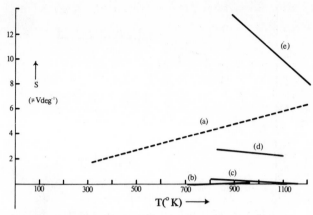

Fig. 7.29. The thermoelectric power of (a) solid Cu, (b) liquid CdSb, (c) liquid InSb, (d) liquid ZnSb, (e) liquid Sb_2Te_3. (Enderby and Walsh 1966).

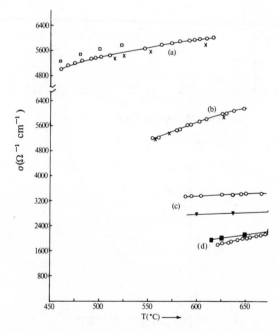

Fig. 7.30. The conductivity of liquid (a) CdSb, (b) ZnSb, (c) Bi_2Te_3, (d) Sb_2Te_3 (for references, see Enderby and Walsh (1966)).

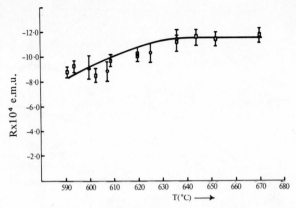

Fig. 7.31. The Hall coefficient of liquid $Bi_2 Te_3$.

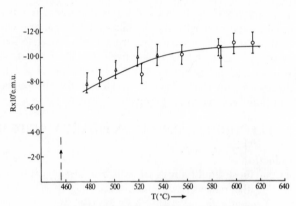

Fig. 7.32. The Hall coefficient of liquid CdSb.

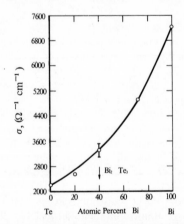

Fig. 7.33. The electrical conductivity of liquid Be-Te at 585°C.

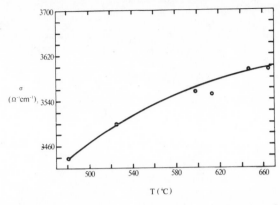

Fig. 7.34. The electrical conductivity of liquid $AuTe_2$ as a function of temperature.

localization occurs; however, the bound states that might be envisaged here concern electrons in essentially atomic-like states, separated energetically from the continuum states, and arising from the incomplete ionization of one or more of the constituents. Alternatively, the deviation from unity of R/R_o might indicate an incipient band gap (Ziman, (1967)). Until an adequate theory of the Hall effect is developed for disordered systems this important point will remain unresolved.

7.7   ELECTRON TRANSPORT IN BINARY LIQUIDS II

7.7.1   Introduction

We now consider those liquid alloys for which the transport parameters are definitely outside the range characteristic of metals for at least for some compositions. We can refer to such alloys as 'true liquid semiconductors' because they exhibit most of the properties which characterise conventional solid semiconductors.

As we have pointed out earlier, it was originally thought that a wide variety of materials could be classed as liquid semiconductors. However, the evidence discussed in Section 7.6 seems to rule out this classification for systems like $AuTe_2$, Bi_2Te_3, Sb_2Te_3, etc. in spite of the fact that $d\sigma/dT$ is positive. Other systems, however, behave in a qualitatively different way. These include liquid Cu-Te, Ag-Te, Tl-Te, In-Te, Mg-Bi and all liquid alloys involving Se as one component. A list showing which alloy systems fall into this category together with the composition(s) of particular interest is given in Table 7.4. In order to make matters more definite we shall focus attention on five groups of liquid alloy systems which constitute alloys of this type. Let M, S and SC refer to pure liquids which have respectively metallic, semi-metallic and semiconducting (or insulating) properties. The five combinations M-M, M-S, M-SC, S-SC, and SC-SC can all, in certain circumstances, be classed as liquid semiconductors. We therefore outline the current experimental position in respect of each of them.

TABLE 7.4 *Type 2 Liquid Alloys*

Liquid Alloy	Composition	Conductivity $\Omega^{-1}\,cm^{-1}$
S – Ag	$Ag_2\,S$	200
S – Pb	PbS	110
S – Cu	$Cu_2\,S$	50
S – Sn	SnS	24
S – Ge	GeS	1.35
S – Tl	$Tl_2\,S_3$	1.7×10^{-2}
	$Tl_4\,S_3$	6.5×10^{-3}
S – Sb	$Sb_2\,S_3$	1.5×10^{-2}
Te – Cu	CuTe	1920
	$Cu_2\,Te$	200
Te – Ag	AgTe	600
	$Ag_2\,Te$	150
Te – Fe	$FeTe_2$	400
Te – Tl	$Tl_2\,Te$	70
Te – Se	$2000 - 10^{-6}$	(range)
Te – Cd	CdTe	40
Te – Zn	ZnTe	40
Te – In	$In_2\,Te_3$	25
Te – Ga	$Ga_2\,Te_3$	10
Se – Pb	PbSe	400
Se – Cu	$Cu_2\,Se$	200
Se – Hg	HgSe	25
Se – Tl	$Tl_2\,Se$	3.0
	$Tl_2\,Se_3$	1.6
	TlSe	1.1
Se – Sb	$Sb_2\,Se_3$	~ 2
Se – In	InSe	0.3
Se – As	$As_2\,Se_3$	
Bi – Mg	$Mg_3\,Bi_2$	45

References for Table 7.4

Allgaier (1969).

Enderby and Collings (1970).

7.7.2 M-M Systems

Although little experimental work has been done on liquids of this type, they are of particular interest because it is possible to follow continually the transition from metallic behaviour to semiconducting behaviour. Liquid Mg-Bi, Mg-Sb and Li-Pb, represent some of the alloys which are known to fall into this group (Enderby, 1974). Experimental results for σ and S have been reported for liquid Mg-Bi by Ilschner and Wagner (1953) and by

Enderby and Collings (1970) and a selection of the data is given in Figures 7.35 to 7.37. Around the critical composition Mg_3Bi_2, S changes sign and σ falls to a very low value. It is not yet possible to say exactly what the minimum value of σ is although it is quite clear that $d\sigma/dT$ is positive for several compositions around the stoichiometric liquid Mg_3Bi_2.

Phase diagrams for two of the systems are given in Figures 7.38 and 7.39.

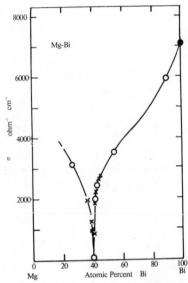

Fig. 7.35. The conductivity of liquid Mg-Bi
x x x Ilschner and Wagner (1953)
o o o Enderby and Collings (1970).

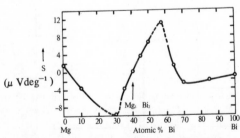

Fig. 7.36. The thermoelectric power of liquid Mg-Bi as a function of composition at temperatures 100°C above the liquidus temperature.

Information concerning the thermodynamic behaviour of Mg-Bi is to be found in the compilation of Hultgreen *et al.* (1963) and it is clear from this evidence that a major change in the bonding characteristics takes place as we proceed from pure liquid Bi to pure liquid Mg. There is evidence (Epstein, 1972) that substantial electromigration occurs in liquid Mg/Bi, with the Mg drifting towards the cathode and the Bi drifting towards the anode.

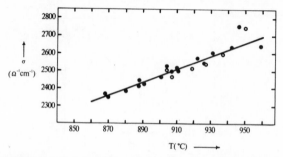

Fig. 7.37. The temperature dependence of the electrical conductivity for liquid Bi + 57.5 at % Mg.

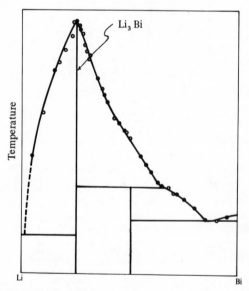

Fig. 7.38. Simplified form of the phase diagram for Li-Bi. A detailed diagram has been given by Hanson (1958).

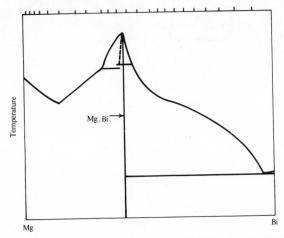

Fig. 7.39. Simplified form of the phase diagram for Mg-Bi. A detailed diagram has been given by Hanson (1958).

TABLE 7.5 *Properties of Type II Liquids*

Liquid Alloy	Temp. ($^{\circ}$C)	σ (Ω^{-1} cm^{-1})	Ref.	R (10^{-4} e.m.u.)	Ref.	S (μV/deg)	Ref.
Te-30 at % Tl	535	1080	ES	$-$ 32.5	ES	$+$ 50	CM
Te-50 at % Tl	535	540	ES	$-$ 72.0	ES	$+$ 95	CM
Te-60 at % Tl	535	300	ES	$-$ 132	ES	$+$ 138	CM
Te-65 at % Tl	535	180	ES	$-$ 217	ES	$+$ 140	CM
Te-67 at % Tl	535	156	ES	$-$ 160	ES	$-$ 100	CM
Te-21.8 at % In	667	110	C	$-$	$-$	$+$ 68	C
Te-70 at % In	667	3300	C	$-$	$-$	$+$ 25	C
CuTe	650	1900	ES	$-$ 16.0	ES	$+$ 50	D
Te-60 at % Cu	1100	1200	D	$-$	$-$	$+$ 75	D
AgTe	550	715	ES	$-$ 125	ES	$+$ 120	D
Te-60 at % Ag	800	500	D	$-$	$-$	$+$ 120	D

References for Table 7.5

ES: Enderby and Simmons (1969).
D: Dancy (1965).
CM: Cutler and Mallon (1966).
C: Cutler (private communication).

7.7.3 M-S Systems

These include Ag-Te, Cu-Te (*not* Au-Te), Ga-Te, In-Te and Tl-Te and represent the most widely studied systems. Phase diagrams for Cu-Te and Tl-Te are shown at Figures 7.40 and 7.41, and a selection of the experimental

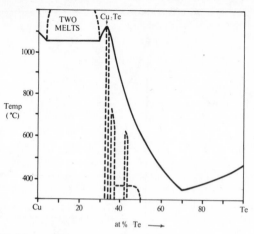

Fig. 7.40. Simplified form of the phase diagram for Cu-Te. A detailed diagram has been
given by Hanson (1958).

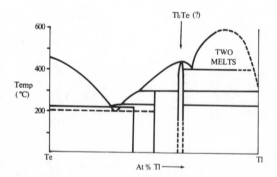

Fig. 7.41. Simplified form of the phase diagram for Tl-Te. A detailed diagram has been
given by Hanson (1958).

data due to Dancy (1965) Cutler and Mallon (1966) and Enderby and
Simmons (1969) are given at Figures 7.42 to 7.48 and in Table 7.5.

Although the system Tl-Te has received special attention, the following
remarks appear to be valid for all alloys within this group.

(i) Alloys within this type possess a two-liquid phase region, often in the
     range $70 \lesssim X_M \gtrsim 100$, where X_M is the atomic percent of the metallic
     component.

(ii) R/R_o is significantly different from unity; R itself is negative at all
     compositions and apparently achieves a maximum absolute value at
     composition corresponding to the minimum conductivity.

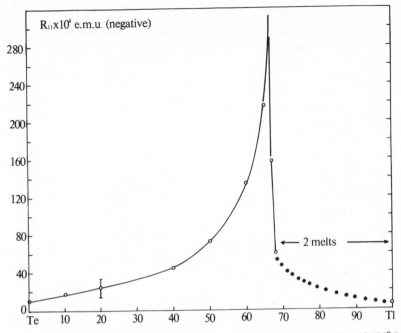

Fig. 7.42. The Hall coefficient of liquid Te-Tl as a function of composition of 585°C. The dotted portion refers to the two phase region.

(iii) S varies rapidly close to, but not necessarily precisely at, the critical composition. In some cases a change in sign is observed.

Interesting new data for the NMR (Warren, 1970) in liquid $Ga_2 Te_3$, indicate that the Knight shift is strongly temperature dependent and that close to the melting point, the Korringa relationship breaks down completely. Work of this type is of potential importance in deciding the extent to which structure in the density of states can be correlated with electronic localisation (see page 420).

7.7.4 M-SC Systems

These include Ni-S and Co-S (Dancy and Derge, (1963)), Tl-Se (Nakamura and Shimoji, (1969)) and In-Se and Bi-Se (Regel et al., (1970)). In general these systems closely resemble those of the M-S group as regards their electrical and thermodynamic properties.

7.7.5 S-SC Systems

There is a considerable literature relating to liquids of these types. Liquid Te-Se has been investigated extensively by Cutler and Mallon (1962) and by Perron (1967). The latter's data are given at Figures 7.49 and 7.50. The

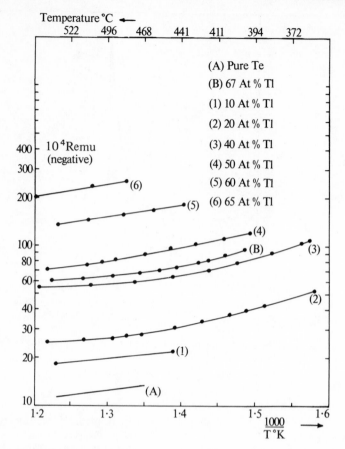

Fig. 7.43. The Hall coefficient as a function of temperature for several liquid Te-Tl alloys.

electrical properties appear to vary in a smooth way from those character-istic of pure Se to those of pure Te. The Hall effect in liquid Te-Se has, recently been investigated by Perron (1972) who finds that R is negative at all compositions.

7.7.6 SC-SC Systems

These include liquid As-Se(Te) for which $R\sigma$ and S have all been deter-mined. The data, due to Edmond (1966) and Male (1967) are given in Figures 7.51 and 7.52. There is insufficient work to draw any general conclusions about alloys of this type, although it appears that a change in sign of S does *not* occur around the stoichiometric compound $As_2 Se_3$.

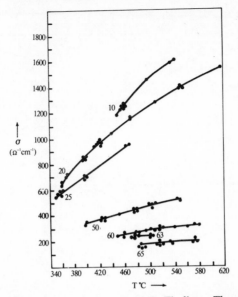

Fig. 7.44. The electrical conductivity of liquid Te-Tl alloys. The number in front of each set of data points represents the atomic percentage of Tl.

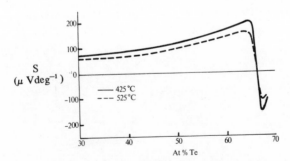

Fig. 7.45. Thermoelectric power for liquid Te-Tl
————— 425°C
- - - - - - 525°C
(After Cutler and Mallon (1966)).

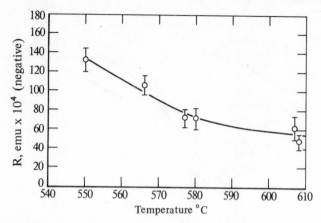

Fig. 7.46. The Hall coefficient of liquid AgTe as a function of temperature.

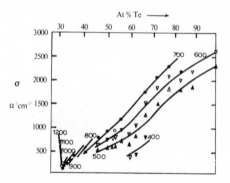

Fig. 7.47. Conductivity of liquid Ag-Te (after Dancy (1965)).

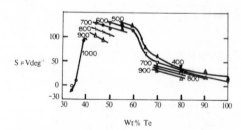

Fig. 7.48. Thermoelectric power of liquid Ag-Te (after Dancy (1965)).

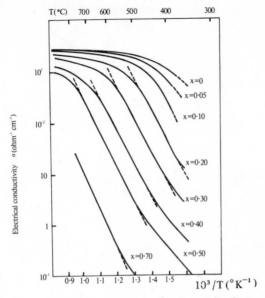

Fig. 7.49. Electric conductivity of liquid $Te_{1-x} Se_x$ alloys (after Perron (1967)).

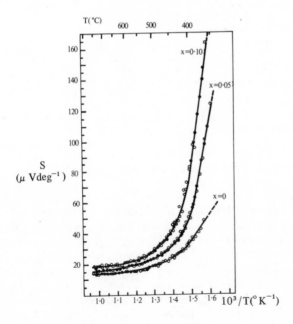

Fig. 7.50. Thermoelectric power of liquid $Te_{1-x} Se_x$ alloys (after Perron (1967)).

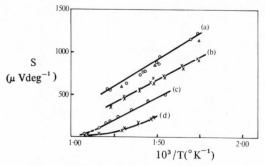

Fig. 7.51. Thermoelectric power of liquid As-Se-Te alloys (a) As_2Se_3, (b) As_2Se_2Te, (c) As_2SeTe_2, (d) As_2Te_3. (after Edmond (1966)).

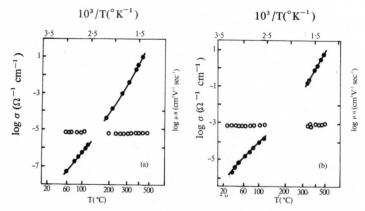

Fig. 7.52. The Hall mobility (o) and the electrical conductivity (•) of solid and liquid As-Se-Te (a) As_2Se_2Te, (b) As_2SeTe_2 (after Male (1967)).

7.7.7 Interpretation of the Experimental Data

(a) *Mott's Approach*

The most comprehensive attack on the theoretical problems posed by the existence of liquid semiconductors has been made by Mott (1966, 1967, 1970, 1971). The essential features of his model are shown in Figure 7.53 and have been described in detail by Cohen, Fritzsche and Ovshinsky (1969) and by Davies and Mott (1970). E_v and E_c separate the extended from the localized states (shown shaded) and mark the energy at which a substantial drop in the mobility occurs. An additional hypothesis is that the energy range of localised states is somewhat greater in the conduction band than in the valence band, thereby explaining the tendency for liquid semi-conductors to exhibit positive thermoelectric powers. All the liquids

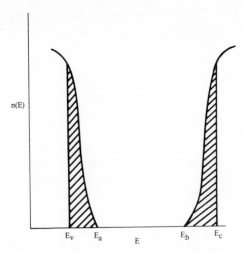

Fig. 7.53. The Mott-CFO model for the density of states. The shaded region indicates the extent of the localisation.

described in this section have a composition range for which $\sigma \leqslant 200\ \Omega^{-1}$ cm^{-1}, which implies that Eqs. (7.29) and (7.34) should be universally used to calculate the electrical properties. Mott identifies the characteristic energy E in Eq. (7.34) through

$$E_F - E_v = E - \beta T \quad \text{(holes)}$$

(7.45)

or

$$E_C - E_F = E - \beta T \quad \text{(electrons)}$$

Evidently the value of β is of crucial significance. If we insert Eq. (7.45) into (7.34) we find that

$\sigma = \sigma_0 \exp{(\beta/k_B)} \exp{(-E/k_B T)}$. An elementary calculation by Mott (1970) indicates that σ_0 is $\sim 200\ \Omega^{-1}$ cm^{-1} which implies that β/k_B is in the region of 2–3. Fortunately, a more accurate method for the determination of β is to measure the thermoelectric power as a function of temperature. It follows from the work of Cutler and Mott (1969) that provided the current is carried by electrons (holes) excited to $E_c(E_v)$ then:

$$S(\text{holes}) \quad = \frac{k_B}{e} \left\{ \frac{E_F - E_V}{k_B T} + 1 - \frac{\beta}{k_B} \right\}$$

(7.46)

$$S(\text{electrons}) \quad = \frac{k_B}{e} \left\{ \frac{E_c - E_F}{k_B T} + 1 - \frac{\beta}{k_B} \right\}$$

Thus a plot of S against 1/T should be a straight line with an intercept of $\left(\dfrac{k_B}{e} - \dfrac{\beta}{e} \right)$.

If both holes and electrons are present, the conductivity and thermo-electric power are given by

$$\sigma = \sigma_h + \sigma_e \, ,$$

$$S = \frac{\sigma_h S_h + \sigma_e S_e}{\sigma_h + \sigma_e}$$

where σ_h, σ_e, S_h, S_e are respectively the hole and electron conductivities and the hole and electron thermoelectric powers.

It is obviously of interest to compare Eqs. (7.34) and (7.36) with experiment and also to test the data for internal consistency. The thermo-electric data on liquid $As_2 Se_3$ and Te-Se are, in fact, consistent with Eq. (7.46). Edmond's data, for example, suggest that β/k_B is 11.6 whilst those of Perron yield a value of 10.4. Cutler and Mallon's (1966) data on the p-type Te-Tl alloys ($1000 < \sigma < 200 \ \Omega^{-1} \, cm^{-1}$) can also be fitted to a curve of the form $S = \frac{A}{T} + B$ although the value of β/k_B deduced from such a plot is less than one half of that for $As_2 Se_3$. There appears to be no way at the present time of establishing whether these values of β/k_B are reasonable or indeed what controls their relative magnitude in one material versus another.

A further difficulty is that the values of the coefficient of the $\frac{1}{T}$ term in σ and in S are not entirely consistent (Hindley 1970), at least so far as liquid $As_2 Se_3$ is concerned. Mott (1971) points out, however, that the mechanism of conduction can change very rapidly with temperature and this makes a detailed comparison of theory and experiment extremely difficult.

One major success of the theory is that in this regime, the Hall mobility should be temperature independent (Friedman (1971)). This is observed in practice as the data given in figure 7.52 show.

In liquid alloys of the M-M, M-S and S-SC it is possible to make measure-ments at compositions or temperatures in which a metallic description of the electron states begins to be valid. Mott (1971) advocates the use of Eq. (7.31) in these situations. The electron states at the Fermi level are no longer fully localised but both S and σ respond to the value chosen for g. Warren's NMR work enables experimental nuclear relaxation rates to be compared with those predicted by the Korringa relationship. The ratio of the experimental to the predicted rate (denoted by Warren as η) turns out to be $\geqslant 10$ for liquids in which $\sigma \leqslant 200^{-1} \, cm^{-1}$ and this result is consistent with the onset of Mott localisation. For metallic liquids, on the other hand, $\eta \sim 1$, so that in the intermediate regime η should fall in the range $1 < \eta < 10$. This behaviour has, in fact, been observed for a variety of liquids (Warren (1971)), including $In_2 Te_3$, $Ga_2 Te_3$ and $Sb_2 Te_3$.

(b) *Cutler's Approach*
In an important series of papers Cutler and co-workers (Cutler and Mallon (1965, 1966) Cutler and Field (1967) Cutler and Paterson (1970)) have applied the ideas of *conventional* semiconductor physics to liquid Tl-Te. By

analogy with the situation for solids, alloys on the Te rich side of the composition Tl_2Te are referred to as p-type whilst those on the Tl side are referred to as n-type, liquid Tl_2Te itself being regarded as an intrinsic semiconductor. Defining

$$\xi_c = (E_F - E_C)/k_BT$$

$$\xi_v = (E_v - E_F)k_BT$$

where now E_v and E_C correspond to band (rather than mobility) edges, we can write, quite generally within the framework of parabolic energy bands,

$$n = \frac{2}{\pi^{\frac{1}{2}}} N_c F_{\frac{1}{2}}(\xi_c)$$

$$p = \frac{2}{\pi^{\frac{1}{2}}} N_v F_{\frac{1}{2}}(\xi_v)$$

where

$$N_c = \frac{1}{4\pi^3}\left(\frac{2\pi m_c^*}{\hbar^2}k_BT\right)^{3/2}$$

$$N_v = \frac{1}{4\pi^3}\left(\frac{2\pi m_v^*}{\hbar^2}k_BT\right)^{3/2}$$

and

$$F_n(x) = \int_o^\infty \frac{u^n du}{\exp(u-x)+1}$$

In order to deduce the transport coefficients, Cutler assumes the existence of a relaxation time, τ, given by

$$\tau = \tau_o(E\text{-}E_c)^r$$

The conductivity and thermopower can now be evaluated at once. For electrons in the conduction band

$$\sigma = e^2 K_o$$

and

$$S = \frac{1}{eT}\frac{K_1}{K_0}$$

where

$$K_n = \frac{4N_c}{3m_c^{*\frac{1}{2}}(k_BT)^{3/2}}\int (E-E_F)^n (E-E_c)^{3/2}\frac{\partial f_o}{\partial E}\,dE$$

and

$$f_o = \frac{1_1}{\exp[(E-E_F)k_BT]+1}.$$

It follows that

$$\sigma_e = \frac{4Nc\,\tau_o\,(r + 3/2)\,(k_BT)^{r|}\,F_{r+1/2}(\xi_c)e^2}{3m_c^*\,\pi^{1/2}}$$ (7.47)

and

$$S_e = \frac{k_B}{e}\left\{\frac{(r + 5/2)F_{r + 3/2}(\xi_c)}{(r + 3/2)F_{r +1/2}(\xi_c)} - \xi_c\right\}$$ (7.48)

Similar expressions can be obtained for the conductivity and thermo-electric power of holes.

Cutler and Field (1968) have analysed their experimental results for liquid Tl-Te in the composition range 66.7 to 72 at % Tl (the so-called p-type liquids) with the help of Eqs. (7.47) and (7.48). They find that S and σ can be correlated very well (Figure 7.54) by a single choice of r, which turns out to be −0.5, provided τ_o is independent of temperature. Within the model used by Cutler and Field these data imply that the electrons are scattered by the disorder in the liquid rather than by thermal vibrations.

Cutler and Paterson (1970) have made measurements of σ and S in a temperature range in which the two liquid phases are mutually soluble in order to study the transition between metallic Tl and semiconducting Tl_2Te. As the concentration of Te in liquid Tl is increased, ρ and the magnitude of S also increase, and their temperature coefficients change

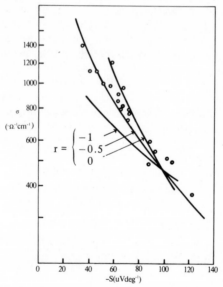

Fig. 7.54. Thermoelectric power versus conductivity for liquid Tl_2Te + excess Tl. Full curves are derived from theory with r = −1, −0.5 and 0. The open circles represent the experimental points.

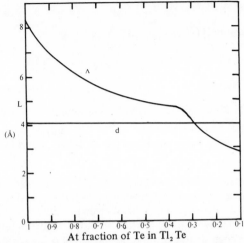

Fig. 7.55. The mean free path (Λ) and the average interatomic spacing (d) for the pseudobinary $Tl_2\,Te$-Tl (after Cutler and Paterson (1970)).

from weakly positive values typical of metals to weakly negative values typical of liquid semiconductors with low resistivity.

They found that the transport behaviour can be accounted for very well by a model which treats the alloy as a solution containing Tl_2 Te molecules plus metallic Tl atoms*. As Te is added to Tl, electrons appear to be removed from the conduction band, and scattering centres due to Tl_2 Te are added. Analysis of the behaviour of S and σ in these terms allow estimates to be made for the electronic mean free path (Figure 7.55). At compositions approaching Tl_2 Te, the 'impurity' contribution to the scattering is consistent with the theoretical analysis of Cutler and Field (1968). It is found that the product of $\Lambda\,k_F$ is in general greater than unity, but approaches unity near the composition Tl_2 Te. It remains an open question, however, whether it is permissible to regard Tl as a trivalent metal at all compositions — an assumption made by Cutler and Paterson.

For the p-type solutions Cutler and Mallon (1966) have used a similar analysis assuming that now the liquid can be thought of as Tl_2 Te + Te. They postulate that the Te atoms give rise to deep lying acceptor levels so that

$$\rho = p_1 \exp\left(-E_1/k_B T\right)$$

where E_1 is a characteristic activation energy and p_1 is a constant for a given composition. Noting that in the non-degenerate case, $\xi = -\ln N_v/p$ and $F_n(\xi) = (\exp \xi)\,\Gamma\,(n+1)$ they obtain from Eq. (7.48)

---

*A theoretical justification for treating Te-Tl as a pseudo-binary Tl_2 Te-Tl has been given by Schaich and Ashcroft (1970).

$$(e/k_B)S = (r + 5/2) + \ln N_v/p \tag{7.49}$$

If N_v is written as $BT^{3/2}$, equation (7.49) reads

$$(e/k_B)S = \tfrac{3}{2} \ln T = [r + 5/2 + \ln B - \ln p_1] + E_1/k_B T \tag{7.50}$$

which suggests that $(e/k)S - \tfrac{3}{2} \ln T$ versus $\dfrac{1}{T}$ should give a straight line plot with slope E_1/T. This was found to be the case and E_1 turns out to be 0.25 eV and is independent of composition. A further analysis of the conductivity data with this value of E_1 showed that the mobility, μ, has the form

$$\mu \alpha (T)^{-x}$$

where x is close to, but not exactly, 3/2. Cutler and Mallon concluded that the scattering mechanism for 'holes' is predominantly of a phonon type, and thus contrasts with the behaviour of electrons in the n-type liquids.

The approach suggested by Cutler is of particular interest because it employs, with a high degree of internal consistency, ideas which are derived from conventional solid state theory. The notions of localisation or of mobility gaps do not appear explicitly although they are probably implicit in some of the arguments used.

In more recent papers, Cutler (1971a, b) has somewhat modified the original model. On the Te side of the stoichiometric composition the structure of the alloy is considered to be chain like with configurations of the form $Te\text{-}(Tl)_n\text{-}Te$. The carriers are generated by the thermal rupture of bonding orbitals. A detailed structural investigation of liquid Te-Tl, in order to validate this model, has not yet been carried out. The evidence for related materials discussed in Section 7.3.5. argues *against* the existence of well defined chains in type II alloys.

(c) *Enderby's Approach*
Enderby and co-workers (Enderby and Walsh (1966) Enderby and Simmons, (1969) Enderby and Collings (1970)) assume that the Hall coefficient rather than the thermopower remains a useful measure of both the sign and the density of the current carriers. This is an entirely different starting point from either that of Mott or of Cutler and since no justification for this drastic assumption is offered, the approach adopted by Enderby is clearly a speculative one. In any case, it applies only to M-M or M-SC systems in which a relatively smooth change from metallic to semi-conducting behaviour is observed.

There are two structural models for liquids whose composition is close to $Tl_2 Te$ which are consistent with the thermodynamic evidence (Terpilowski and Zaleska (1963)). The first of these, suggested by Cutler and Field (1968) and by Enderby and Simmons (1969) has been referred to already, and involves the notion of $Tl_2 Te$ molecules together with excess Te or Tl. An equally plausible description, particularly in view of the electromigration data for liquid Mg-Bi is one in which liquid $Tl_2 Te$ is assumed to be essentially *ionic* in character. The detailed structure experiments referred to in Section 7.3.5 will, it is hoped, help to resolve this uncertainty in due course.

The occurrence of a fairly well-defined molecular group Tl_2Te or of the ionic assembly Tl^+, Tl^+, Te^- cannot be approached in terms of perturbation theory because it involves the formation of bound states. Stern (1966) and others have shown how these can arise if the atomic potentials of the two species are very different. A rough measure of the differences in potentials can be obtained from published tables of electronegativities or stability ratios (Sanderson (1960)), or the well depths which characterise the Heine-Abarenkov-Animalu model potential (see, for instance, Animalu and Heine (1965)). The connection between pseudopotentials and electronegativity implied by this remark has recently been justified by Heine and Weaire (1970).

If the bound states (of whatever character) form a band which is separated in energy from the conduction band the number of free electrons will be depleted and this will give rise to the peak in R as the critical composition is approached. A model for R like that used in Section 7.6.3.(b) when discussing the case of Cu-Sn, yields remarkably good agreement between theory and experiment at least so far as Te-Tl is concerned (Figure 7.56).

From the observed Hall coefficients and within the free electron theory k_F and E_F may be calculated and are given for Tl-Te in Table 7.6; we note that the condition $E_F \gg kT$ is satisfied so that liquid Tl-Te is degenerate at all compositions, except, possibly, at compositions very close to Tl_2Te where the measurements are not yet sufficiently accurate to decide. The thermoelectric power data of Cutler and Mallon (1966) may, therefore, be expressed in terms of the dimensionless parameter ξ with the result shown

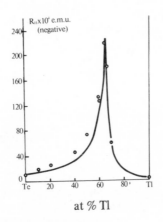

at % Tl

Fig. 7.56. The Hall coefficient of liquid Te-Tl
    o o o Experimental results
                ———— theoretical curve calculated by assuming that the effective valences of Tl_2Te, Te, Tl are 0, 2 and 3 respectively.

TABLE 7.6 *Free Electron Fermi Wave Numbers for Type II Liquids*

Liquid Alloy	Temp °C	R (× 10^4 e.m.u.)	Ref.	k_F (Å^{-1})	E_F (eV)
Te−30 at % Tl	535	− 32.5	ES	0.83	2.65
Te−50 at % Tl	535	− 72.0	ES	0.64	1.56
Te−60 at % Tl	535	− 132	ES	0.52	1.04
Te−65 at % Tl	535	− 217	ES	0.44	0.75
Te−67 at % Tl	535	− 160	ES	0.49	0.92
CuTe	650	− 16.0	ES	1.05	4.16
AgTe	565	− 100	ES	0.58	1.39

Reference for Table 7.6

ES: Enderby and Simmons (1969).

in Figure 7.57. It is important to emphasise that ξ is practically unchanged over the range 10 to 100 at % excess of Tl which suggests that in this composition range g^2 is independent of Tl content. The model described for pure liquid Te in Section 7.5.3. should therefore retain its validity in the alloy. A similar conclusion is suggested by the work of Cutler and Paterson (1970) for $Tl_2 Te + x$ at % excess Tl provided x is greater than ~ 30.

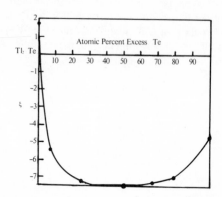

Fig. 7.57. The thermoelectric power parameter ξ as a function of composition for the pseudo-binary $Tl_2 Te + Te$.

Three models to explain the rapid change in ξ around the critical composition $Tl_2 Te$ have been proposed:

(a) The existence of strongly energy dependent scattering mechanisms for excess Te or Tl has been suggested by Enderby and Simmons (1969). These workers point out that a virtual bound state of width ~ 0.2 eV if suitably located with respect to E_F could generate the observed thermopowers. A weakness of the suggestion is that the observed sign change around the stoichiometric compound is a very general phenomenon (it occurs, for example, in liquid Mg-Bi and liquid Li-Pb) and there is no *à priori* reason for supposing that such virtual bound states are always located conveniently close to E_F.

(b) Faber (1972) has suggested that the density of states of the system has an entirely different character depending on whether Te or Tl represents the excess component. A schematic form for the density of states is shown in Figure 7.58 for Tl-Te and Mg-Bi at compositions close to the critical one. In general, the elements with more than four valence electrons (Bi, Te, Se) should, as the excess components, give rise to positive thermopowers; this is in fact observed (Enderby and Collings, 1970). Conversely, elements whose valency is less than four should tend to give rise to negative thermopowers. Apart from the interesting case of excess In in liquid In_2Te (Blakeway 1970), the experimental data which are at present available confirm this prediction (Dancey (1965) Enderby and Collings (1970)).

(c) Enderby (1973) has offered a variation on (b) above but one that may be more attractive from a theoretical point of view. Let us focus attention on an M-M system like Mg-Bi. In pure liquid magnesium, the density of states n(E) will be essentially free electron like. Suppose now that we add a few atoms of bismuth, a strongly electronegative element. Modifications to n(E) will follow the general pattern predicted by the Koster-Slater (1954) theory. If the bismuth potential is strong either a bound state appears below the bottom of the band or else a virtual bound state is created which enhances n(E) around the resonant energy and reduces n(E) just beyond. It seems reasonable to assume that this process continues as the bismuth content increases, the dip at the

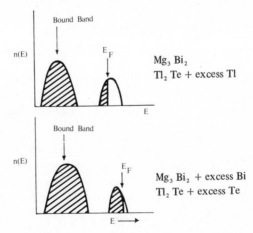

Fig. 7.58. Model density of states for liquid semiconductors.

anti-resonance in n(E) developing into a Mott pseudogap and becoming most marked at the composition Mg_3Bi_2. This model, together with the transport theories discussed earlier, explains
 (i) the rapid variation of S around the critical composition;
 (ii) the reduction in σ as Bi is added to Mg;
 (iii) the peak in R as the critical composition is approached;
 (iv) the rather smooth way in which the electronic transport parameters vary from the metallic values to those characteristic of liquid semiconductors;
 (v) the observation that the Knight shift in type II liquid alloys is progressively reduced as the critical composition is approached (Warren (1972a)).

(d) *Allgaier's Approach*

In two recent papers Allgaier (1969; 1970) has emphasised the dangers in extrapolating ideas particular to ordered solids into the liquid and amorphous states. For example, suppose that a liquid exists in which it is permissible to use an energy-momentum description of the electron state. Since the E-k relationship is the same in all directions, the Fermi surface is a sphere enclosing all the valence (free) electrons. Within this framework, the Hall coefficient is given by

$$R = -\frac{1}{ne} \left\{ \frac{dE}{dk} \middle/ \left| \frac{dE}{dk} \right| \right\}_{k_F}$$

a formula discussed in detail by Jan (1962) and by Enderby (1963). Thus, provided $\left(\frac{dE}{dk}\right)$ is positive, R *will always remain negative even though the Fermi energy is close to a band (or indeed mobility) edge* as the example given in Figure 7.59 shows. The difference between this behaviour and that

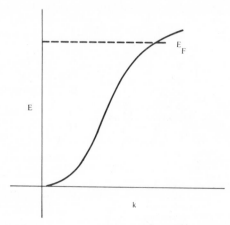

Fig. 7.59. E-k relationship for an isotropic liquid. At the Fermi energy dE/dk is positive so that Hall coefficient is negative even though the band is more than half full.

TABLE 7.7 *Hall Mobilities in Electronically Conducting Liquids*

Liquid	$|\mu_H|$ (cm^2/V sec)	Classification Allgaier	Enderby
Ge	0.50	A	Metallic
Pb	0.39	A	Metallic
Sb	0.39	A	Metallic
Bi	0.23	B	Metallic
CdSb	0.42	B	I
ZnSb	0.28	B	I
AuTe$_2$	0.28	B	I
GeTe	0.36	B	I
Bi$_2$Te$_3$	0.29	B	I
CuTe	0.31	B	II
Sb$_2$Te$_3$	0.26	B	I
Te	0.23	B	I
PbTe	0.15	C	II
Te$_{92}$Sb$_8$	0.23	C	I
AgTe	0.9	C	II
Tl$_{68}$Te$_{32}$	0.55	C	II
TlTe	0.42	C	II
Tl$_3$Te$_2$	0.45	C	II
Ag$_2$Te	15	C	II
Tl$_2$Te	0.25	C	II
As$_2$SeTe$_2$	0.1	C	II
As$_2$Tl$_2$Se$_3$Te	0.1	C	II
AsTlSe$_2$	0.1	C	II
As$_2$Se$_2$Te	0.1	C	II
Se	0.03	C	II

Reference for Table 7.7

Allgaier (1970).

associated with the crystal, where the second derivative of the (E-k) curve dominates the sign of R, is explained in the paper by Jan.

Although Allgaier's papers do not contain any detailed calculations his views about liquid alloys of type I (which he refers to as type B) are in close accord with those expressed in this present review and in the papers by Enderby and co-workers. The essentially metallic character of these materials is now widely accepted.

As regards type II alloys (type C in Allgaier's notation) Allgaier regards the situation as a good deal less clear cut. In Table 7.7, a list of Hall mobilities is given for a variety of liquids of the three main types. Allgaier points out that there is considerable overlap between μ_H values for each class of liquid. For example, the Hall mobility of liquid Bi, which is undoubtedly a metallic conductor, is almost the same as that for Tl_2Te. Allgaier speculates that 'it seems possible . . . . that some kind of straight-forward extension of the simple [metallic] model of range B will eventually provide a basis for understanding the behaviour in range C above.'

7.8 CONCLUSIONS

Most authors have distinguished the existence of at least three types of liquid conductors and it has been an object of this review to suggest ways whereby a unified theory of conducting liquids can be approached. The need for more work both of a theoretical and experimental nature will now be apparent to the reader and in many ways, this article should be regarded only as a progress report.

One field which deserves special attention concerns the structure of liquid semiconductors — particularly those which we have regarded as type II alloys. It is still a matter of dispute whether an 'ionic' or a 'covalent' model is the most appropriate starting point. Carefully planned experiments involving the techniques of neutron diffraction seem to offer the most promising line of attack. Fortunately, sufficient isotopes of different neutron scattering cross-sections, but of interest in connection with liquid semiconductors, exist; studies involving a range of Type II liquids are at present underway at the University of Leicester.

ACKNOWLEDGEMENTS

I wish to thank Professor Sir Nevill Mott for his continued interest in this work, for allowing me to see unpublished material and for generous guidance and sympathetic criticism. I should also like to acknowledge helpful conversations with Dr. E. W. Collings, Dr. C. J. Simmons and Professor N. H. March. Finally, Miss Marion Hancock and Mrs. Angela McQuire are to be thanked for their patient assistance in the preparation of the manuscript.

REFERENCES

Abowitz, G. and Gordon, R. B., (1962), *J. Chem. Phys.,* **37**, 125.
Aldea, A., (1967), *Phys. Stat. Sol.,* **22**, 377.
Aliev, G. M., Abdinov, D.Sh. and Mekhtieva, S. I., (1966), *Sov. Phys. Doklady,* **11**, 305.
Allgaier, R. S., (1969), *Phys. Rev.,* **185**, 227.
Algaier, R. S., (1970), *Phys. Rev.,* **2**, 2257.
Anderson, P. W., (1958), *Phys. Rev.,* **109**, 1492.
Animalu, A. O. E. and Heine, V., (1965), *Phil. Mag.,* **12**, 1249.
Ashcroft, N. W., (1966), *Phys. Letts.,* **23**, 529.
Ashcroft, N. W. and Langreth, D. C., (1967), *Phys. Rev.,* **159**, 500.
Ashcroft, N. W. and Lekner, J., (1966), *Phys. Rev.,* **145**, 83.
Bacon, G. E., (1962), Neutron Diffraction, Clarendon Press, Oxford.
Banyai, L. and Aldea, A., (1966), *Phys. Rev.,* **143**, 652.
Blakeway, R., (1969), *Phil. Mag.,* **20**, 965.
Blech, I. A. and Averbach, B. L., (1965), *Phys. Rev.,* **137**, A1113.

Bond, W. L., (1962), The International Tables for X-ray Crystallography, Vol. II Kynoch Press, Birmingham.

Borland, R. E., (1963), *Proc. Roy. Soc.*, A274, 529.

Busch, G. and Tieche, Y., (1962), *Helv. Phys. Acta*, 35, 273.

Busch, G. and Guntherodt, H. J., (1967), *Phys. Kondens Mat.*, 6, 325.

Cabane, B. and Friedel, J., (1971), *Jnl. de Physique*, 32, 73.

Cabane, B. and Froidevaux, C., (1969), *Phys. Letts.*, 29A, 512.

Cohen, M. H., Fritzsche, H. and Ovshinsky, S. R., (1969), *Phys. Rev. Letts.*, 22, 1065.

Cohen, M. H. and Sac, J., (1972), *J. Non-Cryst. Solids*, 8–10, 696.

Cocking, S. J. and Heard, C. R. T., (1965), A.E.R.E. Report R5016, H.M.S.O., London.

Cusack, N. E., Enderby, J. E., Kendall, P. W. and Tieche, Y., (1965), *J. Sc. Insts.*, 42, 256.

Cusack, N. E. and Kendall, P. W., (1958), *Proc. Phys. Soc.*, 72, 898.

Cusack, N. E., Kendall, P. W. and Fielder, M., (1964), *Phil. Mag.*, 10, 871.

Cutler, M., (1971a), *Phil. Mag.*, 24, 381.

Cutler, M., (1971b), *Phil. Mag.*, 24, 401.

Cutler, M. and Field, M. B., (1968), *Phys. Rev.*, 169, 632.

Cutler, M. and Mallon, C. E., (1962), *J. Chem. Phys.*, 37, 2667.

Cutler, M. and Mallon, C. E., (1965), *J. App. Phys.*, 36, 201.

Cutler, M. and Mallon, C. E., (1966), *Phys. Rev.*, 144, 642.

Cutler, M. and Mott, N. F., (1969), *Phys. Rev.*, 181, 1369.

Cutler, M. and Paterson, R. L., (1970), *Phil. Mag.*, 21, 1033.

Dancy, E. A., (1965), *Trans. Met. Soc.*, AIME, 233, 270.

Dancy, E. A. and Derge, G. J., (1963), *Trans. Met. Soc. AIME*, 227, 1034.

Davies, H. A., (1969), *Phys. Chem. Liquids*, 1, 191.

Davies, E. A. and Mott, N. F., (1970), *Phil. Mag.*, 22, 903.

Edmond, J. T., (1966), *Brit. J. App. Phys.*, 17, 979

Edwards, S. F., (1962), *Proc. Roy. Soc.*, A267, 518.

Enderby, J. E., (1963), *Proc. Phys. Soc.*, 81, 772.

Enderby, J. E., (1973), Band Structure Spectroscopy of Metals and Alloys (D. J. Fabian and L. M. Watson, editors) Academic Press, New York

Enderby, J. E. and Collings, E. W., (1970), *J. Non-Cryst: Solids*, 4, 161.

Enderby, J. E., Hasan, S. B. and Simmons, C. J., (1967), *Adv. Phys.*, 16, 667.

Enderby, J. E. and Hawker, I., (1972), *J. Non-Cryst. Solids*, 8–10, 687.

Enderby, J. E. and Howe, R. A., (1968), *Phil. Mag.*, 18, 923.

Enderby, J. E. and March, N. H., (1965), *Adv. Phys.*, 14, 453.

Enderby, J. E. and North, D. M., (1968), *Phys. Chem. Liquids*, 1, 1.

Enderby, J. E., North, D. M. and Egelstaff, P. A., (1966), *Phil. Mag.*, 14, 961.

Enderby, J. E., North, D. M. and Egelstaff, P. A., (1967), *Adv. Phys.*, 16, 171.

Enderby, J. E. and Simmons, C. J., (1969), *Phil. Mag.*, 20, 125.

Enderby, J. E., Van Zytveld, J. B., Howe, R. A. and Mian, A. J., (1968), *Phys. Letts*, 28A, 144.

Enderby, J. E. and Walsh, L., (1966), *Phil. Mag.*, 14. 991.

Epstein, S. G., (1972), Physics and Chemistry of Liquid Metals (S. Z. Beer, editor) Marcel Dekker, New York.

Faber, T. E., (1972), Introduction to the Theory of Liquid Metals. Cambridge University Press.

Faber, T. E. and Ziman, J. M., (1964), *Phil. Mag.*, 11, 153.

Friedman, L., (1971), *J. Non-Cryst. Solids*, 6, 329.

Fukuyama, H., Ebisawa, H. and Wada, Y., (1969), Proc. Theor. Phys., 42, 494.

Gehlen, P. C. and Enderby, J. E., (1969), *J. Chem. Phys.*, 51, 547.

Gingrich, N. S. and Heaton, L., (1961), *J. Chem. Phys.*, 34, 873.

Greenfield, A. J., (1964), *Phys. Rev.*, A135, 1589.

Gubanov, A. I., (1963), Quantum Electron Theory of Amorphous Semiconductors, (English Translation 1965) Consultants Bureau, New York.

Halder, N. C., Metzger, R. J. and Wagner, C. N. J., (1966), *J. Chem. Phys.*, 45, 1259.

Halder, N. C. and Wagner, C. N. J., (1967a), *Phys. Lett.*, **24A**, 345.
Halder, N. C. and Wagner, C. N. J., (1967b), *J. Chem. Phys.*, **47**, 4385.
Hamer, H., (1962), *Semiconductor Prod.*, **5**, 35.
Hanson, H. P., (1958),, Constitution of Binary Alloys, McGraw-Hill, New York.
Hanson, H. P., Herman, F., Lea, J. D. and Skillman, S., (1964), Acta Cryst., **17**, 1040.
Harrison, W., (1966), Pseudopotentials in the Theory of Metals, Benjamin, New York.
Hawker, I., Howe, R. A. and Enderby, J. E. (1973a), Proceedings of the 5th International Conference on Amorphous and Liquid Semiconductors (to be published).
Hawker, I., Howe, R. A. and Enderby, J. E., (1973b), to be published.
Heine, V. and Weaire, D., (1970), *Solid State Phys.*, **24**, 249.
Hendus, H., (1947), *Zeit. Naturf.*, **A2**, 205.
Henniger, E. H. and Buschert, R. C., (1967), *J. Chem. Phys.*, **46**, 586.
Henniger, E. H., Buschert, R. C. and Heaton, L., (1966), *J. Chem. Phys.*, **44**, 1758.
Hindley, N. K., (1970), *J. Non-Cryst. Solids*, **5**, 17.
Hodgkinson, R. J., (1970), *Phil. Mag.*, **22**, 1187.
Hodgson, J. N., (1961), *Phil. Mag.*, **6**, 509.
Holstein, T., (1961), *Phys. Rev.*, **124**, 1329.
Holstein, T. and Friedman, L., (1968), *Phys. Rev.*, **165**, 1019.
Howe, R. A. and Enderby, J. E., (1967), *Phil. Mag.*, **16**, 141, 467.
Howells, W. S. and Enderby, J. E., (1972), *J. Phys. C.*, **5**, 1277.
Hultgreen, R., Orr, R. L., Anderson, P. D. and Kelley, K. K., (1963),, Selected Values of Thermodynamic Properties of Metals and Alloys, J. Wiley, New York.
Ilschner, B. R. and Wagner, C. N., (1953), *Acta Met.*, **6**, 712.
Jan, J. P., (1962), *Amer. J. Phys.*, **30**, 497.
Joffe, A. F. and Regel, A. R., (1960), Progress in Semiconductors, (A. F. Gibson, editor), J. Wiley, New York.
Kendall, P. W. (1964), *J. Sci. Inst.*, **41**, 485.
Kim, Y. S., Standley, C. L., Kruh, R. F. and Clayton, G. T., (1961), *J. Chem. Phys.*, **34**, 1464.
Lebowitz, J. L., (1964), *Phys. Rev.*, **A133**, 895.
Lomer, W. M. and Low, G. C., (1965), Thermal Neutron Scattering, (P. A. Egelstaff, editor), Academic Press, New York.
Male, J. C., (1967), *Brit. J. Appl. Phys.*, **18**, 1543.
Matsurbara, T. and Kaneyoshi, T., (1968), *Pros. Theor. Phys.*, **40**, 1257.
Mikolaj, G. and Pings, C. N. J., (1967), *J. Chem. Phys.*, **46**, 1412.
Mott, N. F., (1966), *Phil. Mag.*, **13**, 989.
Mott, N. F., (1967), *Adv. Phys.*, **16**, 49.
Mott, N. F., (1970), *Phil. Mag.*, **22**, 1.
Mott, N. F., (1971), *Phil. Mag.*, **24**, 1.
Nakamura, Y. and Shimoji, M., (1969), *Trans. Farad. Soc.*, **65**, 1509.
North, D. M., Enderby, J. E. and Egelstaff, P. A., (1968a), *J. Phys.*, *C*, **1**, 784.
North, D. M., Enderby, J. E. and Egelstaff, P. A., (1968b), *J. Phys. C*, **1**, 1075.
North, D. M. and Wagner, C. N. J., (1969), *J. App. Cryst.*, **2**, 149.
Orton, B. R. and Williams, G. I., (1960), *Acta. Met.*, **8**, 177.
Paalman, H. H. and Pings, C. J., (1962), *J. App. Phys.*, **33**, 2635.
Page, D. I., Egelstaff, P. A., Enderby, J. E. and Wingfield, B. R., (1969), *Phys. Letts.*, **29A**, 6, 296.
Pearson, F. J. and Rushbrooke, G. S., (1957), *Proc. Roy. Soc., Edin.*, **A64**, 305.
Perron, J. C., (1967), *Adv. Phys.*, **16**, 657.
Perron, J. C., (1972), *J. Non-Cryst. Solids*, **8–10**, 272.
Pings, C. N. J., (1968), Physics of Simple Liquids, (H. N. V. Temperley *et al*, editors), North Holland, Amsterdam.
Placzek, G., (1952), *Phys. Rev.*, **86**, 377.
Regel, A. R., Andreev, A. A., Kotov, B. A., Mamaclaliev, M., Okuneva, N. M., Smirnov, I. A. and Shadrichev, E. V., (1970), *J. Non-Cryst. Solids*, **4**, 151.
Roll, A. and Motz, H., (1957), Z. Metallkunde, **48**, 272.
Russel, B. R. and Wahlig, C., (1950), *Rev. Sci. Inst.*, **21**, 1028.

Rzewuski, H. and Werner, Z., (1965), *Electronics Letts.*, 1, 86.
Sanderson, E. A., (1960), Chemical Periodicity, Reinhold, New York.
Schaich, W. and Ashcroft, N. W., (1970), *Phys. Letts.*, 31A, 174.
Shaw, R. W., (1969), *J. Phys. C*, 2, 2335.
Shyu, W. M. and Gaspari, G. D., (1967), *Phys. Rev.*, 178, 985.
Stern, E. A., (1966), *Phys. Rev.*, 144, 545.
Straub, W. D., Roth, H., Bernard, W., Goldstein, S. and Mulhern, J. E., (1968), *Phys. Rev. Lett.*, 21, 752.
Stuke, J., (1969), *Festkörperprobleme*, 9, 46.
Terpilowski, J. and Zaleske, E., (1963), *Roczn. Chem.*, 37, 193.
Thiele, E., (1963), *J. Chem. Phys.*, 39, 474.
Throop, G. J. and Bearman, R. J., (1965). *J. Chem. Phys.*, 42, 2838.
Tourand, G., Cabane, B. and Breuil, M., (1972), *J. Non-Cryst. Solids*, 8–10, 676.
Vineyard, G. H., (1954), *Phys. Rev.*, 96, 93.
Warren, W. W., (1970), *Solid State Comm.*, 8, 1269.
Warren, W. W., (1971), *Phys. Rev.*, B 3, 3708.
Warren, W. W., (1972a), *J. Non-Cryst. Solids*, 8–10, 241.
Warren, W. W., (1972b), *Phys. Rev.*, B 6, 2522
Wertheim, M. S., (1963), *Phys. Rev. Letts.*, 10, 321.
Ziman, J. M., (1960), Electrons and Phonons, Clarendon Press, Oxford.
Ziman, J. M., (1961), *Phil. Mag.*, 6, 1013.
Ziman, J. M., (1967), *Adv. Phys.*, 16.

Subject Index